Handbook of Research on New Solutions and Technologies in Electrical Distribution Networks

Baseem Khan
Hawassa University, Hawassa, Ethiopia

Hassan Haes Alhelou
Tishreen University, Syria

Ghassan Hayek
Tishreen University, Syria

IGI Global
DISSEMINATOR OF KNOWLEDGE

A volume in the Advances in Computer and Electrical Engineering (ACEE) Book Series

Published in the United States of America by
IGI Global
Engineering Science Reference (an imprint of IGI Global)
701 E. Chocolate Avenue
Hershey PA, USA 17033
Tel: 717-533-8845
Fax: 717-533-8661
E-mail: cust@igi-global.com
Web site: http://www.igi-global.com

Library of Congress Cataloging-in-Publication Data

Names: Khan, Baseem, 1987- editor. | Alhelou, Hassan Haes, 1988- editor. | Hayek, Ghassan, 1958- editor.
Title: Handbook of research on new solutions and technologies in electrical distribution networks / Baseem Khan, Hassan Haes Alhelou, and Ghassan Hayek, editors.
Description: Hershey, PA : Engineering Science Reference, [2020] | Includes bibliographical references and index. | Summary: "This book examines the major issues and technological advancements in the electrical distributor sector"-- Provided by publisher.
Identifiers: LCCN 2019026574 (print) | LCCN 2019026575 (ebook) | ISBN 9781799812302 (h/c) | ISBN 9781799812326 (eISBN)
Subjects: LCSH: Electric power distribution.
Classification: LCC TK3001 .N49 2020 (print) | LCC TK3001 (ebook) | DDC 621.319/2--dc23
LC record available at https://lccn.loc.gov/2019026574
LC ebook record available at https://lccn.loc.gov/2019026575

This book is published in the IGI Global book series Advances in Computer and Electrical Engineering (ACEE) (ISSN: 2327-039X; eISSN: 2327-0403)

British Cataloguing in Publication Data
A Cataloguing in Publication record for this book is available from the British Library.

For electronic access to this publication, please contact: eresources@igi-global.com.

Advances in Computer and Electrical Engineering (ACEE) Book Series

Srikanta Patnaik
SOA University, India

ISSN:2327-039X
EISSN:2327-0403

Mission

The fields of computer engineering and electrical engineering encompass a broad range of interdisciplinary topics allowing for expansive research developments across multiple fields. Research in these areas continues to develop and become increasingly important as computer and electrical systems have become an integral part of everyday life.

The **Advances in Computer and Electrical Engineering (ACEE) Book Series** aims to publish research on diverse topics pertaining to computer engineering and electrical engineering. **ACEE** encourages scholarly discourse on the latest applications, tools, and methodologies being implemented in the field for the design and development of computer and electrical systems.

Coverage

- Digital Electronics
- Analog Electronics
- Sensor Technologies
- Applied Electromagnetics
- Optical Electronics
- Chip Design
- Computer Hardware
- Computer Science
- Programming
- VLSI Design

IGI Global is currently accepting manuscripts for publication within this series. To submit a proposal for a volume in this series, please contact our Acquisition Editors at Acquisitions@igi-global.com or visit: http://www.igi-global.com/publish/.

The Advances in Computer and Electrical Engineering (ACEE) Book Series (ISSN 2327-039X) is published by IGI Global, 701 E. Chocolate Avenue, Hershey, PA 17033-1240, USA, www.igi-global.com. This series is composed of titles available for purchase individually; each title is edited to be contextually exclusive from any other title within the series. For pricing and ordering information please visit http://www.igi-global.com/book-series/advances-computer-electrical-engineering/73675. Postmaster: Send all address changes to above address. ©

Titles in this Series

For a list of additional titles in this series, please visit: https://www.igi-global.com/book-series/advances-computer-electrical-engineering/73675

Handbook of Research on Emerging Applications of Fuzzy Algebraic Structures
Chiranjibe Jana (Vidyasagar University, India) Tapan Senapati (Padima Janakalyan Banipith (H.S.), India) and Madhumangal Pal (Vidyasagar University, India)
Engineering Science Reference • © 2020 • 300pp • H/C (ISBN: 9781799801900) • US $285.00

Electrical Insulation Breakdown and Its Theory, Process, and Prevention Emerging Research and Opportunities
Boxue Du (Tianjin University, China)
Engineering Science Reference • © 2020 • 230pp • H/C (ISBN: 9781522588856) • US $215.00

Challenges and Applications for Implementing Machine Learning in Computer Vision
Ramgopal Kashyap (Amity University, Raipur, India) and A.V. Senthil Kumar (Hindusthan College of Arts and Science, India)
Engineering Science Reference • © 2020 • 293pp • H/C (ISBN: 9781799801825) • US $195.00

Handbook of Research on Recent Developments in Electrical and Mechanical Engineering
Jamal Zbitou (University of Hassan 1st, Morocco) Catalin Iulian Pruncu (Imperial College London, UK) and Ahmed Errkik (University of Hassan 1st, Morocco)
Engineering Science Reference • © 2020 • 553pp • H/C (ISBN: 9781799801177) • US $255.00

Architecture and Security Issues in Fog Computing Applications
Sam Goundar (The University of the South Pacific, Fiji) S. Bharath Bhushan (Sree Vidyanikethan Engineering College, India) and Praveen Kumar Rayani (National Institute of Technology, Durgapur, India)
Engineering Science Reference • © 2020 • 205pp • H/C (ISBN: 9781799801948) • US $215.00

Handbook of Research on Advanced Applications of Graph Theory in Modern Society
Madhumangal Pal (Vidyasagar University, India) Sovan Samanta (Tamralipta Mahavidyalaya, India) and Anita Pal (National Institute of Technology Durgapur, India)
Engineering Science Reference • © 2020 • 591pp • H/C (ISBN: 9781522593805) • US $245.00

Novel Practices and Trends in Grid and Cloud Computing
Pethuru Raj (Reliance Jio Infocomm Ltd. (RJIL), India) and S. Koteeswaran (Vel Tech, India)
Engineering Science Reference • © 2019 • 374pp • H/C (ISBN: 9781522590231) • US $255.00

IGI Global
DISSEMINATOR OF KNOWLEDGE

701 East Chocolate Avenue, Hershey, PA 17033, USA
Tel: 717-533-8845 x100 • Fax: 717-533-8661
E-Mail: cust@igi-global.com • www.igi-global.com

Editorial Advisory Board

List of Contributors

Table of Contents

Detailed Table of Contents

With the advancement of technology, the dependency on the electrical energy has been increased greatly. Computer and telecommunication networks, railway network banking, post offices, and life support systems are a few applications that cannot function without electricity. At the same time, these applications demand qualitative energy. However, the quality of power supplied is affected by various internal and external factors of the power system. Harmonics, voltage, and frequency variations deteriorate the performance of the system. Voltage sag/dip is the most frequent problem and there are many methods to overcome this problem. The use of FACT devices is an efficient one. This chapter discusses an overview of the FACT device known as dynamic voltage restorer (DVR) in mitigating voltage sag. The strategy to control this device is also presented. The proposed control strategies are simulated in MATLAB SIMULINK environment and analyzed. The method is utilized and discussed briefly.

Fsaha Mebrahtu, Hawassa University, Hawassa, Ethiopia

Voltage dip in the distribution network is caused by disturbance at different voltage levels and experienced by low voltage customers are established. Voltage dips are those disturbances which damage the power quality of the distribution network and causing heavy economic damage to the customers. This chapter investigates procedures of mitigating the voltage dip by reducing the number of faults due to short circuits, lowering the fault clearing time, and changing the power system design and DSTATCOM Compensator with DG and dynamic voltage restorer.

Youssef Krim, National Engineering School of Monastir, Tunisia
Saber Krim, National Engineering School of Monastir, Tunisia
Mohamed Faouzi Mimouni, National Engineering School of Monastir, Tunisia

In this chapter, a control strategy for a Renewable Distribution Generator (RDG) operates in grid-connected and standalone mode is suggested. This RDG is made up of a wind generator associated with a Super-Capacitors (SC) considered as a storage system. The study investigates a control scheme for RDG integrated into power electrical system to maintain the voltage and the frequency of the grid in an allowable range and to ensure the continuity of power supply in case of grid faults. The proposed control strategy has three parts: a vector control of the wind generator to extract the maximum power; the control of the DC bus voltage by inserting the SC; and a droop control loop proposed to ensure the grid stability. The simulation results demonstrate the reliability of the control system.

Fsaha Mebrahtu, Hawassa University, Hawassa, Ethiopia

In this chapter various harmonic sources and their effect on the distribution network and its mitigation procedures are discussed. In the distribution network, electrical power is mostly utilized for our daily activity. However, the quality of power in the distribution network is affected by different disturbances. The distribution power quality problems deteriorate the performance of the system. One of the disturbances of the distribution network is harmonic distortion. Disturbances not only produce excessive heat in the devices and appliances used in the daily life of human beings, but also reduce the life of the appliances.

Finally, the harmonic distortion mitigation by using active power filter, space vector pulse width modulation, dynamic voltage restorer, voltage phase shift, and fuzzy logic controller is discussed.

Power loss reduction is an important problem that needs to be addressed with respect to generating electrical power. It is important to reduce power loss using locally generated power sources and/or compensations. This chapter brings a method of presents a method of maximizing energy utilization, feeder loss reduction, and voltage profile improvement for radial distribution network using the active and reactive power sources. Distributed Generation (DG) (wind and solar with backup by biomass generation) and shunt capacitor (QG) for reactive power demand are used. Integrating DG and QG at each bus might reduce the loss but it is economically unaffordable, especially for developing countries. Therefore, the utilization optimization method is required for finding an optimal size and location to feeders for placing QG and DG to minimize feeder loss.

Due to the transition of traditional power system to smart structure, integration of renewable energy sources is of great importance. It is performed by using distributed generators and Micro grid. Integration of renewable sources is very useful for the reliability enhancement of distribution system as energy can be supplied locally at distributed level. Therefore, this chapter provides an overview of various reliability methods which are utilized for finding the impact of renewable energy generation on distribution system reliability assessment. These renewable energy sources are integrated in the distribution system at distributed and Micro grid level. Various characteristics of renewable energy sources are discussed to model them. A general problem of reliability assessment in terms of reliability indices is also discussed.

Electric power should deliver a predicable per condition for the technological, economic, and political development of any countries and it is vital for each individual. Power outage is series problem in Ethiopia at the hole of distribution network. This is due to frequent interruptions and much time service restoration. That is why most customers of Ethiopia have their day-to-day activities highly affected and they are strongly complaining to Ethiopia electric utility daily. But this power outage affected customer cost and the Ethiopian utility. In most developing countries including Ethiopia, distribution systems have received considerably less of the attention to reliability modeling and evaluation than have generating and transmitting systems. Life is directly or indirectly dependent on electric power so a utility should deliver reliable power every day for 24 hours and each year for 8,760 hours to satisfy human needs and to perform their works as much as possible with less economy.

With increasing population and urbanization, the demand of electricity also increases day by day; to fulfill this demand, clean and environment-friendly distributed generations are being installed, but these have some issues in power section. With the integration of DG load curve is levelized, feeder voltage is improved; loading effect on the transformer and branches is reduced, and provides electricity with no pollution. This chapter investigates impacts of DGs to the power system; distributed generation means to generate electric power near the power consumption point. Power quality and reliability can be enhanced by the interconnection of distribution generation to an existing distribution system. However, there are so many effects of distributed generation e.g. changing of load losses, increasing of short circuit levels, voltage transient, congestions in the system branches, power quality, and reliability and network protection issues such as false tripping, nuisance tripping, unintentional islanding, neutral shifting is mainly affected.

With the development of smart grid technology, residents can schedule their power consumption pattern in their home to minimize electricity expense, reducing peak-to-average ratio (PAR) and peak load demand. The two-way flow of information between electric utilities and consumers in smart grid opened new areas of applications. In this chapter, the general architectures of the home energy management systems (HEMS) are introduced in a home area network (HAN) based on the smart grid scenario. Efficient scheduling methods for home power usage are discussed. The energy management controller (EMC) receives the demand response (DR) information indicating the Time-of use electricity price (TOUP) through the home gateway (HG). With the DR signal, the EMC achieves an optimal power scheduling scheme that can be delivered to each electric appliance by the HG.

Electric utilities over the domain affected with ecological issues associated with standard fossil fuel-established plants are examining more within the potentiality of interposing energy sources type of plants into the system as an alternative. Integration of Demand Side Management (DSM) and Supply Side Management (SSM) is required in a rational power system planning that implies concurrent deliberation of both qualitative and quantitative problems like costs, fuel mix, and reliability of power supply. This chapter examines the economic and environmental ability of power supplies initiation into an existing peak deficit power system, incorporating both DSM and SSM plans. The Generation Expansion Planning (GEP) study is carried out in the power system for the period of 24 years planning horizon.

Foreword

FOREWORD

I am delighted to write the foreword to this book, as its scope and content provide commercial and technical enterprises with the essential ingredients for implementing and managing power and energy systems techniques and management schemes. The recent advances in the power and energy technology, especially in the distribution power systems operation, control, optimization, renewable, energy management through analysis and simulation allowed a potentially enormous market for energy management and conversion. With advances in the latest field of electrical engineering including: operation and control, energy system security, energy efficiency, optimization and renewable energy resources have contributed to the solution of recent energy problems for any nation and the environmental issues. However, now the major challenge is to maintain the high energy efficiency during operation and to explore sustainable energy resources. To deal with the technical challenges, two major areas in power and energy technologies are being identified:

1. Modern Distribution Power System Operation
2. Modern and Smart Distribution Power System Control

This book provides answers to many challenging questions dealing power and energy system optimization. It addresses a variety of issues related to the energy management and the recently developed optimization techniques. This book comprised of 18 chapters divided into two parts as per the information providing according to the above areas. I recommend this book to researchers and practitioners in the field, and for scientists and engineers involved in power system operation and control. I really appreciate the efforts of all the Editors to compile this book. The managing editors, Dr. Baseem Khan & Dr. Hassan Haes Alhelou, and their team have meticulously collected the chapters, reviewed and place them in appropriate way for better in depth understanding. I believe the readers of power and energy system will be benefited from the work presented in this book.

Akhtar Kalam
Victoria University, Australia

Preface

For the development of smart city, a Smart grid is considered to be the future infrastructure of electrical power generation for secure and sustainable energy development. Active electric power distribution networks (ADNs) is a main element and large part of the power grid infrastructure. They will become even more critical and should be given priority in developing future smart grids to be used in smart city. There are certain issues associated with distribution sector such as distribution losses, inadequate billing, improper monitoring of system, huge reactive power demand under excessive industrialization, harmonics mitigation under excessive use of power electronics circuitry, renewable energy integration with micro-grid and distributed energy sources, exact state determination for the incorporation of intermittency of renewable energy generation, and smart home energy management system modelling. These issues must be addressed to make the distribution system active and smart.

The conventional distribution system depends upon a huge generation margins to meet out the load demand under uncertain operating conditions. Moreover, conventional distribution system planning only takes into account the security criteria with peak load conditions which is not suitable in the scenario of distributed energy sources integration. Therefore, active distribution networks are very important for the integration of renewable energy sources in the system. Moreover, the incorporation of distributed energy sources is the main characteristic of a smart and active distribution network. Thus, an active distribution system is a primary requirement for smart distribution networks. According to the International Council on Large Electric Systems definition, active electrical distribution networks have systems in place to control a combination of distributed energy sources, which is defined as generators, loads, and storage. Further, the future electric grid not only consists of a network, but also consists of a complete system, which has active control approaches for distributed generation, electric vehicle technology, smart metering system, smart monitoring devices, and various storage systems. Distribution network deals with the conventional distribution system in the grid, whereas active distribution grid involves distributed generation too. The basic characteristics of active distribution network include local and coordinated control of power flows, voltage, and fault levels. However, advanced demand side management techniques can also be incorporated for supervision, operational planning, and control. Different issues such as-active and reactive power support, islanding, and black-start capabilities can also be addressed with the use of active distribution network.

Micro-grid consists of different renewable energy sources like wind, solar, and micro turbines are few of them. It also incorporates latest generation technologies such as-fuel cell technologies and combined heat and power technology. To address the aforementioned issues, renewable energy, storage devices, for example, battery energy storage system, electric vehicles and flywheel storage systems can be used. Micro-grid provides a better solution compared to the distributed generation sources in terms of coordination and control. It can be used in islanded mode and grid connected mode as required. Hence, inside micro-grid, operation, control, and coordination problems are of great importance which can be solved with the use of active distribution networks.

In the last decade, use of smart grid technologies increases due to development of smart cities in almost every part of the world, where various smart gird technologies for the distribution sector are developed and installed to provide better uninterrupted services to the users. However, Installation of such systems requires excessive analysis and modelling before implementation. Therefore, different techniques are used by the researchers both from academia and Industry to meet out the challenges faced during their successful implementation.

DESCRIPTION OF THE BOOK

This book is about how to guide and address the major issues inside the most overlooked sector of the electrical Industry, the Electrical Distribution. Various aforementioned issues are discussed in detail with respect to present smart grid scenario. Therefore, the book "New Solutions and Technologies in Electrical Distribution Networks" is to provide a platform to share up-to-date scientific achievements in the core as well as related fields. There are three main objectives underlying this book:

1. Identifying and exploring the scope of different electrical distribution system operation and control methods.
2. Identifying the scope of operation and control in modern distribution power systems with high share of renewable energy resources.
3. Identifying the various operation and control schemes/algorithms/approaches/techniques for the implementation of future power distribution network.

This book aims to be an essential reference source, building on the available literature in the field of modern and future power distribution system operation and control, providing further research opportunities in this field. This specific text is expected to provide the primary and major resources necessary for researchers, academicians, students, faculties, and scientists, across the globe, to adopt and implement new inventions in thrust area of power distribution and energy management. To deal with the technical challenges, two major areas in power and energy technologies are being identified:

1. Modern Distribution Power System Operation
2. Modern and Smart Distribution Power System Control

From recent trends, it can be observed that research is mainly focused on electric distribution system operation and control, renewable energy resource, and the computational intelligence. The domains covered in this book will provide the vast knowledge to the researcher and it will also give common source of information to the power engineers. Energy efficiency and quality is the burning issue nowadays. Researchers are highly involved in this area and this is due to the fact that energy demand is exponentially growing whereas, energy resources are depleting day by day. This requires optimal use of electrical power.

In order to find the optimal operation of the electrical system, we have to explore several optimization techniques. This book will be a source of motivation to the researcher and can exchange the knowledge of energy efficiency, Power quality, reliability assessment, optimization techniques and integration of DG for better utilization of the resources. In order to meet the load demand and to improve the energy efficiency researchers are working in the area of integration of distributed generation, micro-grid, automation in power distribution, hybrid electric vehicle, and synchronized operation of solar power, battery and the grid supply for the development of smart power system for the home. They have developed several optimization techniques involving extensive search, analytical approach, and the computational or artificial intelligence.

In recent years, the computational intelligence based on nature inspired meta-heuristic approaches are finding wider acceptance for optimization of discrete problems particularly for large systems with no fixed solution under different operating conditions. A brief description of the chapters in this book is given as follows.

CHAPTER 1

Distribution system is the final stage of electric power system. It can be classified based on the voltage level, location, number of wires and types of customers. The various classifications of distribution system are explained in detail. System reliability is one of the important design concerns for any distribution system. The various methods of design concept for improving the reliability are explained in detail. Reactive power compensation is one of the main concerns in distribution system. Further, the various methods of reactive power compensation are explained in detail.

CHAPTER 2

As electric distribution system is supplied the energy to end user, various issues are available which directly affected the distribution system due to the operation and activities of power system and customers respectively. The common problems existing in electric distribution systems are under voltage, overloading of distribution system components, unbalanced loading, transformer without online tap changer operation, improper reactive power compensation, power theft, conversion of 3 phase supply into 2 phase supply, voltage sag, harmonics and system resonance condition, voltage fluctuations, problem in fault identification, transients and renewable energy penetration, which result in both functional problems to both distribution power supply company as well as end user. Therefore this chapter provides an overview of these issues exist in the electric distribution system.

CHAPTER 3

With the advancement of technology, the dependency on the electrical energy has been increased greatly. Computer and telecommunication networks, railway network, banking, post office, life support system are few application that just cannot function without electricity. At the same time these applications demand qualitative energy. However, the quality of power supplied is affected by various internal and external factors of the power system. The presence of harmonics, voltage and frequency variations deteriorate the performance of the system. Voltage sag/dip is the most frequently occurring problem. There are many methods to overcome this problem. In this chapter, Dynamic Voltage Restorer (DVR) is utilized to mitigate voltage sag and maintained the power quality of supply. Also the control strategy to control this device is presented. The adopted control strategy is simulated in MATLAB/SIMULINK environment.

CHAPTER 4

Voltage dip in the distribution network is caused by disturbance at different voltage levels and experienced by low voltage customers. Voltage dips are those disturbances which damage the power quality of the distribution network and causing heavy economic damage to the customers. In this chapter, procedures of mitigating the voltage dip is investigated by different means such as reducing the number of faults due to short circuits, lowering the fault clearing time, changing the power system design and installing DSTATCOM Compensator with DG and dynamic voltage restorer, etc.

CHAPTER 5

In this chapter, a control strategy for a Renewable Distribution Generator (RDG) operates in grid-connected and standalone mode is suggested. This RDG is made up of a wind generator associated with a Super-Capacitors (SC) considered as a storage system. The objective of this study is to investigate a control scheme for RDG integrated into power electrical system in order to maintain the voltage and the frequency of the grid in an allowable range and to ensure the continuity of power supply in case of grid faults. The proposed control strategy is composed of three parts. The first one is a vector control of the wind generator to extract the maximum power. The second one is the control of the DC bus voltage by inserting the SC. In the third part, a droop control loop is proposed to ensure the grid stability. The simulation results demonstrate the reliability of the control system.

CHAPTER 6

In the distribution network, electrical power is mostly utilized for our daily activity. However, the quality of power in the distribution network is affected by different disturbances. The power quality problems deteriorate the performance of the system. One of the disturbances of the distribution network is harmonic distortion. These disturbances are not only producing excessive heat in the devices or appliances, but also reduce the lifetime of the appliances. Therefore, this chapter discussed the various harmonic sources and their effect on the distribution network. Further the harmonic distortion mitigation by using active power filter, space vector pulse width modulation, dynamic voltage restorer, voltage phase shift and fuzzy logic controller are discussed.

CHAPTER 7

Power loss reduction is an important problem that needs to be addressed as generating electrical power. It is important to reduce power loss using locally generated power sources and/or compensations. This chapter developed a method of maximizing energy utilization, feeder loss reduction and voltage profile improvement for radial distribution network by using the active and reactive power sources. Distributed Generation (DG) (wind and solar with backup by biomass generation) and shunt capacitor (QG) for reactive power demand are used. Integrating DG and QG at each bus might reduce the loss but it is economically unaffordable especially for developing countries. Therefore, the optimization method is utilized for finding an optimal size and location of feeders for placing QG and DG, so that the feeder loss can be minimized.

CHAPTER 8

Due to the transition of traditional power system to smart structure, integration of renewable energy sources is of the great importance. It is performed by using distributed generators and Micro grid. Integration of renewable sources is very much useful for the reliability enhancement of distribution system as energy can be supplied locally at distributed level. Therefore, this chapter provides an overview of various reliability methods, which are utilized for finding the impact of renewable energy generation on distribution system reliability assessment. These renewable energy sources are integrated in the distribution system at distributed and Micro grid level. Moreover, various characteristics of renewable energy sources are discussed to model them. A general problem of reliability assessment in terms of reliability indices is also discussed.

CHAPTER 9

Electrical power should deliver in a reliable manner for the technological, economic and political development of any country. Power outage is a common problem in various distribution networks. This is due to the fact that most of the interruptions are frequent and required more time for service restoration. Due to this various customers' activities are highly affected. Further, this power outage affected the revenue of electric utility. In most of the developing countries, distribution systems received considerably less attention to reliable modeling and evaluation of the distribution system. Therefore, this chapter discussed the various methods utilized in the power system for the reliability assessment of distribution system.

CHAPTER 10

Now a days, with the increasing population and urbanization, the demand of electricity also increasing day by day; to fulfill this demand clean and environment friendly (renewable energy resources) distributed generations are being installed, but these have some issues in power system. With the integration of distributed generation load curve is levelized, feeder voltage is improved; loading effect on the transformer and branches is reduced, and provide electricity without any pollution. In this chapter, author investigated impacts of DGs on the power system. Distributed generation means to generate electric power near the power consumption point. Power quality and reliability can be enhanced by the interconnection of distribution generation to an existing distribution system. However, there are so many effects of distributed generation e.g. changing of load losses, increasing of short circuit levels, voltage transient, congestions in the system branches, power quality and reliability and network protection issues such as false tripping, nuisance tripping, unintentional islanding, neutral shifting which mainly affected the system.

CHAPTER 11

With the development of smart grid technology, residents have the opportunity to schedule their power consumption pattern in their home by themselves for the purpose of minimizing electricity expense, reducing the peak-to-average ratio (PAR) and peak load demand. The two-way flow of information between electric utilities and consumers in smart grid opened new areas of applications. In this chapter, the general architectures of the home energy management systems (HEMS) are introduced in a home area network (HAN) based on the smart grid scenario and then efficient scheduling methods for home power usage are discussed. The energy management controller (EMC) receives the demand response (DR) information indicating the Time-of use electricity price (TOUP) through the home gateway (HG). With the DR signal, the EMC achieves an optimal power scheduling scheme that can be delivered to each electric appliance by the HG.

CHAPTER 12

Electric utilities over the domain affected with ecological issues associated with standard fossil fuel established plants are examining more within the potentiality of interposing energy sources type of plants into the system as an alternative. Integration of both Demand Side Management (DSM) and Supply Side Management (SSM) are required in a rational power system planning that which imply concurrent deliberation of both qualitative and quantitative problems like costs, fuel mix, and reliability of power supply. In this chapter, an attempt is created to examine the economic and environmental ability of power supplies initiation into an existing peak deficit power system, incorporate both DSM and SSM plans. The Generation Expansion Planning (GEP) study is carried out in the power system for the period of twenty-four years planning horizon.

CHAPTER 13

Microgrid is widely used in real world for advanced forecasting and demand response of renewable energy source, grid integration and operations. Microgrid is consists of conventional and nonconventional energy source such as wind energy, solar energy, biomass energy, hydro power, diesel power, fuel cell, thermal power and many more. It is the combination of AC and DC power such as wind, solar, fuel cell, biomass and Hydro power, which is mostly used in microgrids. It can be operated either grid connected or islanded modes. Therefore, this chapter discussed the different types of microgrids with their mode of operation along with the background of the power system

CHAPTER 14

Building's maintenance has received increasing international attention in various fields of scientific research. As a result, there has been a change in the maintenance of buildings from the preventive to the predictive approach. This is done through an evaluation model to support and assist the management of the facility in selecting alternatives and making appropriate decisions in maintenance according to building status and maintenance budget. This chapter investigated the reasons for the electrical maintenance of the university buildings and the degree of importance of each element of electrical maintenance through the design of a questionnaire in which the electrical components were divided into elements and then each element was linked to all maintenance items that related to it. At the end of the research mathematical models were developed, these models help to forecast the electrical maintenance items and distribution of the maintenance budget, and to verify the validity of these models, they have been applied to study the case of dorm buildings in Tishreen University.

CHAPTER 15

Monitoring and controlling the electrical distribution system for real time becoming very important to improve its operating performance after transition to active distribution systems. So, a huge number of sensors are needed to monitor all part in the systems, But if the sensor installed at all buses, investment cost become huge. To reduce the number of sensors, state estimation approach can be used to estimate the voltage of buses, which do not have sensors. State Estimation (SE) algorithms are broadly classified into Static State Estimator (SSE) and Dynamic State Estimator (DSE). This chapter focused on the classification of the most algorithms that used in active distribution networks, also State estimation in unbalanced distribution systems and Role of PMU in Distribution System State Estimation.

CHAPTER 16

Energy storage system is the vital part of present renewable energy integrated power system as it removes the problem of renewable energy sources' intermittency. Therefore, this chapter examines the modeling and simulation of energy storage (battery, flywheel, etc.) systems interfaced to the power grid by using power electronic device, like chopper module, Rectifier module and filter circuits, which are essential to the load balance between supply and demand, and to eliminate harmonics and ensure efficient, cost effective and reliable operation. Energy storage system in power grid is similar of memory in computer system. Energy efficiency is a key performance indicator for energy storage system. The energy storage system is the most promising component to enhance the system reliability and flexibility.

CHAPTER 17

Power Electronics is the study of switching electronic devices and circuits in order to convert and control the flow of electrical energy. Power Electronics is the basic technology of switching power supplies, power converters, power inverters, motor drives, and motor soft starters. Inverters are devices that transform DC input into fixed AC output through changing the trigger angle, in the ideal statues the output wave form is perfect sine wave, however in the practical case it has some higher harmonics that needs to be reduced. Inverters are an important part of UPS systems, where it would be fully explained in this chapter, where UPSs are uninterruptible Power Supply in which the inverters are included as an essential part that can provide emergency power to a load when the input power source or main power source fails.

CHAPTER 18

Most of the electric machines had a conventional design for speed –control; previously the speed regulation of these motors was done via traditional or mechanical contacts, for example: inserting resistors to the armature circuit or controlling the excited circuit of DC motor, and other methods of control. These classical methods, however, lead to non-linearity in mechanical or electromechanical characteristics, which in turn lead to increased power losses as the result of the non-soft regulation of speed, as well as the great inertia of classical control methods that rely on mechanical and electromagnetic devices.

The book "New Solutions and Technologies in Electrical Distribution Networks" contains 18 chapters of high-quality contributions from international leading researchers in the field of power and energy systems. The book chapters consist of electrical distribution system's voltage and frequency control, stability and reliability improvement, operation, control and planning related issues. Recently, several heuristic and meta-heuristic optimization techniques are presented by the researchers. Simulation results have also appeared in some chapters. Management, maintenance and enhancement techniques for the improvement of power quality, reliability and stability, energy efficiency and security are presented in the chapters. Since energy utilization is mainly by motors load and lighting. It is observed that the energy management and efficiency is the major concern in the present scenario when conventional resources are depleting day by day.

Acknowledgment

This book of research on New Solutions and Technologies in Electrical Distribution Networks is an outcome of the inspiration and encouragement given by many individuals for whom these words of thanks are only a token of our gratitude and appreciation for them.

Our sincere gratitude goes to the peoples who contributed their time and expertise to this book. We highly appreciate their efforts in achieving this project. The editors would like to acknowledge the help of all the people involved in this project and, more specifically, the editors would like to thank each one of the authors for their contributions and the editorial board/reviewers regarding the improvement of quality, coherence and the content presentation of this book.

Second, Editors would like to express their sincere thanks to Jordan Tepper, Jan Travers, Mariah Gilbert, Lindsay Wertman, Courtney Tychinski, and other individuals of IGI Global for their continuous support and giving us an opportunity to edit this book.

The editors are thankful to our family members for their prayers, encouragement, and care shown towards us during the completion of this handbook of research on power system operation and control. Thank you all!!!

We also express our gratitude to the GOD for all the blessings!!!

Baseem Khan
Hawassa University, Hawassa, Ethiopia

Hassan Haes Alhelou
Tishreen University, Syria

Ghassan Hayek
Tishreen University, Syria

Chapter 1
Introduction to Electric Distribution System

Sivaraman P.
https://orcid.org/0000-0002-0591-2266
TECH Engineering Service, Chennai, India

Sharmeela C.
Department of Electrical and Electronics Engineering, College of Engineering, Anna University, Chennai, India

ABSTRACT

Distribution system is the final stage of electric power system, and can be classified based on voltage level, location, number of wires, and types of customers. This chapter explains the various classifications of the distribution system in detail. System reliability is one of the important design concerns for any distribution system. The various methods of design concept to improve reliability are clarified. Reactive power compensation is another main concern in a distribution system. Methods of reactive power compensation are also detailed.

INTRODUCTION

Overview

An electric power system is encompassing the electric power generation in power plants, electric power transmission system is to transmit the generated power from the power plant to load center (over several hundred kilo meters) where the actual load is located, power distribution and utilisation. The Electric Distribution System (EDS) is the final stage of electric power system where the generated power is converted into useful work through distribution lines and feeders. It deals with lower voltage magnitude as reverse of transmission system high voltage (Gonen, 2014, Gers, 2013).

DOI: 10.4018/978-1-7998-1230-2.ch001

The IEEE std. 141-1993 describes the various configuration of distribution based on the reliability requirement. These configurations are simple radial systems, expanded radial system, primary selective system, primary loop system, secondary selective system, ring bus system and spot network (IEEE Std, 1994; Khan et al., 2019; Khan et al., 2018; Khan et al., 2017; Banteywalu et al., 2019; Anteneh et al., 2019; Molla et al., 2019, Molla et al., 2018, Jariso et al. 2018). The reliability of radial system is low as compared with ring bus system. If the distribution is designed for low reliability requirement, the number of equipment's required to the system is less and it requires less economics as compared with distribution is designed for high reliability (Kersting, 2001). Based on the reliability requirement, type of distribution system to be selected based on both technical as well as economic considerations (Alhelou et al., 2019; Makdisie et al., 2018; Alhelou et al., 2018; Alhelou et al., 2016; Haes Alhelou et al., 2019; Njenda et al., 2018).

Based on voltage level, power distribution systems are classified into primary distribution system and secondary distribution system (Khan, 2007, Mehta and Mehta, 1982).

- Typically the various voltage levels of primary distribution systems are 6.6 kV, 11 kV, 22 kV and 33 kV.
- Typically the various voltage levels of secondary distribution are 415 V three phase and 240 V single phase.

Based on the type of consumer, power distribution systems are classified into residential, commercial and industrial distribution system (Sivaraman et al., 2017).

Based on the location, power distribution systems are classified into urban distribution system and rural distribution system.

- Urban distribution system comprising the residential, commercial and industrial power systems.
- Rural distribution system comprising the residential, commercial, industrial power systems and also comprising the agricultural usage.

Based on the no of wires used, power distribution systems are classified into 1ϕ 2W system, 3ϕ 3W system and 3ϕ 4W system (Sivaraman et al., 2017).

- Residential system loads are connected in 1ϕ 2W system and 3ϕ 4W system.
- Commercial system loads are connected in 1ϕ 2W system, 3ϕ 3W system and 3ϕ 4W system.
- Industrial system loads are connected in 1ϕ 2W system, 3ϕ 3W system and 3ϕ 4W system.

The concept of Distributed Generation (DG) introduced in last decade to generate the localised power nearer to the load center or end user loads to reduce the energy losses in transmission & distribution lines, improve the voltage profile across the distribution system (Jenkins et al. 2010). Renewable energy resources play an important role in DG especially solar PV systems and wind turbine systems. The power from DG may integrate at either SDS or PDS. For an example, 100kW rooftop solar PV system integration at SDS and 1000kW ground mounted solar PV system at PDS (Bollen and Hassan, 2011).

Figure 1. Typical 3phase, 3W 11kV primary distribution feeder

DISTRIBUTION SYSTEM BASED ON VOLTAGE LEVEL

Primary Distribution System (PDS)

Primary distribution system is the distribution between main distribution substation and distribution transformers or end user loads. The feeders used between distribution substation and distribution transformers or end user loads are called as primary distribution feeders (Mehta and Mehta, 1982). The figure 1 shows the typical 3phase 3W, 11 kV primary distribution feeder in PDS and SLD for typical industrial system distribution is shown in figure 2.

Voltage Level of PDS

The typical voltage level used in PDS is between 3.3 kV and 33 kV. The primary distribution system is used to provide the power supply to

1. Higher capacity end user loads
2. Secondary distribution system

Figure 2. Typical PDS arrangement

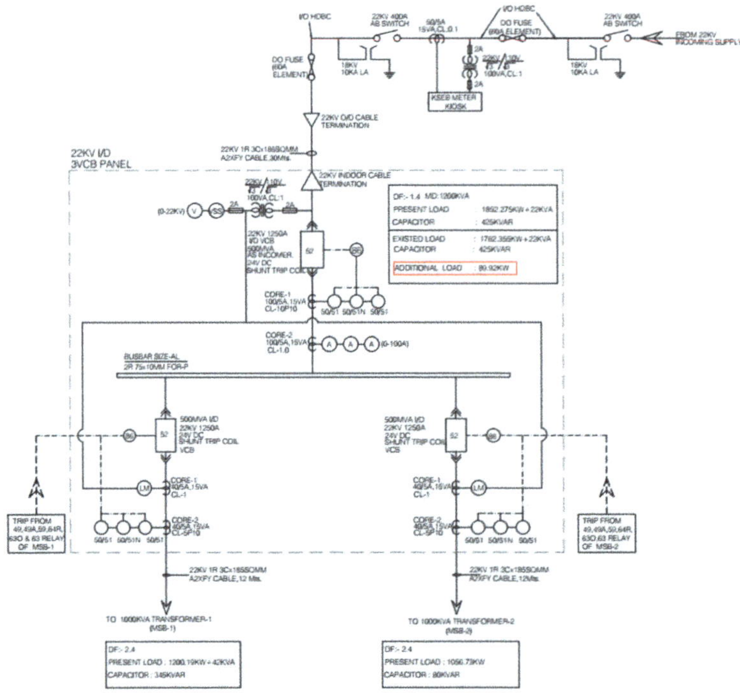

Major Equipment's of PDS

The following equipment's are commonly used in PDS and these are

1. Overhead conductors
2. Underground cables
3. Distribution transformers
4. Series and shunt reactors
5. Capacitor banks
6. Switch gears like VCBs, SF6, fuses, relays, etc.
7. Surge protective devices or surge arrester
8. Isolators
9. Insulators
10. CTs and PTs
11. End user loads like induction motors, drives, arc and induction furnaces, etc.,
12. Medium voltage generators
13. Grounding systems

No. of Wires

The 3 phase 3 wire systems are commonly used in PDS (Mehta and Mehta, 1982).

Figure 3. Typical SDS arrangement

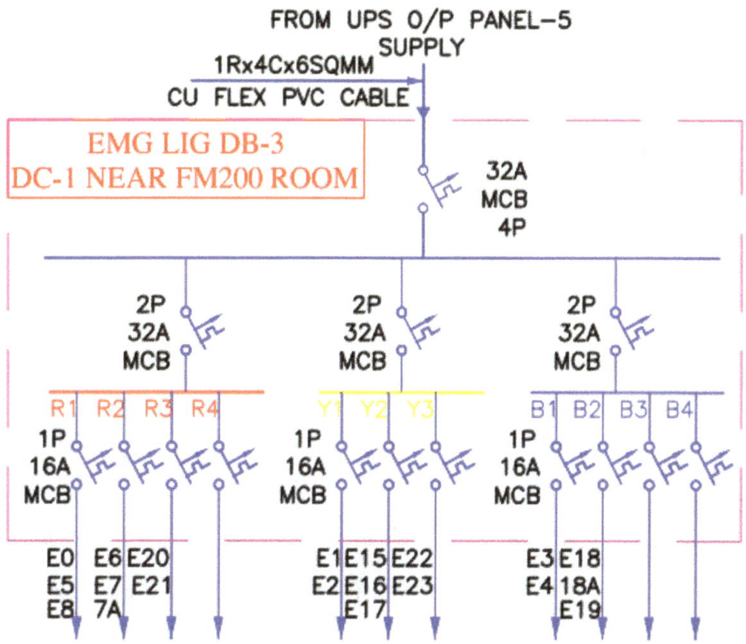

Secondary Distribution System (SDS)

Secondary distribution system is the distribution between distribution transformers and end user loads. The feeders used between distribution transformer and end user loads are called as secondary distribution feeders (Mehta and Mehta, 1982; Khan et al. 2014; Khan et al. 2013; Khan et al. 2012; Negash et al., 2017; Negash et al., 2016; Jariso et al. 2017; Kifle et al. 2018; Yeshalem et al. 2017; Singh et al. 2017; Gupta et al., 2015). The typical SLD for emergency lighting distribution load powered through an UPS system in commercial system is shown in figure 3 and secondary distribution transformer used for powering the residential system loads shown in figure 4.

Voltage Level of SDS

The typical voltage level used in SDS is single phase 230 V/240 V and three phase 400 V / 415 V.

Employment of the SDS

The secondary distribution system is used to provide the power supply to

1. Residential system loads
2. Commercial system loads
3. Industrial system loads

Figure 4. Secondary distribution transformer in SDS

Major Equipment's of SDS

The following equipment's are commonly used in SDS and these are

1. Overhead conductors
2. Underground cables
3. Bus ducts
4. Distribution transformers
5. Series and shunt reactors
6. Capacitor banks
7. Switch gears like MCB, MCCB, ACB, fuses, relays, etc.
8. Low voltage surge protective devices
9. Contactors
10. Isolators
11. Insulators
12. CTs and PTs

Figure 5. Schematic diagram of single phase 2W circuit

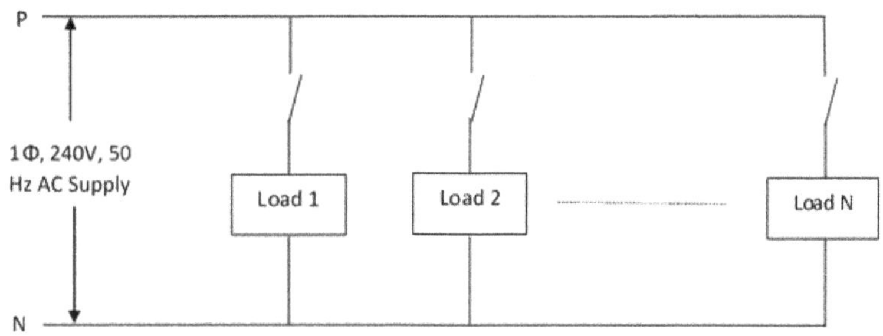

13. End user loads like induction motors, drives, luminaries, UPS systems, kitchen equipment's, electronic gadgets, etc.,
14. Low voltage generators
15. Grounding systems
16. Metering instruments
17. Harmonic filters
18. Solar PV systems

No of Wiress

The commonly used wiring configurations in SDS are (Mehta and Mehta, 1982)

1. Single phase 2W system
2. Three phase 3W system
3. Three phase 4W system

Distribution System Based on Number of Wires

Based on the no. of wires used, distribution systems are generally classified into 1ϕ 2W system, 3ϕ 3W system and 3ϕ 4W system (Mehta and Mehta, 1982, Sivaraman et al., 2017).

Single Phase 2W System

The single phase 2W system commonly used to power the single-phase loads connected in the distribution system. The single phase 2W system is mostly used in SDS. The typical schematic diagram of single phase 2W system is shown in figure 5 and secondary distribution feeder is shown in figure 6.

In figure 5, single phase, 240V, 50Hz AC power supply is used to power the connected loads in the SDS. For an example, luminaries, fans, television, personal computers, etc, in residential or commercial systems are connected in single phase 2W system.

Figure 6. Typical single phase, 2W, 240V secondary distribution feeder

Three Phase 3W System

The three phase 3W system is used to power the three phase loads without neutral in the both primary and secondary distribution system. The typical schematic diagram of three phase 3W, 415V system is shown in figure 7 and three phase 3W, 11kV system primary distribution feeder is shown in figure 1.

In figure7, three phase, 415V, 50Hz AC power supply is used to power the connected motor loads in the SDS. For an example, induction motors in commercial systems are connected in three phase 3W system.

Three Phase 4W System

The three phase 4W system is commonly used to power the combination of both single phase and three phase loads with/without neutral in the SDS. The typical schematic diagram of three phase 4W system is shown in figure 8 and secondary distribution feeder is shown in figure 9.

In figure 8, three phase, 415V, 50Hz AC power supply is used to power the connected three phase 4W loads in the SDS. For an example, VFDs, UPS systems, etc, in commercial systems are connected in three phase 4W system.

Figure 7. Schematic diagram of three phase 3W circuit

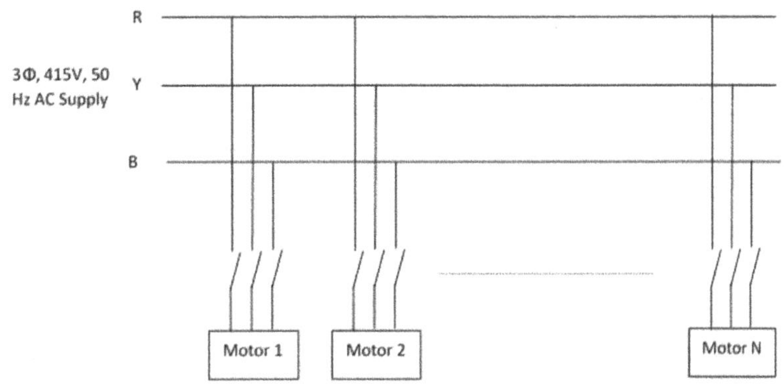

Figure 8. Schematic diagram of three phase 4W circuit

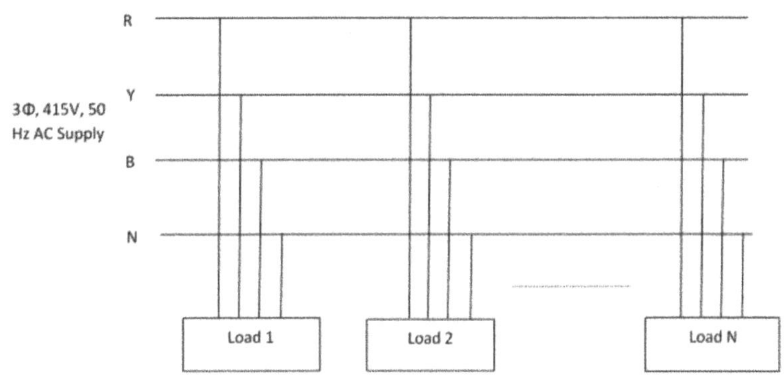

Both single phase 2W system and three phase 3W system can be derived from three phase 4W system, Figure 10 shows, how the single phase 2W circuit and three phase 3W circuit derived from three phase 4W circuit (Sivaraman et al., 2017).

Single phase 2W circuit can be derived from either of R or Y or B phases to the neutral and three phase 3W can be derived from all the phases except neutral in three phase 4W system as shown in figure 10.

Configurations of Distribution System for Different System Reliability

The IEEE std 141-1993 describes the various configuration of distribution based on the reliability requirement as follows (IEEE Std, 1994).

1. Simple radial systems
2. Expanded radial system
3. Primary selective system
4. Primary loop system
5. Secondary selective system

Figure 9. Secondary distribution feeder

Figure 10. Way to derive single phase 2W and three phase 3W circuit from three phase 4W circuit

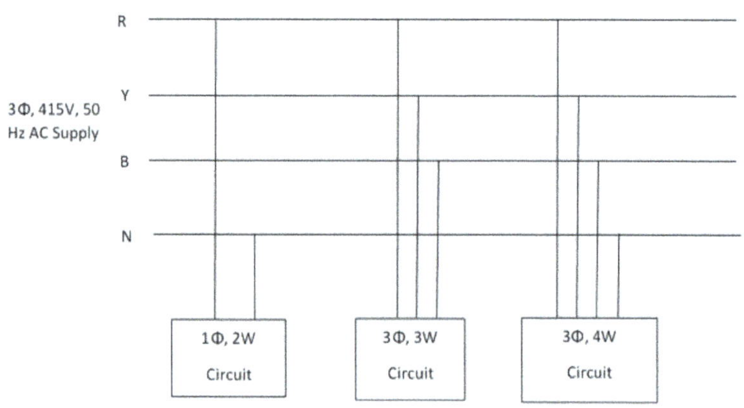

Figure 11. Schematic diagram of simple radial system

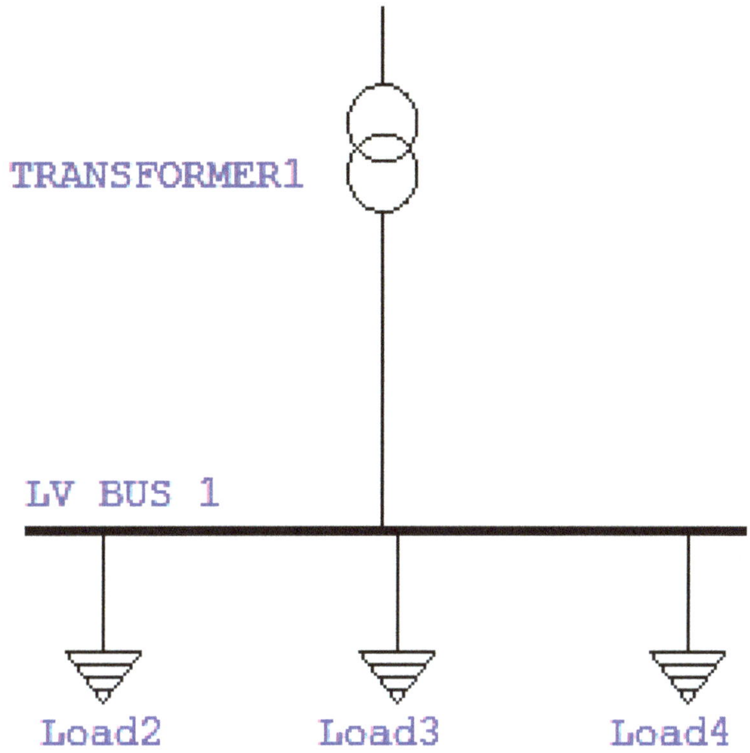

Simple Radial System

The simple radial system receives single power supply input from PDS and distribution transformer supplying all connected feeders in the system. Figure 11 shows the typical schematic diagram of simple radial system.

From figure 11, transformer 1 is powering the connected loads in LV bus 1. In case of failure/outage of transformer or power supply to the transformer, the connected loads in LV bus 1 will not receive the power supply until the problem is resolved.

The simple radial system has no redundant power supply as an alternative in case of emergency and this system can be selected where outage / unavailability of power supply for an extended period is not affect the system process (Khan, 2007). Where the system process gets affected due to power outage / unavailability, simple radial system is not suitable and primary selective or secondary selective system can be used (Fini et al., 2016; Alhelou et al., 2018; Zamani et al., 2018; Alhelou et al., 2015; Njenda et al., 2018; Haes Alhelou et al., 2018; Haes Alhelou et al., 2019).

Advantages:

1. This system requires lesser investment compared to others.
2. Simple operation and lesser complexity.
3. Easy maintenance.

Figure 12. Schematic diagram of expanded radial system

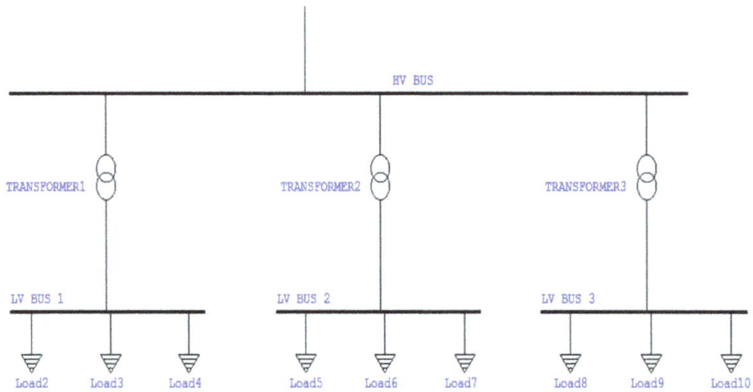

Disadvantages:

1. Failure of transformers, cable, switchgear or primary power supply resulting in power outage to the end user equipment's.
2. Any maintenance and servicing in live condition needs shutdown.

Expanded Radial System

The expanded radial system receives single power supply input from PDS and powering multiple distribution transformers supplying all connected loads in the system. Figure 12 shows the typical schematic diagram of expanded radial system.

Like simple radial system, failure / outage of transformer or power supply to transformer leading to forced outage to the connected loads until the problem is resolved.

The expanded radial system also not having redundant power supply as an alternative in case of emergency and this system can be selected where outage / unavailability of power supply for an extended period is not affect the system process (Khan, 2007). Where the system process gets affected due to power outage / unavailability, simple radial system is not suitable and primary selective or secondary selective system can be used.

Advantages:

1. Multiple transformer power by single power source in a same place.
2. This system requires lesser investment compare to others.
3. Easy maintenance.
4. Simple operation and lesser complexity.

Disadvantages:

1. Failure of transformers, cable or primary power supply resulting in power outage to the end user equipment's.
2. Maintenance and servicing needs equipment shutdown.

Figure 13. Schematic diagram of primary selective system

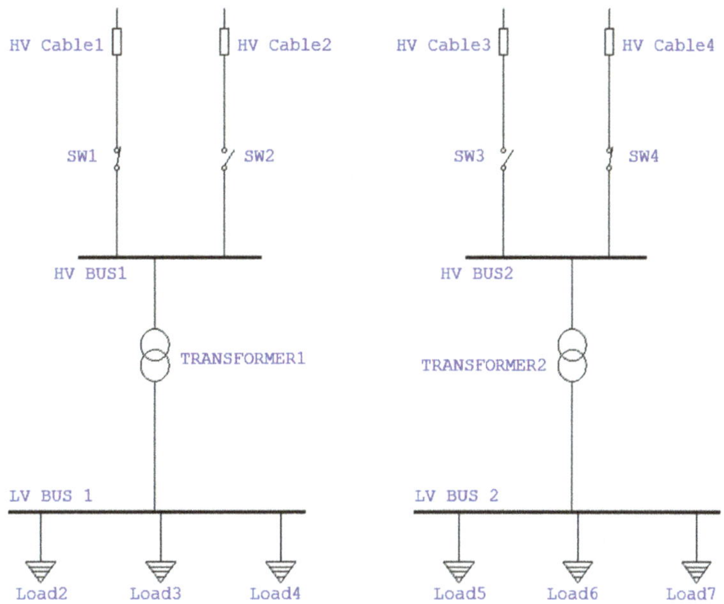

Table 1. Various operating condition primary selective system

S. No	Operating Condition				Remarks
	SW1	SW2	SW3	SW4	
	Transformer 1		Transformer 2		
1	Close	Open	Close	Open	1) HV supply 1 through SW1 is powering transformer 1 2) HV supply 2 through SW3 is powering transformer 2
2	Open	Close	Open	Close	1) HV supply 1 through SW2 is powering transformer 1 2) HV supply 2 through SW4 is powering transformer 2

Primary Selective System

The primary selective system provides the better reliability and power supply continuity as compared with simple and expanded radial system. It receives two power supply input from PDS and powering the single or multiple distribution transformer supplying all connected loads in the system. At a same time, any one of the power supply is used to power the distribution transformer. Suitable interlock to be provided to avoid the two sources getting parallel (IEEE Std, 1991). During the failure of one power supply, other power supply shall be used to power the distribution transformer. In this method, two separate primary distribution circuit is used to power the single transformer. If the plant requires / uses multiple

Figure 14. Schematic diagram of primary loop system

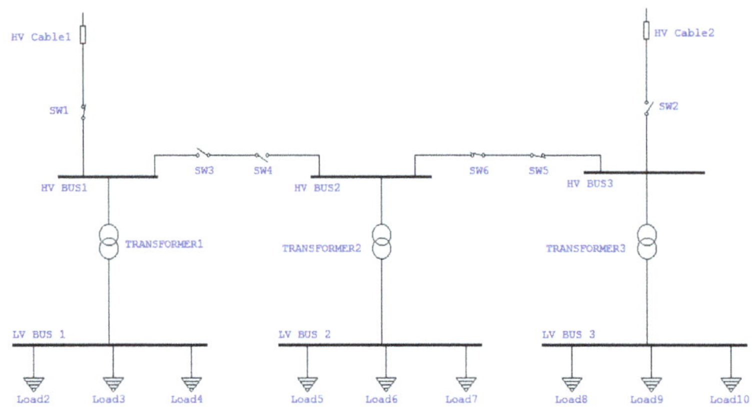

numbers of transformers, two separate circuit shall be used for all transformers (IEEE Std, 1994). Figure 13 shows the typical schematic diagram of primary selective system.

Table 1 describes the various operating condition of primary selective system powering three distribution transformers.

Case 1: HV supply 1 through SW1 is powering transformer 1 and HV supply 2 through SW3 is powering transformer 2

Case 2: HV supply 2 through SW2 is powering transformer 1 and HV supply 1 through SW4 is powering transformer 2

Failure of any one primary power supply will not affect the power supply continuity to the end user loads.

Advantages:

1. High reliability compared with radial system.
2. Maintenance of any one power supply will not affect the service to end user loads.

Disadvantages:

1. Higher capital cost as compared with radial systems.

Primary Loop System

The primary loop system provides the better reliability and power supply continuity as compared with simple and expanded radial system. It receives two power supply input from PDS and powering the multiple distribution transformer supplying all connected loads in the system. Figure 14 shows the typical schematic diagram of primary loop system.

Based on the location and requirement, no of switches in the loop can be selected. In Figure 14, two nos. of switches are selected (SW3 & SW4) between HV bus 1 and HV bus 2, (SW5 & SW6) between HV bus 2 and HV bus 3. This can be reduced to one switch if the location of both HV bus 1 and HV bus 2, HV bus 2 and HV bus 3 are same. Table 2 describes the various operating condition of primary loop system powering three distribution transformers.

Table 2. Various operating condition primary loop system

S. No	Operating Condition				Remarks
	HV supply 1	HV supply 2	SW3 & SW4	SW5 & SW6	
1	Available	Available	Open	Closed	1) HV supply 1 is powering transformer 1 2) HV supply 2 is powering transformer 2 and transformer 3
2	Available	Available	Closed	Open	1) HV supply 1 is powering transformer 1 and transformer 2 2) HV supply 2 is powering transformer 3
3	Available	Not available	Closed	Closed	HV supply 1 is powering transformer 1, transformer 2 and transformer 3
4	Not available	Available	Closed	Closed	HV supply 2 is powering transformer 1, transformer 2 and transformer 3

Normal Operation Time: Either one switch SW3 or SW5 in closed condition:

Case 1: If SW5 is closed, HV supply 1 is powering distribution transformer 1, while HV supply 2 is powering the distribution transformer 2 and 3.

Case 2: If SW3 is closed, HV supply 1 is powering distribution transformer 1 and 2 while HV supply 2 is powering the distribution transformer 3.

Failure of Any One HV Power Supply

Case 3: In the event of failure of HV supply 1, HV supply 2 is powering the distribution transformer 1, 2 and 3 till the resumption.

Case 4: In the event of failure of HV supply 2, HV supply 1 is powering the distribution transformer 1, 2 and 3 till the resumption.

Failure of any one primary power supply will not affect the power supply to end user loads. In the event of failure of two power supply will affect all distribution transformers operations. The system reliability of the system is improved compared with radial system and comparing with primary selective system reliability is less. Detailed operating procedure shall be prepared and followed for operation because both the direction electric power can flow.

Advantages:

1. High reliability compared with radial system.
2. Maintenance of any one power supply will not affect the service to end user loads.
3. No of switches and circuit is less as compared with primary selective system.

Disadvantages:

1. Higher capital cost compared with radial systcms.
2. Maintenance and servicing of distribution transformer needs equipment shutdown.

Figure 15. Schematic diagram of secondary selective system

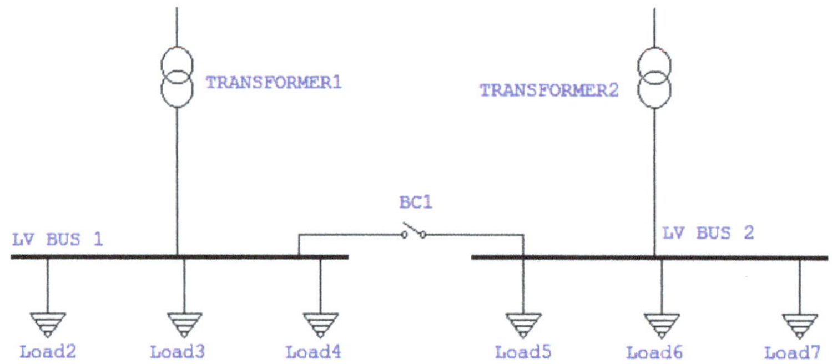

Table 3. Various operating condition of two distribution transformer

S. No	Operating Condition			Remarks
	Transformer 1	Transformer 2	Bus coupler	
1	ON	ON	OFF	1) Transformer 1 is powering LV BUS 1 2) Transformer 2 is powering LV BUS 2
2	OFF	OFF	ON	Transformer 2 is powering LV BUS 1 & LV BUS 2
3	ON	OFF	ON	Transformer 1 is powering LV BUS 1 & LV BUS 2

Secondary Selective System

Figure 15 shows the typical schematic diagram of secondary selective system for two distribution transformer. The secondary selective system receives single power supply input from PDS powering transformer 1. Similarly another distribution transformer 2 receiving the primary power supply from same PDS or different PDS. The provision (bus coupler) is made in secondary side of both distribution transformer to supplying all connected feeders in the system. During the normal operating condition, bus coupler (BC1) is normally open condition. If the failure or outage of transformer and/or primary distribution feeder, power supply is extended through the secondary side bus coupler (BC1).

Table 3 describes the various operating condition of two distribution transformer powering the connected loads.

During the normal operation time, bus coupler is kept open and transformer 1 is powering the LV BUS 1 loads & transformer 2 is powering LV BUS 2 loads. In the event of failure of transformer 1 or power supply to transformer 1, bus coupler is closed and power is extended from transformer 2 to powering the LV BUS 1 load till the resumption. In the event of failure of transformer 2 or power supply to transformer 2, bus coupler is closed and power is extended from transformer 1 to powering the LV BUS 2 load till the resumption.

Figure 16 shows the schematic diagram of secondary selective system for three distribution transformer.

Figure 16. Schematic diagram of secondary selective system

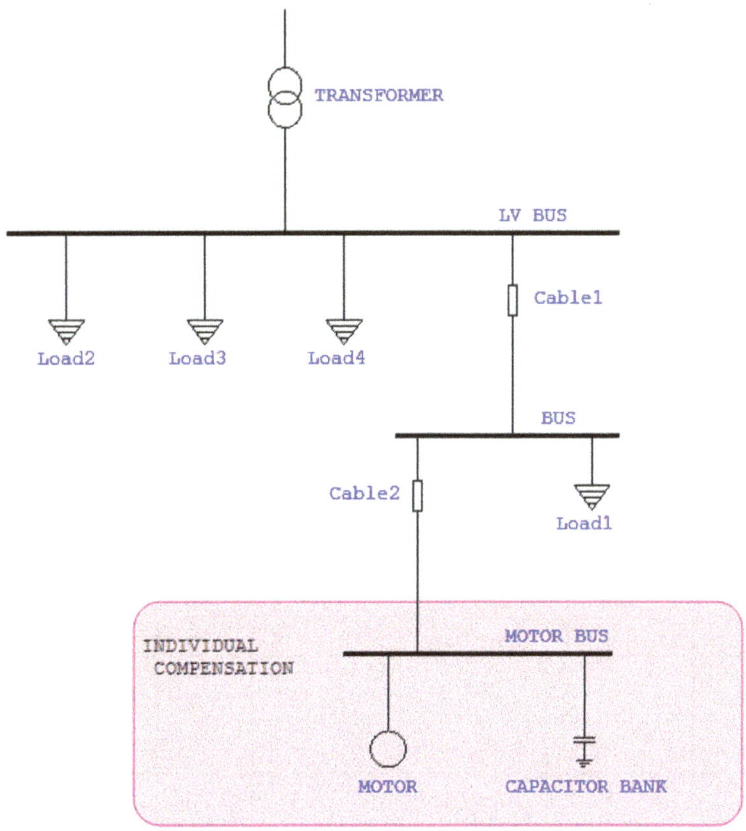

Table 4 describes the various operating condition of two distribution transformer powering the connected loads.

Normal operation time:

Both bus coupler 1 & bus coupler 2 is kept open and transformer 1 is powering the LV BUS 1 loads, transformer 2 is powering LV BUS 2 loads and transformer 3 is powering LV BUS 3 loads.

Failure of transformer 1:

In the event of failure of transformer 1 or power supply to transformer 1, bus coupler 1 is closed and power is extended from transformer 2 to powering the LV BUS 1 load till the resumption.

Failure of transformer 2:

In the event of failure of transformer 2 or power supply to transformer 2, power can be extended from either transformer 1 or transformer 3 to powering the LV BUS 2 load till the resumption. At a time only one bus coupler can be closed either bus coupler 1 or bus coupler 2.

Failure of transformer 3:

In the event of failure of transformer 3 or power supply to transformer 3, bus coupler 2 is closed and power is extended from transformer 2 to powering the LV BUS 3 load till the resumption.

In this method, all the distribution transformer and its circuits shall be rated for total load capacity. The system reliability of the system is improved and providing better supply continuity compared with radial systems.

Table 4. Various operating condition of three distribution transformer

S. No	Operating Condition					Remarks
	Transformer 1	Transformer 2	Transformer 3	Bus coupler 1	Bus coupler 2	
1	ON	ON	ON	OFF	OFF	1) Transformer 1 is powering LV BUS 1 2) Transformer 2 is powering LV BUS 2 3) Transformer 3 is powering LV BUS 3
2	OFF	ON	ON	ON	OFF	1) Transformer 2 is powering LV BUS 1 & LV BUS 2 2) Transformer 3 is powering LV BUS 3
3	ON	OFF	ON	ON	OFF	1) Transformer 1 is powering LV BUS 1 & LV BUS 2 2) Transformer 3 is powering LV BUS 3
4	ON	OFF	ON	OFF	ON	1) Transformer 1 is powering LV BUS 1 2) Transformer 3 is powering LV BUS 2 & LV BUS 3
5	ON	ON	OFF	OFF	ON	1) Transformer 1 is powering LV BUS 1 2) Transformer 2 is powering LV BUS 2 & LV BUS 3
6	ON	OFF	OFF	ON	ON	Transformer 1 is powering LV BUS 1, LV BUS 2 & LV BUS 3
7	OFF	ON	OFF	ON	ON	Transformer 2 is powering LV BUS 1, LV BUS 2 & LV BUS 3
8	OFF	OFF	ON	ON	ON	Transformer 3 is powering LV BUS 1, LV BUS 2 & LV BUS 3

Advantages:

1. High reliability compared with radial system.
2. Maintenance of any one power supply will not affect the service to end user loads.
3. Simple control.

Disadvantages:

1. Distribution transformer and its circuit should be sized for full load capacity.
2. Higher capital cost due to sizing the equipment's for full load capacity.

Table 5. Power electronics base loads

S. No	Name of the loads
1	LEDs
2	CFLs
3	Laptops and mobile phone chargers
4	Televisions
5	Photo copy machine
6	DVD players
7	Personal computers
8	UPS systems
9	Rectifiers
10	VFDs
11	Inverters, etc.,

REACTIVE POWER COMPENSATION AND POWER FACTOR

Reactive Power

Conventionally the power factor is defined as cosine of angle between voltage and current when the system is operating with linear loads without presence of harmonics. Now a days, most of the loads used in residential, commercial and industrial power systems are power electronics-based loads offering nonlinear voltage – current characteristics during their operation. These loads are listed as in table 5.

Due to nonlinear in voltage – current characteristics offered by power electronics based devices, cosine of angle between voltage and current (conventional power factor) is no longer valid in this nonlinear environment. The definition for power factor under nonlinear environment is the ratio of real power to apparent power and this power factor is called as true power factor as expressed in Eq. (1) and (2) (IEEE Std, 1994).

$$PF = \frac{W}{VA} \tag{1}$$

$$= \frac{\sqrt{3}V_L I_l \cos\phi}{\sqrt{3}V_L I_L} \tag{2}$$

Where
PF is true power factor
W is active power
VA is apparent power
V_L is RMS value line to line voltage

I_L is RMS value of line current including harmonics

I_1 is current at fundamental frequency

$\cos\phi$ is displacement power factor

Also true power factor can be defined as multiplication of displacement power factor and distortion power factor is expressed in Eq. (3).

$$\text{True PF} = \text{Displacement PF x Distortion PF} \tag{3}$$

The distortion power factor is equal to unity when the system is operating with linear loads (current is not distorted due to harmonics). Reactive power supports of the major electric equipment's are as follows (IEEE Std, 1994).

1. An induction motor requires both real and reactive power in same direction for its operation.
2. A synchronous motor operating in over excited condition, it requires real power flow into the motor and it delivers the reactive power flow in opposite direction into the system.
3. Capacitors are supplying only the reactive power flow into the system.
4. A generator supplies both real and reactive power flow into the system in same direction.

Due to lesser economics, capacitors are widely used in reactor power compensation (IEEE Std, 1994). The amount of reactive power in VAr is supplied by the capacitors are directly proportional to square of the applied input voltage in Eq. (4).

$$VAr = \frac{V^2}{X_c} \tag{4}$$

Where,

$$X_c = \frac{1}{2\Pi FC}$$

V is applied voltage

$$VAr = 2\Pi FCV^2 \tag{5}$$

Need for Reactive Power Compensation

Reactive power compensation is essential for AC distribution system because most of the loads are RL types and it requires reactive power for their operation. The flow of reactive power from the supply is reducing the system operating power factor. If the power factor is reduced below the specified value of distribution supply company, then distribution supply company imposing the penalty for reduction of power factor value as per norms. For an example, Tamilnadu Generation and Distribution Company in India is penalizing its customer for not maintaining power factor.

Benefits of Reactive Power Compensation

The followings are the direct benefits of reactive power compensation (IEEE Std, 1994)

1. Supplying the localised reactive power demanded by the load.
2. It improves the system power factor at PCC or billing location.
3. It reduces the losses in the system.
4. It avoids the levy / penalty imposed by DISCOMS for not maintaining the power factor at billing point or PCC.
5. It improves system voltage regulation.

Types of Reactive Power Compensation

The types of reactive power compensation are classified into two types.

1. Based on location of connection of reactive power compensating device.
2. Based on reactive power output needed from the device

Generally shunt type reactive power compensation is widely used in commercial and industrial systems. It means, reactive power compensating devices (capacitors banks) are connected parallel to the loads and supplying the reactive power to the loads.

Based on Location of Connection of Reactive Power Compensating Device

Based on the location of placement of reactive power compensating device in the system, reactive power compensation is classified into five types as follows

1. Individual compensation.
2. Group compensation.
3. Feeder compensation.
4. Total compensation at LV side.
5. Total compensation at HV side.

Figure 17. Typical schematic diagram of individual reactive power compensation

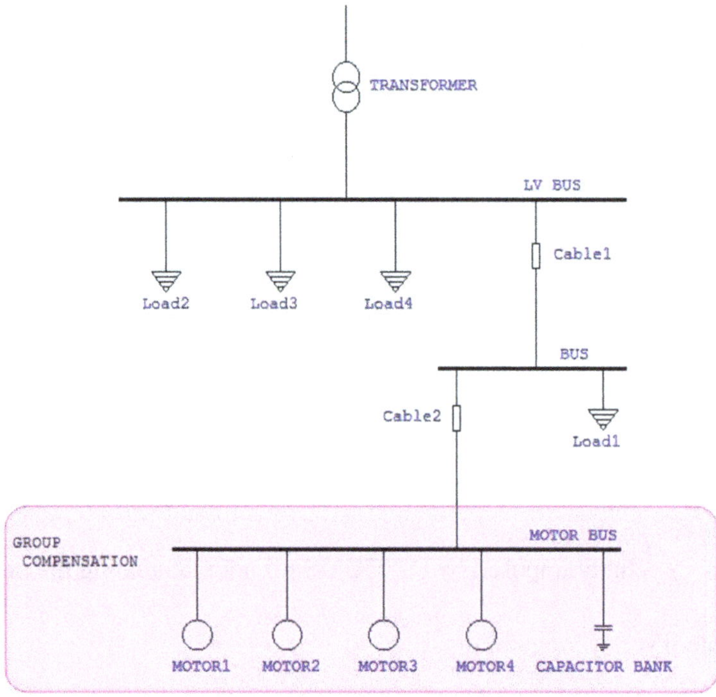

Individual Compensation

The individual reactive power compensation is connecting the reactive power compensating devices nearer to the individual loads in the system. This type of reactive power compensation is suitable for individual higher capacity loads with constant power and not suitable for lower capacity loads due to economics (ABB, 2018). For smaller capacity loads, group compensation or feeder compensation is more suitable. The typical schematic diagram of individual reactive power compensation is shown in figure 17.

Group Compensation

The group reactive power compensation is connecting the reactive power compensating devices to the group of loads in the system. This type of reactive power compensation is suitable for multiple smaller capacity loads connected in the bus (ABB, 2018). Motor control center is the best example for group compensation. If the connected loads are constant power and running continuously throughout the process then fixed reactive power compensation is suitable. If the connected loads are varying and not running through the process then varying reactor power compensation by means of automatic power factor correction is suitable. The typical schematic diagram of group reactive power compensation is shown in figure 18.

Figure 18. Typical schematic diagram of group reactive power compensation

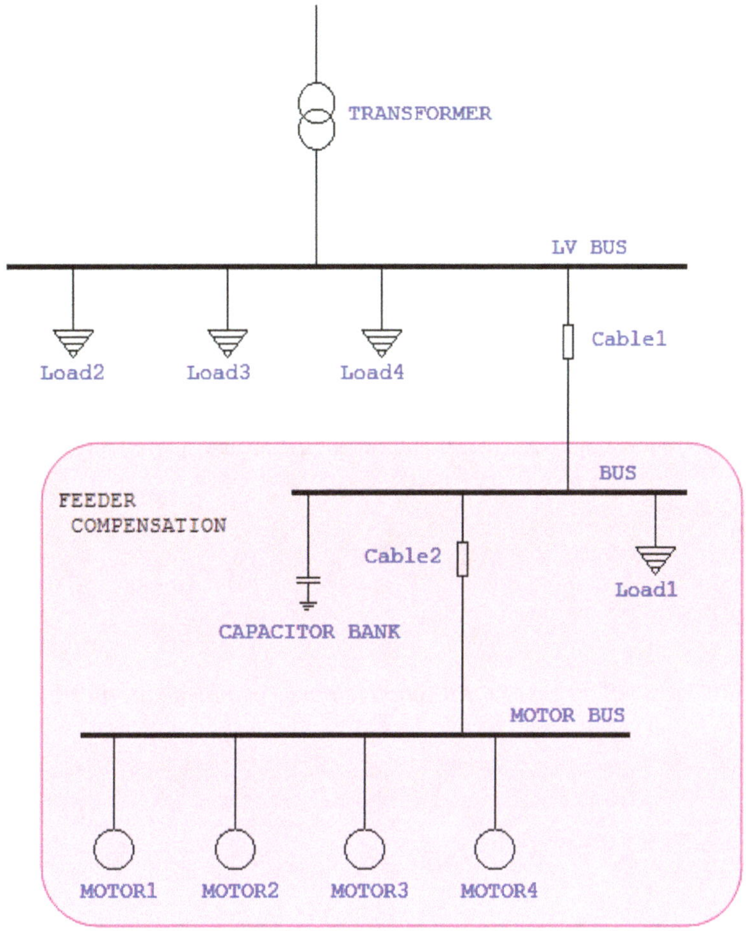

Feeder Compensation

The feeder reactive power compensation is connecting the reactive power compensating devices to the group of loads in the feeder. This type of reactive power compensation is suitable for multiple smaller capacity loads connected in the feeder. If the connected loads are constant power and running continuously throughout the process then fixed reactive power compensation is suitable. If the connected loads are varying and not running through the process then varying reactor power compensation by means of automatic power factor correction is suitable. Feeder compensation and group compensation type of reactive power compensation is used for multiple smaller capacity loads. The typical schematic diagram of feeder reactive power compensation is shown in figure 19.

Figure 19. Typical schematic diagram of feeder reactive power compensation

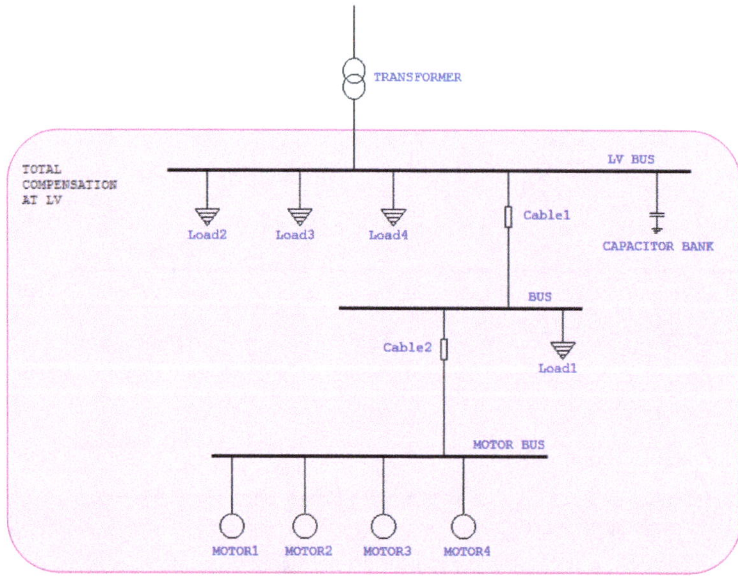

Figure 20. Typical schematic diagram of total reactive power compensation at LV side

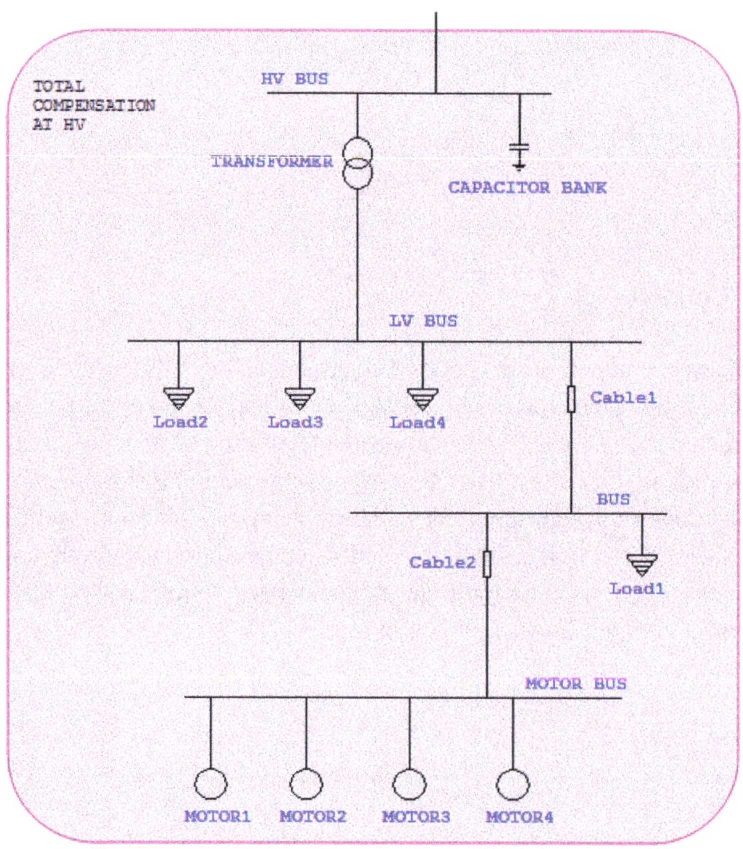

Figure 21. Typical schematic diagram of total reactive power compensation at HV side

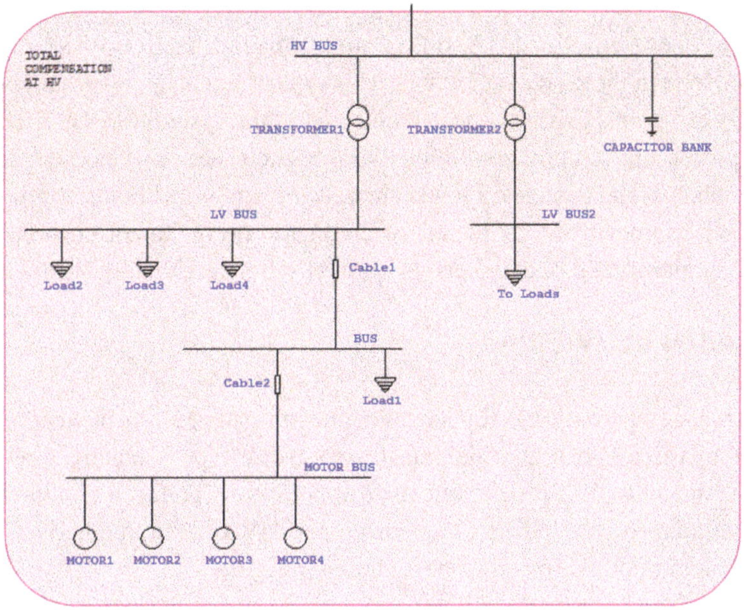

Figure 22. Typical schematic diagram of total reactive power compensation at HV side

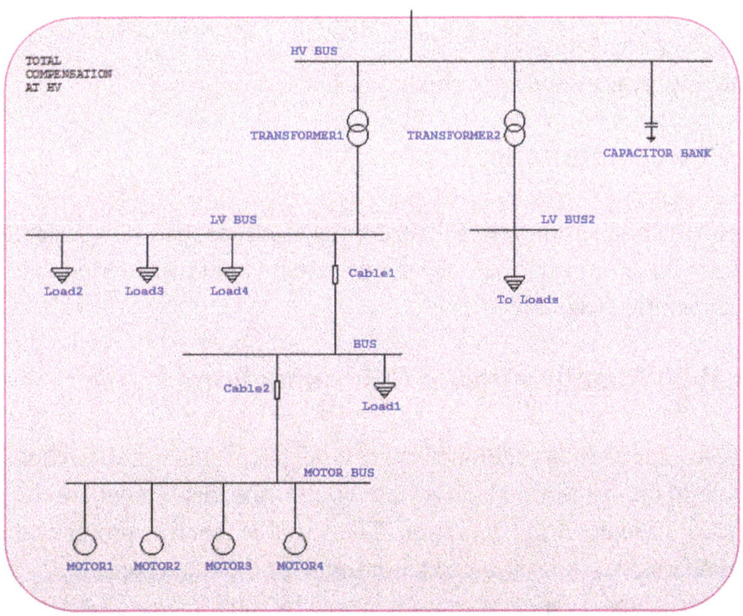

Total Compensation at LV Side

The total reactive power compensation at LV side is connecting the reactive power compensating devices to the total connected load in the system at LV bus. This type of reactive power compensation is suitable for multiple numbers of various capacity loads connected in the system (ABB, 2018). If the connected loads are constant power and running continuously throughout the process then fixed reactive power compensation is suitable. If the connected loads are varying and not running through the process then varying reactor power compensation by means of automatic power factor correction is suitable. The typical schematic diagram of total reactive power compensation at LV side is shown in figure 20.

Total Compensation at HV Side

The total reactive power compensation at HV side is connecting the reactive power compensating devices to the HV bus powering the single or multiple transformers. This type of reactive power compensation is suitable for powering the multiple transformers of various capacity connected in the system. The typical schematic diagram of total reactive power compensation at HV side for single distribution transformer is shown in figure 21 and multiple transformer is shown in figure 22.

Based on Reactive Power Output Needed From the Device

Based on the reactive power output from the compensating device, reactive power compensation is classified into two types (SE, 2015) as follows,

1. Fixed reactive power compensation.
2. Automatic (variable) reactive power compensation.

Fixed Reactive Power Compensation

The fixed reactive power compensation is delivering the fixed reactive power demanded by the loads. This type of compensation is more suitable for the location where the fixed amount of reactive power required through the operation (SE, 2015).

Automatic (Variable) Reactive Power Compensation

The automatic (variable) reactive power compensation is delivering the variable reactive power demanded by the loads. This type of compensation is more suitable for the location where the variable amount of reactive power required throughout the operation. The variable reactive power compensation achieved by using multiple numbers of various size capacitor units for the rated capacity.

The automatic power factor compensation is achieved by the following (ABB, 2018)

1. Voltage and current detection by sensors to measure the power factor.
2. Power factor controller an intelligent unit that compares the actual measured power factor with desired power factor.

a. If the measured power factor is less than the desired power factor, then some capacitor units are switched ON to maintain the power factor to the desired value.
b. If the measured power factor is higher than the desired power factor, then some capacitor units are switched OFF to maintain the power factor to the desired value.

CONCLUSION

This chapter discussed the various types of distribution systems available in the power system. Various classifications are utilized to classify the distribution system. These are based on reliability criteria; number of wires, types of customers. Further different design concepts for improving the reliability is also presented in this chapter. Further various reactive power compensation methods also presented in this chapter.

REFERENCES

ABB. (2018, July). Power factor correction and harmonic filtering in electrical plants, Technical application papers.

Alhelou, H., Hamedani-Golshan, M. E., Zamani, R., Heydarian-Forushani, E., & Siano, P. (2018). Challenges and opportunities of load frequency control in conventional, modern and future smart power systems: A comprehensive review. *Energies*, *11*(10), 2497. doi:10.3390/en11102497

Alhelou, H. H. (2018). *Fault detection and isolation in power systems using unknown input observer. In Advanced condition monitoring and fault diagnosis of electric machines* (p. 38). Hershey, PA: IGI Global.

Alhelou, H. H., Golshan, M., & Fini, M. (2018). Wind driven optimization algorithm application to load frequency control in interconnected power systems considering GRC and GDB nonlinearities. *Electric Power Components and Syst.*

Alhelou, H. H., & Golshan, M. E. H. (2016, May). Hierarchical plug-in EV control based on primary frequency response in interconnected smart grid. In *2016 24th Iranian Conference on Electrical Engineering (ICEE),* (pp. 561-566). IEEE. 10.1109/IranianCEE.2016.7585585

Alhelou, H. H., & Golshan, M. E. H. (2020). Decision-making-based optimal generation-side secondary-reserve scheduling and optimal LFC in deregulated interconnected power system. In *Decision making applications in modern power systems* (pp. 269–299). Academic Press. doi:10.1016/B978-0-12-816445-7.00011-6

Alhelou, H. H., Golshan, M. E. H., & Hatziargyriou, N. D. (2019). Deterministic dynamic state estimation-based optimal LFC for interconnected power systems using unknown input observer. *IEEE Transactions on Smart Grid*, 1. doi:10.1109/TSG.2019.2940199

Alhelou, H. H., Golshan, M. E. H., & Hatziargyriou, N. D. (2019). A decentralized functional observer based optimal LFC considering unknown inputs, uncertainties, and cyber-attacks. *IEEE Transactions on Power Systems.*

Alhelou, H. H., Golshan, M. E. H., Zamani, R., Moghaddam, M. P., Njenda, T. C., Siano, P., & Marzband, M. (2019, June). An improved UFLS scheme based on estimated minimum frequency and power deficit. In 2019 IEEE Milan PowerTech (pp. 1-6). IEEE. doi:10.1109/PTC.2019.8810497

Alhelou, H. H., Golshan, M. H., & Askari-Marnani, J. (2018). Robust sensor fault detection and isolation scheme for interconnected smart power systems in presence of RER and EVs using unknown input observer. *International Journal of Electrical Power & Energy Systems*, *99*, 682–694. doi:10.1016/j.ijepes.2018.02.013

Alhelou, H. H., Hamedani-Golshan, M. E., Heydarian-Forushani, E., Al-Sumaiti, A. S., & Siano, P. (2018, September). Decentralized fractional order control scheme for LFC of deregulated nonlinear power systems in presence of EVs and RER. In *Proceedings 2018 International Conference on Smart Energy Systems and Technologies (SEST)* (pp. 1-6). IEEE. 10.1109/SEST.2018.8495858

Alhelou, H. S. H., Golshan, M. E. H., & Fini, M. H. (2015, December). Multi agent electric vehicle control based primary frequency support for future smart micro-grid. In *Proceedings Smart Grid Conference (SGC)* (pp. 22-27).

Anteneh, D., & Khan, B. (2019). Reliability enhancement of distribution substation by using network reconfiguration a case study at Debre Berhan distribution substation. *International Journal of Economy, Energy, & Environment*, *4*(2), 33.

Banteywalu, S., Khan, B., De Smedt, V., & Leroux, P. (2019). A novel modular radiation hardening approach applied to a synchronous buck converter. *Electronics (Basel)*, *8*(5), 513. doi:10.3390/electronics8050513

Bollen, M., & Hassan, F. (2011). *Integration of distributed generation in the power system*. IEEE Press. doi:10.1002/9781118029039

Fini, M. H., Yousefi, G. R., & Alhelou, H. H. (2016). Comparative study on the performance of many-objective and single-objective optimisation algorithms in tuning load frequency controllers of multi-area power systems. *IET Generation, Transmission, & Distribution*, *10*(12), 2915–2923. doi:10.1049/iet-gtd.2015.1334

Gers, J. M. (2013). *Distribution system analysis and automation*. IET. doi:10.1049/PBPO068E

Gonen, T. (2014). Electric power distribution engineering, Third Ed. Boca Raton, FL: CRC Press.

Gupta, K., Khan, B., & Mubeen, S. E. (2015, February). Available transfer capability enhancement by unified power flowcontroller. In *Proceedings IEEE International Conference on Signal Processing, Informatics, Communication, and Energy Systems (SPICES)*, 2015, pp. 1, 4, 19-21.

Haes Alhelou, H., Hamedani-Golshan, M. E., Njenda, T. C., & Siano, P. (2019). A survey on power system blackout and cascading events: Research motivations and challenges. *Energies*, *12*(4), 682. doi:10.3390/en12040682

Haes Alhelou, H., Hamedani-Golshan, M. E., Njenda, T. C., & Siano, P. (2019). Wide-area measurement system-based optimal multi-stage under-frequency load-shedding in interconnected smart power systems using evolutionary computing techniques. *Applied Sciences*, *9*(3), 508. doi:10.3390/app9030508

Haes Alhelou, H., Hamedani Golshan, M. E., Njenda, T. C., & Siano, P. (2019). WAMS-based online disturbance estimation in interconnected power systems using disturbance observer. *Applied Sciences, 9*(5), 990. doi:10.3390/app9050990

IEEE recommended practice for electric power distribution for industrial plants. (1994, April). In IEEE Std. 141-1993, pp. 1-768.

IEEE recommended practice for electric power systems in commercial buildings. (1991, September). In IEEE Std. 241-1990, pp. 1-768.

Jariso M., Khan B, Tanwar S., Tyagi S., & Rishiwal, V. (2018). Hybrid energy system for upgrading the rural environment, *IEEE GlobeCom 2018.*

Jariso, M., Khan, B., Tesfaye, D., & Singh, J. (2017). Modelling and designing of stand-alone photovoltaic system (case study: Addis Boder Health Center south west Ethiopia). In *Proceedings IEEE International Conference on Electronics, Communication, and Aerospace Technology, ICECA 2017,* pp. 168-173, Coimbatore, India.

Jenkins, N., Ekanayake, J. B., & Strbac, G. (2010). *"Distributed generation", Renewable Energy Series 1.* IET.

Kersting, W. H. (2001). Distribution system modelling and analysis, Fourth Ed. Boca Raton, FL: CRC Press. doi:10.1201/9781420041736

Khan, B., & Agnihotri, G. (2012). A novel transmission loss allocation method based on transmission usage. In *Proceedings IEEE Fifth Power India Conference*, Haryana, India. 10.1109/PowerI.2012.6479479

Khan, B., Agnihotri, G, & Gupta, G. (2013). A multipurpose matrices methodology for transmission usage, loss, and reliability margin allocation in restructured environment. *Electrical & Computer Engineering: An International Journal, 2*(3).

Khan, B., Agnihotri, G., Rathore, P., Mishra, A., & Naidu, G. (2014). A cooperative game theory approach for usage and reliability margin cost allocation under contingent restructured market. *International Review of Electrical Engineering*, *9*(4), 854–862.

Khan, B., Alhelou, H. H., & Mebrahtu, F. (2019). A holistic analysis of distribution system reliability assessment methods with conventional and renewable energy sources. *AIMS Energy*, *7*(4), 413–429. doi:10.3934/energy.2019.4.413

Khan, B., & Singh, P. (2017). The current and future states of Ethiopia's energy sector and potential for green energy: A comprehensive study. *International Journal of Engineering Research in Africa, 33,* 115–139. doi:10.4028/www.scientific.net/JERA.33.115

Khan, B., & Singh, P. (2017). Optimal power flow techniques under characterization of conventional and renewable energy sources: A comprehensive analysis, *Journal of Engineering, 2017,* Article ID 9539506.

Khan, B., & Singh, P. (2017). Selecting a meta-heuristic technique for smart micro-grid optimization problem: A comprehensive analysis. In IEEE Access, 5, pp. 13951-13977.

Khan, B., & Singh, P. (2019). Economic operation of smart micro-grid: A meta-heuristic approach. In Handbook of research on smart power system operation and control (pp. 330-346). Hershey, PA: IGI Global. doi:10.4018/978-1-5225-8030-0.ch014

Khan, B., & Singh, P. (2019). Economic operation of smart micro-grid: A meta-heuristic approach. In Handbook of research on smart power system operation and control (pp. 330-346). Hershey, PA: IGI Global. doi:10.4018/978-1-5225-8030-0.ch014

Khan, B., & Tanwar, S. (2019). Issues associated with microgrid integration. In Handbook of research on smart power system operation and control (pp. 252-264). Hershey, PA: IGI Global. doi:10.4018/978-1-5225-8030-0.ch010

Khan, S. (2007). Industrial power systems. Boca Raton, FL: CRC Press.

Kifle, Y., Khan, B., & Singh, J. (2018). Designing and modeling grid connected photovoltaic system: (Case study: EEU building at Hawassa City). *International Journal of Convergence Computing*, *3*(1), 20–34. doi:10.1504/IJCONVC.2018.091113

Kifle, Y., Khan, B., & Singh, P. (2018). Assessment and enhancement of distribution system reliability by renewable energy sources and energy storage, *Journal of Green Engineering, 8*(3), 2, pp. 219-262.

Makdisie, C., Haidar, B., & Alhelou, H. H. (2018). An optimal photovoltaic conversion system for future smart grids. In Handbook of research on power and energy system optimization (pp. 601–657). Hershey, PA: IGI Global. doi:10.4018/978-1-5225-3935-3.ch018

Metha, V. K., & Metha, R. (1982). *Principle of power system, S.* Chand Publications.

Molla, T., Khan, B., Moges, B., Alhelou, H. H., Zamani, R., & Siano, P. (2019). Integrated optimization of smart home appliances with cost-effective energy management system. *CSEE Journal of Power and Energy Systems*, *5*(2), 249–258.

Molla, T., Khan, B., & Singh, P. (2018). A comprehensive analysis of smart home energy management system optimization techniques, *Journal of Autonomous Intelligence, 1(1)*, 15–21.

Negash, K., Khan, B., Tesfaye, D., & Gayathri, M. (2017). Optimal placement of phasor measurement unit for system observability. In *Proceedings of IEEE International Conference on Computing Methodologies and Communication, (ICCMC 2017),* pp. 73-77, July 18, 19, 2017, Erode India.

Negash, K., Khan, B., & Yohannes, E. (2016). Artificial intelligence versus conventional mathematical techniques: a review for optimal placement of phasor measurement units [Springer]. *Journal of Technology and Economics of Smart Grids and Sustainable Energy*, *1*(1), 10. doi:10.100740866-016-0009-y

Njenda, T. C., Golshan, M. E. H., & Alhelou, H. H. (2018, November). WAMS based intelligent under frequency load shedding considering online disturbance estimation. In *Smart Grid Conference (SGC)* (pp. 1-5). 10.1109/SGC.2018.8777779

Njenda, T. C., Golshan, M. E. H., & Alhelou, H. H. (2018, November). WAMS based under frequency load shedding considering minimum frequency prediction and extrapolated disturbance magnitude. In *Smart Grid Conference (SGC)* (pp. 1-5).

Schneider electric (SE). (2015). Guide for the design and production of LV power factor correction cubicles, Panel builder guide.

Singh, P., & Khan, B. (2017). Smart microgrid energy management using a novel artificial shark optimization, Complexity, Vol. 2017, Article ID 2158926. doi:10.1155/2017/2158926

Singh, P., Khan, B., Vidyarthi, A., Haes Alhelou, H., & Siano, P. (2019). Energy-aware online non-clairvoyant scheduling using speed scaling with arbitrary power function. *Applied Sciences*, *9*(7), 1467. doi:10.3390/app9071467

Sivaraman, P., Sharmeela, C., & Kothari, D. P. (2017, September). Enhancing the voltage profile in distribution system with 40 GW of solar PV rooftop in Indian grid by 2022: a review. 1st International Conference on Large Scale Grid Integration Renewable Energy in India, New Delhi, India.

Sivaraman, P., & Sharmeela, C. (2020). Solar Micro-Inverter. In J. Zbitou, C. Pruncu, & A. Errkik (Eds.), *Handbook of Research on Recent Developments in Electrical and Mechanical Engineering* (pp. 283–303). Hershey, PA: IGI Global.

Yeshalem, M. T., & Khan, B. (2017). Design of an off-grid hybrid PV/wind power system for remote mobile base station: A case study. *AIMS Energy*, *5*(1), 96–112. doi:10.3934/energy.2017.1.96

Zamani, R., Golshan, M. E. H., Alhelou, H. H., & Hatziargyriou, N. (2019). A novel hybrid islanding detection method using dynamic characteristics of synchronous generator and signal processing technique. *Electric Power Systems Research*, *175*, 105911. doi:10.1016/j.epsr.2019.105911

Zamani, R., Hamedani-Golshan, M. E., Haes Alhelou, H., Siano, P., & Pota, R., H. (2018). Islanding detection of synchronous distributed generator based on the active and reactive power control loops. *Energies*, *11*(10), 2819. doi:10.3390/en11102819

KEY TERMS AND DEFINITIONS

Electric Distribution System (EDS): It is the final stage of electric power system where the generated power is converted into useful work through distribution lines and feeders.

Primary Selective System: It receives two power supply input from PDS and powering the single or multiple distribution transformer supplying all connected loads in the system.

Reliability: A measure of power supply provided to the end user loads without power interruption.

Chapter 2
Existing Issues Associated With Electric Distribution System

Sivaraman P.
https://orcid.org/0000-0002-0591-2266
TECH Engineering Service, Chennai, India

Sharmeela C.
Anna University, Chennai, India

ABSTRACT

The common problems existing in electric distribution systems are: under voltage; overloading of distribution system components; unbalanced loading; transformer without OLTC operation; improper reactive power compensation; power theft; conversion of 3phase supply into 2phase supply; voltage sag; harmonics and system resonance condition; voltage fluctuations; problem in fault identification; transients; and renewable energy penetration, resulting in functional problems to both distribution power supply company as well as end user. This chapter discusses the various problems in the electric distribution system.

INTRODUCTION

Overview

The major components used in distribution systems are distribution transformers, distribution overhead lines, cables, switch gears like Air Circuit Breaker (ACB), Moulded Case Circuit Breaker (MCCB), Miniature Circuit Breakers (MCB), Residual Current Circuit Breaker (RCCB), fuses, etc., isolators, contactors, capacitor banks, harmonic filters, end user loads like motors, UPS systems, luminaries (incandescent lamps, florescent lamps, CFLs, LEDs, etc.), resistive loads (iron boxes, water heaters, etc.), electronic gadgets (laptops, mobile phones, digital watches, etc.) and diesel generators. Diesel generators in low voltage EDS are mostly standby generators and it will be in operation only if power grid supply is not available (Alhelou et al., 2019; Makdisie et al., 2018; Alhelou et al., 2018; Alhelou et al., 2016;

DOI: 10.4018/978-1-7998-1230-2.ch002

Haes Alhelou et al., 2019; Njenda et al., 2018). These standby diesel generators are not operating parallel with the power grid at any point of time.

The distribution system is affected by various problems like voltage sag & swell, transients, under voltage & over voltage, unbalance, voltage fluctuations, harmonics, etc. Fault in transmission system have an impact in reduction of voltage in distribution system as voltage sag because transmission system prone to more numbers of transient faults (Sivaraman et al. 2017; Khan et al., 2019; Khan et al., 2018; Khan et al., 2017; Banteywalu et al., 2019; Anteneh et al., 2019; Molla et al., 2019, Molla et al., 2018, Jariso et al. 2018). The detailed explanation of problems in distribution systems are given in this chapter.

PROBLEMS IN ELECTRIC DISTRIBUTION SYSTEM

The problems existing in EDS are

- Under voltage
- Overloading of distribution system components
- Unbalanced loading
- Transformer without OLTC operation
- Improper reactive power compensation
- Power theft
- Conversion of 3 phase supply into 2 phase supply
- Voltage sag
- Harmonics and system resonance condition
- Voltage fluctuations
- Problem in fault identification
- Transients
- Renewable energy penetration

These problems are resulting in functional problems to both distribution power Supply Company as well as end user.

1. **Under Voltage:** The under voltage is the one of the main problem of many low voltage public distribution systems. As per IEEE 1159-2009, under voltage is defined as reduction of voltage magnitude from 0.9pu to 0.1pu for the time duration greater than one minute. The causes of under voltage are sustained overloading of distribution circuit components, failure and inadequate reactive power compensation, switching on back to back higher capacity loads (IEEE Std, 2009, Chattopadhyay et al. 2011, Zobaa and Aleem, 2017). In India, there is no active voltage control in 66kV or 33kV below voltage level (GIZ, 2017).

The impact of under voltage in distribution system are poor performance of equipment's like incandescent lamps, induction motors, etc. In addition to these problems, the power generation from grid connected solar PV system also gets affected in EDS. Solar inverters are designed to operate parallel with the distribution system and most of the solar inverters available in the markets are having the voltage tolerance limits of ±20% from the nominal voltage of the system (Fini et al., 2016; Alhelou et al.,

Figure 1. Single line diagram and location of monitoring

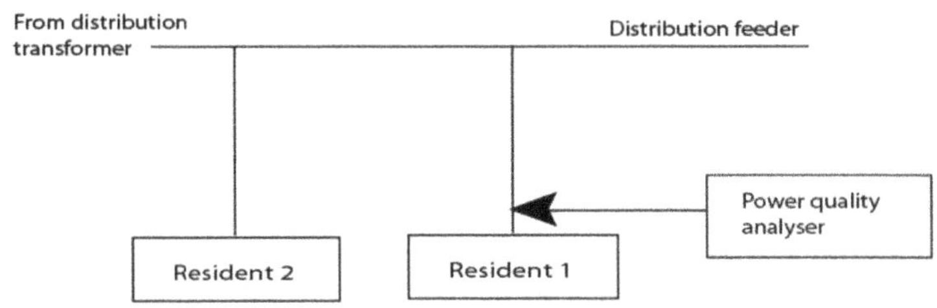

2018; Zamani et al., 2018; Alhelou et al., 2015; Njenda et al., 2018; Haes Alhelou et al., 2018; Haes Alhelou et al., 2019). Whenever, the EDS operating voltage is exceeding the inverter voltage tolerance limits, then the solar inverter automatically disconnected from the EDS and will get connected after the EDS voltage become within the inverter tolerance limits (Sivaraman et al. 2017).

Example: The voltage profile monitoring has been carried out in 1phase, 240V, residential system by using Fluke 435-II power quality analyser at Vellalur village in Tamilnadu, India from 01/07/2017 06:33:41 hours to 02/07/2017 17:13:00 hours. The block diagram of EDS and location of power quality measurement is shown in figure 1.

The summary of the voltage monitoring is listed in table 1 and voltage trend is shown in figure 2.

From table 1, the minimum recorded voltage is 168.3V and maximum recorded voltage is 237.5V between 01/07/2017 06:33:41 hours and 02/07/2017 17:13:00 hours.

The voltage profile on 01/07/2017 daytime monitoring is shown in figure 3. The summary of the voltage monitoring is listed in table 2.

From table 2, the minimum recorded voltage is 180.8V and maximum recorded voltage is 237.5V between 01/07/2017 06:33:41 hours and 01/07/2017 17:54:00 hours.

The voltage profile trend on 02/07/2017 daytime monitoring is shown in figure 4. The summary of the voltage monitoring is listed in table 3.

From table 3, the minimum recorded voltage is 190.9V and maximum recorded voltage is 230V between 02/07/2017 06:50:11 hours and 02/07/2017 17:13:00 hours.

Nominal voltage of 1ϕ 2W distribution system in India is 240 V. Solar inverters are designed to operate $\pm$20% of nominal voltage of 240 V. Inverter operating voltage is listed in table 4.

Comparing the table 2 and 3 with table 4, voltage profile in distribution system is exceeding the solar inverter operation voltage tolerance limits and in this case, inverter automatically disconnected from the grid. During this time, output power from the solar inverter is zero till the distribution system voltage become within the voltage tolerance limits of the inverter.

Table 1. Summary of voltage monitoring

Voltage	Min (V)	Max (V)	Avg (V)
Phase to Neutral	168.3	237.5	206.3

Figure 2. Voltage profile in trend

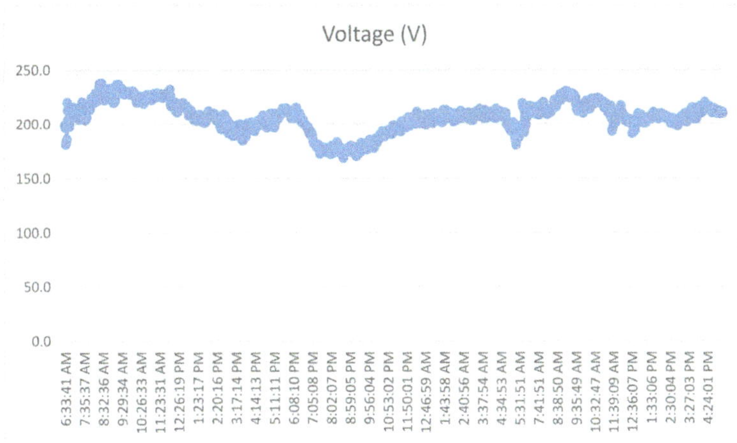

Table 2. Summary of voltage monitoring on 01/07/2017 day time

Voltage	Min (V)	Max (V)	Avg (V)
Phase to Neutral	180.8	237.5	212.8

Figure 3. Voltage profile trend on 01/07/2017 day time

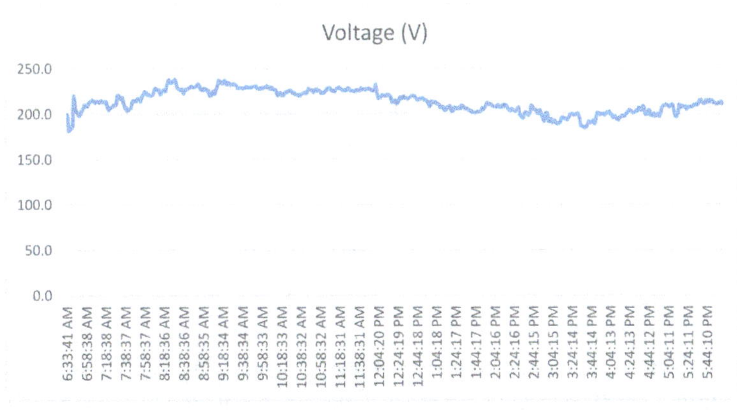

Table 3. Summary of voltage monitoring on 02/07/2017 day time

Voltage	Min (V)	Max (V)	Avg (V)
Phase to Neutral	190.9	230	211.3

Figure 4. Voltage profile trend on 02/07/2017

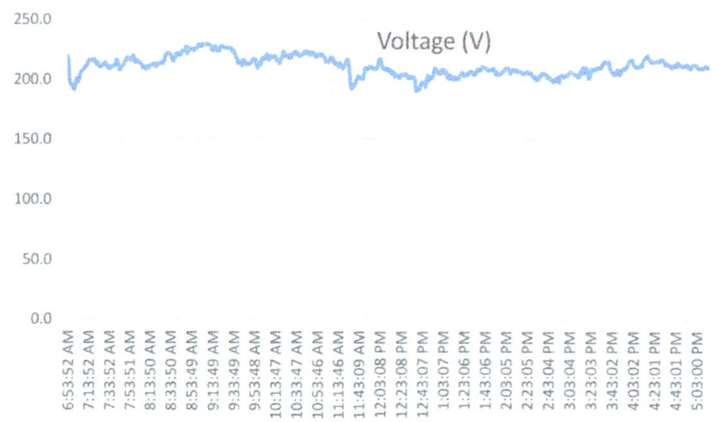

Table 4. Inverter operating voltage

Nominal Voltage	Min (V)	Max (V)
240 V	192	288

2. **Overloading of Distribution Components:** The distribution systems are designed certain criteria (present load and expected future load growth). Due to various reasons like unbalancing loadings amongst three phases, under sized components, improper reactive power compensation, power theft, harmonics, etc. cause the over loading in distribution network components like transformers, distribution lines and cables, etc. This resulting in higher voltage drop in the over loaded phase. Overloading of distribution components shall overcome by selecting the proper equipment rating, balancing the loads whenever single phase loads are connected in three phase 4W systems, by installing the suitable reactive power compensating devices at one or multiple locations in the distribution systems (Sivaraman et al. 2017; Khan et al. 2014; Khan et al. 2013; Khan et al. 2012; Negash et al., 2017; Negash et al., 2016; Jariso et al. 2017; Kifle et al. 2018; Yeshalem et al. 2017; Singh et al. 2019; Gupta et al., 2015).
3. **Unbalanced Loading:** One of the major problems in most of the 3ϕ 4W distribution system is unbalanced loading. Unbalance loading means, the distribution system operating in unequal load impedance connected in all three phases leading to unequal current drawl in all three phases. The main cause of unbalanced loading is mainly due to 1ϕ loads. This problem exists mainly due to same distribution feeder powering the combination both 1ϕ & 3ϕ loads in residential, commercial and industrial power system. These 1ϕ loads are majorly used in residential and commercial system powered through 3ϕ 4W distribution network. However, all 1ϕ loads needs to be balanced amongst three phases but if these loads are not balanced then the distribution system become unbalanced. In this condition, current flows in all three phases are different from each other. Due to this, voltage

drop/phase in the EDS components shall varying. That is lesser voltage drop in lightly loaded phase and higher voltage drop in highly loaded phase. Due to this, receiving end loads will get unequal voltage magnitude between three phases and leading to poor performance of the equipment's connected in this network (Sivaraman et al. 2017, Chattopadhyay et al. 2011, Zobaa and Aleem, 2017).

Causes:

- Unequal load impedance of single-phase loads connected into the system
- Unbalanced voltage from incoming utility power supply
- Unequal impedance offered by cables or conductors

Impacts:

- Increased power losses in the distribution circuit
- Increased current flow in neutral conductor
- Reduced performance of equipment's like motors
- Overheating of equipment's

4. **Transformer Without OLTC Operation:** Most of the distribution transformers present in the rural Indian power systems are deployed without an OLTC. Due to various reasons like reduced primary voltage from the nominal voltage, over loading, more reactive power drawl etc. leading to voltage at the distribution system is reduced below the nominal voltage (under voltage). Without OLTC operation, the distribution system is not achieved voltage control in the secondary side (Sivaraman et al. 2017). Active voltage control shall be achieved by using the secondary distribution transformers with OLTC (GIZ, 2017).
5. **Improper Power Factor Compensation:** Drawl of more reactive power from the distribution system shall reduce the active power consumption capacity. The poor reactive power compensation or no reactive power compensation shall lead to consumption of more reactive power from the grid (ABB, 2018). Presently most of the Indian DISCOMs are mandated the minimum power factor requirement at the billing location or PCC for medium scale and large scale commercial systems, industrial systems. DISCOMs are imposing the penalty for not maintaining the minimum power factor requirement at billing location or PCC. However, for residential and small-scale commercial systems, maintaining the power factor correction is not mandatory (Sivaraman et al. 2017). Due to various reasons like, lack of awareness, low cost, easy availability, people are opting the equipment's operating at low power factor (0.4 to 0.7 lag PF).

This can be avoided by using the power factor corrected equipment's, using the properly sized reactive power compensation devices and installing the reactive power compensating devices at suitable location.

6. **Power Theft:** Distribution systems are highly affected by drawing of power in temporary or permanent illegal connection. This is called as power thefts. The person who are conversant in operation of electrical system can make the illegal connections at the distribution lines easily. These illegal connections are drawing the uncounted or undocumented real and reactive power from the distribution line. That means illegal connections are drawing un-accountable real and reactive power from the distribution system and this leading to over loading of distribution circuits like transformer, distribution lines, cables (Sivaraman et al. 2017). It also resulting to unbalanced loading in all the three phase and it directly affect the voltage profile of the distribution system. Finally financial burden to DISCOMs. It can be avoided by improving the grid visibility. That means, grid shall be monitored continuously by grid operator and disconnect the particular feeder/section if power theft is found.
7. **Conversion of 3 Phase Supply into 2 Phase Supply:** Some distribution supply company having the restrictions to providing the continuous 3ϕ supply to agricultural usage. This condition still exist in some countries like India. Distribution Supply Company provides the 3ϕ power supply for defined time period and after this period, distribution Supply Company change the 3ϕ power supply into 2ϕ supply. By doing this, distribution Supply Company avoids the agricultural load operating on the particular time. Whenever utility company changing the 3ϕ supply into 2ϕ supply, 3ϕ 4W distribution becomes to 2ϕ 3W system and the loads connected in the third phase will not receive the supply voltage. Many of the rural distribution systems in India facing this problem as changing 3ϕ to 2ϕ supply to reduce the agricultural uses and have an impact in entire distribution, because there is no separate feeder is used for agriculture and residential application (Sivaraman et al. 2017). A common feeder is used for powering all the residential, commercial (small scale) and agricultural loads. So all the 1ϕ loads in the particular distribution system is forced to connect at any two phase (R & Y) or (R & B) or (Y & B) except the absent phase.

Due to this, loading in these two phase (R & Y) or (R & B) or (Y & B) keep increasing and one absent phase (R or Y or B) is not loaded at all. This leads to unbalanced loading of all the three phase, resulting to higher voltage drop in the loaded phase, no voltage drop in the absent phase so unbalanced voltage at the receiving end.

This condition severely affect the solar PV system integration into the distribution system. A 3ϕ solar PV system is connected parallel to the distribution grid can delivers the power when all three phases are available and it cannot deliver the power when any one phase is absent till the absent phase recovered to nominal voltage. So for this condition, 3ϕ solar PV system cannot deliver the power whenever 3ϕ supply is converted into 2ϕ supply.

In this condition 3ϕ solar PV system is not viable for continuous operation and forced to get 1ϕ solar PV systems. This 1ϕ solar PV systems can connect at the particular two (available) phase only and absent phase in this distribution grid becomes not used. For this reason, voltage profile of these two phases exceed the inverter voltage tolerance limits (typically ±20%), then 1ϕ solar PV system gets disconnected from the grid and connected after the grid voltage within ±20% voltage tolerance limit (Sivaraman et al. 2017).

8. **Voltage sag:** The IEEE 1159-2009, definition for voltage sag is short duration reduction of RMS voltage from 0.9pu to 0.1pu for time duration from 0.5 cycles to less than one minute (IEEE Std, 2009).

The main causes of voltage sag in the distribution system are (Chattopadhyay et al. 2011, Zobaa and Aleem, 2017).

1. Electrical fault (SLG)
2. Large motor starting
3. Energisation of higher rating transformer
4. Energisation of group of load at a same time
5. Faults in transmission system resulting in voltage sag in distribution system
6. Temporary overloading of circuit components
7. Poor wiring connection or loose connection
8. Operation of SPDs and TVSS
9. De-Energisation of capacitor banks

The impacts of voltage sag in the distribution systems are

1. Mal operation / tripping / failure of sensitive electronic loads
2. Reduced performance of motors, incandescent lamps, etc.

The equipment's are sensitive to reduction of voltage, shall be protected from voltage sag by using the sag fighter or stabilizers.

10. **Harmonics and System Resonance Condition:** Harmonics are the one of the inevitable problems in modern low voltage AC distribution system. As per IEEE 1159-2009, definition of harmonics is sinusoidal current or voltage having frequencies that are integral multiples of fundamental supply frequency (50Hz or 60Hz) (IEEE Std, 2009, Chattopadhyay et al. 2011).

Now a days power electronics based loads are widely used for sophisticated operation and control. Some of these loads are UPS systems, VFDs, SMPS, LEDs, etc., and these loads are exhibit nonlinear voltage-current characteristics (harmonics) during their operation.

The impacts of harmonics in EDS are (Wakileh, 2001, Das, 2015).

- Premature failure of equipment's
- Overloading of cables
- Overloading of neutral conductors due to triplen harmonics
- Overloading of distribution transformers
- Overheating of equipment's like transformers, motors, cables, etc, resulting to insulation failure
- Increased audible noise in transformers, motors, etc.
- Skin effect and proximity effect in power cables
- Mal operation of power electronic loads
- Higher neutral to ground voltage resulting in failure of sensitive electronics loads
- Poor voltage and current quality affecting the performance of other connected loads in the distribution
- Current imbalance in transformer, excess loading of transformer in individual phases
- EMI on communication systems

- System resonance condition

Resonance is a condition in power system. In this condition network inductive reactance is equal to capacitive reactance at some frequency due to harmonics. During this condition, cyclic energy transfer between the inductive and capacitive elements in the power system. This phenomenon is called as resonance and at which frequency resonance phenomenon is occurring is called as resonance frequency (Wakileh, 2001, Chattopadhyay et al., 2011).

The condition for resonance occurs in the system is

$$X_L = X_C$$

Where, X_L is inductive reactance and X_C is capacitive reactance.

The major electrical equipment's in the systems are inductive by theirs characteristics. Use of power factor correction capacitors and harmonic filters are introducing the capacitive characteristics into the system. Resonance condition occurs only due to multiple frequencies generated by non-linear loads. At any one of multiple frequencies, system inductive reactance is equal to capacitive reactance. Based on the system configuration, resonance conditions are classified into two types (IEEE Std, 1998).

- Series resonance
- Parallel resonance

One of the easiest ways to overcome resonance condition is by means of adding the reactor. Adding the reactor in series with each capacitor shall change the system to inductive at resonant frequencies and capacitive at fundamental frequency.

11. **Voltage Fluctuations:** Voltage fluctuation is continuous change in instantaneous voltage (cycle to cycle) due to variation in load resistance in every cycle (IEEE Std, 2014). The continuous voltage changes are called as voltage fluctuation. Definition of voltage fluctuation by IEC 61000-3-3 is series of changes of RMS voltage evaluated as a single value for each successive half period between zero crossings of the source voltage (IEC, 2017). The major cause of voltage fluctuation is arc furnace in the industrial plant. Voltage fluctuations have an impact in illumination intensity from lamp. That is continuous variation in voltage have an impact in illumination density resulting in noticeable change in illumination by normal human eye. This phenomenon is called as flicker or voltage flicker.

Definition of flicker by IEC 61000-3-3 is impression of unsteadiness of visual sensation induced by a light stimulus whose luminance or spectral distribution fluctuates with time.

Flicker is classified into two categories as short-term flicker (P_{st}) and long-term flicker (P_{lt}). Short term flicker is measured and evaluated over a short time period in minutes. Long term flicker is measured and evaluated over a long time period in hours. As per IEC 61000-3-3, the limits of short-term flicker shall not be more than 1.0 and long-term flicker shall not be more than 0.65 (IEC, 2017).

Voltage fluctuations can be caused by any of the conditions of which few are listed below

Table 5. Classifications of transients

Parameter	Category	Type	Duration / Frequency
Transients	Impulsive	Nano second	< 50 ns
		Micro second	50 ns - 1 ms
		Milli second	> 1 ms
	Oscillatory	Low frequency	< 5 kHz
		Medium frequency	5 - 500 kHz
		High frequency	500 kHz - 5 MHz

Table 6. Sources of transients

Category	Type	Sources of problems
Transients	External sources	Lighting strike
		Opening and closing of energized lines
		Breaker opening and reclosing
		Transients from other users in shared sources
	Internal sources	Switching operation
		High resistance fault
		Photocopy machines
		Welding machine
		Capacitor bank switching
		High frequency switching in inverter or SMPS

1. Arc furnace
2. Arc welding
3. Arching load

The impacts of voltage fluctuations are few are listed below

1. Reduced performance of the equipment's connected in same bus
2. Flickering in incandescent lamp

12. **Problem in Fault Identification:** Fault identification by manual identification in distribution system is the complex task for DISCOMs. The country like India, distribution system is a large network, manual fault identification is complex and more challenging. In manual fault identification, group of people or team has to investigate the particular section of distribution manually to identify the faults. This method takes more time to identify the fault locations and this will reduce the supply reliability to the end users.

13. **Transients:** Transients are momentary changes in voltage or current or both resulting due to switching operation, lightning, faults (Das, 2015, Chattopadhyay et al. 2011, Zobaa and Aleem, 2017). The time duration for a transient vary from nano-second to milli-second. The classifications of transients as listed in IEEE 1159 – 2009 are shown under table 5 (IEEE Std, 2009).

Transient over voltage is the result of rapid change of current in an inductive circuit.
The sources of transients are listed in below table 6.
The impacts of transients in the distribution systems are

1. Failures of insulation in inductive loads if the transients exceed the impulsive voltage withstand capability.
2. Failure of equipment's due to lightning induced over voltage.
3. Failure of semiconductor devices and its associated circuits due to peak inverse voltage.
4. Failure of control wiring in electronic circuits.

Renewable Energy Penetration

Now a days, distributed generation plays an important role in localised energy generation nearer to the loads. Renewable energy sources like solar PV and wind energy conversion systems are widely used for distributed generation for localized energy generation (Sivaraman et al., 2017). The following are limiting the renewable energy integration into the distribution system

1. Under voltage: The impact of under voltage in solar PV generation is explained in under voltage.
2. Voltage sag: Solar PV inverter will get disconnected from the grid due to reduction of voltage during voltage sag event and connected to grid after the voltage sag. Voltage sag may occur at many times in a day due to various reasons like faults, switching ON higher capacity loads & transformers, etc. Tripping of inverter during voltage sag can be avoided by LVRT features of the inverter. Modern inverters having the inbuilt LVRT features within the inverters.
3. Unbalanced loading amongst three phases: The impact of unbalanced loading amongst three phases will affect the 3phase solar PV system integration into the system.
4. Conversion of 3 phase supply into 2 phase supply: The impact of conversion of 3 phase supply into 2 phase supply is explained in conversion of 3 phase supply into 2 phase supply.
5. Rating of distribution transformer: The rating of the distribution transformer is limiting the renewable energy integration capacity into the system.

CONCLUSION

As distribution system faces so many issues, this chapter provides the overview of all these issues. Different issues such as under voltage, overloading of distribution system components, unbalanced loading, transformer without online tap changer operation, improper reactive power compensation, power theft, conversion of 3 phase supply into 2 phase supply, voltage sag, harmonics and system resonance condition, voltage fluctuations, problem in fault identification, transients and renewable energy penetration are discussed in this chapter.

REFERENCES

ABB. (2018). Power factor correction and harmonic filtering in electrical plants, Technical application papers, July, 2018.

Alhelou, H., Hamedani-Golshan, M. E., Zamani, R., Heydarian-Forushani, E., & Siano, P. (2018). Challenges and opportunities of load frequency control in conventional, modern and future smart power systems: A comprehensive review. *Energies*, *11*(10), 2497. doi:10.3390/en11102497

Alhelou, H. H. (2018). *Fault detection and isolation in power systems using unknown input observer. Advanced condition monitoring and fault diagnosis of electric machines* (p. 38). Hershey, PA: IGI Global.

Alhelou, H. H., Golshan, M., & Fini, M. (2018). Wind driven optimization algorithm application to load frequency control in interconnected power systems considering GRC and GDB nonlinearities. *Electric Power Components and Syst.*

Alhelou, H. H., & Golshan, M. E. H. (2016, May). Hierarchical plug-in EV control based on primary frequency response in interconnected smart grid. In *2016 24th Iranian Conference on Electrical Engineering (ICEE),* (pp. 561-566). IEEE. 10.1109/IranianCEE.2016.7585585

Alhelou, H. H., & Golshan, M. E. H. (2020). Decision-making-based optimal generation-side secondary-reserve scheduling and optimal LFC in deregulated interconnected power system. In *Decision making applications in modern power systems* (pp. 269–299). Academic Press. doi:10.1016/B978-0-12-816445-7.00011-6

Alhelou, H. H., Golshan, M. E. H., & Hatziargyriou, N. D. (2019). Deterministic dynamic state estimation-based optimal LFC for interconnected power systems using unknown input observer. *IEEE Transactions on Smart Grid*, 1. doi:10.1109/TSG.2019.2940199

Alhelou, H. H., Golshan, M. E. H., & Hatziargyriou, N. D. (2019). A decentralized functional observer based optimal LFC considering unknown inputs, uncertainties, and cyber-attacks. *IEEE Transactions on Power Systems*.

Alhelou, H. H., Golshan, M. E. H., Zamani, R., Moghaddam, M. P., Njenda, T. C., Siano, P., & Marzband, M. (2019, June). An improved UFLS scheme based on estimated minimum frequency and power deficit. In 2019 IEEE Milan PowerTech (pp. 1-6). IEEE. doi:10.1109/PTC.2019.8810497

Alhelou, H. H., Golshan, M. H., & Askari-Marnani, J. (2018). Robust sensor fault detection and isolation scheme for interconnected smart power systems in presence of RER and EVs using unknown input observer. *International Journal of Electrical Power & Energy Systems*, *99*, 682–694. doi:10.1016/j.ijepes.2018.02.013

Alhelou, H. H., Hamedani-Golshan, M. E., Heydarian-Forushani, E., Al-Sumaiti, A. S., & Siano, P. (2018, September). Decentralized fractional order control scheme for LFC of deregulated nonlinear power systems in presence of EVs and RER. In *2018 International Conference on Smart Energy Systems and Technologies (SEST)* (pp. 1-6). IEEE. 10.1109/SEST.2018.8495858

Alhelou, H. S. H., Golshan, M. E. H., & Fini, M. H. (2015, December). Multi agent electric vehicle control based primary frequency support for future smart micro-grid. In *Smart Grid Conference (SGC)* (pp. 22-27).

Anteneh, D., & Khan, B. (2019). Reliability enhancement of distribution substation by using network reconfiguration a case study at Debre Berhan distribution substation. *International Journal of Economy, Energy, & Environment*, *4*(2), 33.

Banteywalu, S., Khan, B., De Smedt, V., & Leroux, P. (2019). A novel modular radiation hardening approach applied to a synchronous buck converter. *Electronics (Basel)*, *8*(5), 513. doi:10.3390/electronics8050513

Chattopadhyay, S., Sengupta, S., & Mitra, M. (2015). *Electric power quality*. Springer.

Das, J. C. (2015). Power system harmonics and passive filter designs, IEEE Press.

Fini, M. H., Yousefi, G. R., & Alhelou, H. H. (2016). Comparative study on the performance of many-objective and single-objective optimisation algorithms in tuning load frequency controllers of multi-area power systems. *IET Generation, Transmission & Distribution*, *10*(12), 2915–2923. doi:10.1049/iet-gtd.2015.1334

GIZ. (2017). *Analysis of Indian electricity distribution systems for the integration of high shares of rooftop PV, final report*. GIZ.

Gupta, K., Khan, B., & Mubeen, S. E. (2015). Available transfer capability enhancement by unified power flowcontroller. *IEEE International Conference on Signal Processing, Informatics, Communication, and Energy Systems (SPICES),* 2015, pp. 1, 4, 19-21.

Haes Alhelou, H., Hamedani-Golshan, M. E., Njenda, T. C., & Siano, P. (2019). A survey on power system blackout and cascading events: Research motivations and challenges. *Energies*, *12*(4), 682. doi:10.3390/en12040682

Haes Alhelou, H., Hamedani-Golshan, M. E., Njenda, T. C., & Siano, P. (2019). Wide-area measurement system-based optimal multi-stage under-frequency load-shedding in interconnected smart power systems using evolutionary computing techniques. *Applied Sciences*, *9*(3), 508. doi:10.3390/app9030508

Haes Alhelou, H., Hamedani Golshan, M. E., Njenda, T. C., & Siano, P. (2019). WAMS-based online disturbance estimation in interconnected power systems using disturbance observer. *Applied Sciences*, *9*(5), 990. doi:10.3390/app9050990

IEC 61000-3-3. (2017). Limitation of voltage changes, voltage fluctuations and flicker in public low-voltage supply systems, for equipment with rated current ≤16A per phase and not subject to conditional connection, 2017.

IEEE. (1998). Recommended practice for industrial and commercial power systems analysis (Brown book). In IEEE Std 399-1997, pp. 1-488.

IEEE. (2009) Recommended practice for monitoring electric power quality. In IEEE Std 1159-2009 (Revision of IEEE Std 1159-1995), pp. 1-94.

IEEE. (2014) Recommended practice and requirements for harmonic control in electric power systems. In IEEE Std 519-2014 (Revision of IEEE Std 519-1992), pp. 1-29.

Jariso, M., Khan, B, Tanwar, S., Tyagi, S., & Rishiwal, V. (2018). Hybrid energy system for upgrading the rural environment, *IEEE GlobeCom 2018.*

Jariso, M., Khan, B., Tesfaye, D., & Singh, J. (2017). Modelling and designing of stand-alone photovoltaic system (Case study: Addis Boder Health Center southwest Ethiopia). In *Proceedings IEEE International Conference on Electronics, Communication, and Aerospace Technology, ICECA 2017,* pp. 168-173, Coimbatore, India.

Khan, B., & Agnihotri, G. (2012). A novel transmission loss allocation method based on transmission usage. In *Proceedings IEEE Fifth Power India Conference*, 2012, Haryana, India. 10.1109/PowerI.2012.6479479

Khan, B., Agnihotri, G., & Gupta, G. (2013). A multipurpose matrices methodology for transmission usage, loss and reliability margin allocation in restructured environment, *Electrical & Computer Engineering: An International Journal, 2*(3), September 2013.

Khan, B., Agnihotri, G., Rathore, P., Mishra, A., & Naidu, G. (2014). A cooperative game theory approach for usage and reliability margin cost allocation under contingent restructured market. *International Review of Electrical Engineering*, *9*(4), 854–862.

Khan, B., Alhelou, H. H., & Mebrahtu, F. (2019). A holistic analysis of distribution system reliability assessment methods with conventional and renewable energy sources. *AIMS Energy*, *7*(4), 413–429. doi:10.3934/energy.2019.4.413

Khan, B., & Singh, P. (2017). The current and future states of Ethiopia's energy sector and potential for green energy: A comprehensive study. *International Journal of Engineering Research in Africa*, *33*, 115–139. doi:10.4028/www.scientific.net/JERA.33.115

Khan, B., & Singh, P. (2017). Optimal power flow techniques under characterization of conventional and renewable energy sources: A comprehensive analysis, *Journal of Engineering, vol. 2017,* Article ID 9539506.

Khan, B., & Singh, P. (2017). Selecting a meta-heuristic technique for smart micro-grid optimization problem: A comprehensive analysis, in IEEE Access, 5, pp. 13951-13977.

Khan, B., & Singh, P. (2019). Economic operation of smart micro-grid: a meta-heuristic approach. In Handbook of research on smart power system operation and control (pp. 330-346). Hershey, PA: IGI Global. doi:10.4018/978-1-5225-8030-0.ch014

Khan, B., & Singh, P. (2019). Economic operation of smart micro-grid: a meta-heuristic approach. In Handbook of research on smart power system operation and control (pp. 330-346). Hershey, PA: IGI Global. doi:10.4018/978-1-5225-8030-0.ch014

Khan, B., & Tanwar, S. (2019). Issues associated with microgrid integration. In Handbook of research on smart power system operation and control (pp. 252-264). Hershey, PA: IGI Global. doi:10.4018/978-1-5225-8030-0.ch010

Kifle, Y., Khan, B., & Singh, J. (2018). Designing and modeling grid connected photovoltaic system: (Case study: EEU building at Hawassa city). *International Journal of Convergence Computing*, *3*(1), 20–34. doi:10.1504/IJCONVC.2018.091113

Kifle, Y., Khan, B., & Singh, P. (2018). Assessment and enhancement of distribution system reliability by renewable energy sources and energy storage, Journal of Green Engineering, 8(3), 2, pp. 219-262.

Makdisie, C., Haidar, B., & Alhelou, H. H. (2018). An optimal photovoltaic conversion system for future smart grids. In Handbook of research on power and energy system optimization (pp. 601–657). Hershey, PA: IGI Global. doi:10.4018/978-1-5225-3935-3.ch018

Molla, T., Khan, B., Moges, B., Alhelou, H. H., Zamani, R., & Siano, P. (2019). Integrated optimization of smart home appliances with cost-effective energy management system. *CSEE Journal of Power and Energy Systems*, *5*(2), 249–258.

Molla, T., Khan, B., & Singh, P. (2018). A Comprehensive Analysis of Smart Home Energy Management System Optimization Techniques, *Journal of Autonomous Intelligence, 1(1)*, 15–21.

Negash, K., Khan, B., Tesfaye, D., & Gayathri, M. (2017). Optimal placement of phasor measurement unit for system observability. In *Proceedings of IEEE International Conference on Computing Methodologies and Communication, (ICCMC 2017),* pp. 73-77, Erode, India.

Negash, K., Khan, B., & Yohannes, E. (2016). Artificial intelligence versus conventional mathematical techniques: A review for optimal placement of phasor measurement units [Springer]. *Journal of Technology and Economics of Smart Grids and Sustainable Energy*, *1*(1), 10. doi:10.100740866-016-0009-y

Njenda, T. C., Golshan, M. E. H., & Alhelou, H. H. (2018, November). WAMS based intelligent under frequency load shedding considering online disturbance estimation. In *Smart Grid Conference (SGC)* (pp. 1-5). 10.1109/SGC.2018.8777779

Njenda, T. C., Golshan, M. E. H., & Alhelou, H. H. (2018, November). WAMS based under frequency load shedding considering minimum frequency prediction and extrapolated disturbance magnitude. In *Smart Grid Conference (SGC)* (pp. 1-5).

Singh, P., & Khan, B. (2017). Smart microgrid energy management using a novel artificial shark optimization, Complexity, vol. 2017, Article ID 2158926. doi:10.1155/2017/2158926

Singh, P., Khan, B., Vidyarthi, A., Haes Alhelou, H., & Siano, P. (2019). Energy-aware online non-clairvoyant scheduling using speed scaling with arbitrary power function. *Applied Sciences*, *9*(7), 1467. doi:10.3390/app9071467

Sivaraman, P., Sharmeela, C., & Kothari, D. P. (2017). Enhancing the voltage profile in distribution system with 40 GW of solar PV rooftop in Indian grid by 2022: A review. *1st International Conference on Large Scale Grid Integration Renewable Energy in India*, New Delhi, India.

Sivaraman, P., & Sharmeela, C. (2020). Solar Micro-Inverter. In J. Zbitou, C. Pruncu, & A. Errkik (Eds.), *Handbook of Research on Recent Developments in Electrical and Mechanical Engineering* (pp. 283–303). Hershey, PA: IGI Global.

Wakileh, G. J. (2001). Power systems harmonics: Fundamentals, analysis, and filter design. Springer.

Yeshalem, M. T., & Khan, B. (2017). Design of an off-grid hybrid PV/wind power system for remote mobile base station: A case study. *AIMS Energy*, *5*(1), 96–112. doi:10.3934/energy.2017.1.96

Zamani, R., Golshan, M. E. H., Alhelou, H. H., & Hatziargyriou, N. (2019). A novel hybrid islanding detection method using dynamic characteristics of synchronous generator and signal processing technique. *Electric Power Systems Research*, *175*, 105911. doi:10.1016/j.epsr.2019.105911

Zamani, R., Hamedani-Golshan, M. E., Haes Alhelou, H., Siano, P., & Pota, H. R. (2018). Islanding detection of synchronous distributed generator based on the active and reactive power control loops. *Energies*, *11*(10), 2819. doi:10.3390/en11102819

Zobaa, A. F., & Aleem, S. H. E. A. (2017). Power quality in future electrical systems, IET, 2017.

KEY TERMS AND DEFINITIONS

Transient: It is a momentary variation in current, voltage, or frequency.

Unbalanced Loading: The imbalance occurs when an open or short circuit appears at the load.

Voltage Sag: A voltage sag or voltage dip is a short duration reduction in RMS voltage which can be caused by a short circuit, overload or starting of electric motors.

Chapter 3
Power Quality Improvement in Distribution System Using Dynamic Voltage Restorer

Tesfahun Molla
Hawassa University, Hawassa, Ethiopia

ABSTRACT

With the advancement of technology, the dependency on the electrical energy has been increased greatly. Computer and telecommunication networks, railway network banking, post offices, and life support systems are a few applications that cannot function without electricity. At the same time, these applications demand qualitative energy. However, the quality of power supplied is affected by various internal and external factors of the power system. Harmonics, voltage, and frequency variations deteriorate the performance of the system. Voltage sag/dip is the most frequent problem and there are many methods to overcome this problem. The use of FACT devices is an efficient one. This chapter discusses an overview of the FACT device known as dynamic voltage restorer (DVR) in mitigating voltage sag. The strategy to control this device is also presented. The proposed control strategies are simulated in MATLAB SIMULINK environment and analyzed. The method is utilized and discussed briefly.

INTRODUCTION

Modern society is fully dependent on the Power generated by generating station. Traditional power system comprises of three parts i.e. generation, transmission and distribution of electrical power in the form of AC. The generated power should have good quality so that it can energize all equipments or appliances equally and satisfactorily (Papic, 2000). Due to heavy loads or any abnormal conditions or faults on the line reduces the quality of the power, becomes less suitable for further applications.

The power provided by generating station must be improved for delivering pure and clean power to the end users. For delivering a good quality of power, Flexible AC Transmission System (FACTS) devices like static synchronous series compensator (SSSC), static synchronous compensator (STATCOM), interline power flow controller (IPFC), unified power flow controller (UPFC) etc. were used.

DOI: 10.4018/978-1-7998-1230-2.ch003

Generally, FACTS devices are modified to be used in electrical distribution system known as Custom Power Devices. Some of the widely used custom power devices are Distribution Static Synchronous Compensator (DSTATCOM), Dynamic Voltage Restorer (DVR), Active filter (AF) and Unified power quality conditioner (UPQC) (Pal and Gupta, 2015; Khan et al., 2019; Khan et al., 2018; Khan et al., 2017; Banteywalu et al., 2019; Anteneh et al., 2019; Molla et al., 2019, Molla et al., 2018, Jariso et al. 2018). These devices are used to reduce power quality problems. DVR is one of the most efficient and effective custom power devices due to its fast response, lower cost and smaller size. Power electronics technology had played an important role in power flow control and utilization of electrical energy. Consumers need constant sine wave, constant frequency and symmetrical voltage with a constant root mean square (RMS) value to continue the production (Alhelou et al., 2019; Makdisie et al., 2018; Alhelou et al., 2018; Alhelou et al., 2016; Haes Alhelou et al., 2019; Njenda et al., 2018). So to satisfy these demands, of course the disturbances must be eliminated from the system. Some of the typical Power quality issues in the system such as sags, swells, these disturbances change the shape of the sine waveform of the supply voltage and adversely affect the performance of equipment connected to the system (Ansal et al. 2016). The voltage sag's magnitude mainly ranged from 10% to 90% of nominal voltage and with duration from half a cycle to 1 min and swell is defined as an increase in RMS voltage or current at the power frequency for durations from 0.5 cycles to 1 min. (Singh et al. 2016).

There are two general approaches to mitigate power quality disturbances. One approach is to ensure that the process equipment is less responsive to disturbances, allowing it to ride-through the disturbances (Pal and Gupta, 2015; Khan et al. 2014; Khan et al. 2013; Khan et al. 2012; Negash et al., 2017; Negash et al., 2016; Jariso et al. 2017; Kifle et al. 2018; Yeshalem et al. 2017; Singh et al. 2017; Gupta et al., 2015). The other approach to suppress or neutralize the disturbance at the customer end is installing custom power devices. The DVR is one of most effective and efficient power electronic custom power device, which is used to inject voltage in series with distribution feeder in order to compensate for voltage sag/swell. In order to restore the load voltage, active and/or reactive power should be injected into the distribution feeder.

Literature Review

Voltage quality is the most important portion of power quality. The quality of voltage can be affected by several events including voltage sags and voltage swells. These events can cause malfunction of voltage sensitive loads and automated process disruption. As a result, the voltage sags and swells can lead to huge financial and technical losses. Therefore, the voltage sags and swells should be avoided as far as possible. One of the solutions is to install a proper device at sensitive load location to mitigate the voltage sags and swells (Alhelou et al., 2019; Makdisie et al., 2018; Alhelou et al., 2018; Alhelou et al., 2016; Haes Alhelou et al., 2019; Njenda et al., 2018). Power electronic based solutions have been widely applied for solving the problems (Babaei and Kangarlu, 2015). There are varieties of custom power devices available each with its own benefits and boundaries: Active Power Filters (APF), Battery Energy Storage Systems (BESS), Distribution Series Capacitors (DSC), Solid-State Transfer Switches (SSTS), Surge Arresters (SA), Super conducting Magnetic Energy Systems (SMES), Uninterruptible Power Supply (UPS), Static Electronic Tap Changers (SETC), Solid State Fault Current Limiter (SSFCL), Static VAR Compensator (SVC) and Thyristor Switched Capacitors (TSC), unified power-quality conditioner (UPQC), Distribution-STATCOM (DSTATCOM) and dynamic voltage restorer (DVR).

DVR is considered as an effective and efficient custom power device for mitigating the impact of voltage disturbances on sensitive load. In addition, DVR also has functions such as reactive power compensation and harmonic compensation (Mohammed Shazly et al. 2013). The first DVR was installed at 12.47 kV substations in 1996 in North America, Anderson. The Dynamic Voltage Restorer (DVR) is fast, flexible, efficient and cost-effective solution to voltage sag and voltage swell problems. It is used to protect sensitive loads and regulate the load voltage. The modeling and the operating principles are discussed briefly. Different control strategy to mitigate voltage sag is presented. The fuzzy logic controller is used to control the injection of voltage when the fault occurs (Kalia et al., 2014). The DVR injects the voltage by Multi level inverter which converts dc voltage in the Energy Storage System (ESS) into ac and the amount of the injection of voltage is controlled by the fuzzy logic controller. The various control techniques are discussed in (Rovai and Doorwar, 2014). The compensation methods and the operating principles of DVR are discussed. SVPWM based DVR is presented in (Ding et al. 2002). In this control algorithm, the three Phase supply is converted into synchronously rotating d-q reference frame (Fini et al., 2016; Alhelou et al., 2018; Zamani et al., 2018; Alhelou et al., 2015; Njenda et al., 2018; Haes Alhelou et al., 2018; Haes Alhelou et al., 2019). The d-component gives information for depth of sag and q-component tells us about phase shift information. The error generated is given to SVPWM for DVR operation.

Power Quality

"Power quality is described as the variation of voltage, current and frequency in the power system. It refers to a wide variety of electromagnetic phenomena that characterized the voltage and current at the given time and location in the power system" (Mohammed Shazly et al., 2013). Broadly speaking, a power system can be grouped into three sub system such as:

- Electricity generation system, also known as a power resource, is the process of generating electrical energy from other forms of energy.
- Electric-power transmission system is the bulk transfer of electrical energy, from generating power plants to electrical substations located near demand centers. This is distinct from the local wiring between high-voltage substations and customers, which is typically referred to as electric power distribution. Transmission lines, when interconnected with each other, become transmission networks. Transmission system is to transmit a large-scale electric power to distribution system.
- Electricity distribution system is the final stage in the delivery of electricity to end users. A distribution system's network carries electricity from the transmission system and delivers it to consumers. Typically, the network would include medium-voltage power lines, substations and pole mounted transformers, low-voltage distribution wiring and sometimes meters.

Power Quality Problems

Power quality and reliability cost the industry large amounts due to mainly sags and short-term interruptions. Distorted and unwanted voltage wave forms, too. And the main concern for the consumers of electricity was the reliability of supply. Here authors define the reliability as the continuity of supply. The problem of distribution lines is divided into two major categories. First group is power quality, second is power reliability (Rani and Sivakumar, 2014). First group consists of harmonic distortions, impulses and

Figure 1. Voltage divider model for voltage sags/swells

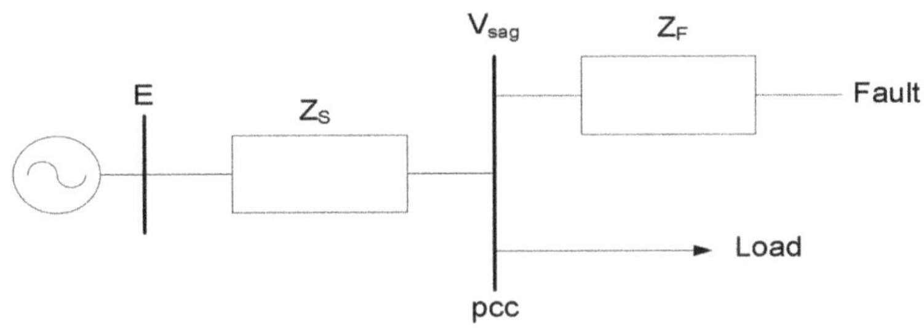

swells. Second group consists of voltage sags and outages. Voltage sags is much more serious and can cause a large amount of damage. If exceeds a few cycle, motors, robots, servo drives and machine tools cannot maintain control of process. The best-known disturbances of the voltage waveform are voltage sags and swells, harmonics, inter-harmonics and voltage imbalances. Voltage-quality problems are as follows (Rauf and Khadkikar, 2015).

- Voltage Sag
- Voltage Swell
- Harmonics
- Voltage transients
    - Impulsive Transients
    - Oscillatory Transient
- Flicker

Computation of Voltage Sags and Swells

To calculate the amount of voltage sag / swell in radial systems, the voltage divider model can be used as shown in Figure 1.

In Figure 1 authors see two impedances: Z_S is the source impedance at the point-of-common coupling; and Z_F is the impedance b00etween the point-of-common coupling and the fault. The point-of-common coupling is the point from which both the fault and the load are fed. In the voltage divider model, the load current before as well as during the fault is neglected. There is thus no voltage drop between the load and the PCC.

The voltage at PCC, and thus the voltage at the equipment terminals, can be found from:

$$Vsag/swell = \frac{ZF}{Zs + ZF} * E \tag{1}$$

Equation (1) can be used to calculate the sag/swell magnitude as a function of thc distance to the fault. Therefore, we have to write $Z_F = Z_L$, with z is the impedance of the feeder per unit length and L the distance between the fault and the pcc, leading to

$$Vsag/swell = \frac{ZL}{Zs + ZL} * E \tag{2}$$

Figure 2. General configuration of Dynamic Voltage Restorer (Pal and Gupta, 2015)

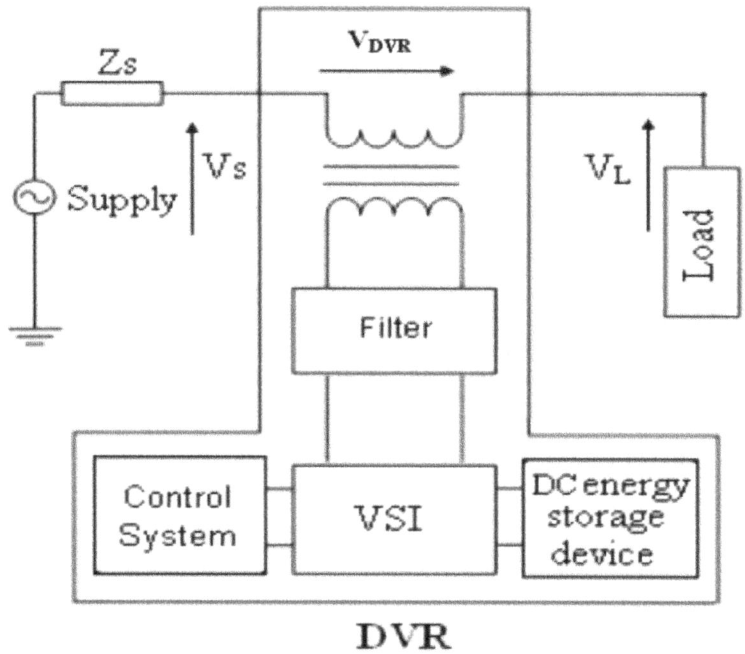

Dynamic Voltage Restorer (DVR)

The DVR (Dynamic Voltage Restorer) is a series connected solid state device that injects additional voltage into the system in order to regulate the load side voltage to the desired magnitude and waveform even when the source voltage is unbalanced or distorted (Pal and Gupta, 2015). It is normally installed in a distribution system between the supply and a critical load feeder a t the so-called point of common coupling (PCC).Its primary function is to rapidly boost up the load -side voltage in the event of voltage sag in order to avoid any power disruption to that load.

DVR Model

The DVR is a custom power device that is connected in series with the distribution system as shown in Figure 2. The main component of the DVR consists of an injection transformer, harmonic filter, VSI, an energy storage and control system. The main function of DVR is to mitigate the voltage sag, although sometimes, additional functions such as harmonics compensation and reactive power compensation are also integrated to the device. A schematic diagram of conventional DVR presented in Figure 2.

1. **The Converter:** It is used to produce required voltage for compensation from fixed voltage. For DC link energy storage voltage source inverter (VSI) is used. A stiff DC voltage supply of low impedance at the input is used to energize the VSI. The output voltage of the converter is independent of the load current. The capacitor used in the VSI reduces the variations in output voltage (González et al. 2014). Graetz bridge inverter and Neutral point clamp inverter are two common

inverter connections used for three phase DVR. H-bridge inverter is the common method used for single phase DVRs.

2. **Energy Storage Device:** The required ac voltage to be injected to the grid is synthesized from a stiff dc-link. The energy storage is required to provide active power injection to the load to restore the supply voltages during deep voltage dips. Lead-acid batteries, Super Conducting Magnetic Energy Storage (SMES), flywheel or Super-capacitors can be used for energy storage (Papic, 2000). The depth and duration of the sag decides the capacity of the energy storage required for the DVR.
3. **Filtering Unit:** The inverter output of the DVR is distorted and contains lots of harmonics due to the nonlinear characteristics of the semiconductor switches used. The filter unit is used to filter higher-order switching harmonics generated by the PWM VSI and improve the quality of the energy supply. Inverter side and line side filtering are the basic types of filtering schemes.
4. **Injection Transformer:** The primary functions of the transformer are to boost the voltage generated by the VSI and to isolate and couple the DVR to the distribution system. The maximum effectiveness and reliability can only be ensured by proper selection of the electrical parameters of the injection transformer. The turns ratio, MVA rating, primary winding voltage and current ratings, and the short-circuit impedance values of transformers are required for proper interconnection of the injection transformer into the DVR.

Protection Modes of DVR

The DVR has three modes of operation, which are protection mode, standby mode (during steady state), and injection/boost mode (during sag).

1. **Protection Mode:** If the current on the load side exceeds a permissible limit due to a short circuit on the load or large inrush current, the DVR will be isolated from the systems by using the bypass switches
2. **Standby Mode: ($V_{DVR} = 0$):** In the standby mode the booster transformer's low-voltage winding is shorted through the converter. No switching of semiconductors occurs in this mode of operation and the full load current will pass through the transformer primary.
3. **Injection/Boost Mode: ($V_{DVR} \neq 0$):** In the Injection/Boost mode, the DVR is injecting a compensating voltage through the booster transformer after the detection of a disturbance in the supply voltage.

Voltage Injection Methods of DVR

The way in which the dynamic voltage restorer (DVR) is used during the voltage injection mode depends upon several limiting factors such as: DVR power rating, load conditions, and voltage-sag type. For example, some loads are sensitive to phase-angel jumps, some others are sensitive to a change in voltage magnitude and some others are tolerant to all these disturbances. Therefore, the control strategies to be applied depend upon the load characteristics. There are four different methods of DVR voltage injection (Mohammed Shazly et al., 2013). These are as follows:

Figure 3. Schematic diagram of a DVR

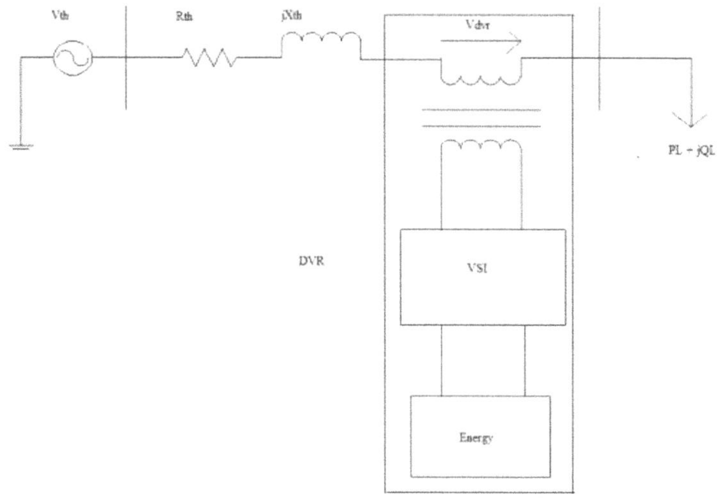

- Pre-sag/dip compensation method.
- In-phase compensation method.
- In-phase advanced compensation method.
- Voltage tolerance method with minimum energy injection.

Calculation of Voltage Injected by DVR

The Thevenin equivalent circuit of the electrical networks is shown in Figure 3. During voltage unbalance conditions, DVR injects the desired voltage level through injection transformer to maintain constant voltage profile. Z_{th} is the Thevenin's equivalent impedance, whose value depends upon the types of fault in the system.

As per the diagram shown in figure 3:

Applying KVL,

$$V_{th} - Z_{th} I_L + V_{DVR} = V_L \tag{3}$$

$$V_{DVR} + V_{th} = V_L + Z_{th} I_L \tag{4}$$

The series voltage injected by DVR can be given as

$$V_{DVR} = V_L + Z_{th} I_L - V_{th} \tag{5}$$

Where, V_{th} = equivalent thevenin voltage of the system

V_L= load voltage

Z_{th} = equivalent thevenin impedance of the system

I_L= Load current and

$$I_L = [\frac{P_L + Q_L}{V_L}]^* \quad (6)$$

Taking V_L as reference, equation (4.3) can be rephrase as

$$V_{DVR} <_{\alpha =} V_L < 0^0 + Z_{th} I_L < (\beta - \alpha) - V_{th} < (\delta) \quad (7)$$

Where, α = angle of V_{DVR}
β = angle of system impedance Z_{th}
δ = angle of system voltage V_{th}
α = Load pf angle and

$$\alpha = \tan^{-1} (\frac{Q_L}{P_L}) \quad (8)$$

The complex power injected by DVR is

$$S_{DVR} = V_{DVR} I_L^* \quad (9)$$

Control Techniques of DVR

To maintain the voltage profile of the load voltage in the time when sensitive loads are connected, there must be appropriate control techniques. The switching pulse generated for VSI is based on SVPWM. It is a simple method and better than other PWM techniques. Normally, three phase inverters use SPWM technique. However, problem like large noise peak at carrier frequency are present in such technique. Hence, SVPWM has an advantage over such technique such as better dc utilization and easy implementation with digital signal processor.

SVPWM is used as a control method for DVR. The Phase Locked Loop (PLL) circuit used here is used to generate a unit sinusoidal wave which is in phase with main supply voltage. Figure 4 shows the control scheme for DVR.

SPECIFICATION AND MODELING OF THE SYSTEM

The necessary data for the calculation of the voltage required to mitigate the voltage sag/swell in the system is provided in the table 1 below.

Voltage sag occurs due to sudden disconnection of load, sudden disconnection of capacitor bank or faults in the system whereas voltage swell occurs due to connection of capacitive load. Voltage unbalance may also occur for certain duration of time in the system due to different faults in the network. During this period voltage disturbance occurs at PCC and DVR operates to restore/maintain the voltage profile.

Figure 4. Control scheme for DVR

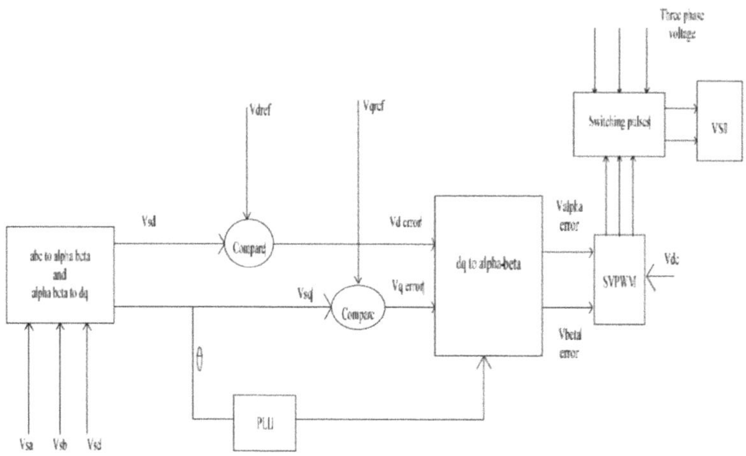

Table 1. Specification parameters of DVR

SN	Specification parameters	Value
1	The main supply voltage per phase	230 V
2	Line resistance	0.1Ω
3	Line inductance	0.5mH
4	Series transformer turns ratio	1:1
4	DC bus voltage	120V
5	DC capacitor	2200μF
6	Filter	LF = 3mH and CF=1μF
7	Load	RL=40ohm and LL=60mH
8	Line frequency	50Hz
9	Switching frequency	10kHz

DVR operates and inject the desired voltage in series to compensate the voltage sag/dip using the transformer connected across the line. Figure 5 presents the Simulink model of DVR.

A three-phase fault is generated in the system to create balanced voltage sag in the time duration between 0.2s to 0.3s and performed the analysis.

CONCLUSION

Voltage sags and voltage swells are the most frequent problems that affect power quality in industries and customers. In order to overcome this problem, Voltage quality can be sufficiently improved by using some custom power devices such as dynamic voltage restorer and distribution static synchronous compensator. The Dynamic Voltage Restorer (DVR) is fast, flexible, efficient and cost-effective solution to

Figure 5. Circuit diagram of DVR in MATLAB Simulink Environment

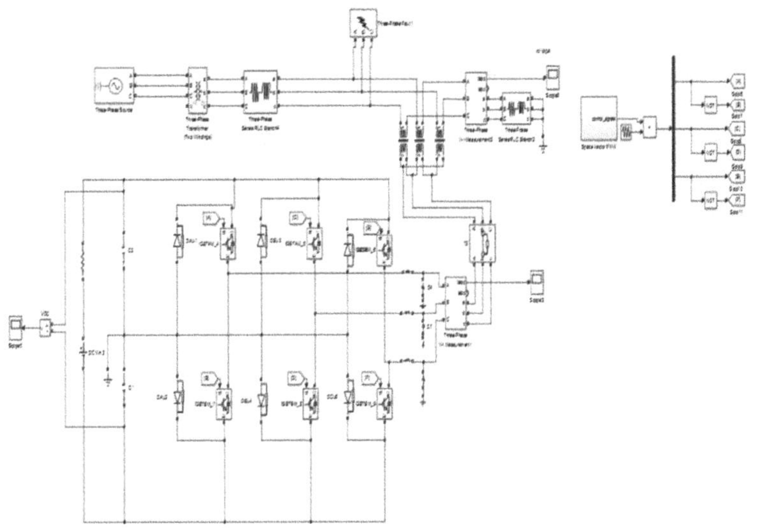

voltage sag and voltage swell problems. The simulation of balanced voltage sag is done. The modeling and the operating principles are discussed.

As far as future work is concerned various artificial intelligence based controlling mechanisms can be adopted to control DVR, which is better than other controlling mechanisms. Further, the transformer used for isolation can be substituted with other mechanisms like power electronic devices to reduce the cost. Moreover, the energy storage used here can be improved with ultra-capacitor design.

REFERENCES

Alhelou, H., Hamedani-Golshan, M. E., Zamani, R., Heydarian-Forushani, E., & Siano, P. (2018). Challenges and opportunities of load frequency control in conventional, modern and future smart power systems: A comprehensive review. *Energies, 11*(10), 2497. doi:10.3390/en11102497

Alhelou, H. H. (2018). *Fault detection and isolation in power systems using unknown input observer. In Advanced condition monitoring and fault diagnosis of electric machines* (p. 38). Hershey, PA: IGI Global.

Alhelou, H. H., Golshan, M., & Fini, M. (2018). Wind driven optimization algorithm application to load frequency control in interconnected power systems considering GRC and GDB nonlinearities. *Electric Power Components and Systems, 46*(11-12), 1223-1238.

Alhelou, H. H., & Golshan, M. E. H. (2016, May). Hierarchical plug-in EV control based on primary frequency response in interconnected smart grid. In *2016 24th Iranian Conference on Electrical Engineering (ICEE),* (pp. 561-566). IEEE. 10.1109/IranianCEE.2016.7585585

Alhelou, H. H., & Golshan, M. E. H. (2020). Decision-making-based optimal generation-side secondary-reserve scheduling and optimal LFC in deregulated interconnected power system. In *Decision making applications in modern power systems* (pp. 269–299). Academic Press. doi:10.1016/B978-0-12-816445-7.00011-6

Alhelou, H. H., Golshan, M. E. H., & Hatziargyriou, N. D. (2019). Deterministic dynamic state estimation-based optimal LFC for interconnected power systems using unknown input observer. *IEEE Transactions on Smart Grid*, 1. doi:10.1109/TSG.2019.2940199

Alhelou, H. H., Golshan, M. E. H., & Hatziargyriou, N. D. (2019). A decentralized functional observer based optimal LFC considering unknown inputs, uncertainties, and cyber-attacks. *IEEE Transactions on Power Systems*.

Alhelou, H. H., Golshan, M. E. H., Zamani, R., Moghaddam, M. P., Njenda, T. C., Siano, P., & Marzband, M. (2019, June). An improved UFLS scheme based on estimated minimum frequency and power deficit. In 2019 IEEE Milan PowerTech (pp. 1-6). IEEE. doi:10.1109/PTC.2019.8810497

Alhelou, H. H., Golshan, M. H., & Askari-Marnani, J. (2018). Robust sensor fault detection and isolation scheme for interconnected smart power systems in presence of RER and EVs using unknown input observer. *International Journal of Electrical Power & Energy Systems*, *99*, 682–694. doi:10.1016/j.ijepes.2018.02.013

Alhelou, H. H., Hamedani-Golshan, M. E., Heydarian-Forushani, E., Al-Sumaiti, A. S., & Siano, P. (2018, September). Decentralized fractional order control scheme for LFC of deregulated nonlinear power systems in presence of EVs and RER. In *2018 International Conference on Smart Energy Systems and Technologies (SEST)* (pp. 1-6). IEEE. 10.1109/SEST.2018.8495858

Alhelou, H. S. H., Golshan, M. E. H., & Fini, M. H. (2015, December). Multi agent electric vehicle control based primary frequency support for future smart micro-grid. In *Proceedings Smart Grid Conference (SGC)* (pp. 22-27). IEEE.

Ansal, V., Ravikumar, K., & Parthiban, P. (2016). Transformer less dynamic voltage restorer for voltage sag mitigation, *2016 Biennial International Conference on Power and Energy Systems: Towards Sustainable Energy (PESTSE)*, Bangalore, India, 2016, pp. 1-4.

Anteneh, D., & Khan, B. (2019). Reliability enhancement of distribution substation by using network reconfiguration A case study at Debre Berhan distribution substation, *International Journal of Economy, Energy, & Environment*, *4*(2), 33.

Babaei, E., & Kangarlu, M. F. (2015). Comparison four topologies for three-phase dynamic voltage restorer. In *Proceedings 2015 International Conference on Renewable Energy Research and Applications (ICRERA)*, Palermo, Sicily. pp. 1527-1532. 10.1109/ICRERA.2015.7418663

Banteywalu, S., Khan, B., De Smedt, V., & Leroux, P. (2019). A novel modular radiation hardening approach applied to a synchronous buck converter. *Electronics (Basel)*, *8*(5), 513. doi:10.3390/electronics8050513

Ding, H., Swangyan, S., Xianzhong, D., & Jun, G. (2002). A novel dynamic voltage restorer and its unbalanced control strategy based on space vector PWM. *Electrical Power and Energy Systems*, *24*(9), 693–699. doi:10.1016/S0142-0615(02)00004-2

Fini, M. H., Yousefi, G. R., & Alhelou, H. H. (2016). Comparative study on the performance of many-objective and single-objective optimisation algorithms in tuning load frequency controllers of multi-area power systems. *IET Generation, Transmission, & Distribution*, *10*(12), 2915–2923. doi:10.1049/iet-gtd.2015.1334

González, M., Cárdenas, V., & Espinosa, G. (2014). Advantages of the passivity-based control in dynamic voltage restorers for power quality improvement. *Simulation Modelling Practice and Theory*, *47*, 221–235. doi:10.1016/j.simpat.2014.06.009

Gupta, K., Khan, B., & Mubeen, S. E. (2015). Available transfer capability enhancement by unified power flowcontroller. In *IEEE International Conference on Signal Processing, Informatics, Communication, and Energy Systems (SPICES),* pp. 1, 4, 19-21.

Haes Alhelou, H., Hamedani-Golshan, M. E., Njenda, T. C., & Siano, P. (2019). A survey on power system blackout and cascading events: Research motivations and challenges. *Energies*, *12*(4), 682. doi:10.3390/en12040682

Haes Alhelou, H., Hamedani-Golshan, M. E., Njenda, T. C., & Siano, P. (2019). Wide-area measurement system-based optimal multi-stage under-frequency load-shedding in interconnected smart power systems using evolutionary computing techniques. *Applied Sciences*, *9*(3), 508. doi:10.3390/app9030508

Haes Alhelou, H., Hamedani Golshan, M. E., Njenda, T. C., & Siano, P. (2019). WAMS-based online disturbance estimation in interconnected power systems using disturbance observer. *Applied Sciences*, *9*(5), 990. doi:10.3390/app9050990

Jariso, M., Khan, B., Tanwar, S., Tyagi, S., & Rishiwal, V. (2018). Hybrid energy system for upgrading the rural environment, IEEE GlobeCom 2018.

Jariso, M., Khan, B., Tesfaye, D., & Singh, J. (2017). Modelling and designing of stand-alone photovoltaic system (Case study: Addis Boder Health Center southwest Ethiopia). In *Proceedings IEEE International Conference on Electronics, Communication, and Aerospace Technology, ICECA 2017,* pp. 168-173, 20-21. Coimbatore, India.

Kalia, C., Kumar, D., & Rana, P. (2014). Analysis of power quality improvement by using BFO-fuzzy controlled DVR, *International Journal of Application or Innovation in Engineering & Management, 3*(8).

Khan, B., & Agnihotri, G. (2012). A novel transmission loss allocation method based on transmission usage. In *Proceedings IEEE Fifth Power India Conference*, Haryana, India. 10.1109/PowerI.2012.6479479

Khan B., Agnihotri G, & Gupta G. (2013). A multipurpose matrices methodology for transmission usage, loss and reliability margin allocation in restructured environment, *Electrical & Computer Engineering: An International Journal, 2*(3).

Khan, B., Agnihotri, G., Rathore, P., Mishra, A., & Naidu, G. (2014). A cooperative game theory approach for usage and reliability margin cost allocation under contingent restructured market. *International Review of Electrical Engineering*, *9*(4), 854–862.

Khan, B., Alhelou, H. H., & Mebrahtu, F. (2019). A holistic analysis of distribution system reliability assessment methods with conventional and renewable energy sources. *AIMS Energy*, *7*(4), 413–429. doi:10.3934/energy.2019.4.413

Khan, B., & Singh, P. (2017). The current and future states of Ethiopia's energy sector and potential for green energy: A comprehensive study. *International Journal of Engineering Research in Africa*, *33*, 115–139. doi:10.4028/www.scientific.net/JERA.33.115

Khan, B., & Singh, P. (2017). Optimal power flow techniques under characterization of conventional and renewable energy sources: A comprehensive analysis, *Journal of Engineering, vol. 2017,* Article ID 9539506.

Khan, B., & Singh, P. (2017). Selecting a meta-heuristic technique for smart micro-grid optimization problem: A comprehensive analysis, in IEEE Access, 5, pp. 13951-13977.

Khan, B., & Singh, P. (2019). Economic operation of smart micro-grid: A meta-heuristic approach. In Handbook of research on smart power system operation and control (pp. 330-346). Hershey, PA: IGI Global. doi:10.4018/978-1-5225-8030-0.ch014

Khan, B., & Singh, P. (2019). Economic operation of smart micro-grid: A meta-heuristic approach. In Handbook of research on smart power system operation and control (pp. 330-346). Hershey, PA: IGI Global. doi:10.4018/978-1-5225-8030-0.ch014

Khan, B., & Tanwar, S. (2019). Issues associated with microgrid integration. In Handbook of research on smart power system operation and control (pp. 252-264). Hershey, PA: IGI Global. doi:10.4018/978-1-5225-8030-0.ch010

Kifle, Y., Khan, B., & Singh, J. (2018). Designing and modeling grid connected photovoltaic system: (Case study: EEU building at Hawassa city). *International Journal of Convergence Computing*, *3*(1), 20–34. doi:10.1504/IJCONVC.2018.091113

Kifle, Y., Khan, B., & Singh, P. (2018). Assessment and enhancement of distribution system reliability by renewable energy sources and energy storage, Journal of Green Engineering, 8(3), 2, pp. 219-262.

Makdisie, C., Haidar, B., & Alhelou, H. H. (2018). An optimal photovoltaic conversion system for future smart grids. In Handbook of research on power and energy system optimization (pp. 601–657). Hershey, PA: IGI Global. doi:10.4018/978-1-5225-3935-3.ch018

Mohammed Shazly, A., Cerrada Aurelio, G., Abdel-Moamen, M. A., & Hasanin, B. (2013, January). Dynamic voltage restorer (DVR) system for compensation of voltage sags, state-of-the-art review. *International Journal of Computational Engineering Research*, *3*(1), 177–183.

Molla, T., Khan, B., Moges, B., Alhelou, H. H., Zamani, R., & Siano, P. (2019). Integrated optimization of smart home appliances with cost-effective energy management system. *CSEE Journal of Power and Energy Systems*, *5*(2), 249–258.

Molla, T., Khan, B., & Singh, P. (2018). A comprehensive analysis of smart home energy management system optimization techniques, *Journal of Autonomous Intelligence, 1*(1), 15–21.

Negash, K., Khan, B., Tesfaye, D., & Gayathri, M. (2017). Optimal placement of phasor measurement unit for system observability. In *Proceedings of IEEE International Conference on Computing Methodologies and Communication, (ICCMC 2017),* pp. 73-77, Erode, India.

Negash, K., Khan, B., & Yohannes, E. (2016). Artificial intelligence versus conventional mathematical techniques: A review for optimal placement of phasor measurement units [Springer]. *Journal of Technology and Economics of Smart Grids and Sustainable Energy, 1*(1), 10. doi:10.100740866-016-0009-y

Njenda, T. C., Golshan, M. E. H., & Alhelou, H. H. (2018, November). WAMS based intelligent under frequency load shedding considering online disturbance estimation. In *Proceedings Smart Grid Conference (SGC)* (pp. 1-5). 10.1109/SGC.2018.8777779

Pal, R., & Gupta, S. (2015, June). State of the art: Dynamic voltage restorer for power quality improvement. *Electrical & Computer Engineering: An International Journal, 4*(2), 79–98.

Papic, I. (2000). Power quality improvement using distribution static compensator with energy storage system. In Proceedings *9th International Conference on Harmonics and Quality of Power*, Orlando, FL, pp. 916-920, vol. 3. 10.1109/ICHQP.2000.896851

Rani, P. A., & Sivakumar, R. (2014, January). Improvement of power quality using DVR in distribution systems. *International Journal of Innovative Research in Science, Engineering, and Technology, 3*(Special Issue), 1.

Rauf, A. M., & Khadkikar, V. (2015, May). An enhanced voltage SAG compensation scheme for dynamic voltage restorer. *IEEE Transactions on Industrial Electronics, 62*(5), 2683–2692. doi:10.1109/TIE.2014.2362096

Rovai, C. H., & Doorwar, A. (2014). An overview of various control techniques of DVR. In *Proceedings 2014 International Conference on Circuits, Power, and Computing Technologies (ICCPCT-2014),* Nagercoil, India. pp. 53-57. 10.1109/ICCPCT.2014.7054844

Sadigh, A. K., & Smedley, K. M. (2016). Fast and precise voltage sag detection method for dynamic voltage restorer (DVR) application. *Electric Power Systems Research, 130*, 192–207. doi:10.1016/j.epsr.2015.08.002

Singh, P., & Khan, B. (2017). Smart microgrid energy management using a novel artificial shark optimization, Complexity, vol. 2017, Article ID 2158926. doi:10.1155/2017/2158926

Singh, P., Khan, B., Vidyarthi, A., Haes Alhelou, H., & Siano, P. (2019). Energy-aware online non-clairvoyant scheduling using speed scaling with arbitrary power function. *Applied Sciences, 9*(7), 1467. doi:10.3390/app9071467

Singh, S., Rai, V., Kumar, A., & Sahay, K. B. (2016). Simulation and comparison of DVR and D-STATCOM for voltage sag mitigation. In *Proceedings 2016 IEEE 6th International Conference on Power Systems (ICPS),* New Delhi, India. pp. 1-6. 10.1109/ICPES.2016.7584238

Yeshalem, M. T., & Khan, B. (2017). Design of an off-grid hybrid PV/wind power system for remote mobile base station: A case study. *AIMS Energy*, *5*(1), 96–112. doi:10.3934/energy.2017.1.96

Zamani, R., Golshan, M. E. H., Alhelou, H. H., & Hatziargyriou, N. (2019). A novel hybrid islanding detection method using dynamic characteristics of synchronous generator and signal processing technique. *Electric Power Systems Research*, *175*, 105911. doi:10.1016/j.epsr.2019.105911

Zamani, R., Hamedani-Golshan, M. E., Haes Alhelou, H., Siano, P., & Pota, H. R. (2018). Islanding detection of synchronous distributed generator based on the active and reactive power control loops. *Energies*, *11*(10), 2819. doi:10.3390/en11102819

KEY TERMS AND DEFINITION

Dynamic Voltage Restorer: It is a series connected solid state device that injects additional voltage into the system in order to regulate the load side voltage to the desired magnitude and waveform even when the source voltage is unbalanced or distorted.

Power Quality: It is described as the variation of voltage, current and frequency in the power system.

Voltage Sag (Dip): It is a brief reduction in RMS voltage of 10% or more below specified (rated) equipment voltage for a period of 1/2cycle to 1min.

Voltage Swells: These are the opposite of voltage sags (dips) and they are defined as a momentary increase in RMS voltage of 10% or more above specify (rated) equipment voltage for a period of 1/2 cycle to 1 min.

APPENDIX

List of Abbreviations

D-STATCOM Distribution static Compensator
DVR Dynamic Voltage Restorer
FACTS Flexible AC Transmission System
IPAC In-Phase Advanced Compensation
IPC In-Phase Compensation
PCC Point of common coupling
PDC Pre-Sag/Dip Compensation
PWM Pulse width Modulation
SMES Super Conducting Magnetic Energy Storage
SV Space Vector
SVPWM Space vector pulse width modulation
VSI Voltage source Inverter

Chapter 4
Voltage Drop Mitigation in Smart Distribution Network

Fsaha Mebrahtu
Hawassa University, Hawassa, Ethiopia

ABSTRACT

Voltage dip in the distribution network is caused by disturbance at different voltage levels and experienced by low voltage customers are established. Voltage dips are those disturbances which damage the power quality of the distribution network and causing heavy economic damage to the customers. This chapter investigates procedures of mitigating the voltage dip by reducing the number of faults due to short circuits, lowering the fault clearing time, and changing the power system design and DSTATCOM Compensator with DG and dynamic voltage restorer.

INTRODUCTION

Voltage dip is the unexpected reduction of the root mean square (RMS) voltage, and the cause of voltage dip in power system distribution is due to short circuit, overload or starting of motors. According to the IEEE the variation of the voltage sags is between the ranges of 10% to 90% i.e. 0.1 pu to 0.9 pu of the nominal voltage followed by the recovery within a short time 10 ms to 60 sec. It starts when voltage sags occur to low voltage than threshold voltage (V_{thr}) usually 0.9 pu voltage dip and temporary interruption of power are possibly the main power quality problems affecting the end user or customers. These actions are mostly allied with a disturbance anywhere on the power system source. If the disturbances are faced by the end users than these disturbances will be exist. Voltage dip is the most common meanwhile they can be interconnected with the end users. Even the voltage dip is 4 to 5 cycles, it affects the sensitive distribution materials to drop out fuses (Alhelou et al., 2019; Makdisie et al., 2018; Alhelou et al., 2018; Alhelou et al., 2016; Haes Alhelou et al., 2019; Njenda et al., 2018).

DOI: 10.4018/978-1-7998-1230-2.ch004

There are various causes for power quality problems, which affect the distribution network. These are temporary faults; transient, harmonic distortion, notch voltage, flicker, voltage swells and voltage sages. Sensitive electronic materials, series and nonlinear loads including accuracy industrial procedures are unfavorably affected by power quality problems and results in momentary losses, damaged equipment and product quality degradation (Brumsickle, et al. 2001).

According to IEEE standard 1159-1995, the voltage dip variation is between 0.9 pu to 0.1 pu for the duration of 50.5 cycles, which is smaller than a minute. Generally facing due to the disturbance of distribution and transmission networks and it is the most dangerous to the power quality problems (Brumsickle, et al. 2001; Khan et al., 2019; Khan et al., 2018; Khan et al., 2017; Banteywalu et al., 2019; Anteneh et al., 2019; Molla et al., 2019, Molla et al., 2018, Jariso et al. 2018).

The main cause of voltage sags are exchange processes related with momentary cut of power source, the movement of huge current allied with the initial huge current requirement of electric motors or the flow of current to the distribution networks from the power source. So these conditions originated from the end users due to the lack of experience. The main reason of temporary voltage sags is probably the lightning strike. Each of these cases may cause a dip with unusual features (Fini et al., 2016; Alhelou et al., 2018; Zamani et al., 2018; Alhelou et al., 2015; Njenda et al., 2018; Haes Alhelou et al., 2018; Haes Alhelou et al., 2019).

Voltage dip can be damaged electric machines, especially the power electronics devices, which can be trip because of their construction to prevent circuits. Voltage dips can also put stress on the personal computers and it will reduce their life time. In the standard, voltage sag events are divided into three subclasses in terms of time durations:

- Instantaneous voltage sag: 0.5 to 30 cycles
- Momentary voltage sag: 30 cycles to 3 seconds
- Temporary voltage sag: > 3 to 60 seconds

Indices for Voltage Sag Quantification

Indices for voltage sag are formed to quantify the performance of an entire power system or a single measurement point in terms of sag events. In the IEEE Std. 1564-2014 (IEEE std. 2014), the definitions of the following indices are provided. The index variation of system average in RMS:

- Voltage Sag Tables
- Voltage Sag Severity
- Voltage Sag Energy

SARFI index is divided into two subgroups as follows:

- SARFI- X
- SARFI- curve

SARFI-X index gives the amount of voltage dip actions regarding specified threshold values. For example, SARFI-70 indicates that the amount of actions where reserved voltage is below 70% of the reference voltage value.

SARFI – Curves provide equipment compatibility limits for voltage disturbances. The most considered curves are CBEMA, ITIC and SEMI F47. These curves demonstrate the compatibility of sensitive equipment such as PLC, ASD and PC in terms of retained voltages and durations.

Voltage sag tables present a count of sag events by means of retained voltages and durations in a table format. The rows symbolize ranges of retained voltage whereas the column of the tables represented within the range dip voltage period. There are several tables suggested, including UNIPEDE, IEC 61000-4-11 and IEC 61000-2-8.

Voltage dip severity is evaluated using retained voltage and sag duration by comparing the reference curves such as ITIC and SEMI F47 for each single event. The index is greater than one if the sag event is longer or deeper than the allowed reference value.

The energy of voltage dip can be defined as the period of disturbance that delivers similar energy loss during the voltage dip(Fini et al., 2016; Alhelou et al., 2018; Zamani et al., 2018; Alhelou et al., 2015; Njenda et al., 2018; Haes Alhelou et al., 2018; Haes Alhelou et al., 2019) .

The magnitude of sag voltage in radial network can be calculated by using the voltage divider models. It is a simple and adequate model to characterize a sag event. The dip voltage before the fault voltage value, Z_s is impedance of the source at the point of common coupling (PCC), and Z_f is the total impedance between the fault location and PCC (Dugan, et al, 1996). During a fault at an adjacent feeder, the voltage at the PCC can be evaluated by using Equation (1).

$$V_{sag} = \frac{Z_f}{Z_s + Z_f} E \qquad (1)$$

Cause of Voltage Dip

The dip voltage can be commonly categorized by the dip magnitude, period and frequency (ITIC, 2000). Network impedances regulate the dip magnitude. When considering dips produced by faults, the protection practices specify the dip period, and the fault frequencies determine the amount of voltage sags.

In distribution power system the dip voltage is caused due to the short circuit, over loads and starting of machines. The importance of voltage dips has improved due to the problems sags cause to many type of equipment. To be able to estimate the damage voltage sags cause to customer also new tools and analysis program are needed. The causes of faults are follows:

- The incensement of current in the system
- Increase of system impedance
- Lightning
- Heavy winds
- Trees falling across the line
- Vehicles colliding with towers or poles
- Birds shorting lines
- Aircraft colliding with the transmission lines
- Destruction

MITIGATION OF VOLTAGE DIP IN ELECTRICAL DISTRIBUTION NETWORK

Voltage sag events have adverse effects on equipment that is sensitive to short duration supply voltage variations. In the textile and plastic industries in particular, sensitive equipment may malfunction or trip resulting in significant financial loss due to voltage sag.

There are several mitigation methods to minimize the negative impacts of voltage sag events for both utility and end-user sides. On the other hand, end-users can implement mitigation equipment such as AVC, DVR, CVT, STS and UPS against voltage sag events to keep the production process running. In addition, D-STATCOM and Distribution Static VAR Compensator (DSVC) can be used by utilities and large industrial consumers. Utility side mitigation methods can be applied in three ways as follows.

- Reducing the number of faults due to short circuits
- Lowering the fault clearing time
- Changing the power system design

Reducing the Number of Faults Due to Short Circuits

Short circuit faults occur for various reasons and cannot be avoided completely in electrical power systems. Moreover, faults in the adjacent feeders generally cause voltage sags. Hence, by decreasing the short circuit disturbance will be reduce the event of voltage sags. For any scheduling and operation of power system goal is to decrease blackouts or voltage sags in its transmission and distribution networks. But if electrical faults are faced in the power system dip voltage will be occurred at the distribution networks (Ipinnimo, et al. 2013; Khan et al. 2014; Khan et al. 2013; Khan et al. 2012; Negash et al., 2017; Negash et al., 2016; Jariso et al. 2017; Kifle et al. 2018; Yeshalem et al. 2017; Singh et al. 2019; Gupta et al., 2015).

Mitigation of Voltage Sags by Using DG

Due to the variation of loads at the distribution side variation of voltage will be occurred, and the dip voltage is the function of reactive components of the load currents and resistance of transformers. In electrical power network the reactive power is either consumed or generated in almost every parameters of the system. This is the result of voltage profile on the distribution networks.

Figure 1. Flow diagram of voltage compensation detection method

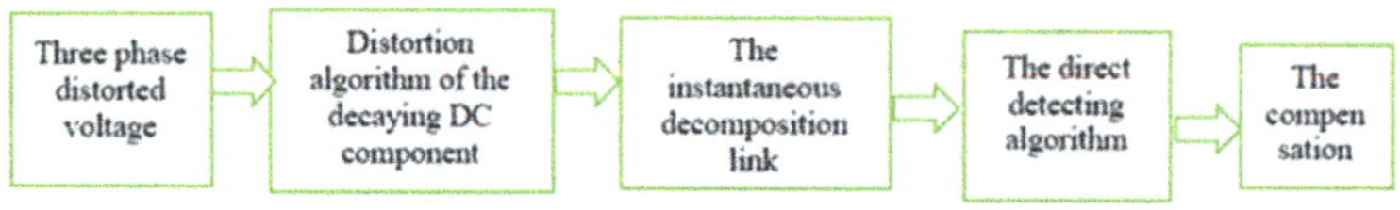

DG has made it economical to power supply this reactive power closer to the load and also improve the voltage profile in the power distribution system (Gaikwad, et al. 2012). The following subtitles are used to mitigate the DG based schemes for voltage variation.

Application of Converter Based DG Asynchronous and Synchronous Generator

The dip voltage can be controlled by using the three distribution generator units. These are asynchronous generator, synchronous generator and converters linked into generators. The occurrence of power electronics, synchronous equipment induction generators have positive influence on the maintained voltage at the time of voltage sags. At this time reactive power will be injected to the system or to the grid by using the machine connected there, consequently voltage sags is certainly increases the gain voltage at the time sags. The performance of power electronics converters based on DG, synchronous generators, and asynchronous generators during voltage dips is corroborated and examined by (Renders et al. 2008).

On the other side the reported effects of asynchronous and synchronous generators on the voltage sags in high to medium voltage distribution system, further the effect on voltage sags in low voltage distribution network is better than negligible. Power electronics converters-based DG is established with the same result on the voltage dip in lower voltage system in opposition to high voltage networks.

Series Compensation

The grid voltage dip can be mitigated by introducing parallel current or a series voltage to the existing network. This could do by using power electronics converters at the DG systems which involve huge active current, if the voltage has to be retaining in both single phase and magnitude to its pre disturbance values. The power electronic converter of a distributed generation system can be stretched with a succession compensator as shown in figure 1. DVR is used in series to compensate or to eliminate the sag voltage with the distribution generation depend on balanced components estimation.

This technique warrants the improvement of the positive sequence amplitude at a time of sags to sensitive loads (Bajpai, et al. 2011). From the power electronics converters device one of them is DVR is used to protect from different disturbances. DVR is attached in series with the distribution networks and it has the ability of absorbing and producing active and reactive power at the terminal of AC power supply. Different parts of DVR are as flows:

- Controlling diagram or switching control scheme
- Energy storage device in the form of DC
- Voltage source converter filter
- Injection/coupling transformer is connected in series.

The working principle of DVR is to the addition of voltage is the wanted magnitude and can returning to the sensitive load voltage to the needed sinusoidal and amplitude as explained in (Liu. Et al. 2007). The application of DVR in power distribution network is to compensate the voltage in order to balance the system line current.

The protection procedure of the voltage reimbursement is very important for removing the dip voltage and it also controls at different disturbances, like transient voltage, excessive voltage and variation of voltage. But it has the following disadvantage:

- It can't use at the time of voltage sags that exceeds the rating of series converters.
- It needs a number of power electronics materials.

Distribution generation is very important for both micro grid and grid connected operations. During the working of grid connected mode the converter controller of distribution generation scheme normalizes the value of current that is current model control. The DG scheme, which is allied through the converters into the grid by the three-static switch of transformers is introduced between the load side and utility supply. When the sag voltage is faced the static transformer switch will detach the load side, where sensitive materials are placed, at the utility side. The static transformer switch involves back to back thyristors. The distribution generation regulates voltage load side and the load will be disconnected from the utility supply. During this time sensitive apparatus are not damaged by the sag voltage. Micro grid operation has numerous benefits such as during the dip voltage and disturbances then it is also inexpensive solution as the quantity of power electronic element.

Shunt Active Filter with Energy Storage (SAFES)

In the DG micro grid the dip voltage can also mitigate by using the parallel active power filter with its energy storage mechanisms. The planned control approach is dependent on the state space pole placement design for a parallel active power filter with it storages in the micro grid which is guaranteed voltage regulation and elimination of harmonics on the load side. At the time of dip voltage the fundamental reference current for the SAFES is evaluated and it works in combination with the point of common coupling. In the micro grid power generation SAFES is used to improve the power improvement in the existence distribution generation. The hypothetical investigation has inspected how to control a power electronic converters energy source in organization with an energy source, with a slow dynamic response. It is seen that this scheme is able to quickly recover the grid voltage after a considerable load step change.

D-STATCOM Compensator with DG

D-STATCOM mechanism is connected in the system in parallel of six pulses Semiconductor Bridge like conventional STATCOM is applied to the HV transmission networks. The two group devices are differing in power. The controller of D-STATCOM is planned for the grid of distributions and it has the advantage of compensating the voltage dips or swells. The working principle of D-STATCOM is to vaccinate a set of three unbalanced compensating current to the system, such as the system current become pure sinusoidal balanced and the voltages are in phase. The current control mode performs the compensators in order to balance the system voltage. Also there is a drawback consist in unbalanced compensation of four wire network needs four leg device to be applied, however the last function is to perform the three legs compensators effectively. The compensator of D-STATCOM is a flexible device, which can operate in current control mode to eliminate the voltage variation, unbalanced and reactive power and in voltage control mode as stabilizer of voltage. Based on the grid power source D-STATCOM is efficient to mitigate the voltage dips in the distribution networks. Nevertheless, sometimes the connection of single phase of DG to the grid may deteriorate power quality considerably.

Additionally in D-STATCOM, three phase voltage source converter is applied as a front end of a DG unit. D-STATCOM is a voltage source inverter (VSI) used to regulate and balance the voltage at the distribution network bus by injecting reactive power. The controller of D-STATCOMM consists of two types of controllers those are outer voltage controller and inner current. The outer controller is designed both negative and positive sequence frames to be able to recompense for voltage imbalance.

The VSC controller is added to the voltage regulation in order to keep the voltage at the load bus equal to the normal voltage even in case of inverter unbalanced or balanced dip voltage. During the faults of higher order harmonics of the high current on grid side power source the VSC is connected at the weak bus to compensate. They are two type of DC link voltages are elaborated. Those are constant or stiff DC voltage source and regulated or weak DC voltage with a constant current source.

The controlling of the DG system is implemented in rotating dq frame that is synchronized with the voltage at the point of common coupling by using phase locked loop. The key controller contains three types of controller in order to improve the stability margin and at the same time to damp the oscillations at the resonance frequency of the LCL filter. The advantage of voltage source converter in the power distribution system is to enhance the system stability and to improve the voltage profile at the weak bus. Nonetheless the limited rating of the DG result is limiting the voltage regulation margin.

Intelligent Approach for Voltage Sags Mitigation by

The cause of voltage sags is transient procedure in an electrical power distribution system and it random in nature. When disturbance faced in the network the magnitude of voltage will be changed and it can be solved the dip voltage by artificial intelligence. A mutual drawback in existing system is that an indepth knowledge of the method is prerequisite to eliminate the sag voltage (Ipinnimo et al., 2012). In procedural works, an intelligent scheme is observed as a scheme that observes its environment and consequently takes actions to maximize its chances of achievement. Like this systems and methods are mutually called artificial intelligence. The common AI tools that are mostly used for power system control are fuzzy logic control, neural network, genetic algorithm, and expert system and support vector machine. Advantages of intelligent method are as flows:

- The ability to generalize high speed
- It has the ability to model appropriately
- It synthesized the complex map accurately in the power system
- It can handle noisy and incomplete data

Optimal DG Placement and Sizing for Mitigating a Voltage Dip

The behavior of sag voltage in power distribution power system produced by faults is extremely affected by the size and location and of DG. The use of practical swarm optimization method is to launch the placement of optimal DG and sizing on the power distribution system for the mitigation of voltage sags is existing. Meanwhile the sag voltage depends on different equipment's like location of disturbance, the load type and sensitivity and impedances of the system. The advantage of PSO process shows the mitigation of voltage dip, when DG was optimally situated and sized. During the time voltage dip it is important to find the weak bus automatically to correct the voltage variation in the system. Genetic Algorithm was use for optimal placement and sizing of DG for loss reduction and voltage dip mitigation with the present of DG (Amanifar et al., 2012).

Dynamic Voltage Restorer with Artificial Intelligence

Arrangement of and control diagram of dynamic voltage restorer differs depending on the nature and behavior of protection of the load. For a sensitive materials of distribution power system, the load is established how the low energy can be injected by using feed forward neural system for separating the negative sequence components from the basic sequence parameters in balanced sag voltage and also correct the dip voltage in distribution system. The fast acting and efficient controller is recommended for satisfying the dynamic voltage restorer technique with fuzzy logic controller for industrial induction motor loads at the time of voltage sags. The purpose of controlling mechanism in power distribution system is to get constant voltage with constant magnitude at the point, where the sensitive loads are connected under line of disturbance. In fuzzy logic mechanism-controlled PI controller are adjusted online PI gains. The main benefit of adaptive fuzzy PI controller over the classical one (PI) is its capability to recompense notching when the DVR is connected to a weak power system. In addition, adaptive fuzzy PI controller improves performance of the DVR and solves the problem of PI tuning (Banaei et al., 2006).

Tap Changing Transformer

Tap changing transformer is accomplished through the use of back to back transformer with its reasonable response time and it is known for medium applications which are greater than 3 KVA. Nevertheless, high control resolution needs for a large number of silicon-controlled rectifier (60 SCRs for +/-3% regulation with+10/-20% input range), control for fast response will be complex. Additional disadvantage of this structure is its vulnerability to high transient current with motor loads upon tap changing and its weak transient voltage rejection (McGranaghan et al., 1993).

Saturable Reactor Regulators

The diagram of control result voltage by varying the impedance of a saturable reactor is a simple and has a good line transient rejection. The disadvantages of this method comprise slow response usually 10 cycles and high output impedance, which gives high faults with nonlinear loads sensitive to load power factor, will not have surge currents such as motor starting and will not suppress transients generated inside power generation (McGranaghan et al., 1993).

Motorized Variacs

The motor variacs are worked to keep out the output voltage regulation. This structure can have huge surge currents commonly used in industrial setting due to high surge capability. On the other hand, it has a slow response about 30 V/s and is not appropriate for sensitive materials. Also it has considerable maintenance requirements and weak transient suppression. Its slow response limits its efficiency (McGranaghan et al., 1993).

Soft Switching Line Conditioners

Soft switching line conditioner is the fast combine the fast response and high performance of active line conditioner with minor cost of the more conventional solutions. IGBT is the heart of power line conditioner because it is a soft switching inverter technology and the resonance DC link inverter a high efficiency and high tasks inverters. In industrial area there is automatic voltage regulator and its capacity is up to 1 MVA. It realized using the soft switching mechanism to compensate the dip voltages. In order to make cost effective, the power units are based on the hybrid structure by using active and passive materials (McGranaghan et al., 1993). Also the output voltage will maintain within 1% of nominal value with wide variation in input voltage.

Static Voltage Regulator (SVR)

Static VAR voltage regulator through the use of static tap changers basically regulates the voltage materials operational levels. Unlike conventional load tap changer, which are equipped with a time delayed mechanical tap changer, static changer are intended to respond instantaneously by choosing the suitable tap voltage without the need to progress through series of lower tap voltage. The SVR doesn't need the use of energy storage and it has relatively small foot print for the amount of load and it can protect. The SVR is able to precise the dip voltage conditions usually 55% of the normal voltage maximum depth in quarter of a cycle 4ms to allow even the most sensitive manufacture material to remove through the dip voltage conditioner caused by faults in the utility distribution networks (McGranaghan et al. 1993).

Ferro Resonant Transformers (CVT)

A Ferro resonant transformer is also known as constant voltage transformer, and it can be handled during the voltage dip, always below 20 kVA. In fact, they are specifically smart for a constant low power loads. It is more difficult to present the variable load especially with high inrush current for CVT because the tuned circuit on the output. The Ferro resonant transformer core structure is designed, so that the secondary operates in flux saturation and the secondary winding resonates with the capacitor in a tuned circuit. As a result of this saturated operating mode, changing the primary or line voltage may change the current but will not vary the flux or the secondary induced voltage. The output waveforms are not sinusoidal (square wave with a high harmonic content) especially with nonlinear loads. An accurately selected neutralizing winding cancels out most of the harmonic content of the output voltage and yields a satisfactory low distortion sine wave (McGranaghan et al., 1993).

CONCLUSION

As voltage dip is the critical issue in the electrical distribution system, this chapter provides the detailed overview of the various methods, which are utilized in power system to mitigate the voltage dip. Various techniques such as DVR, D-STATCOM, SVR, CVT, tap changing transformers, variacs, conditioners, series compensation, Saturable Reactor Regulators, tap changing transformer, shunt active power filters are discussed to mitigate the voltage dip problem in the distribution network.

REFERENCES

Alhelou, H., Hamedani-Golshan, M. E., Zamani, R., Heydarian-Forushani, E., & Siano, P. (2018). Challenges and opportunities of load frequency control in conventional, modern and future smart power systems: A comprehensive review. *Energies, 11*(10), 2497. doi:10.3390/en11102497

Alhelou, H. H. (2018). *Fault detection and isolation in power systems using unknown input observer. In Advanced condition monitoring and fault diagnosis of electric machines* (p. 38). Hershey, PA: IGI Global.

Alhelou, H. H., Golshan, M., & Fini, M. (2018). Wind driven optimization algorithm application to load frequency control in interconnected power systems considering GRC and GDB nonlinearities. *Electric Power Components and Systems, 46*(11-12), 1223-1238.

Alhelou, H. H., & Golshan, M. E. H. (2016, May). Hierarchical plug-in EV control based on primary frequency response in interconnected smart grid. In *2016 24th Iranian Conference on Electrical Engineering (ICEE),* (pp. 561-566). IEEE. 10.1109/IranianCEE.2016.7585585

Alhelou, H. H., & Golshan, M. E. H. (2020). Decision-making-based optimal generation-side secondary-reserve scheduling and optimal LFC in deregulated interconnected power system. In *Decision making applications in modern power systems* (pp. 269–299). Academic Press. doi:10.1016/B978-0-12-816445-7.00011-6

Alhelou, H. H., Golshan, M. E. H., & Hatziargyriou, N. D. (2019). Deterministic dynamic state estimation-based optimal LFC for interconnected power systems using unknown input observer. *IEEE Transactions on Smart Grid,* 1. doi:10.1109/TSG.2019.2940199

Alhelou, H. H., Golshan, M. E. H., & Hatziargyriou, N. D. (2019). A decentralized functional observer-based optimal LFC considering unknown inputs, uncertainties, and cyber-attacks. *IEEE Transactions on Power Systems.*

Alhelou, H. H., Golshan, M. E. H., Zamani, R., Moghaddam, M. P., Njenda, T. C., Siano, P., & Marzband, M. (2019, June). An improved UFLS scheme based on estimated minimum frequency and power deficit. In Proceedings 2019 IEEE Milan PowerTech (pp. 1-6). IEEE. doi:10.1109/PTC.2019.8810497

Alhelou, H. H., Golshan, M. H., & Askari-Marnani, J. (2018). Robust sensor fault detection and isolation scheme for interconnected smart power systems in presence of RER and EVs using unknown input observer. *International Journal of Electrical Power & Energy Systems, 99,* 682–694. doi:10.1016/j.ijepes.2018.02.013

Alhelou, H. H., Hamedani-Golshan, M. E., Heydarian-Forushani, E., Al-Sumaiti, A. S., & Siano, P. (2018, September). Decentralized fractional order control scheme for LFC of deregulated nonlinear power systems in presence of EVs and RER. In *2018 International Conference on Smart Energy Systems and Technologies (SEST)* (pp. 1-6). IEEE. 10.1109/SEST.2018.8495858

Alhelou, H. S. H., Golshan, M. E. H., & Fini, M. H. (2015, December). Multi agent electric vehicle control based primary frequency support for future smart micro-grid. In *Proceedings Smart Grid Conference (SGC)* (pp. 22-27). IEEE.

Amanifar, O., & Golshan, M. E. H. (2012, May). Optimal DG allocation and sizing for mitigating voltage sag in distribution systems with respect to economic consideration using Particle Swarm Optimization. In 2012 Proceedings of 17th Conference on Electrical Power Distribution (pp. 1-8). IEEE.

Anteneh, D., & Khan, B. (2019). Reliability enhancement of distribution substation by using network reconfiguration, A case study at Debre Berhan Distribution Substation. *International Journal of Economy, Energy, & Environment*, *4*(2), 33.

Bajpai, R. S., & Gupta, R. (2011, June). Series compensation to mitigate harmonics and voltage sags/swells in distributed generation based on symmetrical components estimation, *IEEE International Symposium* Industrial Electronics (ISIE*)*, 1639–1644.

Banaei, M. R., Hosseini, S. H., & Khajee, M. D. (2006, August). Mitigation of voltage sag using adaptive neural network with dynamic voltage restorer. In Proceedings Power Electronics and Motion Control Conference, IPEMC, pp. 1–5.

Banteywalu, S., Khan, B., De Smedt, V., & Leroux, P. (2019). A novel modular radiation hardening approach applied to a synchronous buck converter. *Electronics (Basel)*, *8*(5), 513. doi:10.3390/electronics8050513

Brumsickle, W. E., Schneider, R. S., Luckjiff, G. A., Divan, D. M., & McGranaghan, M. F. (2001). Dynamic sag correctors: Cost-effective industrial power line conditioning. *IEEE Transactions on Industry Applications*, *37*(1), 212–217. doi:10.1109/28.903150

Dugan, R. C., McGranaghan, M. F., & Beaty, H. W. (1996). Electrical power systems quality. New York: McGraw-Hill.

Farhoodnea, M., Mohamed, A., & Shareef, H. (2014, May). A comparative study on the performance of custom power devices for power quality improvement. In *Proceedings 2014 IEEE Innovative Smart Grid Technologies-Asia (ISGT ASIA),* (pp. 153-157). IEEE.

Fini, M. H., Yousefi, G. R., & Alhelou, H. H. (2016). Comparative study on the performance of many-objective and single-objective optimisation algorithms in tuning load frequency controllers of multi-area power systems. *IET Generation, Transmission, & Distribution*, *10*(12), 2915–2923. doi:10.1049/iet-gtd.2015.1334

Gaikwad, D., & Mehraeen, S. (2012, April). Reactive power considerations in reliability analysis of photovoltaic systems. In Proceedings 2012 IEEE Green Technologies Conference, pp. 1–6. doi:10.1109/GREEN.2012.6200982

Gupta, K., Khan, B., & Mubeen, S. E. (2015). Available transfer capability enhancement by unified power flowcontroller. In *Proceedings IEEE International Conference on Signal Processing, Informatics, Communication, and Energy Systems (SPICES)*, pp. 1, 4, 19-21.

Haes Alhelou, H., Hamedani-Golshan, M. E., Njenda, T. C., & Siano, P. (2019). A survey on power system blackout and cascading events: Research motivations and challenges. *Energies*, *12*(4), 682. doi:10.3390/en12040682

Haes Alhelou, H., Hamedani-Golshan, M. E., Njenda, T. C., & Siano, P. (2019). Wide-area measurement system-based optimal multi-stage under-frequency load-shedding in interconnected smart power systems using evolutionary computing techniques. *Applied Sciences*, *9*(3), 508. doi:10.3390/app9030508

Haes Alhelou, H., Hamedani Golshan, M. E., Njenda, T. C., & Siano, P. (2019). WAMS-based online disturbance estimation in interconnected power systems using disturbance observer. *Applied Sciences*, *9*(5), 990. doi:10.3390/app9050990

IEEE guide for voltage sag indices, IEEE 1564, 2014.

Information Technology Industry Council (ITIC), Note, 2000.

Ipinnimo, O., Chowdhury, S., Chowdhury, S. P., & Mitra, J. (2012). Intelligent voltage dip detection in power networks with distributed generation (DG). In Proceedings 2012 North American Power Symposium (NAPS), pp. 1-6. IEEE.

Ipinnimoa, O., Chowdhurya, S., Chowdhurya, S. P., & Mitrab, J. (2012). A review of voltage dip mitigation techniques with distributed generation in electricity networks.

Jariso, M., Khan, B., Tanwar, S., Tyagi, S., & Rishiwal, V. (2018, December). Hybrid energy system for upgrading the rural environment. In *2018 IEEE Globecom Workshops* (pp. 1-6). IEEE.

Jariso, M., Khan, B., Tesfaye, D., & Singh, J. (2017). Modelling and designing of stand-alone photovoltaic system (Case study: Addis Boder Health Center southwest Ethiopia), *IEEE International Conference on Electronics, Communication, and Aerospace Technology, ICECA 2017*, pp. 168-173, Coimbatore, India.

Khan, B., & Agnihotri, G. (2012). A novel transmission loss allocation method based on transmission usage. In *Proceedings IEEE Fifth Power India Conference*, Haryana, India. 10.1109/PowerI.2012.6479479

Khan, B., Agnihotri, G, Gupta G. (2013). A multipurpose matrices methodology for transmission usage, loss, and reliability margin allocation in restructured environment, *Electrical & Computer Engineering: An International Journal, 2*(3), September 2013.

Khan, B., Agnihotri, G., Rathore, P., Mishra, A., & Naidu, G. (2014). A cooperative game theory approach for usage and reliability margin cost allocation under contingent restructured market. *International Review of Electrical Engineering*, *9*(4), 854–862.

Khan, B., Alhelou, H. H., & Mebrahtu, F. (2019). A holistic analysis of distribution system reliability assessment methods with conventional and renewable energy sources. *AIMS Energy*, *7*(4), 413–429. doi:10.3934/energy.2019.4.413

Khan, B., & Singh, P. (2017). The current and future states of Ethiopia's energy sector and potential for green energy: A comprehensive study. *International Journal of Engineering Research in Africa, 33*, 115–139. doi:10.4028/www.scientific.net/JERA.33.115

Khan, B., & Singh, P. (2017). Optimal power flow techniques under characterization of conventional and renewable energy sources: A comprehensive analysis, *Journal of Engineering, vol. 2017,* Article ID 9539506.

Khan, B., & Singh, P. (2017). Selecting a meta-heuristic technique for smart micro-grid optimization problem: a comprehensive analysis. In IEEE Access, 5, pp. 13951-13977.

Khan, B., & Singh, P. (2019). Economic operation of smart micro-grid: a meta-heuristic approach. In Handbook of research on smart power system operation and control (pp. 330-346). Hershey, PA: IGI Global. doi:10.4018/978-1-5225-8030-0.ch014

Khan, B., & Singh, P. (2019). Economic operation of smart micro-grid: A meta-heuristic approach. In Handbook of research on smart power system operation and control (pp. 330-346). Hershey, PA: IGI Global. doi:10.4018/978-1-5225-8030-0.ch014

Khan, B., & Tanwar, S. (2019). Issues associated with microgrid integration. In Handbook of research on smart power system operation and control (pp. 252-264). Hershey, PA: IGI Global. doi:10.4018/978-1-5225-8030-0.ch010

Kifle, Y., Khan, B., & Singh, J. (2018). Designing and modeling grid connected photovoltaic system: (Case study: EEU building at Hawassa city). *International Journal of Convergence Computing*, *3*(1), 20–34. doi:10.1504/IJCONVC.2018.091113

Kifle, Y., Khan, B., & Singh, P. (2018). Assessment and enhancement of distribution system reliability by renewable energy sources and energy storage, *Journal of Green Engineering, 8*(3), 2, pp. 219-262.

Liu, X., & Li, P. (2007). The effect of DVR on distribution system with distributed generation. In *Proceedings of International Conference on Electrical Machines and Systems*, pp. 277–281.

Makdisie, C., Haidar, B., & Alhelou, H. H. (2018). An optimal photovoltaic conversion system for future smart grids. In Handbook of research on power and energy system optimization (pp. 601–657). Hershey, PA: IGI Global. doi:10.4018/978-1-5225-3935-3.ch018

McGranaghan, M. F. (1993). Voltage sags in industrial systems. *IEEE Trans., 29*(2), pp 397-403.

Molla, T., Khan, B., Moges, B., Alhelou, H. H., Zamani, R., & Siano, P. (2019). Integrated optimization of smart home appliances with cost-effective energy management system. *CSEE Journal of Power and Energy Systems*, *5*(2), 249–258.

Molla, T., Khan, B., & Singh, P. (2018). A comprehensive analysis of smart home energy management system optimization techniques, *Journal of Autonomous Intelligence, 1(1)*, 15–21.

Negash, K., Khan, B., Tesfaye, D., & Gayathri, M. (2017). Optimal placement of phasor measurement unit for system observability. In *Proceedings of IEEE International Conference on Computing Methodologies and Communication, (ICCMC 2017),* pp. 73-77, Erode, India.

Negash, K., Khan, B., & Yohannes, E. (2016). Artificial intelligence versus conventional mathematical techniques: A review for optimal placement of phasor measurement units [Springer]. *Journal of Technology and Economics of Smart Grids and Sustainable Energy*, *1*(1), 10. doi:10.100740866-016-0009-y

Njenda, T. C., Golshan, M. E. H., & Alhelou, H. H. (2018, November). WAMS based intelligent under frequency load shedding considering online disturbance estimation. In *Proceedings 2018 Smart Grid Conference (SGC),* (pp. 1-5). IEEE. 10.1109/SGC.2018.8777779

Njenda, T. C., Golshan, M. E. H., & Alhelou, H. H. (2018, November). WAMS based under frequency load shedding considering minimum frequency prediction and extrapolated disturbance magnitude. In *2018 Smart Grid Conference (SGC),* (pp. 1-5). IEEE.

Singh, P., & Khan, B. (2017). Smart microgrid energy management using a novel artificial shark optimization, Complexity, vol. 2017, Article ID 2158926. doi:10.1155/2017/2158926

Singh, P., Khan, B., Vidyarthi, A., Haes Alhelou, H., & Siano, P. (2019). Energy-aware online non-clairvoyant scheduling using speed scaling with arbitrary power function. *Applied Sciences*, *9*(7), 1467. doi:10.3390/app9071467

Yeshalem, M. T., & Khan, B. (2017). Design of an off-grid hybrid PV/wind power system for remote mobile base station: A case study. *AIMS Energy*, *5*(1), 96–112. doi:10.3934/energy.2017.1.96

Zamani, R., Golshan, M. E. H., Alhelou, H. H., & Hatziargyriou, N. (2019). A novel hybrid islanding detection method using dynamic characteristics of synchronous generator and signal processing technique. *Electric Power Systems Research*, *175*, 105911. doi:10.1016/j.epsr.2019.105911

Zamani, R., Hamedani-Golshan, M. E., Haes Alhelou, H., Siano, P., & Pota, H. R. (2018). Islanding detection of synchronous distributed generator based on the active and reactive power control loops. *Energies*, *11*(10), 2819. doi:10.3390/en11102819

KEY TERMS AND DEFINITIONS

Series Compensation: It is the method of improving the system voltage by connecting a capacitor in series with the transmission line.

Transformer Tap Changer: The purpose of a tap changer is to regulate the output voltage of a transformer.

Voltage Dip: It is a short, temporary drop in the voltage magnitude in the distribution or customer's electrical system.

Chapter 5
Improvement of the Electrical Network Stability by Using a Renewable Distributed Generator

Youssef Krim
National Engineering School of Monastir, Tunisia

Saber Krim
National Engineering School of Monastir, Tunisia

Mohamed Faouzi Mimouni
National Engineering School of Monastir, Tunisia

ABSTRACT

In this chapter, a control strategy for a Renewable Distribution Generator (RDG) operates in grid-connected and standalone mode is suggested. This RDG is made up of a wind generator associated with a Super-Capacitors (SC) considered as a storage system. The study investigates a control scheme for RDG integrated into power electrical system to maintain the voltage and the frequency of the grid in an allowable range and to ensure the continuity of power supply in case of grid faults. The proposed control strategy has three parts: a vector control of the wind generator to extract the maximum power; the control of the DC bus voltage by inserting the SC; and a droop control loop proposed to ensure the grid stability. The simulation results demonstrate the reliability of the control system.

DOI: 10.4018/978-1-7998-1230-2.ch005

INTRODUCTION

The development of wind and photovoltaic energy systems enters within the overall framework of the fight against climate changes (Shahryaria et al., 2019). The major contribution would be that of wind power. Moreover, the particular contexts of the island systems lead to a rapid development of wind energy production (Zabihollah et al., 2018). The most of the integrated wind turbines in the world operate either in the grid connected mode or islanded mode (Somnath and Ashok, 2018). The major problem of today, which would make it difficult to integrate wind turbines continuously, is that they generally do not participate in system services, particularly frequency and voltage control because its intermittent form (Jihane and Mohamed, 2019). Thus, to maintain a secure operation and system reliability with high penetration of renewable energy system, Energy Storage System (ESS) is required to mitigate uncertainties of this renewable source in more and more grid codes to deal with their intermittent nature (Youssef et al., 2019). If suitably deployed, ESS can help operators to manage variability effectively in generation and load. However, the energy storage technology faces limitations, such as high cost, the life span is limited by the number of charge/discharge cycles and the inverse proportionality between specific energy and specific power (Jamila et al., 2018). The most suitable storage system is the battery because it is a proven technology and has a good ratio performance/cost and is suitable for constant power utilization (Cabrane et al., 2016). However, a regular maintenance is necessary for this storage system. In other hand, Super-Capacitors (SC) are characterized by their high efficiency, energy transfer in short duration (Perdana et al., 2018). The envisaged applications for supercapacitors include the automobile, urban public transport and renewable energies (Changfu et al., 2018). In the automotive segment, supercapacitors technology satisfies start-up functions, especially low-temperature constraints, and power unit of hybrid traction chain. Also, it can improve the energy efficiency of tramway and metros by allowing the recovery of braking energy, which allows an increase in traffic without grid investment and reduces the ownership cost (Qi et al., 2019). In renewable energies field, supercapacitors are used to compensate the rapid intermittencies of these sources and to participate in the electrical network support (Krim et al., 2019). To valorize the utility of the storage system, a power management algorithm is necessary.

The management of electrical ESS is an important research topic, in particular in combination with intermittent RES, such as wind energy (Alhelou et al., 2019; Makdisie et al., 2018; Alhelou et al., 2018; Alhelou et al., 2016; Haes Alhelou et al., 2019; Njenda et al., 2018). In fact, in addition to real-time energy management, several objectives (constraints and network services, storage levels and availability, lifespan and aging, etc.) are needed to be taken into consideration all together. The management method is introduced to ensures the storage system set-point be applied within its safe operating limits and to calculates the quantities of powers required for grid stability. The Power Management Algorithm (PMA) is already used in many applications like, for poly-generation microgrids, performance evaluation of a grid independent hybrid renewable energy system (Yanjun et al., 2018) and PMA for DC microgrid systems (Srikanth et al., 2019). The power management module represents the top layer in the hierarchy of the management system (Achraf et al., 2018). Its task is to determine the set-point to be applied to each component of the RDG. A component can represent in this study the storage system, the wind source with its conversion chain, the electrical network and the power consumers (Load). The power budget exchanged with the grid at the connection point is then simply calculated, at any instant t, by the management algorithm as a function of the instantaneous variations of the frequency and the voltage. The power injected to the load is the sum of the power produced by the wind source, supplied by the storage system and the power absorbed from the grid in case of over-frequency. The management of the energy distribution between these different components is assured by a Boolean management method.

For the systems services, several researches are explored in the voltage and frequency control. Ref. (Tomás et al., 2018) considers that the application of the model predictive control approach is well adapted to control the frequency and voltage of an isolated wind power system. Ref. (Savaghebi et al., 2012) depicts a control method for selective compensation of main voltage harmonics to an allowable level in a grid-connected operation. Ref. (Andrés et al., 2018) demonstrates using experimental tests that variable speed wind energy can contribute to the primary frequency regulation. Ref. (Priya and Sagar, 2018; Khan et al., 2019; Kifle et al., 2018; Khan et al., 2018; Khan et al., 2017) illustrates the feasibilities of the droop control strategy in controlling the frequency and voltage in both grid-tied and separated mode. In this framework, this study is focused on the sector of the wind power. The wind turbine with the super-capacitors and load constitute a Renewable Distributed Generator (RDG) that is used to exchange the controlled active and reactive powers with the grid to ensure the power quality. The RDG operates according to the stability of the electrical network in two modes. Firstly, a standalone mode is detected in the event of a grid fault. In this mode, the role of the RDG is to ensure a continuous supply of load. Secondly, a grid-connected mode is detected when the variations of the grid frequency and voltage are not severe but do not belong to the optimal margins. In the latter mode, the role of the RDG is to participate in system services by improving the grid voltage and frequency plans and make them within the desired margins. In addition, a control scheme is instigated for the RDG. The first task of this control is the power management system used to manage the power flow and to detect the connected or islanded operation mode of the RDG. Its second control task is the droop control system used to improve the grid stability by controlling the exchanged active and reactive power.

This paper is organized as follows. The first section is the modeling and control of the renewable distribution generator. The second section is the grid side converter control. In this section a droop control is put forward to improve the voltage and frequency plane of the grid, and ensure a continuous supply of the load in case of grid fault. Finally, a simulation results in the last section illustrate the robustness of the proposed control strategy.

MODELING AND CONTROL OF THE RENEWABLE DISTRIBUTED GENERATOR

The renewable distributed generation is schematically illustrated in Figure 1. It decomposes in three main blocks. The first one is a wind energy system coupled to the PMSG. The second block is a SC module. It is connected with the wind energy system through a common DC bus voltage. The third one is a variable load and the electrical network.

Depending on the stability state of the electrical network, the RDG operates in two modes:

- Connected-mode: In this case, the RDG is connected to the grid through a static switch noted Common Connection Point (PCC) (Jundi et al., 2018) (PCC is closed). The PCC is the connection point of the wind generator, the load and the electrical network. He participates in system services by reducing the grid voltage and frequency disturbances at the PCC to ensure its stability. Quantities of active and reactive powers are imposed to reduce these disturbances and make them within the margins mentioned in the standards.

Figure 1. Structure of the renewable distribution generator

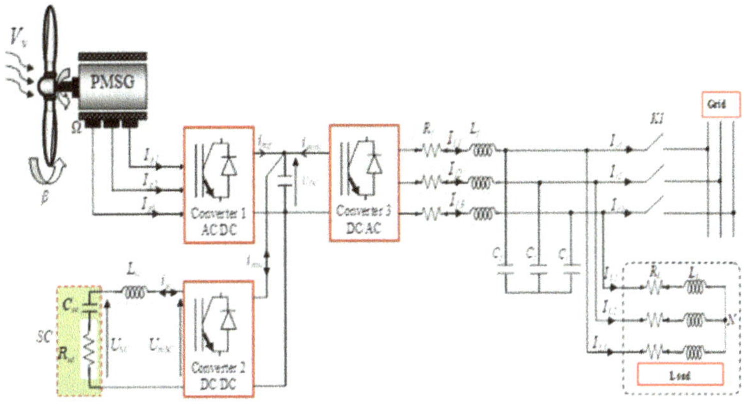

- Standalone mode is detected in case of grid failure (the fluctuations of the frequency and the voltage of the network at the point PCC are more severe than the maximum margins mentioned in the standards). In this mode, the RDG operates in a separated mode to ensure a continuous supply of load during this grid fault. Therefore, the exchange of powers with the grid is stopped.

Wind Turbine Modeling and Control

The modeling of the turbine is inspired by (Youssef et al., 2019a). The available power P_w of the wind crossing a surface S is defined by:

$$P_w = \frac{1}{2}\rho S V_w^3 \tag{1}$$

The captured power "P_{aer}" by the wind turbine is then:

$$P_{aer} = \frac{1}{2}C_p(\lambda,\beta)\rho S V_w^3 \tag{2}$$

where ρ is the density of air and V_w is the wind speed.

The power coefficient C_p is defined by the ratio of the captured power by the turbine on the wind power. C_p is a function of the specific λ speed and a pitch angle β (Howlader and Senjyu, 2016):

$$\begin{cases} C_p(\lambda,\beta) = 0.5179(98\lambda_i - 0.4\beta + 5)e^{-21\lambda_i} + 0.068\lambda \\ \lambda_i = \dfrac{1}{\dfrac{1}{\lambda + 0.089} - \dfrac{0.035}{\beta^3 + 1}} \end{cases} \tag{3}$$

Figure 2. Architecture of Wind Turbine Modeling

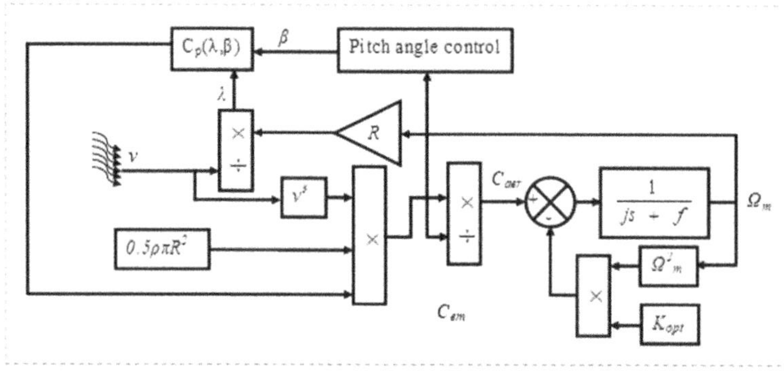

Where the specific speed is defined as:

$$\lambda = \frac{R\Omega_m}{V_w} \tag{4}$$

where R is the radius of the turbine.

Then, the model diagram of the studied wind turbine is depicted in Figure 2.

To ensure the maximum efficiency of the turbine, the power coefficient must be maintained at its maximum. According to Betz theory, the C_p cannot exceed 0.593 (Chaudhary et al., 2019). The evolution of this coefficient depending on λ and β is given by Figure 3, shows an optimum value of the speed ratio (λ_{opt}=8.15) corresponding to a maximum value of the power coefficient (C_{pmax}=0.4794) for β=0.

Eq. (5) and (6) gives the expression of the maximum power obtained by using the MPPT strategy (Youssef et al., 2018):

$$P_{MPPT} = \frac{1}{2}\frac{\rho C_{p\max} R^5 \Omega^3{}_m}{\lambda_{opt}{}^3} \tag{5}$$

Then the electromagnetic torque is:

$$T_{em_{MPPT}} = \frac{1}{2}\frac{\rho C_{p\max} R^5 \Omega^2{}_m}{\lambda_{opt}{}^3} \tag{6}$$

In order to produce the maximum available power, the orientation angle of the blades must be zero. This is the MPPT operation. However, when the wind speed exceeds its nominal value, it is necessary to limit the mechanical speed of the turbine in its nominal value to protect the wind system against the strong variations of the wind speed. In this study, a control system called "Pitch angle control" is inserted to control the power and speed of the wind turbine. It limits them to their nominal values if the wind is too high. The operation of the "Pitch angle control" system is described in Figure 4.

Figure 3. C_p versus λ curve

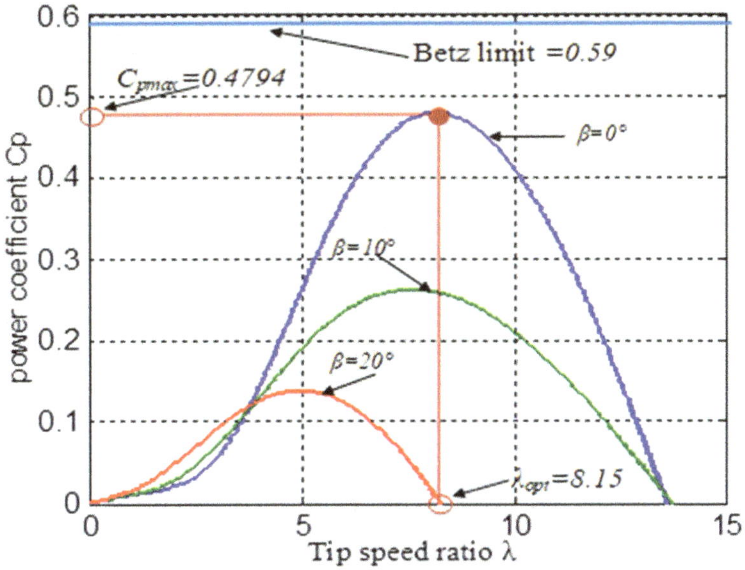

Figure 4. Characteristics of the wind turbine

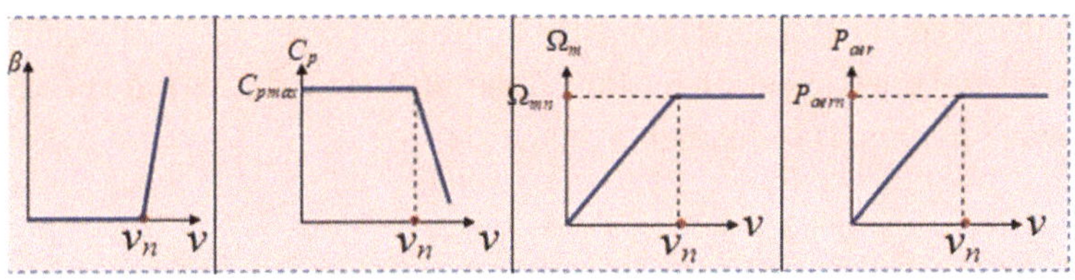

1. Modeling and control of the PMSG

The wind turbine is connected to the common DC bus by means the PMSG and a PWM rectifier. The PMSG is classically modeled in the d-q park reference by the following equations (Krim et al., 2016):

$$\begin{cases} V_{sd} = R_s i_{gd} + L_s \dfrac{di_{gd}}{dt} - p\Omega_m L_s i_{gq} \\ V_{sq} = R_s i_{gq} + L_s \dfrac{di_{gq}}{dt} + p\Omega_m L_s i_{gq} + p\Omega_m \phi_m \\ T_{em} = p i_{gq} \phi_m \end{cases} \tag{7}$$

Figure 5. PMSG and AC/DC converter control

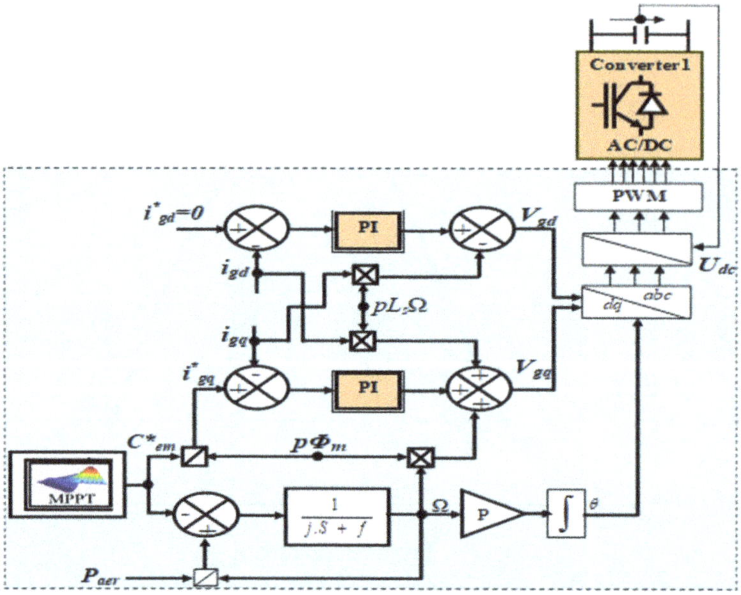

where R_s is the stator winding resistance, L_s is the stator winding inductance, i_{gd} and i_{gq} are the direct and quadratic stator currents, V_{sd} and V_{sq} are the direct and quadratic stator voltages, p is the number of pole pairs and Φ_m is the permanent magnet rotor flux (Wb).

Taking into account the coefficient of friction f, the mechanical rotation speed of the generator is defined by the following dynamics fundamental equation:

$$j\frac{d}{dt}\Omega_m = T_{aer} - T_{em} - f\Omega_m \tag{8}$$

with:

j is the total inertia of the system and T_{aer} is the aerodynamic torque calculated by equation (9) as follows:

$$T_{aer} = \frac{P_{aer}}{\Omega_m} \tag{9}$$

The Figure 5 shows the control strategy of the wind generator. This strategy is based on the vector control applied to PMSG to extract a maximum power. The principle of this control is to impose a reference direct current i^*_{sd} equal to zero and a quadratic current reference i^*_{sq} proportional to the reference torque given by the MPPT algorithm as follows:

Figure 6. Control system of stator currents

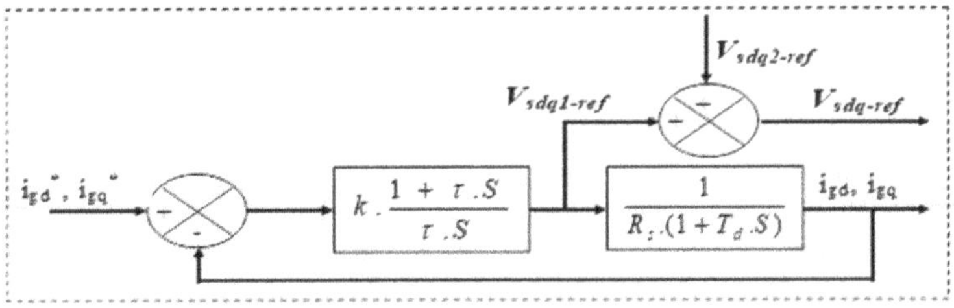

$$\begin{cases} i_{gd}^{*} = 0 \\ i_{gq}^{*} = \dfrac{T_{em_{MPPT}}}{p\phi_m} \end{cases} \tag{10}$$

In order to impose reference currents, two Proportional Integral (PI) controllers are used. These controllers are identical because the transfer function is similar in the d and q axes. Figure 6 represents the regulation loop of the currents in the axes d and q.

Using the poles compensation method, the time constant of the regulator studied can be calculated as follows:

$$\begin{cases} \tau = \dfrac{L_s}{R_s} \\ k = \dfrac{3L_s}{\tau} \end{cases} \tag{11}$$

Modeling and Control of the Super-Capacitors Module

An emerging technology and increasingly applied, super-capacitors are able to store and discharge energy very quickly and efficiently (Fini et al., 2016; Alhelou et al., 2018; Zamani et al., 2018; Alhelou et al., 2015; Njenda et al., 2018; Haes Alhelou et al., 2018; Haes Alhelou et al., 2019). Due to their many advantages, SCs are currently used in many applications and are considered in a range of future applications that are as diverse (Zineb et al., 2017). Unlike a ceramic or electrolytic capacitor, an SC does not have a dielectric. Instead, it uses the "electric double layer" that forms at the interface of the solid (electrode) and the liquid (electrolyte). Figure 7 present the structure of the SC. A typical SC is composed of electrodes and electrolytes (including the electrolyte salt) and a separator (to prevent contact between positive and negative electrodes). The electrodes are positioned on the electricity collectors and coated with activated carbon powder. A double electric layer is formed at each interface where the activated carbon powder comes into contact with the electrolyte. The operating principle of the SC is equivalent to that of the conventional capacitor with the only difference that the SC does not have a dielectric layer in its ionic liquid electrolytic zone.

Figure 7. Super-capacitor structure

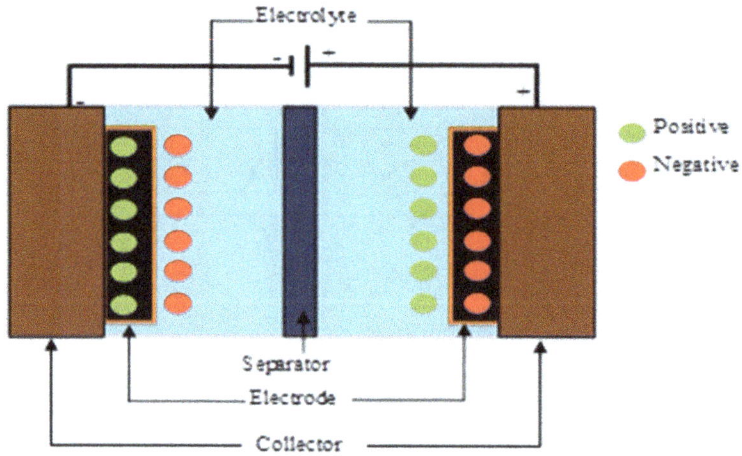

In this context, to ensure a balance between production and consumption and to ensure DC bus voltage stability at its reference value, a super capacitor module as storage system is inserted. A realistic model of the super-capacitor is composed by a series capacitance C_{ssc} with a resistance R_{ssc}.

The maximum power can be provided by associating super-capacitors in series and in parallel. A SC bank of N_c super capacitors in series and N_p super capacitors in parallel is modeled by a capacitor C_{sc} in series with a resistance R_{sc} as follows:

$$\begin{cases} C_{sc} = \dfrac{N_p}{N_c} C_{ssc} \\ R_{sc} = \dfrac{N_c}{N_p} R_{ssc} \end{cases} \tag{12}$$

The SC bank is connected to the DC bus by means a bidirectional DC/DC converter and an inductive filter, as shown in Figure 1.

In order to control the charge and discharge current of the SC, a PI controller is used to adjust the current I_{sc} with the reference value I^*_{sc} as depicted in part 2 of Figure 8:

$$U_{msc} = U_{sc} - PI\left(I^*_{sc} - I_{sc}\right) \tag{13}$$

The reference power exchanged with the super-capacitors '$P_{sc\text{-}ref}$' is generated by the Power Management Algorithm (PMA) according to the SC voltage 'U_{sc}', as shown in Figure 9.

The PMA receives the produced power "P_g", the power demanded by the load "P_L", the necessary power for ancillary services and the State of Charge (SOC) of the SC 'U_{sc}'. Then, it allows managing the RDG into four operating modes:

Figure 8. DC bus, SC, and DC/DC converter control

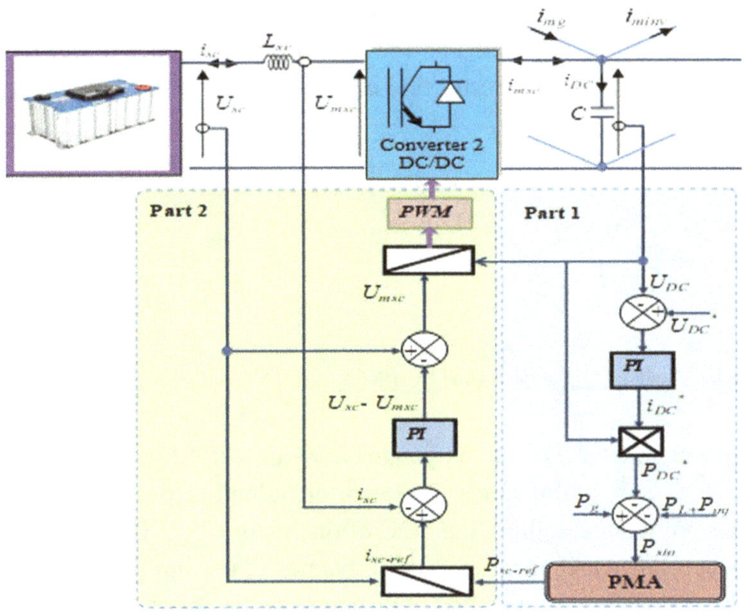

Figure 9. Power management algorithm

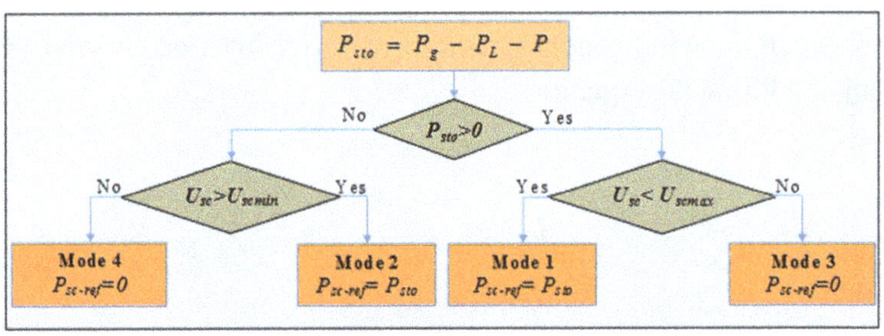

- **Mode 1:** When the power provided by the generator is higher than that demanded by load and system services "$P_g > P_L + P$" and the SC does not reach its maximum SOC "$U_{sc} < U_{scmax}$", it is the charging mode of the SC. The reference power applied to the storage system module is: "$P_{sc\text{-}ref} = P_g - P_L - P$". In this condition the wind turbine operates in MPPT.
- **Mode 2:** If the generated power is not sufficient to cover the grid and load demand power "$P_g < P_L - P$" and the storage system does not reach its minimum SOC "$U_{sc} > U_{sc\,min}$", it is a discharge mode of the SC. In this mode, the SC acts to compensate the shortage of demanded power to ensure the balance between generation and consumption. The reference power applied to the storage system module is: "$P_{sc\text{-}ref} = P_g - P_L - P$".

- **Mode 3:** When the SC reaches its maximum SOC "$U_{sc} > U_{scmax}$", and the produced power exceeds the demanded one. In this case, the wind generator operates in a Limited Power Point Tracking (LPPT) to generate only the demanded power (k_d=1). The reference power applied to the storage system module is: "$P_{sc\text{-}ref}=0$". This restriction is made by the variation of the angle β as follows:

$$\begin{cases} \beta_{ref} = 0 \ if \ k_d = 0 \\ \beta_{ref} = \dfrac{\Delta\beta}{\Delta P}\left(P_g - P_L\right) if \ k_d = 1 \end{cases} \quad (14)$$

where k_d is the degradation factor deduced by PMA.

- **Mode 4:** The power generation system operates in mode 4 if "$P_g<P_L+P$" and SC attends is minimum SOC "$U_{sc} < U_{scmin}$". In this mode, the no-priority load is disconnected to reach the balance production-consumption and to limit the SOC of the SC in U_{scmin}. The disconnection is made by a load management relay, which measures the active power consumption in the installation and, as soon as the defined threshold is exceeded, it disconnects the non-preferential charges and re-feeds them after a pre-programmed period. The reference power applied to the storage system module is: "$P_{sc\text{-}ref}=0$".

Modeling and Control of the DC Bus Voltage

The power transfer between the wind generator and the inverter must be ensured via the DC bus. This DC bus is modeled by the following equation:

$$U_{DC} = \frac{1}{C_{DC}}\int I_{DC}dt \quad (15)$$

From Figure 1, the I_{DC} current is deduced by the following equation:

$$I_{DC} = I_{mg} - I_{\min j} - I_{msc} \quad (16)$$

The powers on the DC bus level are determined by multiplying the equation (16) by U_{DC}:

$$I_{DC}U_{DC} = P_{DC} = P_g - P_L - P_{\text{in} j} - P_{sc} \quad (17)$$

Part 1 of Figure 8 represents the control loop of the DC bus. A PI controller is applied to maintain the measured voltage U_{DC} equivalent to the reference U^*_{DC}.

$$I^*_{DC} = PI\left(U^*_{DC} - U_{DC}\right) \quad (18)$$

Figure 10. Detection algorithm of the grid stability

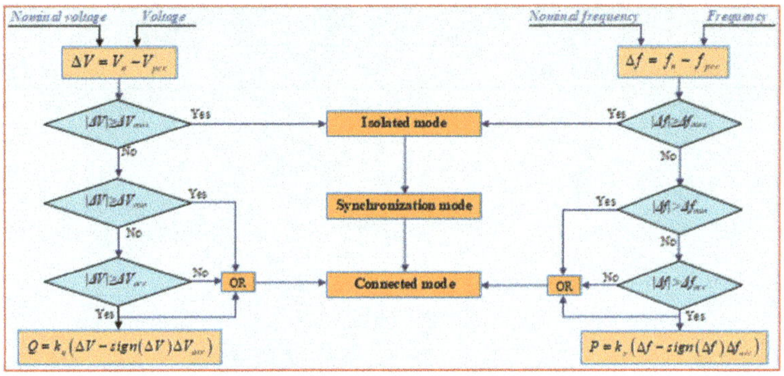

Modeling of the Grid Connection Interface (LC filter)

The architecture of the Voltage Source Inverter (VSI) system related to the grid and the load with an LC filter is depicted in Figure 1. The model of the LC filter can be written, in the d-q rotating frame, by the following equation systems:

$$\frac{d}{dt}\begin{pmatrix} i_{fd} \\ i_{fq} \end{pmatrix} = \begin{pmatrix} -\frac{R_f}{L_f} & \omega \\ -\omega & -\frac{R_f}{L_f} \end{pmatrix}\begin{pmatrix} i_{fd} \\ i_{fq} \end{pmatrix} + \frac{1}{L_f}\begin{pmatrix} V_{id} - V_{cd} \\ V_{iq} - V_{cq} \end{pmatrix} \tag{19}$$

$$\frac{d}{dt}\begin{bmatrix} V_{cd} \\ V_{cq} \end{bmatrix} = \frac{1}{C_f}\begin{bmatrix} i_{fd} \\ i_{fq} \end{bmatrix} + \omega\begin{bmatrix} V_{cd} \\ -V_{cq} \end{bmatrix} - \frac{1}{C_f}\begin{bmatrix} i_{td} \\ i_{tq} \end{bmatrix} \tag{20}$$

where i_{td} and i_{tq} are the direct and quadratic currents transferred to the load and the grid, i_{fd} and i_{fq} are the direct and quadratic currents modulated by VSI, V_{id} and V_{iq} are the direct and quadratic voltages modulated by VSI, V_{cd} and V_{cq} are the direct and quadratic voltages at the terminals of the filter capacitors, and ω is the coordinate angular frequency.

GRID STABILITY CONTROL

The wind generator is used to participate in the system services by improving the voltage and frequency planes of the electrical network and to ensure a continuous supply of load in case of grid fault. Therefore, the studied RDG operates in two modes. Grid connected mode and island mode. Therefore, the participation in the system service requires a detection algorithm to satisfy the requested energy constraints.

Table 1. Standards used for grid stability

Δf_{min}	Δf_{max}	ΔV_{min}	ΔV_{max}
±0.1 Hz	±1 Hz	±5 V	±20 V

Figure 11. Droop control loop

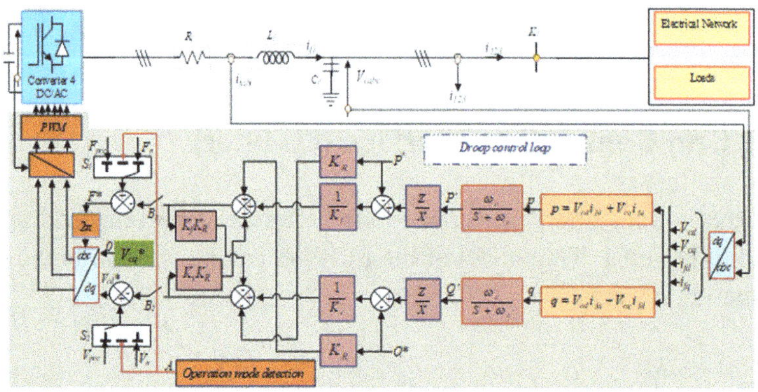

Detection Algorithm

The detection algorithm allows to detect the islanding in case of grid failure and to calculate the necessary active and reactive powers to reduce the variations of the frequency and the voltage at the PCC point. Thus, the different operation modes of the RDG are presented in Figure 10.

- If "$|\Delta f| \geq |\Delta f_{max}|$" or "$|\Delta V| \geq |\Delta V_{max}|$" an islanded mode is active. In this mode the renewable distribution generator operates in standalone mode to protect the electrical equipment against the strong variations of the frequency and voltage. Then the automatic switches B_1, B_2 presented in figure 11 are opened, and S_1 and S_2 are respectively in f_n and V_n states.
- When "$|\Delta f| < |\Delta f_{min}|$" and "$|\Delta V| < |\Delta V_{min}|$" the levels of the frequency and the voltage are acceptable, so the adjustment is not necessary. In this condition a grid-connected mode is active. Then K_1, B_1 and B_2 are closed, and S_1 and S_2 are respectively in f_{pcc} and V_{pcc} states.
- When "$f_n + |\Delta f_{acc}| < f_{pcc} \leq f_{max}$ and $V_n + |\Delta V_{acc}| < V_{pcc} \leq V_{max}$" or "$f_{min} \leq f_{pcc} < f_n - |\Delta f_{acc}|$ and $V_{min} \leq V_{pcc} < V_n - |\Delta V_{acc}|$" a grid-connected mode is active but the levels of the frequency and the voltage are not optimum. For this purpose, the necessary powers P and Q defined by Eq. (21) must be exchanged with the grid to make the grid frequency and voltage in the desired margins. Then K_1, B_1 and B_2 are closed, and S_1 and S_2 are respectively in f_{pcc} and V_{pcc} states.

Where Δf_{min}, Δf_m, ΔV_{min}, and ΔV_m are indicated in the following table:

This algorithm calculates the quantities of the active and reactive powers P and Q in order to reduce the variations of the frequency and the voltage.

$$\begin{cases} P = k_f\left(\Delta f - sign\left(\Delta f\right)\Delta f_{\min}\right) \\ Q = k_V\left(\Delta V - sign\left(\Delta V\right)\Delta V_{\min}\right) \end{cases} \tag{21}$$

where k_f and k_V are adjustable gains.

Droop Control Loop

A droop control is designed to adjust the frequency and voltage at *PCC* point by controlling the injected or absorbed powers *P* and *Q* with the electrical distribution network. This control is depicted in Figure 11.

The active and reactive powers transferred to the load and to the grid are calculated from the output currents and voltages of the inverter as follows (Youssef et al., 2018a):

$$\begin{aligned} p &= \frac{3}{2}(V_{cd}i_{fd} + V_{cq}i_{fq}) \\ q &= \frac{3}{2}(V_{cd}i_{fq} - V_{cq}i_{fd}) \end{aligned} \tag{22}$$

After having used a low-pass filter, the fundamental parts of the transferred active and reactive powers are obtained:

$$\begin{cases} P = \dfrac{\omega_c}{S + \omega_c} p \\ Q = \dfrac{\omega_c}{S + \omega_c} q \end{cases} \tag{23}$$

Notice that there is decoupling between the active power and the frequency "*P-F*", and the reactive power and the voltage "*Q-V*", as follows (Beena et al., 2018).

$$\begin{cases} P = \dfrac{V_c V_{pcc}}{Z}\delta \\ Q = \dfrac{V_c(V_c - V_{pcc})}{Z} \end{cases} \tag{24}$$

Eq. (23) illustrates the relationship between the active power and the frequency and a relationship between the reactive power and the voltage (Rahman et al., 2016). This is the basis of droop control method:

$$\begin{cases} \Delta F = F^{*} - F = -k_p(P - P^{*}) \\ \Delta V = V^{*} - V = k_q(Q - Q^{*}) \end{cases} \tag{25}$$

Figure 12. Classical PI control scheme of VSI system

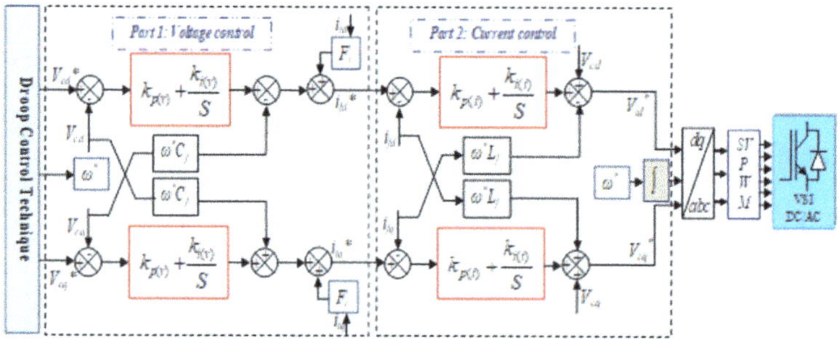

Another expression of the active "P" and reactive "Q" powers transferred to the load and grid:

$$\begin{cases} P = \dfrac{X}{Z}P - \dfrac{R}{Z}Q \\ Q = \dfrac{R}{Z}P + \dfrac{X}{Z}Q \end{cases} \tag{26}$$

where R and L are the resistance and inductance of the line connecting the two voltages V_c and V_{pcc}. By introducing Eq. 25 in Eq. 26, Eq. 26 becomes:

$$\begin{aligned} P &= \frac{X}{Z}\left(\frac{k_p P^{*} - \Delta F}{k_p} - \frac{R}{X}\frac{k_q Q^{*} - \Delta V}{k_q}\right) \\ Q &= \frac{X}{Z}\left(\frac{k_q Q^{*} - \Delta V}{k_q} + \frac{R}{X}\frac{k_p P^{*} - \Delta F}{k_p}\right) \end{aligned} \tag{27}$$

Assuming that $K_R = \dfrac{R}{X}, k_f = -\dfrac{1}{k_p}, k_v = -\dfrac{1}{k_q}$, so the variations of the frequency and the voltage are defined by:

$$\begin{cases} \Delta F = \dfrac{1}{k_f}\left(\dfrac{Z}{X}P - P^{*}\right) + K_R k_v \Delta V + K_R Q^{*} \\ \Delta V = \dfrac{1}{k_v}\left(\dfrac{Z}{X}Q - Q^{*}\right) - K_R k_f \Delta F - K_R P^{*} \end{cases} \tag{28}$$

The outputs of the droop control loop are the tuning variations that will be subtracted from the phase and the amplitude of the grid at the PCC point in order to make them in the desired margins.

Finally, whatever the operation mode of the generator, the droop control system presents as outputs the reference voltages. The phase and the RMS value of these voltages are maintained at each moment within a desirable margin more stable than that mentioned in the standards. Thereafter, the inverter adjusts the voltages measured at the output of the LC filter with those deduced by the droop controller using PI controllers. To start the design of the control technique, it is first necessary to determine the references of the voltages and currents supplied to the load. The reference voltages V_{cdr} and V_{cqr} are given by a park transformation of a balanced three-phase voltage system with a frequency of 50 Hz and amplitude of 230V.

The block diagram of current and voltage control based on PI controllers is shown in Figure 12.

The control structure of the two components of voltages V_{cd} and V_{cq} is shown in part 1 of Figure 12. This command is performed by two conventional PI controllers. The differential algebraic equations of these two controllers are written as follows (Krim et al., 2019):

$$\begin{array}{l} \dot{\phi} = V_{cd}^{\bullet} - V_{cd} \\ \dot{\phi}_q = V_{cq}^{\bullet} - V_{cq} \end{array} \quad (29)$$

By inverting the relation (20), the reference currents i_{fd}* and i_{fq}* modulated by the voltage regulation block are obtained as follows:

$$\begin{cases} i_{fd}^{\bullet} = Fi_d - \omega^{\bullet} C_f V_{cq} + k_p(v)\dot{\phi}_d + k_i(v)\phi_d \\ i_{fq}^{\bullet} = Fi_q - \omega^{\bullet} C_f V_{cd} + k_p(v)\dot{\phi}_q + k_i(v)\phi_q \end{cases} \quad (30)$$

The control structure of the two current components i_{fd} and i_{fq} is shown in part 2 of Figure 12. The differential algebraic equations of these two control loops are given by the following expressions (Krim et al., 2019):

$$\begin{array}{l} \dot{\phi}_d = i_{fd}^{\bullet} - i_{fd} \\ \dot{\phi}_q = i_{fq}^{\bullet} - f_{fq} \end{array} \quad (31)$$

By inverting the relation (19), the reference voltages V_{id}* and V_{iq}*, which represent the control voltages of the inverter are obtained as follows:

$$\begin{cases} V_{id}^{\bullet} = V_{cd} - \omega^{\bullet} L_f i_{fq} + k_p(i)\dot{\phi}_d + k_q(i)\phi_d \\ V_{iq}^{\bullet} = V_{cq} + \omega^{\bullet} L_f i_{fd} + k_p(i)\dot{\phi}_q + k_q(i)\phi_q \end{cases} \quad (32)$$

Figure 13. (a) Wind profil, (b) Active power demanded by the load

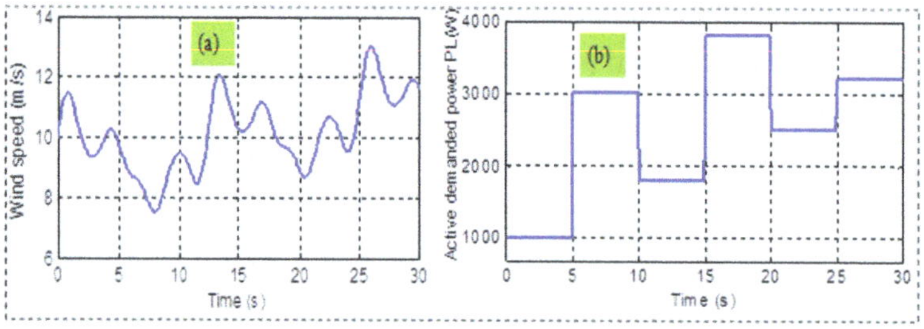

Figure 14. (a) Grid frequency, (b) Grid voltage

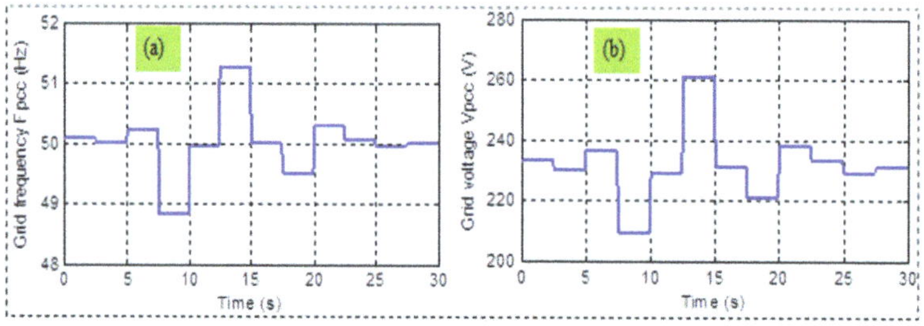

By this way, the possibility of using the power interfaces of the renewable distributed generator to participate in system services is analyzed in this study. In fact, the power management algorithm and the adaptive control strategy are implemented to control the converters of the studied generator simultaneously ensuring:

- Power flow management between the different components
- keeping the DC bus voltage stable at its reference value
- Controlling of the charge and discharge current of the storage system
- Calculating of the necessary active power for adjusting the grid frequency
- Calculating of the necessary reactive power for adjusting the grid voltage
- Controlling of the grid frequency and voltage
- Controlling of the voltages and currents at the LC filter output

SIMULATIONS RESULTS

The control strategy proposed in this chapter has been validated using Matlab/Simulink.

The system is simulated under the following conditions:

-Variable wind profile varies between 7 and 13 m/s presented in Figure 13a.

- ◦ Active power demanded by the load depicted by Figure 13b.

-The variations of the grid frequency and voltage are represented by Figure 14a and Figure 14b respectively.

Figure 15. Mechanical speed of the PMSG

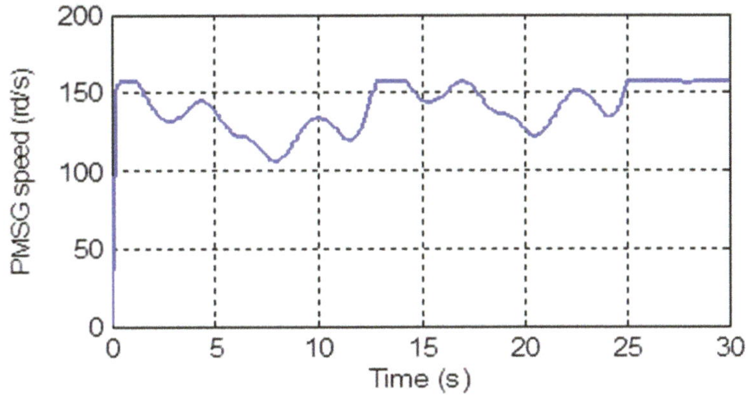

Figure 16. (a) Power coefficient C_p, (b) Pitch angle β

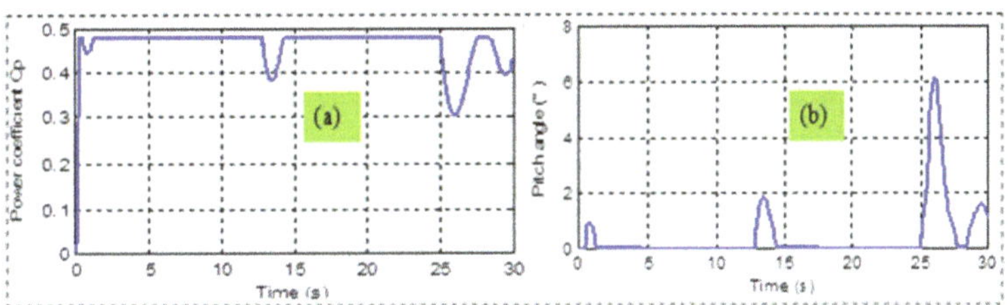

Figure 17. Currents at DC bus level

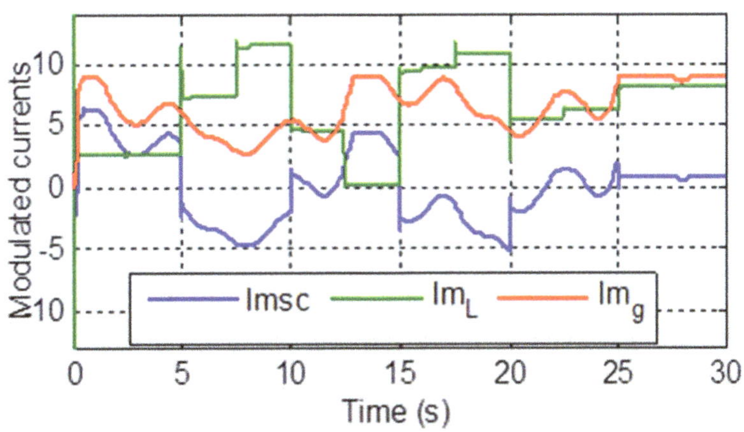

A wind turbine is designed to rotate at a nominal speed and produce a nominal power. For this, a control system called 'pitch angle' was inserted in order to control the power output of the wind turbine and thus to limit its speed when the wind speed is high. The mechanical speed of the PMSG is given by Figure 15.

Figure 18. DC busvoltage U_{dc}

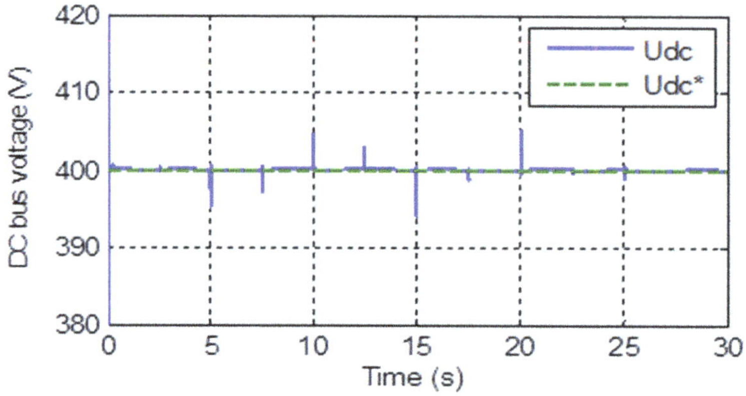

Figure 19. (a) State of K_1, (b) Active power exchanged with grid, (c) Reactive power exchanged with grid.

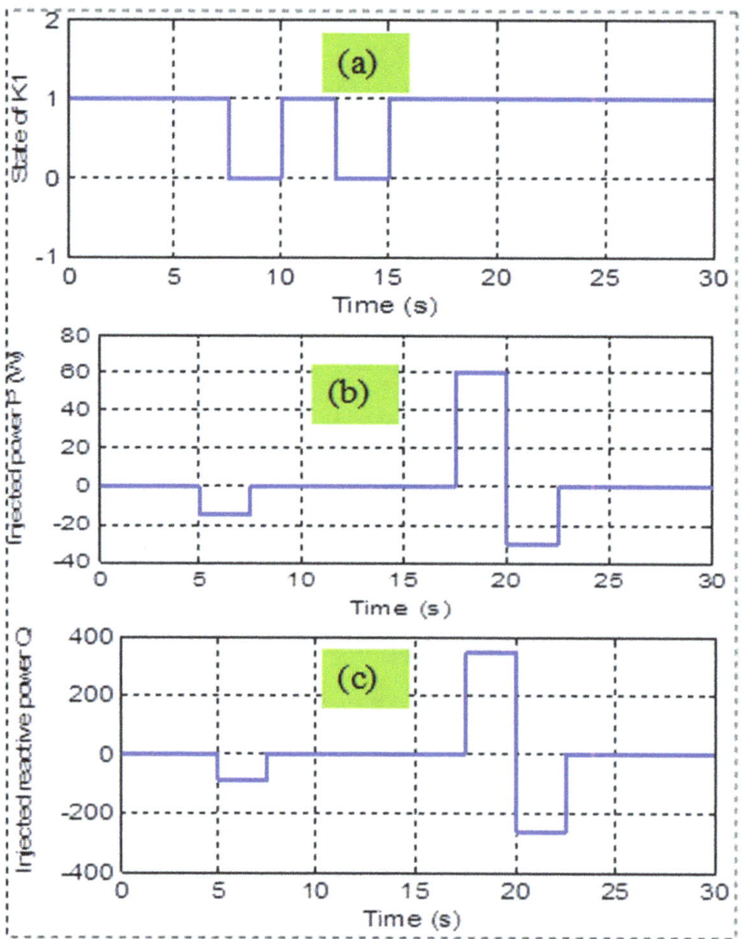

Figure 20. Grid voltage before and after regulation

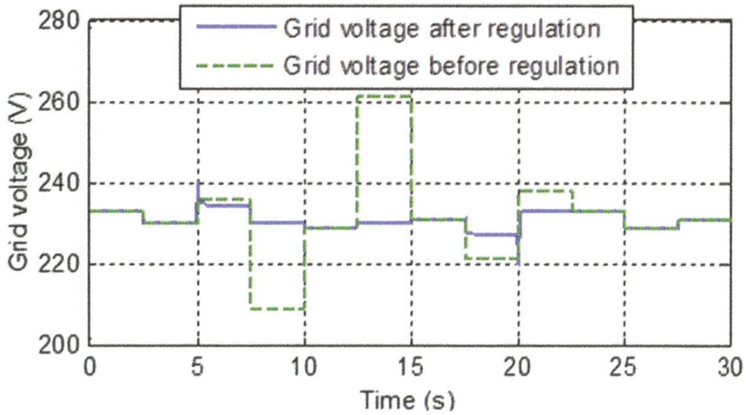

Figure 21. Grid frequency before and after regulation

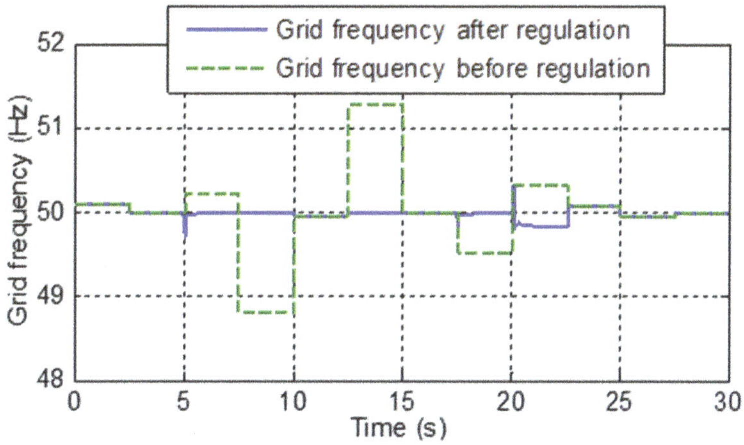

Figure 16b gives the variation of the pitch angle β. The operation of this control system is summarized as follows: when the wind speed exceeds its nominal value this angle β increases, subsequently a power coefficient C_p decrease (Figure 16a).

Figure 17 shows the evolutions of the currents at the DC bus level. It is clear that the production-consumption balance is achieved, which proves the efficiency of the SC control strategy. The balance between these three currents enables us to obtain a constant DC bus voltage at 400 V as presented in Figure 18.

Figure 19a shows the transition from grid-connected mode "*PCC*= 1" to stand alone mode "*PCC*= 0". This transition is accruing to protect the electrical equipments when the voltage or the frequency exceeds the maximum level. Then, the adjustment active and reactive powers injected to the grid are illustrated in Figure 19b and Figure 19c. The role of these powers transferred to the grid is to reduce the fluctuations of the grid voltage and frequency and make them within the acceptable zones. All this is validated by Figure 20 and Figure 21, which respectively represent the grid voltage and frequency after regulation.

CONCLUSION

In this chapter, the improvement of the frequency and voltage planes of the grid by means a renewable distributed generator has been put forward. Three power converters have been used. The control of the PMSG side converter guarantees the operation in MPPT of the wind generator to extract the maximum power. The storage system side converter is responsible for the regulation of the DC bus voltage by storing or requesting energy from the SC module to ensure the power balance. The droop control strategy applied to the grid side converter ensuring a continuous supply of load in case of grid fault and participates in system services by setting the grid frequency and voltage in grid-connected mode. Finally, the simulation results show that the control strategy participates in the grid stability with the help of the SC storage system.

REFERENCES

Achraf, A., Abdelkarim, M., & Lotfi, K. (2018). Assisted power management of a stand-alone renewable multi-source system, *Energy, 145*(15), 195-205.

Alhelou, H., Hamedani-Golshan, M. E., Zamani, R., Heydarian-Forushani, E., & Siano, P. (2018). Challenges and opportunities of load frequency control in conventional, modern and future smart power systems: A comprehensive review. *Energies, 11*(10), 2497. doi:10.3390/en11102497

Alhelou, H. H. (2018). *Fault detection and isolation in power systems using unknown input observer. In Advanced condition monitoring and fault diagnosis of electric machines* (p. 38). Hershey, PA: IGI Global.

Alhelou, H. H., Golshan, M., & Fini, M. (2018). Wind driven optimization algorithm application to load frequency control in interconnected power systems considering GRC and GDB nonlinearities. *Electric Power Components and Systems, 46*(11-12), 1223-1238.

Alhelou, H. H., & Golshan, M. E. H. (2016, May). Hierarchical plug-in EV control based on primary frequency response in interconnected smart grid. In *2016 24th Iranian Conference on Electrical Engineering (ICEE),* (pp. 561-566). IEEE. 10.1109/IranianCEE.2016.7585585

Alhelou, H. H., & Golshan, M. E. H. (2020). Decision-making-based optimal generation-side secondary-reserve scheduling and optimal LFC in deregulated interconnected power system. In *Decision making applications in modern power systems* (pp. 269–299). Academic Press. doi:10.1016/B978-0-12-816445-7.00011-6

Alhelou, H. H., Golshan, M. E. H., & Hatziargyriou, N. D. (2019). Deterministic dynamic state estimation-based optimal LFC for interconnected power systems using unknown input observer. *IEEE Transactions on Smart Grid*, 1. doi:10.1109/TSG.2019.2940199

Alhelou, H. H., Golshan, M. E. H., & Hatziargyriou, N. D. (2019). A decentralized functional observer based optimal LFC considering unknown inputs, uncertainties, and cyber-attacks. *IEEE Transactions on Power Systems*.

Alhelou, H. H., Golshan, M. E. H., Zamani, R., Moghaddam, M. P., Njenda, T. C., Siano, P., & Marzband, M. (2019, June). An improved UFLS scheme based on estimated minimum frequency and power deficit. In Proceedings 2019 IEEE Milan PowerTech (pp. 1-6). IEEE. doi:10.1109/PTC.2019.8810497

Alhelou, H. H., Golshan, M. H., & Askari-Marnani, J. (2018). Robust sensor fault detection and isolation scheme for interconnected smart power systems in presence of RER and EVs using unknown input observer. *International Journal of Electrical Power & Energy Systems*, *99*, 682–694. doi:10.1016/j.ijepes.2018.02.013

Alhelou, H. H., Hamedani-Golshan, M. E., Heydarian-Forushani, E., Al-Sumaiti, A. S., & Siano, P. (2018, September). Decentralized fractional order control scheme for LFC of deregulated nonlinear power systems in presence of EVs and RER. In *2018 International Conference on Smart Energy Systems and Technologies (SEST)* (pp. 1-6). IEEE. 10.1109/SEST.2018.8495858

Alhelou, H. S. H., Golshan, M. E. H., & Fini, M. H. (2015, December). Multi agent electric vehicle control based primary frequency support for future smart micro-grid. In *2015 Smart Grid Conference (SGC),* (pp. 22-27). IEEE.

Beena, V., Jayaraju, M., & Sebin, D. K. (2018). Active and reactive power control of single phase transformerless grid connected inverter for distributed generation system. *International Journal of Applied Engineering Research*, *13*(1), 150–157.

Cabrane, Z, Ouassaid, M, & Maaroufi, M. (2016). Analysis and evaluation of battery-supercapacitor hybrid energy storage system for photovoltaic Installation. *International Journal of Hydrogen Energy - Elsevier*, 1-11.

Changfu, Z., Lei, Z., Xiaosong, H., Zhenpo, W., Torsten, W., & Michael, P. (2018). A review of fractional-order techniques applied to lithium-ion batteries, lead-acid batteries, and supercapacitors. *Journal of Power Sources*, *390*, 286–296. doi:10.1016/j.jpowsour.2018.04.033

Chaudhary, M. K., & Prakash, S. (2019). Investigation of blade geometry and airfoil for small wind turbine blade, *Advanced Science*, *Engineering in Medicine*, *11*(5), 448–452.

Fini, M. H., Yousefi, G. R., & Alhelou, H. H. (2016). Comparative study on the performance of many-objective and single-objective optimisation algorithms in tuning load frequency controllers of multi-area power systems. *IET Generation, Transmission, & Distribution*, *10*(12), 2915–2923. doi:10.1049/iet-gtd.2015.1334

Haes Alhelou, H., Hamedani-Golshan, M. E., Njenda, T. C., & Siano, P. (2019). A survey on power system blackout and cascading events: Research motivations and challenges. *Energies*, *12*(4), 682. doi:10.3390/en12040682

Haes Alhelou, H., Hamedani-Golshan, M. E., Njenda, T. C., & Siano, P. (2019). Wide-area measurement system-based optimal multi-stage under-frequency load-shedding in interconnected smart power systems using evolutionary computing techniques. *Applied Sciences*, *9*(3), 508. doi:10.3390/app9030508

Haes Alhelou, H., Hamedani Golshan, M. E., Njenda, T. C., & Siano, P. (2019). WAMS-based online disturbance estimation in interconnected power systems using disturbance observer. *Applied Sciences*, *9*(5), 990. doi:10.3390/app9050990

Howlader, A. M., & Senjyu, T. (2016). A comprehensive review of low voltage ride through capability strategies for the wind energy conversion systems. *Renewable & Sustainable Energy Reviews*, *56*, 643–658. doi:10.1016/j.rser.2015.11.073

Jamila, S., Seifeddine, B. E., Mohamed, B., & Mohamed, F. M. (2018). Optimal sizing of energy storage systems using frequency-separation-based energy management for fuel cell hybrid electric vehicles. *IEEE Transactions on Vehicular Technology*, *10*, 9337–9346.

Jihane, K., & Mohamed, C. (2019). Study of the different structures of hybrid systems in renewable energies: A review. *Energy Procedia*, *157*, 323–330. doi:10.1016/j.egypro.2018.11.197

Jundi, J., Guangya, Y., Arne, H. N., & Peter, R.-H. (2018). Study of control strategies of power electronics during faults in microgrids, *Hybrid-Renewable Energy Systems in Microgrids,* 109-146.

Khan, B., & Agnihotri, G. (2012). A novel transmission loss allocation method based on transmission usage. In *Proceedings IEEE Fifth Power India Conference*, Haryana, India. 10.1109/PowerI.2012.6479479

Khan, B., Agnihotri, G., & Gupta, G. (2013) A multipurpose matrices methodology for transmission usage, loss, and reliability margin allocation in restructured environment, *Electrical & Computer Engineering: An International Journal, 2*(3).

Khan, B., Agnihotri, G., Rathore, P., Mishra, A., & Naidu, G. (2014). A cooperative game theory approach for usage and reliability margin cost allocation under contingent restructured market. *International Review of Electrical Engineering*, *9*(4), 854–862.

Khan, B., Alhelou, H. H., & Mebrahtu, F. (2019). A holistic analysis of distribution system reliability assessment methods with conventional and renewable energy sources. *AIMS Energy*, *7*(4), 413–429. doi:10.3934/energy.2019.4.413

Khan, B., & Singh, P. (2017). The current and future states of Ethiopia's energy sector and potential for green energy: A comprehensive study. *International Journal of Engineering Research in Africa*, *33*, 115–139. doi:10.4028/www.scientific.net/JERA.33.115

Khan, B., & Singh, P. (2017). Optimal power flow techniques under characterization of conventional and renewable energy sources: a comprehensive analysis, *Journal of Engineering, vol. 2017,* Article ID 9539506.

Khan, B., & Singh, P. (2017). Selecting a meta-heuristic technique for smart micro-grid optimization problem: A comprehensive analysis. In IEEE Access, vol. 5, pp. 13951-13977.

Khan, B., & Singh, P. (2019). Economic operation of smart micro-grid: A meta-heuristic approach. In Handbook of research on smart power system operation and control (pp. 330-346). Hershey, PA: IGI Global. doi:10.4018/978-1-5225-8030-0.ch014

Khan, B., & Singh, P. (2019). Economic operation of smart micro-grid: A meta-heuristic approach. In Handbook of research on smart power system operation and control (pp. 330-346). Hershey, PA: IGI Global. doi:10.4018/978-1-5225-8030-0.ch014

Khan, B., & Tanwar, S. (2019). Issues associated with microgrid integration. In Handbook of research on smart power system operation and control (pp. 252-264). Hershey, PA: IGI Global. doi:10.4018/978-1-5225-8030-0.ch010

Kifle, Y., Khan, B., & Singh, J. (2018). Designing and modeling grid connected photovoltaic system: (Case study: EEU building at Hawassa city). *International Journal of Convergence Computing*, *3*(1), 20–34. doi:10.1504/IJCONVC.2018.091113

Kifle, Y., Khan, B., & Singh, P. (2018). Assessment and enhancement of distribution system reliability by renewable energy sources and energy storage, *Journal of Green Engineering, 8*(3), 2, pp. 219-262.

Krim, Y., Abbes, D., Krim, S., & Mimouni, M. F. (2019). A flexible control strategy of a renewable active generator to participate in system services under grid faults. *Int Trans Electr. Energ. Syst.*, 1-24.

Makdisie, C., Haidar, B., & Alhelou, H. H. (2018). An optimal photovoltaic conversion system for future smart grids. In Handbook of research on power and energy system optimization (pp. 601–657). Hershey, PA: IGI Global. doi:10.4018/978-1-5225-3935-3.ch018

Njenda, T. C., Golshan, M. E. H., & Alhelou, H. H. (2018, November). WAMS based intelligent under frequency load shedding considering online disturbance estimation. In *2018 Smart Grid Conference (SGC)* (pp. 1-5). IEEE. 10.1109/SGC.2018.8777779

Njenda, T. C., Golshan, M. E. H., & Alhelou, H. H. (2018, November). WAMS based Under Frequency Load Shedding Considering Minimum Frequency Prediction and Extrapolated Disturbance Magnitude. In *2018 Smart Grid Conference (SGC)* (pp. 1-5). IEEE.

Peña Asensio, A., Arnaltes Gómez, S., Rodriguez-Amenedo, J., García Plaza, M., Eloy-García Carrasco, J., & Alonso-Martínez de las Morenas, J. (2018). A voltage and frequency control strategy for stand-alone full converter wind energy conversion systems. *Energies, 11*(3), 474.

Perdana, Y. S., Muyeen, S. M., Al-Durra, A., Helmo, K. M-P, & Marcelo, G. S. (2018). Direct connection of supercapacitor-battery hybrid storage system to the grid-tied photovoltaic system, *IEEE Transactions on Sustainable Energy,* 1-10.

Pippia, T., Sijs, J., & De Schutter, B. (2018, December). A parametrized model predictive control approach for microgrids. In *2018 IEEE Conference on Decision and Control (CDC)* (pp. 3171-3176). IEEE.

Priya, S. S., & Sagar, B. K. (2018). An advanced droop control method for proportional reactive power sharing. *International Research Journal of Engineering and Technology*, *5*(7), 1735–1743.

Qi, L., Bo, S., Yuchen, P., Ying, H., Tianhong, W., Liangzhen, Y., & Weirong, C. (2019). A state machine control based on equivalent consumption minimization for fuel cell/ supercapacitor hybrid tramway. *IEEE Transactions on Transportation Electrification*, *5*(2), 552–564. doi:10.1109/TTE.2019.2915689

Rahman, M. S., Hossain, M. J., & Lu, J. (2016). Coordinated control of three-phase AC and DC type EV–ESSs for efficient hybrid microgrid operations. *Energy Conversion and Management*, *122*, 488–503. doi:10.1016/j.enconman.2016.05.070

Savaghebi, M., Vasquez, J. C., Jalilian, A., Guerrero, J. M., & Lee, T. L. (2012). Selective compensation of voltage harmonics in grid-connected microgrids. *Mathematics and Computers in Simulation, 91*, 211–228. doi:10.1016/j.matcom.2012.05.015

Shahryaria, E., Shayeghi, H., Mohammadi-ivatloo, B., & Moradzadeh, M. (2019). A copula-based method to consider uncertainties for multi-objective energy management of microgrid in presence of demand response. *Energy, 175*, 879–890. doi:10.1016/j.energy.2019.03.129

Somnath, D., & Ashok, K. A. (2018). Power flow control of PV-wind-battery hybrid renewable energy systems for stand-alone application. *International Journal of Renewable Energy Research-IJRER, 8*(1), 36–43.

Srikanth, K., & Mahesh, K. M. (2019). Design and stability analysis of DC microgrid with hybrid energy storage system. *IEEE Transactions on Sustainable Energy, 10*(3), 1603–1612. doi:10.1109/TSTE.2019.2891255

Yanjun, H., Hong, W., Amir, K., Bin, L., Jie, J., Kegang, Z., & Chuan, H. (2018). A review of power management strategies and component sizing methods for hybrid vehicles. *Renewable & Sustainable Energy Reviews, 96*, 132–144. doi:10.1016/j.rser.2018.07.020

Youssef, K., Dhaker, A., Saber, K., & Mohamed, F. M. (2018). Power management and second-order sliding mode control for standalone hybrid wind energy with battery energy storage system. In *Proceedings of the Institution of Mechanical Engineers, Part I: Journal of Systems and Control Engineering, 232*(10), 1-22.

Youssef, K., Dhaker, A., Saber, K., & Mohamed, F. M. (2018a). Classical vector, first-order sliding mode and high-order sliding-mode control for a grid-connected variable speed wind energy conversion system: A comparative study. *Wind Engineering, 42*(1), 1–22.

Youssef, K., Dhaker, A., Saber, K., & Mohamed, F. M. (2019). Intelligent droop control and power management of active generator for ancillary services under grid instability using fuzzy logic technology. *Control Engineering Practice, 81*, 215–230.

Youssef, K., Dhaker, A., Saber, K., & Mohamed, F. M. (2019a). A second-order sliding-mode control for a real time emulator of a wind power system synchronized with electrical network, *International Transactions on Electrical Energy Systems,* 1-20.

Youssef, K, Krim, S, Mimouni, M. F. (2016). Control of a wind farm connected to the grid at a frequency and variable voltage. *International Journal of Renewable Energy Research, 6*(3), 747-758.

Zabihollah, N. A., Masoud, G., Misagh, I. S., Erfan, K., & Ahmad, S. (2018). Evaluation of hydrogen production from harvesting wind energy at high altitudes in Iran by three extrapolating Weibull methods. *International Journal of Hydrogen Energy, 43*(6), 3110–3132. doi:10.1016/j.ijhydene.2017.12.154

Zamani, R., Golshan, M. E. H., Alhelou, H. H., & Hatziargyriou, N. (2019). A novel hybrid islanding detection method using dynamic characteristics of synchronous generator and signal processing technique. *Electric Power Systems Research, 175*, 105911. doi:10.1016/j.epsr.2019.105911

Zamani, R., Hamedani-Golshan, M. E., Haes Alhelou, H., Siano, P., & Pota, H. R. (2018). Islanding detection of synchronous distributed generator based on the active and reactive power control loops. *Energies*, *11*(10). doi:10.3390/en11102819

Zineb, C., Mohammed, O., & Mohamed, M. (2017). Battery and supercapacitor for photovoltaic energy storage: A fuzzy logic management. *IET Renewable Power Generation*, *11*(8), 1157–1165. doi:10.1049/iet-rpg.2016.0455

KEY TERMS AND DEFINITIONS

Connected Mode: It is an operation mode of the generator when the variations of the frequency and the voltage at the point PCC are acceptable. In this mode, the RDG is used to participate in system services by improving the variation plans of the frequency and the tension and make them in optimal margins.

Grid Fault: The grid is faulty when the peaks of voltage and frequency are severe and exceeds the maximum margins mentioned in the standards.

Renewable Distributed Generator: It is an electric power generation system based on renewable generators (wind energy, solar energy) associated with storage systems (batteries, super-capacitors).

Standalone Mode: It is an operation mode of the generator when the variations of the frequency and the voltage at the point PCC are severe. In this mode, the RDG is used to ensure a continuous supply of load.

Super-Capacitor: It is a short-term storage system used to store energy in case of excess and provide energy in case of lack in order to ensure the balance between production and consumption.

Chapter 6
Harmonic Mitigation Techniques in Smart Distribution Network

Fsaha Mebrahtu
Hawassa University, Hawassa, Ethiopia

ABSTRACT

In this chapter various harmonic sources and their effect on the distribution network and its mitigation procedures are discussed. In the distribution network, electrical power is mostly utilized for our daily activity. However, the quality of power in the distribution network is affected by different disturbances. The distribution power quality problems deteriorate the performance of the system. One of the disturbances of the distribution network is harmonic distortion. Disturbances not only produce excessive heat in the devices and appliances used in the daily life of human beings, but also reduce the life of the appliances. Finally, the harmonic distortion mitigation by using active power filter, space vector pulse width modulation, dynamic voltage restorer, voltage phase shift, and fuzzy logic controller is discussed.

INTRODUCTION

Current or voltage having frequency that is the numerical multiple of the fundamental frequency at which the supply system is planned to perform i.e. 50 or 60 Hz is called harmonics. Harmonic current produce from the nonlinear loads and it caused harmonic of the drop voltage that is transmitted through the distribution network systems. It depends on the harmonic impedance of the scheme. Inter harmonics are harmonics in currents or voltages and has frequency but they are not the multiple of the source frequency (Pogaku et al. 2006).

Traditionally, the distribution network of electrical power system has to address harmonics because the distribution system users are typically planned for the linear supply (Alhelou et al., 2019; Makdisie et al., 2018; Alhelou et al., 2018; Alhelou et al., 2016; Haes Alhelou et al., 2019; Njenda et al., 2018). With the utilization of different speed drives and the power electronics converters, needed in industries, some order harmonic current are being inserted into the electrical distribution networks (Khan et al.,

DOI: 10.4018/978-1-7998-1230-2.ch006

2019; Kifle et al. 2018; Khan et al., 2018; Khan et al., 2017; Banteywalu et al., 2019; Anteneh et al., 2019; Molla et al., 2019, Molla et al., 2018, Jariso et al. 2018). There are important methods which are broadly utilized to reduce the harmonic currents in the power supply. The harmonics can be described by the following equation depicted below.

$$f_h = hf_{ac} \tag{1}$$

Where, f_h represents the hth harmonics and f_{ac} represents the fundamental frequency of the system.

The following equations are represented for the sinusoidal current and voltage that's depend on the time't' as explained (Singh et al. 2012).

$$V(t) = V * \sin(\omega t) \tag{2}$$

$$i(t) = I * \sin(\omega t) \tag{3}$$

Where $\omega = 2\pi f$ is the angular velocity of the periodic waveform is form and ω is vitiation of phase angle between the voltages and the waveform of current.

If current leads the voltage, the sign of phase angle ω is positive and if current lags the voltage, it is negative. For the periodic sinusoidal wave form the simplified Fourier expression will be as flows.

$$V(t) = V_0 + V_1 \sin(\omega t) + V_2 \sin(\omega t) + V_3 \sin(\omega t) \ldots V_n \sin(n\omega t) + V_{n+1} \sin(n+1)\omega t) \tag{4}$$

The Fourier transform of a series is depicted in equation (4), V0 is the wave form of the DC element. V_1, V_2, V_3... V_n are the maximum value. These are called the periodic wave form of the system harmonics. The frequency of the harmonics are the result of the integer multiple of the fundamental source frequency such as a fundamental of 50 Hz, the *3rd* harmonic would be 150 Hz and the 5th harmonic would be 250Hz (Akagi et al. 2016).

For the power system having 50 Hz (or 60 Hz), the fundamental frequency 150 Hz (180 Hz) and (300 Hz) etc. are the higher order harmonic i.e. 3rd and 5th.

- Odd Harmonics (5th, 7th.....)
- Even Harmonics (2nd, 4th ....)
- Triplen Harmonics (3rd, 9th, 15th...)

Any periodical wave form that is non integer multiple of the basic fundamental frequency is known as the inter harmonics i.e. 2.5th = 125 Hz at 50 Hz base frequency.

TOTAL HARMONIC DISTORTION (THD)

THD of Current

THD of current can be expressed as it shows in equation (5).

$$\mathrm{THD} = \sqrt{\frac{\sum_{H=2}^{\infty} I_h^2}{I_1}} \tag{5}$$

Where, I_h is individual harmonic component, I_1 is represented for fundamental element.

THD of Voltage

THD is a difficult and frequently ambiguous idea to understand. But, when it divided into the fundamental definition of harmonic distortion, it will be easier to know. The electrical distribution network has an AC source and electrical load (Fini et al., 2016; Alhelou et al., 2018; Zamani et al., 2018; Alhelou et al., 2015; Njenda et al., 2018; Haes Alhelou et al., 2018; Haes Alhelou et al., 2019). The linear load produces current which is sinusoidal by nature. The nature of the nonlinear loads is to produce distorted current or it is not pure sinusoidal, therefore at the distribution network distortions are generated. As it shows in equation (6), the total harmonic distortion of voltage wave form are the summation of the harmonic component wave forms related to the fundamental elements of voltage (Sergery et al., 2002).

$$THD = \frac{\sqrt{(V_2^2 + V_3^2 + V_4^2 + \ldots V_n^2)}}{V_1} * 100\% \tag{6}$$

SOURCES OF HARMONICS

Some of the sources of harmonics are nonlinear loads that generate harmonics. The harmonics are created because of the variation of device's resistance with respect to time. The Main Causes of Harmonics in Industrial Power System are as follows (Sergery et al., 2002).

- Non-linear loads
- Variable Speed Drives
- Arc furnace
- Converters (rectifiers and inverters)
- Power Electronic Devices
- Electronic ballasts for fluorescent lighting
- Computers and PCs,
- Chargers of the battery

Figure 1. Non-linear loads positioned as far upstream as possible

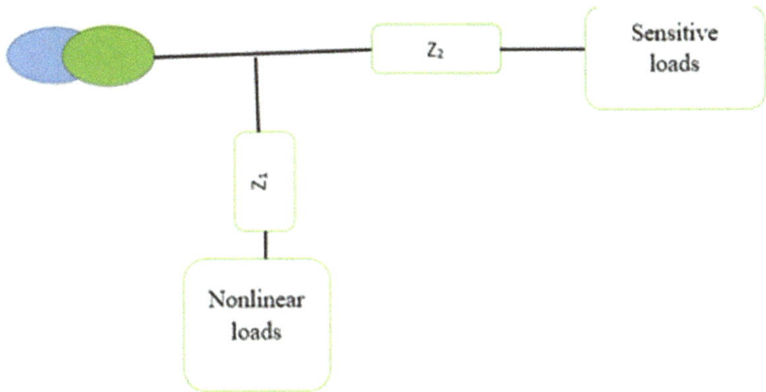

Figure 2. Grouping of non-linear loads and connection as far upstream as possible

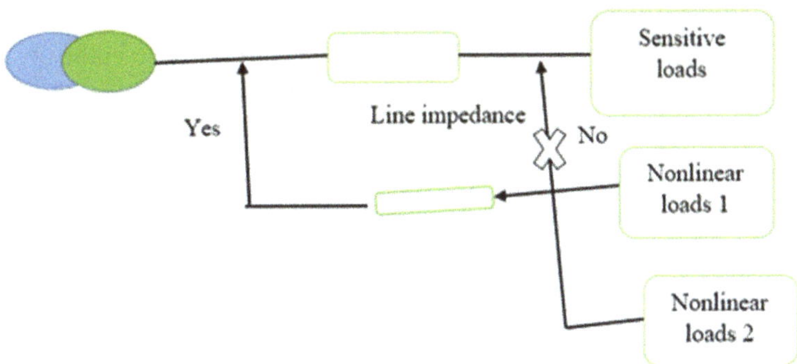

- Controlled rectifiers,
- Traction material
- Uninterruptable power supply and
- Static VAR compensators.

Position the Non-Linear Loads Upstream in the System

As the short circuit power decreases, consequently the value of total harmonic distortion will increase. For the economic considerations, it is superior to connect the non-linear loads as far upstream as possible as shown in figure 1.

Figure 3. Supply of non-linear loads via an isolated transformer

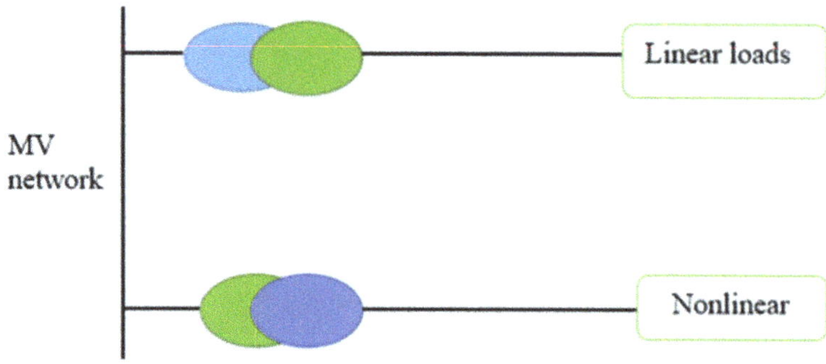

Nonlinear Loads Groups

Nonlinear loads equipment's should be isolated from others as presented in figure 2. The two clusters of devices should be supplied by different sets of bus bars.

In endeavoring to bound harmonics, a further enhancement can be found by creating a source via isolated transformer as indicated in the Figure 3.The drawback is the increase in the cost of the installation.

Identification of the Harmonics in the System

If one observe irregularly involvement in some events like alarms going off, or MCB's, MCCB's, RCD's, flickering lights and earth leakage devices tripping for no seeming reason one can know that in electrical system harmonic distortion is existed. The result of harmonic distortion is heat generation in the the electrical cables, switchboard, and overheating of motors. Substitution of the motor's bearings and insulation often indicates the presence of harmonics.

EFFECTS OF HARMONICS

In the distribution power network, harmonic distortions are not necessary in all the electrical applications and process. They have contrary outcome on the electrical networks as well as on its running conditions. Harmonics generate resonance in the distribution system. The restraining property may convert due to the existence of harmonics. In addition harmonics has some adverse effects on the performance of revolving machines, transformers and transmission systems.

Harmonics that are produced in industrial plant may find their way into the electric utility system. This generates a voltage distortion that disturbs the quality of service for all the customers connected to the same feeder. So consequently it is vital to control the magnitude of the harmonic currents that flow into the electric utility network (Sergery et al. 2002).

The total harmonics distortions (THD) of currents create disapproving effect on the industries system such as electronics devices can't function properly and measurement errors. Some of the significant effects of harmonics are as follows:

1. Distortion of the main source voltage, unwanted currents flowing in the supply network produce extra energy losses.
2. Malfunctioning operation of regulating devices, disturbed operation of fluorescent lamps, television receivers or other equipment.
3. High harmonic amplitudes can excess the power distribution network and the neutral conductor can be overheat and become the reason to burn out.
4. Additional heating in motors and transformers.
5. Increased probability of relay malfunctions.
6. Cable line insulation failure
7. Motor failures
8. Flickering of fluorescent lights
9. Fuses blowing
10. Circuit breakers tripping

Problems Associated with Harmonics

- Due to the following reasons, harmonics are very dangerous for power system:
- Capacitors can be blow from over voltage
- Drive can be trip offline
- Transformer can be overheat possibly to destruction
- Motor will be heat more
- Ballast can drop
- Electronics will be malfunction

MITIGATION OF HARMONIC IN ELECTRICAL DISTRIBUTION NETWORK

They are various methods which are developed to mitigate the harmonic and satisfy the the IEEE standards. The intention of these methods is to make the input current free from the harmonics, in order to decrease the whole THD current. The general procedures can be classified as follows (Akagi et al. 2016).

- Active powers filter systems.
- Space vector pulse width modulation (SVPWM)
- Passive power filters
- Hybrid systems of active and passive filter.
- Distributed Generation
- Fuzzy logic control based on the shunt active power filter
- Artificial neural network-based shunt active filter
- voltage phase shifting concept
- Dynamic voltage restorer

Active Power Filter Systems (Series and Parallel)

Active power filter is one of the methods used to mitigate the harmonics in the power distribution networks, nevertheless, active power filter has a number of power electronics converters and they are more expensive than passive filters. The main idea of active power filter is to inject voltage into the line equivalent to the magnitude of the current and voltage harmonics produced by the nonlinear loads with 180^0 phase angle variations, so they cancel each other. Active power filter can be categorized based on the power converters type, topology, and number of phases. The power converter types can be either current source inverter or voltage source inverters. Based on the current source inverter active power filter inductor acts as energy storage device although voltage source inverter of active power filter capacitor is used as energy storage device. The scheme of the system can be series, shunt or a combination of both. The third classification is based on the number of phases, those are single phase (2-wire), three phase (3 or 7 wire) of the network. For those high-power nonlinear loads three phase active filter are used. Active power filter of many arrangements has been announced and enhanced in Allirani et al. (2015).

Space Vector Pulse Width Modulation (SVPWM)

The multilevel power inverter produces a sinusoidal voltage from discrete voltage level and pulse width modulation (PWM) approaches. Modulation procedures hybrid multilevel inverter can be categorized based on the method of switching frequency. Various pulse width modulation procedures has been established to attain the succeeding. Those are wide linear modulation range, less switching loss, and reduced total harmonic distortion (THD) in the spectrum of switching wave form and easy to accomplishing and less computation time. The most widely used approaches that apply with pulse width modulation (PWM) system are sinusoidal pulse width modulation and space vector pulse width modulation (SVPWM) (Khan et al., 2014; Khan et al., 2013; Khan et al., 2012; Negash et al., 2017; Negash et al., 2016; Jariso et al., 2017; Kifle et al., 2018; Yeshalem et al., 2017; Singh et al., 2019; Gupta et al., 2015). Now days the most utilized PWM technique is space vector pulse width modulation (SVPWM). This type of scheme in the voltage source inverter will be enhanced the bus voltage consumption and less commutation losses (VenkataBramhanaidu et al., 2017).

The most predominant pulse width modulation scheme, which is the most efficient among all the other PWM methods is the space vector pulse width modulation, since it controls to generate large voltage associated with less total harmonic distortion as well as field oriented structures to achieve motor control (Allirani et al., 2017).

Passive Filter

In power distribution network passive power filter is used to dissipate the distorted harmonic current based on the normalized frequency. To remove the harmonics created in the power system, filter impedance is very low. From the various types of filters, single tuned filter is commonly used in industries for dealing the harmonic pollution (Awadalla et al., 2015 and Soomro et al., 2015). Reactive power (Q) and voltage (V) are injected to the weak bus to compensate the power network by using the capacitor bank, meanwhile, inductance and resistance are based on the selected harmonics (n) that need to be decrease. To eliminate the harmonic order of 3rd, 5th, 7th, and 9th in the power distribution network four types of set filters are used (Syed et al., 2018, Halpin et al., 1998).

$$C_{filter} = \frac{Q}{2\pi f v^2}$$

$$L_{filter} = \frac{V^2}{2n^2 Q^2 \pi f}$$

$$R_{filter} = \frac{V^2}{nQ^2}$$

Passive filter are normally the combination of resistance, inductance and capacitance, whose parameters are calculated by various methods to get the removal of harmonics. They either remove the harmonic currents off the line or block them between the systems by tuning the parameters to produce resonance at a selected frequency. They also supplied the reactive power to the network and to increase the power quality. However, they have a drawback of potentially interference unfavorably with the electrical network and the operation of passive filter depends mostly on the source impedance. Additionally these can be used for canceling of a specific harmonic frequency, so the number of passive filters increases with the increasing of the number of harmonics on the network. Classification of passive filter is as follows.

- Shunt passive filter
- Series passive filter

PASSIVE SHUNT FILTER

Single Tuned Filter

Single tuned "notch" is one of the most utilized passive filters and is more economical type filter, which is commonly adequate for the applications. The connection of the filter is series tuned to provide low impedance to specific harmonic current and it is associated in parallel with the electrical system. During the overheating time, the produced harmonic current is diverted from their usual path to the filter. The power factor is also corrected with the help of filter. In fact capacitor bank is used to correct the power factor and it is used to make single tuned filters. They are tuned at low harmonic frequency.

Passive Series Filter

Contrasting notch filter which is allied in parallel with the electrical power system networks, also a series passive filter is coupled in series with the load. The connection of the inductance and capacitance are in shunt and they are tuned to deliver high impedance at a selected harmonic frequency. During this time the flow of harmonic current are blocked with the help of impedance at the tuned frequency only. During the normal condition the fundamental frequency of the filter is designed to provide low impedance, thereby allowing only the fundamental current with a minor additional impedance and losses. The advantage of series filters is to block a single harmonic current (like 3rd harmonic) in a single phase circuit, where it is impossible to take advantage of zero sequence characteristics. To limit multiple harmonic currents, each need a series filter tuned to the harmonics and the system can produce significant losses at the fundamental frequency.

Hybrid Filters

For the application of high electrical power distribution, active power filter topology is expensive due to the high rating and high switching frequency of PWM power converters. The use of passive power filter is done for the filtration of harmonics of the large nonlinear loads. Nevertheless, there is drawback of passive filter that the efficiency of the filter is affected due to the value of impedance of the system and with resonance is produced, which cause the increasing harmonic current in the power supply. Therefore, alternative method for harmonic reduction is utilized such as hybrid active power filter. Hybrid active power filter delivered the combined advantage of active and passive power filter and eliminate the drawbacks.

Distributed Generation

Nowadays different utilities encourage the interconnection of the independent energy sources in the power distribution system. The integration of distributed generator (DG) into the distribution network has the both negative and positive effect on the distribution network. Now a day's number of researchers are studied the effect of DG on the power distribution system. The effect of DG on the power quality depends on the type of DG, and it interfaces with the main power supply system. Further, it depends on the value of distribution generation unit, converter scheme, overall size of the DG relative to the system and load at the connecting point, and feeder voltage regulation practice. Based on the interfacing of the DG with distribution system, it can be grouped in to two types. These are:

- Inverter based DG
- Without inverter DG

Based on the inverter-based DG, a number of renewable energy sources are connected to the distribution networks to satisfy the end users or eliminate the power quality problems. The location of DG is very important. It must be placed at the distribution network level of electrical grid and placed near the load center. The influence of DG based on the energy losses, profile of voltage, short circuit current, harmonics and reliability of power supply. It tested independently before connecting to the distribution network (Jegadeesan et al. 2014).

Voltage Phase Shifting

In electrical power distribution system the harmonic can be minimized by using phase shifting transformer. This type of mitigation method is the originally developed by electromagnetic technology to eliminate the most damaging harmonics like 3rd to 21st. This procedure is known as phase shifting. The transformers can be used to decrease the existing harmonics in the line current. This minimization method can be designed into new way to protect the future harmonic problems. Under the balanced load condition, the line current system has good performance with the help of phase shifting principle. This is due to the shifting of the line by 120 degree between the three phases, when the sign of one phase positive, the other phases will cancel each other.

Dynamic Voltage Restorer (DVR)

Dynamic voltage restorer is the custom power device that is connected in series with the power distribution networks. The key components of the dynamic voltage restorer are injection transformer, voltage source inverter, filter, and energy storage device and control system. There are various advantages of DVR, which are as follows:

- To controls the flow of active power
- It is not expensive compared to the UPS
- Has higher energy capacity and cost effective compared with the SEMES device
- It is smaller in size compared to D-STATCOM
- It is more efficient compare to UPS

The circuit of power converter is being controlled by using pulse width modulation (PWM) pulse according to the system voltage. The function of the capacitor in the dynamic voltage restorer serves as energy storage and it will inject into the power network to compensate the system voltage during the voltage dip and the controller also used to change the DC to AC inverter mode into rectifier mode to change the capacitor in the DC energy link. In dynamic voltage restorer, the use of filter is to convert the inverted pulse width modulation wave form into sinusoidal wave form, by eliminating the unwanted signals generated from the voltage source inverter or power electronics converters action. The main advantage of dynamic voltage restorer is the easy control, fast response time for any detections and it is not expensive.

CONCLUSION

As harmonics is the major issue in the preset power system due to the large utilization of power electronics devices in the system as well as industrial drives and load. Therefore, harmonic mitigation is very important from the power quality point of view. This chapter provides the detailed overview about the various types of harmonics, sources of harmonics and their quantification. Further, various methods to mitigate these harmonics, utilized in preset power system are also discussed. These methods are based on various types of filter, distributed generation, phase shifting transformer and dynamic voltage restorer. Their advantages as well as disadvantages are also discussed in this chapter.

REFERENCES

Akagi, H. (2016). Modern active filters and traditional passive filters. *Bulletin of the Polish Academy of Sciences. Technical Sciences*, *54*(3), 255–269.

Alhelou, H., Hamedani-Golshan, M. E., Zamani, R., Heydarian-Forushani, E., & Siano, P. (2018). Challenges and opportunities of load frequency control in conventional, modern and future smart power systems: a comprehensive review. *Energies*, *11*(10), 2497. doi:10.3390/en11102497

Alhelou, H. H. (2018). *Fault detection and isolation in power systems using unknown input observer. In Advanced condition monitoring and fault diagnosis of electric machines* (p. 38). Hershey, PA: IGI Global.

Alhelou, H. H., Golshan, M., & Fini, M. (2018). Wind driven optimization algorithm application to load frequency control in interconnected power systems considering GRC and GDB nonlinearities. *Electric Power Components and Systems, 46*(11-12), 1223-1238.

Alhelou, H. H., & Golshan, M. E. H. (2016, May). Hierarchical plug-in EV control based on primary frequency response in interconnected smart grid. In *2016 24th Iranian Conference on Electrical Engineering (ICEE),* (pp. 561-566). IEEE. 10.1109/IranianCEE.2016.7585585

Alhelou, H. H., & Golshan, M. E. H. (2020). Decision-making-based optimal generation-side secondary-reserve scheduling and optimal LFC in deregulated interconnected power system. In *Decision making applications in modern power systems* (pp. 269–299). Academic Press. doi:10.1016/B978-0-12-816445-7.00011-6

Alhelou, H. H., Golshan, M. E. H., & Hatziargyriou, N. D. (2019). Deterministic dynamic state estimation-based optimal LFC for interconnected power systems using unknown input observer. *IEEE Transactions on Smart Grid*, 1. doi:10.1109/TSG.2019.2940199

Alhelou, H. H., Golshan, M. E. H., & Hatziargyriou, N. D. (2019). A decentralized functional observer based optimal LFC considering unknown inputs, uncertainties, and cyber-attacks. *IEEE Transactions on Power Systems.*

Alhelou, H. H., Golshan, M. E. H., Zamani, R., Moghaddam, M. P., Njenda, T. C., Siano, P., & Marzband, M. (2019, June). An improved UFLS scheme based on estimated minimum frequency and power deficit. In 2019 IEEE Milan PowerTech (pp. 1-6). IEEE. doi:10.1109/PTC.2019.8810497

Alhelou, H. H., Golshan, M. H., & Askari-Marnani, J. (2018). Robust sensor fault detection and isolation scheme for interconnected smart power systems in presence of RER and EVs using unknown input observer. *International Journal of Electrical Power & Energy Systems*, *99*, 682–694. doi:10.1016/j.ijepes.2018.02.013

Alhelou, H. H., Hamedani-Golshan, M. E., Heydarian-Forushani, E., Al-Sumaiti, A. S., & Siano, P. (2018, September). Decentralized fractional order control scheme for LFC of deregulated nonlinear power systems in presence of EVs and RER. In *2018 International Conference on Smart Energy Systems and Technologies (SEST)* (pp. 1-6). IEEE. 10.1109/SEST.2018.8495858

Alhelou, H. S. H., Golshan, M. E. H., & Fini, M. H. (2015, December). Multi agent electric vehicle control based primary frequency support for future smart micro-grid. In *Smart Grid Conference (SGC)* (pp. 22-27). IEEE.

Allirani, S., & Thurai, R. (2015). *V. B. development of space vector pulse width modulation algorithm for voltage source inverter using dsPIC controller 30f4011(IJARCCE), 2*(3), pp. 01-05.

Anteneh, D., & Khan, B. (2019). Reliability enhancement of distribution substation by using network reconfiguration: A case study at Debre Berhan Distribution Substation. *International Journal of Economy, Energy, & Environment, 4*(2), 33.

Awadalla, M., Omer, M., & Mohamed, A. (2015). Single-tuned filter design for harmonic mitigation and optimizationwith capacitor banks. In *Proceedings of the 1st International Conference on Computing, Control, Networking, Electronics, and Embedded Systems Engineering, (ICCNEEE '15),* pp. 242–247, IEEE, Khartoum, Sudan.

Banteywalu, S., Khan, B., De Smedt, V., & Leroux, P. (2019). A novel modular radiation hardening approach applied to a synchronous buck converter. *Electronics (Basel), 8*(5), 513. doi:10.3390/electronics8050513

Fini, M. H., Yousefi, G. R., & Alhelou, H. H. (2016). Comparative study on the performance of many-objective and single-objective optimisation algorithms in tuning load frequency controllers of multi-area power systems. *IET Generation, Transmission, & Distribution, 10*(12), 2915–2923. doi:10.1049/iet-gtd.2015.1334

Gupta, K., Khan, B., & Mubeen, S. E. (2015). Available transfer capability enhancement by unified power flowcontroller. In *Proceedings IEEE International Conference on Signal Processing, Informatics, Communication, and Energy Systems (SPICES),* pp. 1, 4, 19-21.

Haes Alhelou, H., Hamedani-Golshan, M. E., Njenda, T. C., & Siano, P. (2019). A survey on power system blackout and cascading events: Research motivations and challenges. *Energies, 12*(4), 682. doi:10.3390/en12040682

Haes Alhelou, H., Hamedani-Golshan, M. E., Njenda, T. C., & Siano, P. (2019). Wide-area measurement system-based optimal multi-stage under-frequency load-shedding in interconnected smart power systems using evolutionary computing techniques. *Applied Sciences, 9*(3), 508. doi:10.3390/app9030508

Haes Alhelou, H., Hamedani Golshan, M. E., Njenda, T. C., & Siano, P. (2019). WAMS-based online disturbance estimation in interconnected power systems using disturbance observer. *Applied Sciences, 9*(5), 990. doi:10.3390/app9050990

Halpin, S. M., Ribeiro, P. F., & Dai, J. J. (1998). Frequency-domain harmonic analysis methods, IEEE Power Engineering Society: Tutorial in Harmonics Modeling and Simulation, pp. 49–54.

Jariso, M., Khan, B., Tanwar, S., Tyagi, S., & Rishiwal, V. (2018). Hybrid energy system for upgrading the rural environment, *IEEE GlobeCom 2018.*

Jariso, M., Khan, B., Tesfaye, D., & Singh, J. (2017). Modelling and designing of stand-alone photovoltaic system (Case study: Addis Boder Health Center southwest Ethiopia), *IEEE International conference on Electronics, Communication, and Aerospace Technology, ICECA 2017,* pp. 168-173, Coimbatore, India.

Jegadeesan, M., & Venkatesh, N. (2014, March). Estimation and mitigation of current and voltage harmonics in distribution system using distributed generation, *International Journal of Innovative Research in Science, Engineering, and Technology*, *3*(Special Issue), 3.

Khan, B., & Agnihotri, G. (2012). A novel transmission loss allocation method based on transmission usage. In *Proceedings IEEE Fifth Power India Conference*, Haryana, India. 10.1109/PowerI.2012.6479479

Khan, B., Agnihotri, G, Gupta, G. (2013). A multipurpose matrices methodology for transmission usage, loss and reliability margin allocation in restructured environment, *Electrical & Computer Engineering: An International Journal, 2*(3).

Khan, B., Agnihotri, G., Rathore, P., Mishra, A., & Naidu, G. (2014). A cooperative game theory approach for usage and reliability margin cost allocation under contingent restructured market. *International Review of Electrical Engineering*, *9*(4), 854–862.

Khan, B., Alhelou, H. H., & Mebrahtu, F. (2019). A holistic analysis of distribution system reliability assessment methods with conventional and renewable energy sources. *AIMS Energy*, *7*(4), 413–429. doi:10.3934/energy.2019.4.413

Khan, B., & Singh, P. (2017). The current and future states of Ethiopia's energy sector and potential for green energy: A comprehensive study. *International Journal of Engineering Research in Africa*, *33*, 115–139. doi:10.4028/www.scientific.net/JERA.33.115

Khan, B., & Singh, P. (2017). Optimal power flow techniques under characterization of conventional and renewable energy sources: A comprehensive analysis, *Journal of Engineering, vol. 2017,* Article ID 9539506.

Khan, B., & Singh, P. (2017). Selecting a meta-heuristic technique for smart micro-grid optimization problem: A comprehensive analysis. In IEEE Access, 5, pp. 13951-13977.

Khan, B., & Singh, P. (2019). Economic operation of smart micro-grid: a meta-heuristic approach. In Handbook of research on smart power system operation and control (pp. 330-346). Hershey, PA: IGI Global. doi:10.4018/978-1-5225-8030-0.ch014

Khan, B., & Singh, P. (2019). Economic operation of smart micro-grid: A meta-heuristic approach. In Handbook of research on smart power system operation and control (pp. 330-346). Hershey, PA: IGI Global. doi:10.4018/978-1-5225-8030-0.ch014

Khan, B., & Tanwar, S. (2019). Issues associated with microgrid integration. In Handbook of research on smart power system operation and control (pp. 252-264). Hershey, PA: IGI Global. doi:10.4018/978-1-5225-8030-0.ch010

Kifle, Y., Khan, B., & Singh, J. (2018). Designing and modeling grid connected photovoltaic system: (Case study: EEU building at Hawassa city). *International Journal of Convergence Computing*, *3*(1), 20–34. doi:10.1504/IJCONVC.2018.091113

Kifle, Y., Khan, B., & Singh, P. (2018). Assessment and enhancement of distribution system reliability by renewable energy sources and energy storage, *Journal of Green Engineering, 8*(3), 2, pp. 219-262.

Lyshevski, S. E., Ehawary, M. E., Bolea, L., Nasar, A., Kesrtin, W. H., & Crow, C. S. M. (2002). Power quality. Boca Raton, FL: CRC Press.

Makdisie, C., Haidar, B., & Alhelou, H. H. (2018). An optimal photovoltaic conversion system for future smart grids. In Handbook of research on power and energy system optimization (pp. 601–657). Hershey, PA: IGI Global. doi:10.4018/978-1-5225-3935-3.ch018

Molla, T., Khan, B., Moges, B., Alhelou, H. H., Zamani, R., & Siano, P. (2019). Integrated optimization of smart home appliances with cost-effective energy management system. *CSEE Journal of Power and Energy Systems*, *5*(2), 249–258.

Molla, T., Khan, B., & Singh, P. (2018). A comprehensive analysis of smart home energy management system optimization techniques, *Journal of Autonomous Intelligence, 1(1)*, 15–21.

Negash, K., Khan, B., Tesfaye, D., & Gayathri, M. (2017). Optimal placement of phasor measurement unit for system observability. In *Proceedings of IEEE International Conference on Computing Methodologies and Communication, (ICCMC 2017)*, pp. 73-77, Erode, India.

Negash, K., Khan, B., & Yohannes, E. (2016). Artificial intelligence versus conventional mathematical techniques: a review for optimal placement of phasor measurement units [Springer]. *Journal of Technology and Economics of Smart Grids and Sustainable Energy*, *1*(1), 10. doi:10.100740866-016-0009-y

Njenda, T. C., Golshan, M. E. H., & Alhelou, H. H. (2018, November). WAMS based intelligent under frequency load shedding considering online disturbance estimation. In *2018 Smart Grid Conference (SGC)* (pp. 1-5). 10.1109/SGC.2018.8777779

Njenda, T. C., Golshan, M. E. H., & Alhelou, H. H. (2018, November). WAMS based under frequency load shedding considering minimum frequency prediction and extrapolated disturbance magnitude. In *2018 Smart Grid Conference (SGC)* (pp. 1-5).

Pogaku, N., & Green, T. C. (2006, May 11). Harmonic mitigation throughout a distribution system: A distributed-generator-based solution. *IEE Proceedings. Generation, Transmission, and Distribution*, *153*(3), 350–358. doi:10.1049/ip-gtd:20050086

Nasir, S. S., Jamian, J. J., & Mustafa, M. W. (2018). Minimizing harmonic distortion impact at distribution system with considering large-scale EV load behaviour using modified lightning search algorithm and pareto-fuzzy approach, Vol. 2018, Article ID 6587493.

Singh, B., Al-Haddad, K., & Chandra, A. (2012, October). A review of active filters for power quality improvement. *IEEE Transactions on Industrial Electronics*, *46*(5), 960–971. doi:10.1109/41.793345

Singh, P., & Khan, B. (2017) Smart microgrid energy management using a novel artificial shark optimization, Complexity, vol. 2017, Article ID 2158926. doi:10.1155/2017/2158926

Singh, P., Khan, B., Vidyarthi, A., Haes Alhelou, H., & Siano, P. (2019). Energy-aware online non-clairvoyant scheduling using speed scaling with arbitrary power function. *Applied Sciences*, *9*(7), 1467. doi:10.3390/app9071467

Soomro, D. M., & Almelian, M. M. (2015). Optimal design of a single tuned passive filter to mitigate harmonics in power frequency. *Journal of Engineering and Applied Sciences (Asian Research Publishing Network)*, *10*(19), 9009–9014.

VenkataBramhanaidu, D., & Sreenivasulu, A. (2017). Implementation of space vector pulse width modulation technique on three phase two level inverter, *IOSR-JEEE, 12*(3) Ver. II May – June. 2017.

Yeshalem, M. T., & Khan, B. (2017). Design of an off-grid hybrid PV/wind power system for remote mobile base station: A case study. *AIMS Energy*, *5*(1), 96–112. doi:10.3934/energy.2017.1.96

Zamani, R., Golshan, M. E. H., Alhelou, H. H., & Hatziargyriou, N. (2019). A novel hybrid islanding detection method using dynamic characteristics of synchronous generator and signal processing technique. *Electric Power Systems Research*, *175*, 105911. doi:10.1016/j.epsr.2019.105911

Zamani, R., Hamedani-Golshan, M. E., Haes Alhelou, H., Siano, P., & Pota, H. R. (2018). Islanding detection of synchronous distributed generator based on the active and reactive power control loops. *Energies*, *11*(10), 2819. doi:10.3390/en11102819

KEY TERMS AND DEFINITIONS

Distributed Generation: It is an approach that employs small-scale technologies to produce electricity close to the end users of power.

Hybrid Filter: It combines a number of passive and/or active filters and their structure may be of series or parallel topology or a combination of the two.

Total Harmonic Distortion (THD): It is a measurement of the harmonic distortion present in a signal and is defined as the ratio of the sum of the powers of all harmonic components to the power of the fundamental frequency.

Chapter 7
Loss Minimization in Active Distribution Network

Bawoke Simachew
Hawassa University, Hawassa, Ethiopia

ABSTRACT

Power loss reduction is an important problem that needs to be addressed with respect to generating electrical power. It is important to reduce power loss using locally generated power sources and/or compensations. This chapter brings a method of presents a method of maximizing energy utilization, feeder loss reduction, and voltage profile improvement for radial distribution network using the active and reactive power sources. Distributed Generation (DG) (wind and solar with backup by biomass generation) and shunt capacitor (QG) for reactive power demand are used. Integrating DG and QG at each bus might reduce the loss but it is economically unaffordable, especially for developing countries. Therefore, the utilization optimization method is required for finding an optimal size and location to feeders for placing QG and DG to minimize feeder loss.

INTRODUCTION

Distribution system is the important component of the power system, which supplies power to end users. At this point, system voltage will be step down; as a result, system loss is higher. According to Electric Power Research Institute (EPRI) (Short and Swayne, 2012, Badran et al. 2017, Ali et al. 2017) New York, the distribution power loss is about 70 percent of all the energy loss and this is even higher during peak load conditions. Different researchers (Lalitha et al. 2016, Abdelaziz et al. 2016, Reddy et al. 2017), also indicate that distribution system has more than 13% of the total power generation (Alhelou et al., 2019; Makdisie et al., 2018; Alhelou et al., 2018; Alhelou et al., 2016; Haes Alhelou et al., 2019; Njenda et al., 2018). If distribution network and load arrangement is not balanced, the system loss will be higher. This is happened when the drawn current is higher than the proportion of voltage increase (Xie et al. 2018). The over increase of demands causes heating, which further increases losses and insulation break down for networks (Christeen et al. 2017, Elsheikh et al. 2014, Thang and Minh, 2017).

DOI: 10.4018/978-1-7998-1230-2.ch007

Losses are increased due to improper cable size, increasing demand of reactive power or improper current flow to satisfy the demand of active and reactive power. It might be economical to configure the system with new proper sized lines; however, cable sizing only does not bring the required loss reduction and voltage profile improvement. To reduce power losses and improve the voltage profile, one should consider the active and reactive demands and load characteristics at each buses (Khan et al., 2019; Khan et al., 2018; Khan et al., 2017; Banteywalu et al., 2019; Anteneh et al., 2019; Molla et al., 2019, Molla et al., 2018, Jariso et al. 2018). To fix the loss issues and improve the voltage profile, both DG for active power and shunt capacitors for unbalanced system would be required. It might be difficult to optimize the varying load and find the exact location (Alhelou et al., 2019; Makdisie et al., 2018; Alhelou et al., 2018; Alhelou et al., 2016; Haes Alhelou et al., 2019; Njenda et al., 2018). Hence, it is required to consider the load under different loading conditions i.e. during off peak, during peak and in average state. Furthermore, it is important to consider the output probability density of DG (wind and solar) and load distribution probability of the distribution network.

Several definitions are given for DG based on plant rating, technology used, generated voltage levels, point of connection, etc. EPRI defined distributed generation as the generation from "a few kilo-watts to 50 MW" (Lalitha et al. 2016). For proper sizing and allocation of shunt capacitor and DG meta-heuristic optimization technique called whale optimization algorithm is used. Both active and reactive power demands are analyzed. For this DG availability, probability densities are considered (Fini et al., 2016; Alhelou et al., 2018; Zamani et al., 2018; Alhelou et al., 2015; Njenda et al., 2018; Haes Alhelou et al., 2018; Haes Alhelou et al., 2019; Khan et al. 2014; Khan et al. 2013; Khan et al. 2012; Negash et al., 2017; Negash et al., 2016; Jariso et al. 2017; Kifle et al. 2018; Yeshalem et al. 2017; Singh et al. 2017; Gupta et al., 2015). The biggest issue in DG design is the resource availability. In according to resource availability per hour or per day solar availability and wind speed variation would affect the power output.

The integration of DG and shunt capacitor has the following additional contributions:

- Improve the network load carrying capacity without any abnormal voltage drop
- Decrease stress on feeders
- Safe guard for insulation breakdown
- Bring clean city and promote carbon free energy
- Increases the diversity of energy source
- Entrepreneurship city youths

DESIGN PROCEDURE

Figure 1 presented the design procedure for loss minimization algorithm.

Shunt Capacitor and Renewable Energy Assessment

Shunt Capacitor Modeling: The discrete size of capacitors could be taken for optimal planning of Renewable-based DGs in the distribution system. The capacitors available on the market are smaller units (150 kVAr), which are further integer multiples of factor U using Equation (1) as reported in (Kumar, 2017):

Figure 1 Design Procedure

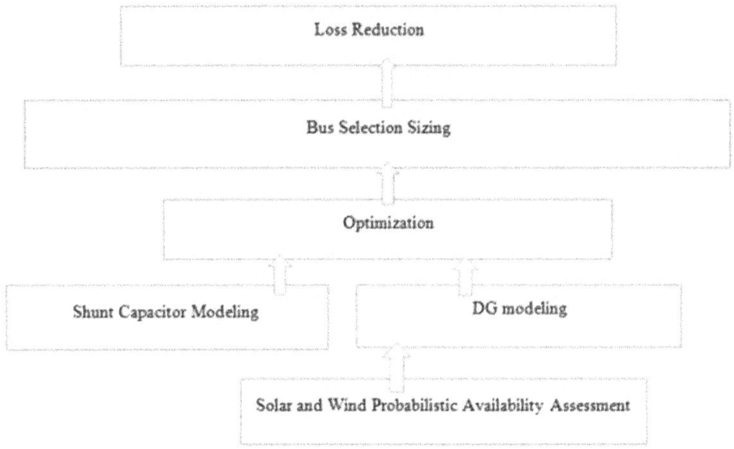

$$Q_{max} = U \times Q_o \tag{1}$$

Where, U is an integer.

Renewable DG Modeling: Different types of renewable DGs are used in the distribution system. Here wind speed-and solar irradiance-based renewable DGs as interconnected with radial distribution system using an hourly multi-state modeling (Nallagownden et al. 2017).

Wind Speed Modeling: Wind speed using Weibull probability density function for particular wind speed at t^{th} time hour, given as follows (Kumar, 2017, Nallagownden et al. 2017):

$$f^t(v) = \left(\frac{k^t}{c^t}\right)\left(\frac{v^t}{c^t}\right)^{k^t - 1} \times \exp[-\left(\frac{v^t}{c^t}\right)^{k^t}] \tag{2}$$

Where $f^t(v)$ is the probability of wind speed at t^{th} time hour, k^t and c^t are the shape and scale parameter respectively. The k^t and c^t of the wind speed can be measured by the following equations (3) and (4),

$$k^t = \left(\frac{\sigma^t}{\mu^t}\right)^{-1.086} \tag{3}$$

$$c^t = \frac{\mu^t}{``\left(1+\frac{1}{k^t}\right)} \tag{4}$$

Where $\mu^t and\, \sigma^t$ and shows the mean and standard deviation of the wind speed at t[th] time hour. The probability of wind speed at any specific hour for ¡ states can be calculated by integrating the probabilities of each state during that hour as given in Equation (5):

$$p\left(v_p^t\right) = \begin{cases} \int_0^{\frac{\left(v_p{}^t + v_p{}^t + 1\right)}{2}} f^t\left(v\right).dv,\, for\ p = 1 \\ \int_{\frac{\left(v_{p-1}{}^t + v_p{}^t\right)}{2}}^{\frac{\left(v_p{}^t + v_p{}^t + 1\right)}{2}} f^t\left(v\right).dv,\, for\ p = 2\ldots\left(nbv, s-1\right) \\ \int_{\frac{\left(v_{p-1}{}^t + v_p{}^t\right)}{2}}^{\infty} f^t\left(v\right).dv,\, for\ p = nbv, s \end{cases} \tag{5}$$

Power Generation from Wind Turbines: Mathematically output power of wind turbine during t[th] time hour at each state p can be calculated by the following Equation (6):

$$p^t{}_{wind} = \sum_{p=1}^{nbv,s} PDG_{wind,p} \times P\left(v_p{}^t\right) \tag{6}$$

where P^t_{wind} is the output power of wind turbine at time t hour, $PDG_{wind,p}$ is the power output of wind turbine given from the manufacturer at state p can be calculated by Equations (7)–(9) as given below:

$$PDG_{wind,p} = \begin{cases} 0, 0 \le v_{avg,p \le v_{ci}} \\ \alpha \times v_{avg,p}{}^3 + \beta \times p_{rated, v_{ci} \le v_{avg,p} \le v_r} \\ p_{rated, v_r \le v_{avg,p} \le v_{co}} \\ 0, v_{avg,p \ge v_{co}} \end{cases} \tag{7}$$

$$\alpha = \left(\frac{p_{rated}}{v_r^3 - v_{ci}^3} \right) \tag{8}$$

$$\beta = \left(\frac{v_{ci}^3}{v_r^3 - v_{ci}^3} \right) \tag{9}$$

Where $PDG_{wind,p}$ is the characteristic of power generation taken from the power performance curve provided by the manufacturer at each of pth state. P_{rated} is the rated output power from the wind turbine, υ_{aw} is the average wind speed, υ_{ci} is the cut-in speed, υ_{cr} is the rated speed and υ_{co} is the cut-out speed.

Solar Irradiance Modeling: The stochastic characteristics of solar irradiance can be modeled using Beta probability distribution function for particular solar irradiance at t^{th} time as:

$$f^t(s) = \begin{cases} \frac{\Gamma(\alpha^t + \beta^t)}{\Gamma(\alpha^t) \times \Gamma(\beta^t)} \times (s^t)^{\alpha^t - 1} \times (1 - s^t)^{\beta^t - 1}, 0 \le s \le 1; \alpha, \beta \ge 0 \\ 0, otherwise \end{cases} \tag{10}$$

where $f^t(t)$ is the Beta distribution function of solar irradiance s at t^{th} time hour, s is the solar irradiance measured in (kW/m^2), α and β are the statistical parameter of $f^t(t)$ and can be calculated by the mean (μ) and standard deviation (σ) of the random variable s as follows:

$$\alpha^t = \frac{\mu^t \times \beta^t}{(1 - \mu^t)} \tag{11}$$

$$\beta^t = (1 - \mu^t) \times \left(\frac{\mu^t (1 + \mu^t)}{(s^t)^2} - 1 \right) \tag{12}$$

The probability of solar irradiance at any specific hour for many states can be calculated by integrating the probabilities of each state during that hour as given in Equation (13):

$$p\left(v_p^t\right)=\begin{cases}\displaystyle\int_{0}^{\frac{\left(s_p{}^t+s_p{}^t+1\right)}{2}} f^t\left(s\right).ds,\ for\ p=1\\ \displaystyle\int_{\frac{\left(s_{p-1}{}^t+s_p{}^t\right)}{2}}^{\frac{\left(s_p{}^t+s_p{}^t+1\right)}{2}} f^t\left(s\right).ds,\ for\ p=2\ldots\left(nbv,s-1\right)\\ \displaystyle\int_{\frac{\left(s_{p-1}{}^t+s_p{}^t\right)}{2}}^{\infty} f^t\left(s\right).ds,\ for\ p=nbv,s\end{cases} \tag{13}$$

Power Generation from Solar PV: Mathematically output power from solar PV module during t[th] time hour at each state p can be calculated from the following Equation (14):

$$p^t{}_{wind}=\sum_{p=1}^{nbv,s} PDG_{solar,p}\times P\left(s_p{}^t\right) \tag{14}$$

where p_{solar}^t is the output power of solar PV module at time t hour, $PDG_{solar,p}$ is the power output of solar PV module given from manufacturer at state p can be calculated by Equations (15)–(19) as given below:

$$T_{Cp}=T_A+S_{ap}\left(\frac{N_{OT}-20}{0.8}\right) \tag{15}$$

$$I_P=S_{ap}\left[I_{SC}+K_i\left(T_C-25\right)\right] \tag{16}$$

$$V_P=V_{OC}-K_V T_{cp} \tag{17}$$

$$FF=\left(\frac{V_{MM}\times I_{MMP}}{V_{OC}\times I_{SC}}\right) \tag{18}$$

$$PDG_{solar,P} = N \times FF \times I_P \times V_P \tag{19}$$

Where T_{CP} and T_A are the cell temperature and ambient cell temperature during state p, both are measured in (°C). S_{ap} is the mean irradiance of state P, N_{OT} is the nominal operating temperature measured in (°C). I_P and I_V are the total current and voltage of state p, whereas I_{sc} and V_{oc} are the short circuit current measured in (amps) and open circuit voltage measured in (volts). K_i and K_v are the current and voltage temperature coefficient measured in (amps/°C) and (volts/°C) respectively. I_{MMP} and V_{MMP} are the current and voltage at maximum power point, measured in amps and in volts respectively. N is the total number of PV modules and FF is the fill factor.

Biomass Modeling: In modern era, energy storing mechanisms is the unsolved issue. Biomass this case can be an alternative solution since energy from the sun is stored by photosynthesis during plants food production.

Physical Properties of Biomass: The chemical and energy output characterization is directly related with its moisture and density during conversion process of biomass (Yazie et al. 2017). The moisture content can be evaluated on dry (1.5) and wet basis wb (1.6) the latter is largely used (PP, 2017).

Wet basis

$$moisture\,content\left(MC\right) = \frac{mass\,of\,water}{mass\,of\,wet\,biomass} = h_{wet} = \frac{m_{tot} - m_{dry}}{m_{tot}} \times 100\% \tag{20}$$

Dry basis

$$moisture\,content = \frac{mass\,of\,water}{mass\,of\,dry\,wood} = h_{dry} = \frac{m_{tot} - m_{dry}}{m_{dry}} \times 100\% \tag{21}$$

Where: $mtot$ = the total mass, thus including moisture,

$mdry$ = the mass of the dry Substance,

$m_{tot} - m_{dry}$ =the moisture mass.

The moisture content of wood is different in seasons and its species and typical values 40 to 50 percent. This can be further decrease to below 20 percent using natural drying.

Energy Properties: From an energy point of view, the most important feature of a fuel is its heating value that quantifies the heat generated by complete combustion of a unit mass of the material. Heating value (calorific value) is the heat released by the fuel when completely burnt, and may be determined at constant volume or constant pressure, and flue gas is cooled back to the initial temperature (ambient temperature).

Heat Balance in a Complete Combustion: Heat released from combustion (Heat of Combustion) = heat to vaporize existing water + heat to vaporize water product + heat loss to environment

Table 1. Co-firing and plant efficiency

Co-firing rate	Drop in efficiency (% points)
0%	0.0%
5%	0.5%
10%	0.8%
15%	0.9%

Higher/Gross Heating/Calorific Value(HHV, Hs) - assumes that the water vapor in the products condenses and thus includes the latent heat of vaporization of the water vapor in the products.

- HHV or GHV = Heat of Combustion per unit mass

Lower/Net Heating/Calorific Value (LHV, Hi) - does not contain the latent heat, the water in flue gas remain in steam form at the initial temperature.

- LHV or NHV = (Heat of Combustion - heat to vaporize existing water - heat to vaporize water product) per unit mass
- Possible dimensions: MJ/kg, MJ/m^3, kWh/kg, kWh/m^3

$$LHV_{dry} = HHV_{dry} - 9Hq \quad (22)$$

Where, H is hydrogen content in dry biomass(5 to7 percent) and q is water condensation heat, that is equal to 2.4 MJ/kg; the subscript dry obviously indicates that the values are referred to dry basis, while the factor 9 relates to the fact that the produced water quantity is nine times higher than the hydrogen content. It must be noted that, precisely because discussion has been done on dry basis, here water vapor is only related to the presence of hydrogen in the structure of the fuel that gives water in the combustion process, and is not associated to the moisture content: thus the formula. The difference between HHV and LHV is typically equal to 1 to 1.5 MJ/kg (Yazie et al. 2017).

The moisture content HHV and LHV value of some biomass is shown in the glossary (RETSCREEN, nd). The ratio of dry matter at anthesis and final grain was between 1.29-1.50 t/ha of dry matter to each ton of grain. The light grazing had the highest conversion ratio, while the heavy grazing had the lowest (PP, 2017).

The efficiency and capacity of power plant decrease as co firing ratio of biomass increases:

$$eff\ drop = -0.4\left(co - fire\,ratio\right)^2 + 0.12\left(co - fire\,ratio\right)$$

Table 1 presents the Co-firing and plant efficiency. To calculate the mass of the biomass and coal, the percent of coal fired with biomass is required (RETSCREEN, nd).

$$\frac{t.biomass}{year} = power\ plant\ size * \%cofire * capacity\ factor * 8760\ {}^{hrs}\!/\!_{yr} * heat\,rate * \left[HHV\right]_{biomass}^{-1} \tag{23}$$

Load Flow

Forward/ Backward Sweep Load Flow: RDS have some hard conditions such as radial meshed grids, instable operation, high R/X ratios and distributed generation. Due to these reasons, the Newton Raphson, Gauss Siedel and other conventional load flow algorithms are failed with the distribution networks. Therefore, a modern method called forward/backward sweep is used (Boktor et al. 2017). The forward and/ or backward sweep processes distribution power flow developed using Kirchhoff's law is computed as.

- Branch currents are calculated using the backward sweep
- The bus voltage will be computed in forward sweep

Meta-heuristic Optimization Methods: The Meta heuristic method is an iterative generation process that helps the search process to efficiently locate near-optimal solutions using learning strategies and intelligently combining different concepts that will help exploit and scour the search space. This strategy is utilized to find the exact or near exact optimal solutions. This method can be divided into two distinct categories: Single (unique) solution and Population solution (Badran et al. 2017). Different methods of optimization technique could be used. The most popular optimization methods are GA, PSO, WOA etc. all are derived from the natural movement of insects, birds and fish for hunting their food or protecting themselves from pray.

Objective Function and Constrained Equations

Copper losses are predominant in distribution system. This can be calculated as follows (Reddy et al. 2017):

Objective Function

The objective function is the task that should be considered hence for best loss minimization active and reactive power consideration or total power loss consideration is important.

Constraint Equations

The constraints are limiting conditions for achieving the required level of optimization

1. 1. Voltage constraints

$$V_{min} \leq V_i \leq V_{max} \tag{24}$$

2. 2. DG constrained the value of DG or QG might not infinitely abundant hence limiting the limit is important

$$P^t_{DG,min} \leq P^t_{DG} \leq P^t_{DG,max} \tag{25}$$

3. 3. Capacitor constrained

$$Q^t_{cap,min} \leq Q^t_{cap} \leq Q^t_{cap,max} \tag{26}$$

4. 4. Line Capacity Constraints: the line capacity constraints of line i is limited by its maximum thermal rating limit as:

$$S^t_{li} \leq S_{li,rated} \tag{27}$$

Candidate Selection Methods

Due to economic reasons integrating DG and QG is no longer applicable hence different methods would be exercised among this:

Loss Sensitivity Factor Method: This is either with respect to active or reactive power loss to find the most affective area in the loss. The loss sensitivity factors are calculated from the base case load flow (that is, without compensation) and the values are arranged in descending order for all the transmission lines of the given system. A vector that holds the respective ''end buses'' of the lines arranged in descending order of the values of the loss sensitivity factors is stored. The bus with inflowing power is the one considered for capacitor placement (Elsheikh et al. 2014).

Voltage Normalization: The other is normalizing voltage magnitudes are calculated by considering the base case voltage magnitudes given as below.

$$\text{Norm}[i] = |V[i]|/0.95 \tag{28}$$

$$\text{Norm}[i] = |V[i]|/0.95 \tag{29}$$

Where, V[i] is the base voltages of the corresponding bus

The "Norm[i]" decides whether the buses need reactive compensation or not. The buses whose Norm[i] value is less than 1.01 can be selected as the candidate buses for capacitor placement.

Voltage sensitivity index sensitivity index or load probability methods could be used but loss sensitivity method is the most popular one.

CONCLUSION

The chapter is discussed the problem of distribution power loss minimization and energy maximization. To do this researcher discussed different analytical, artificial intelligent and meta-heuristic techniques. To make the distribution load flow optimum using optimal sizing and location of shunt capacitors, DG and/or the combination of both.

Optimization of the distribution system is an open research that is undergoing on the current era. Unbalanced distribution network computing methods are ongoing to find optimum solution for radial and mesh distribution systems. Power Engineers are keep finding computation algorithms to make the system more optimum. To do this proper shunt capacitor placement developing better load flow solution, network reconfiguration, maintains feeder losses and DG placement are some of the tasks developed by researchers.

REFERENCES

Abdelaziz, A. Y., Ali, E. S., & Abd Elazim, S. M. (2016). Optimal sizing and locations of capacitors in radial distribution systems via flower pollination optimization algorithm and power loss index, *Engineering Science and Technology, an International Journal, Vol. 19*(1), pp. 610-618.

Alhelou, H., Hamedani-Golshan, M. E., Zamani, R., Heydarian-Forushani, E., & Siano, P. (2018). Challenges and opportunities of load frequency control in conventional, modern and future smart power systems: A comprehensive review. *Energies, 11*(10), 2497. doi:10.3390/en11102497

Alhelou, H. H. (2018). *Fault detection and isolation in power systems using unknown input observer. In Advanced condition monitoring and fault diagnosis of electric machines* (p. 38). Hershey, PA: IGI Global.

Alhelou, H. H., Golshan, M., & Fini, M. (2018). Wind driven optimization algorithm application to load frequency control in interconnected power systems considering GRC and GDB nonlinearities. *Electric Power Components and Systems, 46*(11-12), 1223-1238.

Alhelou, H. H., & Golshan, M. E. H. (2016, May). Hierarchical plug-in EV control based on primary frequency response in interconnected smart grid. In *2016 24th Iranian Conference on Electrical Engineering (ICEE),* (pp. 561-566). IEEE. 10.1109/IranianCEE.2016.7585585

Alhelou, H. H., & Golshan, M. E. H. (2020). Decision-making-based optimal generation-side secondary-reserve scheduling and optimal LFC in deregulated interconnected power system. In *Decision making applications in modern power systems* (pp. 269–299). Academic Press. doi:10.1016/B978-0-12-816445-7.00011-6

Alhelou, H. H., Golshan, M. E. H., & Hatziargyriou, N. D. (2019). Deterministic dynamic state estimation-based optimal LFC for interconnected power systems using unknown input observer. *IEEE Transactions on Smart Grid*, 1. doi:10.1109/TSG.2019.2940199

Alhelou, H. H., Golshan, M. E. H., & Hatziargyriou, N. D. (2019). A decentralized functional observer based optimal LFC considering unknown inputs, uncertainties and cyber-attacks. *IEEE Transactions on Power Systems.*

Alhelou, H. H., Golshan, M. E. H., Zamani, R., Moghaddam, M. P., Njenda, T. C., Siano, P., & Marzband, M. (2019, June). An improved UFLS scheme based on estimated minimum frequency and power deficit. In 2019 IEEE Milan PowerTech (pp. 1-6). IEEE. doi:10.1109/PTC.2019.8810497

Alhelou, H. H., Golshan, M. H., & Askari-Marnani, J. (2018). Robust sensor fault detection and isolation scheme for interconnected smart power systems in presence of RER and EVs using unknown input observer. *International Journal of Electrical Power & Energy Systems*, *99*, 682–694. doi:10.1016/j.ijepes.2018.02.013

Alhelou, H. H., Hamedani-Golshan, M. E., Heydarian-Forushani, E., Al-Sumaiti, A. S., & Siano, P. (2018, September). Decentralized fractional order control scheme for LFC of deregulated nonlinear power systems in presence of EVs and RER. In *2018 International Conference on Smart Energy Systems and Technologies (SEST)* (pp. 1-6). IEEE. 10.1109/SEST.2018.8495858

Alhelou, H. S. H., Golshan, M. E. H., & Fini, M. H. (2015, December). Multi agent electric vehicle control based primary frequency support for future smart micro-grid. In *2015 Smart Grid Conference (SGC)* (pp. 22-27).

Ali, A., Nor, N. M., Ibrahim, T., & Romlie, M. F. (2017). Sizing and placement of battery-coupled distributed photovoltaic generations. *Journal of Renewable and Sustainable Energy*, *9*(5). doi:10.1063/1.4995531

Anteneh, D., & Khan, B. (2019). Reliability enhancement of distribution substation by using network reconfiguration a case study at Debre Berhan Distribution Substation. *International Journal of Economy, Energy, & Environment*, *4*(2), 33.

Badran, O., Mekhilef, S., Mokhlis, H., & Dahalan, W. (2017). Optimal reconfiguration of distribution system connected with distributed generations: A review of different methodologies. *Renewable & Sustainable Energy Reviews*, *73*, 854–867. doi:10.1016/j.rser.2017.02.010

Banteywalu, S., Khan, B., De Smedt, V., & Leroux, P. (2019). A novel modular radiation hardening approach applied to a synchronous buck converter. *Electronics (Basel)*, *8*(5), 513. doi:10.3390/electronics8050513

Bhullar, S., & Ghosh, S. (2018). Optimal integration of multi distributed generation sources in radial distribution networks using a hybrid algorithm. *Energies*, *12*(4-5), 2–15.

Boktor, C. G., Youssef, A. R., Ali, A. H., & Kamel, S. (2017). Optimal distribution power flow including shunt capacitor allocation based on voltage deviation and power loss minimization. In *Proceedings 2017 Nineteenth International Middle East Power Systems Conference (MEPCON)*, Menoufia University, 2017. 10.1109/MEPCON.2017.8301288

Christeen, G. B., Abdel-Raheem, Y., Ali, A. H., & Kamel, S. (2017). Optimal distribution power flow including shunt capacitor allocation based on voltage deviation and power loss minimization. In *2017 Nineteenth International Middle East Power Systems Conference (MEPCON)*, Menoufia University, 2017.

Elsheikh, A., Helmy, Y., Abouelseoud, Y., & Elsherif, A. (2014, November). Optimal capacitor placement and sizing in radial electric power systems. *Alexandria Engineering Journal*, *10*(4), 809–816. doi:10.1016/j.aej.2014.09.012

Elsheikh, A., Helmy, Y., Abouelseoud, Y., & Elsherif, A. (2014). Optimal capacitor placement and sizing in radial electric power systems. *Alexandria Engineering Journal, 53*(4), 809–816. doi:10.1016/j.aej.2014.09.012

Fini, M. H., Yousefi, G. R., & Alhelou, H. H. (2016). Comparative study on the performance of many-objective and single-objective optimisation algorithms in tuning load frequency controllers of multi-area power systems. *IET Generation, Transmission, & Distribution, 10*(12), 2915–2923. doi:10.1049/iet-gtd.2015.1334

Gupta, K., Khan, B., & Mubeen, S. E. (2015). Available transfer capability enhancement by unified power flow controller. In *Proceedings IEEE International Conference on Signal Processing, Informatics, Communication, and Energy Systems (SPICES),* 2015, pp. 1, 4, 19-21.

Haes Alhelou, H., Hamedani-Golshan, M. E., Njenda, T. C., & Siano, P. (2019). A survey on power system blackout and cascading events: Research motivations and challenges. *Energies, 12*(4), 682. doi:10.3390/en12040682

Haes Alhelou, H., Hamedani-Golshan, M. E., Njenda, T. C., & Siano, P. (2019). Wide-area measurement system-based optimal multi-stage under-frequency load-shedding in interconnected smart power systems using evolutionary computing techniques. *Applied Sciences, 9*(3), 508. doi:10.3390/app9030508

Haes Alhelou, H., Hamedani Golshan, M. E., Njenda, T. C., & Siano, P. (2019). WAMS-based online disturbance estimation in interconnected power systems using disturbance observer. *Applied Sciences, 9*(5), 990. doi:10.3390/app9050990

Jariso, M., Khan, B., Tanwar, S., Tyagi, S., & Rishiwal, V. (2018). Hybrid energy system for upgrading the rural environment, *IEEE GlobeCom 2018.*

Jariso, M., Khan, B., Tesfaye, D., & Singh, J. (2017). Modelling and designing of stand-alone photovoltaic system (Case study: Addis Boder Health Center southwest Ethiopia). In *Proceedings IEEE International Conference on Electronics, Communication, and Aerospace Technology, ICECA 2017,* pp. 168-173, Coimbatore, India.

Khan, B., & Agnihotri, G. (2012). A novel transmission loss allocation method based on transmission usage. In *Proceedings IEEE Fifth Power India Conference*, Haryana, India. 10.1109/PowerI.2012.6479479

Khan, B., Agnihotri, G., & Gupta, G. (2013). A multipurpose matrices methodology for transmission usage, loss and reliability margin allocation in restructured environment, *Electrical & Computer Engineering: An International Journal, 2*(3).

Khan, B., Agnihotri, G., Rathore, P., Mishra, A., & Naidu, G. (2014). A cooperative game theory approach for usage and reliability margin cost allocation under contingent restructured market. *International Review of Electrical Engineering, 9*(4), 854–862.

Khan, B., Alhelou, H. H., & Mebrahtu, F. (2019). A holistic analysis of distribution system reliability assessment methods with conventional and renewable energy sources. *AIMS Energy, 7*(4), 413–429. doi:10.3934/energy.2019.4.413

Khan, B., & Singh, P. (2017). The current and future states of ethiopia's energy sector and potential for green energy: A comprehensive study. *International Journal of Engineering Research in Africa, 33*, 115–139. doi:10.4028/www.scientific.net/JERA.33.115

Khan, B., & Singh, P. (2017). Optimal power flow techniques under characterization of conventional and renewable energy sources: A comprehensive analysis, *Journal of Engineering, vol. 2017,* Article ID 9539506.

Khan, B., & Singh, P. (2017). Selecting a meta-heuristic technique for smart micro-grid optimization problem: a comprehensive analysis. In IEEE Access, 5, pp. 13951-13977.

Khan, B., & Singh, P. (2019). Economic operation of smart micro-grid: a meta-heuristic approach. In Handbook of research on smart power system operation and control (pp. 330-346). Hershey, PA: IGI Global. doi:10.4018/978-1-5225-8030-0.ch014

Khan, B., & Singh, P. (2019). Economic operation of smart micro-grid: a meta-heuristic approach. In Handbook of research on smart power system operation and control (pp. 330-346). Hershey, PA: IGI Global. doi:10.4018/978-1-5225-8030-0.ch014

Khan, B., & Tanwar, S. (2019). Issues associated with microgrid integration. In Handbook of research on smart power system operation and control (pp. 252-264). Hershey, PA: IGI Global. doi:10.4018/978-1-5225-8030-0.ch010

Kifle, Y., Khan, B., & Singh, J. (2018). Designing and modeling grid connected photovoltaic system: (Case study: EEU building at Hawassa city). *International Journal of Convergence Computing*, *3*(1), 20–34. doi:10.1504/IJCONVC.2018.091113

Kifle, Y., Khan, B., & Singh, P. (2018). Assessment and enhancement of distribution system reliability by renewable energy sources and energy storage, *Journal of Green Engineering, 8*(3), 2, pp. 219-262.

Kumar, M., Nallagownden, P., & Elamvazuthi, I. (2017). Multi-objective PSO based optimal placement of solar power DG in radial distribution system. *J. Electrical Systems*, *13*(2), 322–331.

Lalitha, M. P., Suresh Babu, P., & Adivesh, B. (2016). Optimal distributed generation and capacitor placement for loss minimization and voltage profile improvement using symbiotic organisms search algorithm. *International Journal of Electrical Engineering*, *9*(3), 250–262.

Makdisie, C., Haidar, B., & Alhelou, H. H. (2018). An optimal photovoltaic conversion system for future smart grids. In Handbook of research on power and energy system optimization (pp. 601–657). Hershey, PA: IGI Global. doi:10.4018/978-1-5225-3935-3.ch018

Molla, T., Khan, B., Moges, B., Alhelou, H. H., Zamani, R., & Siano, P. (2019). Integrated optimization of smart home appliances with cost-effective energy management system. *CSEE Journal of Power and Energy Systems*, *5*(2), 249–258.

Molla, T., Khan, B., & Singh, P. (2018). A comprehensive analysis of smart home energy management system optimization techniques, *Journal of Autonomous Intelligence, 1*(1), 15–21.

Nallagownden, P., Kumar, M., & Elamvazuthi, I. (2017). Optimal placement and sizing of renewable distributed generations and capacitor banks into radial distribution systems.

Negash, K., Khan, B., Tesfaye, D., & Gayathri, M. (2017). Optimal placement of phasor measurement unit for system observability. In *Proceedings of IEEE International Conference on Computing Methodologies and Communication, (ICCMC 2017)*, pp. 73-77, Erode, India.

Negash, K., Khan, B., & Yohannes, E. (2016). Artificial intelligence versus conventional mathematical techniques: A review for optimal placement of phasor measurement units [Springer]. *Journal of Technology and Economics of Smart Grids and Sustainable Energy*, *1*(1), 10. doi:10.100740866-016-0009-y

Njenda, T. C., Golshan, M. E. H., & Alhelou, H. H. (2018, November). WAMS based intelligent under frequency load shedding considering online disturbance estimation. In 2018 *Smart Grid Conference (SGC)* (pp. 1-5). 10.1109/SGC.2018.8777779

Njenda, T. C., Golshan, M. E. H., & Alhelou, H. H. (2018, November). WAMS based under frequency load shedding considering minimum frequency prediction and extrapolated disturbance magnitude. In *2018 Smart Grid Conference (SGC)* (pp. 1-5).

Power, P. (2009). *The cost of using biomass for power generation and potential for decreased greenhouse gas emissions.* Retrieved from https://sequestration.mit.edu/pdf/AmandaCuellar_Thesis_June2012.pdf

Reddy, P. D. P., Reddy, V. C. V., & Manohar, T. G. (2017). *Whale optimization algorithm for optimal sizing of renewable resources for loss reduction in distribution systems* (pp. 1–13). Renewables Wind, Water, and Soloar.

RETSCREEN International, Renewable Energy Project Analysis: Ret screen Engineering and Cases Textbook. Available at http://www.retscreen.net

Short, T., & Swayne, T. (2012). Assessment of Transmission and Distribution Losses in New York PID071178 (NYSERDA 15464), The New York State Energy Research and Development Authority, New York.

Singh, P., & Khan, B. (2017). Smart microgrid energy management using a novel artificial shark optimization, Complexity, vol. 2017, Article ID 2158926. doi:10.1155/2017/2158926

Singh, P., Khan, B., Vidyarthi, A., Haes Alhelou, H., & Siano, P. (2019). Energy-aware online non-clairvoyant scheduling using speed scaling with arbitrary power function. *Applied Sciences*, *9*(7), 1467. doi:10.3390/app9071467

Thang, V. V., & Minh, N. D. (2017). Optimal allocation and sizing of capacitors for distribution systems reinforcement based on minimum life cycle cost and considering uncertainties. *The Open Electrical & Electronic Engineering Journal*, *11*(18), 165–176. doi:10.2174/1874129001711010165

Xie, J., Liang, C., & Xiao, Y. (2018). Reactive power optimization for distribution network based on distributed random gradient-free algorithm. *Energies*, *11*(3), 534. doi:10.3390/en11030534

Yazie, C., Akalu, T., Yalfal, T., & Baye, B. (2017). *Characterization of potato production, marketing and utilization in North Western Amhara* Region, Ethiopia. *Journal of Horticulture and Forestry*, *9*(3), 17-25.

Yenesew, M. (2017). Enhanced particle swarm optimization (PSO) algorithm for reactive power optimization in the distribution system [IJERME]. *International Journal of Engineering Research and Modern Education*, *2*(2), 56–64.

Yeshalem, M. T., & Khan, B. (2017). Design of an off-grid hybrid PV/wind power system for remote mobile base station: A case study. *AIMS Energy*, *5*(1), 96–112. doi:10.3934/energy.2017.1.96

Zamani, R., Golshan, M. E. H., Alhelou, H. H., & Hatziargyriou, N. (2019). A novel hybrid islanding detection method using dynamic characteristics of synchronous generator and signal processing technique. *Electric Power Systems Research*, *175*, 105911. doi:10.1016/j.epsr.2019.105911

Zamani, R., Hamedani-Golshan, M. E., Haes Alhelou, H., Siano, P., & Pota, H. R. (2018). Islanding detection of synchronous distributed generator based on the active and reactive power control loops. *Energies*, *11*(10), 2819. doi:10.3390/en11102819

KEY TERMS AND DEFINITIONS

Active Distribution Networks: These are distribution networks that have systems in place to control a combination of distributed energy source.

Distributed Generation: It is an approach that employs small-scale technologies to produce electricity close to the end users of power.

Distribution Loss: It includes losses in between points of distribution to consumers.

APPENDIX

List of Acronyms

DG Distributed Generations
EPRI Electric Power Research Institute
GA Genetic Algorithm
MC Moisture content
MC Moisture Content
PSO Particle Swarm Optimization
PV Photo Voltaic

Chapter 8
Reliability Assessment of Microgrid-Integrated Electrical Distribution System

Baseem Khan
https://orcid.org/0000-0002-0562-0933
Hawassa University, Hawassa, Ethiopia

Mesfin Fanuel
Hawassa University, Hawassa, Ethiopia

ABSTRACT

Due to the transition of traditional power system to smart structure, integration of renewable energy sources is of great importance. It is performed by using distributed generators and Micro grid. Integration of renewable sources is very useful for the reliability enhancement of distribution system as energy can be supplied locally at distributed level. Therefore, this chapter provides an overview of various reliability methods which are utilized for finding the impact of renewable energy generation on distribution system reliability assessment. These renewable energy sources are integrated in the distribution system at distributed and Micro grid level. Various characteristics of renewable energy sources are discussed to model them. A general problem of reliability assessment in terms of reliability indices is also discussed.

INTRODUCTION

The primary purpose of electrical energy system is to provide reliable and quality energy to its consumers within permissible limits (Adegboye and Ekundayo, 2011). Probability of the power system to operate adequately without any outage in specified time duration under the normal operating constraints is called Reliability of the power system (Kifle et al. 2018). The performance of distribution system can be assessed with the help of available data of distribution system. Mal-functional elements of the system can also be identified with the help of reliability assessment study of distribution system. Further, elements, which require emergency replacement, can be replaced from the system. Moreover, reliability

DOI: 10.4018/978-1-7998-1230-2.ch008

assessment study is also helpful to recommend new components in the system, so that the overall reliability of the system will enhance (Muhaini et al. 2010). Reliability methods have technical as well as economical characteristics. Therefore, for designing and operation of power system at all the stages, these techniques have been adopted as benchmark standards (Billinton and Allan, 1996). There is the transition of conventional distribution system into the smart distribution system with the integration of renewable energy sources and smart sensors (Adefarati and Bansal, 2016, Shah et al. 2015). It supports smart technologies and distributed generation units in the electrical power system, which enhances the reliability of power system by reducing the outage duration and minimizing the interruption cost. In smart grid structure, micro grid technology provides the holistic approach for the integration of renewable energy sources. It has several benefits over conventional grid system as it's minimise energy losses, improve reliability and enhance energy management. Further, at distribution level, micro grid technology provides better solution of energy scarcity, generation coordination and control problems due to its better performance with respect to distributed generation technology (Kifle et al. 2018, Jariso et al. 2017). Further fast restoration and enhanced capacity utilization of distribution system raise the income of the utilities. By analysing the technical, environmental and economic aspects of renewable energy sources integration, reliability of distribution system can be enhanced by the power utilities. Based on the energy output, renewable distributed generation units can be integrated into the conventional distribution system. Output of these units depends on the solar irradiance of selected area, wind speed availability, repair rate, network configuration, customer load profile, etc. By utilizing the historical meteorological data of wind speed and solar irradiance of selected site, performance of wind turbine and photovoltaic panels can be easily assessed. Therefore, the availability of renewable energy sources on hourly basis can be simply predicted. Therefore, this chapter explores the integration of renewable energy sources in the form of Micro grid, for enhancing the reliability of distribution system by reducing the power outage cost (Alhelou et al., 2019; Makdisie et al., 2018; Alhelou et al., 2018; Alhelou et al., 2016; Haes Alhelou et al., 2019; Njenda et al., 2018).

Renewable energy sources can be performed following functions (Singh and Khan, 2017):

1. Enhance the reliability of distribution system
2. Minimize the greenhouse gases emissions
3. Reduce the fuel consumption and consumption cost of traditional power plants

This chapter discuss the reliability assessment of distribution system in the presence of renewable distributed energy sources, Micro grid and energy storage system for minimizing the energy outage in the system. For this purpose renewable energy sources such as wind turbine, solar photo-voltaic can be integrated into conventional distribution system in the form of distributed generators or micro grid. These sources are generally designed in such a way that whenever they are available they will supply the energy.

RELATED WORK

As distribution system mainly contributes the outages at the load points due to its radial structure, distribution system reliability problem has great importance in current smart grid structure (Billinton and Allan, 1996). Many literatures reported the problem of distribution system reliability enhancement by different methods such as reconfiguration, utilization of advanced protection system and underground cables, renewable energy sources integration etc. Out of these different methods, renewable energy sources integration method has several benefits over other conventional methods such as incorporation of green energy sources; utilization of available local resources etc. Therefore, many researchers addressed the problem of reliability assessment and enhancement of distribution system by incorporating renewable energy sources such as wind turbine, photovoltaic and energy storage (Billinton and Allan, 1996). In Saboori et al. (2015) and Li et al. (2012), reliability of conventional distribution system is studied in the presence of renewable distributed generation and battery energy storage system. For optimizing the efficiency, losses and reliability of power system, the characteristics of energy storage system is explored by Arifujjaman (2015). For estimating the impacts of wind, photovoltaic and micro hydro systems on the sustainability of environment, Bhandari *et al.* (2014) presented the methods of economic hybridizations. For optimizing the life cycle cost, CO_2 emission and dump energy of hybrid power system, which incorporated wind, photovoltaic, diesel and battery storage, a genetic algorithm-based optimization method is proposed by Ogunjuyigbe *et al.* (2016). Different characteristics of renewable energy sources are analysed by authors to minimize the objective function. A novel method is proposed by Ahadi *et al.* (2016) to minimize the overall cost of the standalone system, which is optimally combined with renewable energy sources. A network reconfiguration technique is proposed by Duan *et al.* (2015) for enhancing the reliability of traditional distribution system by calculating the improved reliability indices. Two major factors such as availability of power supply from utility and power system components' capacity are the basis of this proposed method. For estimating the reliability assessment's impact on distribution system a pseudo dynamic planning method is proposed by Ouza *et al.* (2015). A distribution system expansion planning technique is proposed by Abbasi and Seifi (2015) for enhancing the reliability of traditional distribution system. The main objective of this method is to optimize investment, operation cost and reliability of the system. An optimal switch allocation method is presented by Locatelli *et al.* (2015) for improving the reliability of traditional distribution system. The cost associated with energy not supplied is minimized by this method. For calculating reliability worth and system cost, different methods are presented by the Ghajar *and* Billinton (2006). The basic requirements for these techniques are customer damage functions and assessment rates for interrupted energy. The main objective of this work is to optimize the benefits and costs, which are associated with distributed generation.

Monte Carlo Simulation (MC)

Monte Carlo simulation technique is a versatile method, which is utilized by various researchers for computing reliability of traditional distribution network as well as hybrid distribution system, associated with distributed generation and battery storage by calculating reliability indices (Alkuhayli, 2012, Gao and Billinton, 2009, Billinton and Gao, 2008). MC simulation technique considered the stochastic behaviour of renewable energy sources and calculates their impacts on hybrid distribution system reliability. Stochastic nature of renewable energy sources has great impact on the calculation of reliability indices of hybrid distribution system, which is also confirmed by the simulation results of MC technique. MC simulation technique can be very much useful in various power system problems as presented in references (Borges, 2012, Zhanga et al., 2011, Aghaebrahimi et al., 2011). As MC is a simulation technique, sometimes the results are not accurate as compare to other exact technique. The major drawback of MC simulation technique is the large simulation time, which makes it difficult to apply on large power systems because they have multiple objective functions with many system constraints. Therefore, MC simulation technique is most suitable for small power systems.

Analytical Method

Reliability analysis of traditional as well as hybrid power system with large number of generating units can also be estimated with the help of analytical methods. In Zhang and Sun (2015) reliability analysis of traditional and hybrid system with distributed generation resources is performed with the help of analytical method. The basic condition adopted in this technique is the supply of power from distributed generators to loads when there is an outage from utilities. Generally for estimating the reliability of non-intermittent generating units, analytical methods are utilized. It can also used for small power systems. Li and Zio (2012) more accurate results can be obtained with the help of analytical methods as compare to MC simulation technique with less computational time. Analytical methods' results are effective in many power system problems, but this effectiveness is restricted due to their inability to handle the stochastic behaviour of major elements in power system under various operating conditions.

Detailed Model Analysis

For multi objective function, a more accurate and realistic solutions can be obtained by incorporating a multi-state model of power system with various elements (wind turbine, photovoltaic and energy storage system), which include stochastic characteristics of these elements at respective states. This model can very useful for representing multiple operating states of different elements of renewable distributed generators (Al-Muhaini et al., 2013, Billinton and Gao, 2006, Ding et al., 2011, Esau et al., 2014). As compared to analytical and MC simulation technique, this model provides more insight on load buses and reliability of system.

Electrical Distribution system

It is a final destination of the electrical power system which makes a link among generation and transmission with customer demands (Alhelou et al., 2019; Makdisie et al., 2018; Alhelou et al., 2018; Alhelou et al., 2016; Haes Alhelou et al., 2019; Njenda et al., 2018). It has simple design and low cost of installation. Distribution utilities utilized this network to supply the electricity for customer demands. The distribution system must be assessed based on its reliability criteria for finding the effects of power system outages and their frequency and duration. This will helpful for averting the irregular interruptions in the distribution system (Wang and Billinton, 2002). Various factors on which the reliability of any distribution system depends are as follows (Adefarati and Bansal, 2017):

1. Corresponding environment
2. System configuration (radial or mesh)
3. Type of installed loads
4. Location of demands

Figure 1 shows the circuit diagram of distribution system with renewable energy generation (Micro grid). As shown in the figure, during faulted condition circuit breaker of feeder 1 is opened and the system is disconnected from the utility grid. This is called an islanded mode of operation. In this operation mode, loads on the feeder will be energised by MG. Loads on second feeder are supplied either by

Figure 1. Distribution system with renewable energy sources

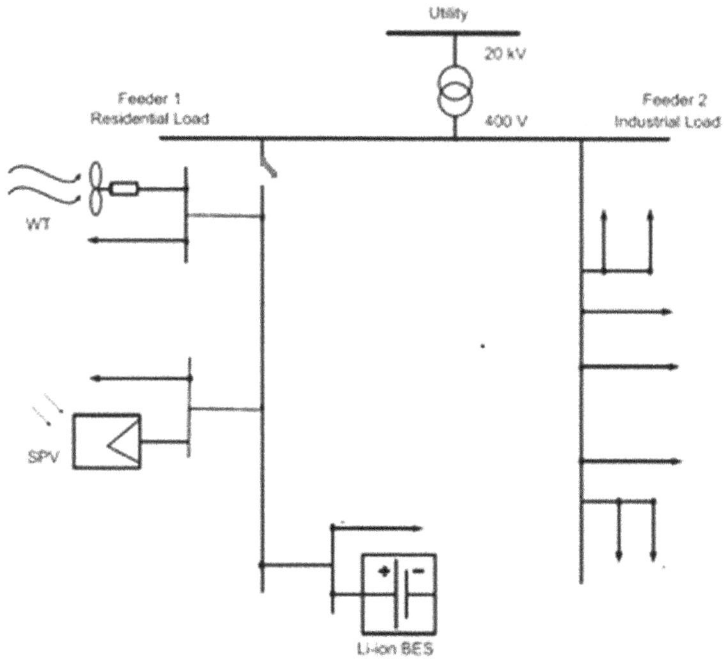

utility grid or MG. An optimal rated capacity will be achieved if there is an adequate correlation between generation and loads.

Hybrid Distribution System

Figure 1 shows the hybrid distribution system, which incorporates photovoltaic, wind turbine and battery storage system. Due to various technical and financial constraints, utilities are not able to meet the electricity demand of customers. Therefore, these constraints encouraged power utilities for the integration of renewable energy sources into existing system. For increasing the reliability of power system, distributed energy generators can be integrated at various levels of the power system. Due to numerous advantages of wind and solar photo voltaic energy sources, they are normally integrated by the utilities to enhance the reliability of power system (Kifle et al. 2018). As wind and solar energy sources are intermittent in their nature, a fraction of their generated energy must be stored in the battery storage system. It will enhance the efficiency of system by reducing the outage time (Khan and Singh, 2017, Mulualem and Khan, 2017, Zhou et al. 2011). The modelling of solar photovoltaic, wind turbine and energy storage system is discussed in the next section.

Modelling of Renewable Distributed Generators in MG

The modelling of solar photovoltaic, wind turbine and energy storage system is discussed in this section.

Figure 2. Solar photo voltaic cell equivalent circuit

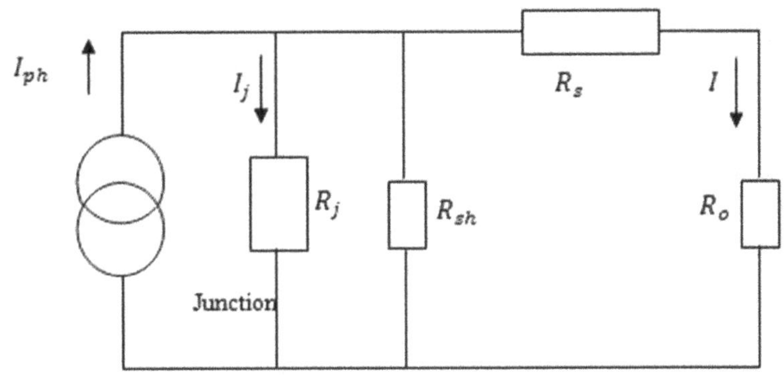

Solar Photo Voltaic System

The sunlight energy can be directly converted into electricity with the help of solar photovoltaic system (Khan et al. 2014; Khan et al. 2013; Khan et al. 2012; Negash et al., 2017; Negash et al., 2016; Jariso et al., 2017; Kifle et al., 2018; Yeshalem et al., 2017; Singh et al., 2019; Gupta et al., 2015). A number of photovoltaic modules are joined together either in series or parallel or combined ways for generating the desired output of energy in the solar photo-voltaic system. This energy output depends on the current and voltage ratings provided by the manufacturer. Different configurations such as grid-tied and islanded mode can be utilized for energy generation from photo-voltaic energy sources. There are several applications, which can be energized with the help of photovoltaic system. For example vehicles, television, radio, street lights etc. can easily supply with help of photovoltaic. Features like low operating, repair and fuel cost, green energy and no air pollution as well as noise pollution makes rapid growth of photovoltaic technology (Kifle et al., 2018). Factors on which the output of the photovoltaic system depends are solar irradiance and operating temperature of the selected location, where this system will be installed. Photovoltaic system can be expressed as presented in figure 2 (Kifle et al., 2018, Jariso et al., 2017, Singh and Khan, 2017). The basic component of solar photo voltaic system is solar cell. It is basically a P-N junction and its equivalent circuit presented in figure 2.

In this Figure 2, I_{ph} represents the photo current of cell i.e. current source. The non linear impedance of P-N junction is represented by R_j. The intrinsic series and shunt resistances of the cell is represented by R_s and R_{sh}, respectively. Generally, R_{sh} has a large value in comparison to R_s, therefore, R_s can be neglected for simplification of the calculation. Reverse saturation current of cell is represented by I_{rs}. Generally, I_{rs} is assumed zero (Chowdhury et al., 2013). Equations (1) to (5) represent the solar photo voltaic mathematical model. To simplify this model, it is assumed that $R_s = 0$, $R_{sh} = \infty$ and $I_{rs} = 0$. By applying Kirchhoff law on the figure 1, the energy output of solar photovoltaic array, consisted N number of modules is computed by equations (1) to (5).

$$T_C = T_A + \frac{(N_{oct} - 20)S}{80} \tag{1}$$

$$I = I_{ph} = \left[I_{sc} + K_i \left(T_c - 25 \right) \right] \frac{S}{100} \tag{2}$$

$$V = V_{OC} - K_V T_C \tag{3}$$

$$FF = \frac{V_{mpp} * I_{mpp}}{V_{oc} I_{sc}} \tag{4}$$

$$P_{out} = N * FF * V * I = n_s * n_p * FF * V * I \tag{5}$$

Here, output current of photo voltaic array is represented by I. Array output voltage is V. Number of cells in series and parallel are denoted by n_s and n_p, respectively. Number of modules is presented by N. Effective and ambient temperatures of cell are represented by T_C and T_A, respectively in $^{\circ}C$. Short circuit current capacity of cell is denoted by I_{sc}. Open circuit voltage of cell is presented by V_{oc}. Current and voltage at maximum power point are represented by I_{mpp} and V_{mpp}, respectively. Short circuit current and open circuit voltage temperature coefficients are denoted by K_i and K_v, respectively. S represents the solar radiation in mW/cm^2 and FF denotes the fill factor.

Wind Turbine System

Wind velocity is the key factor on which the output of wind turbine system depends. In this system, kinetic energy of wind is converted into electrical energy by wind turbines (Khan et al., 2019; Kifle et al., 2018; Khan et al., 2018; Khan et al., 2017; Banteywalu et al., 2019; Anteneh et al., 2019; Molla et al., 2019, Molla et al., 2018, Jariso et al., 2018). Areas, where wind speed comes into the range of operating speeds of wind turbine, specified by the manufacturer, wind turbine system can be installed (Granado et al., 2016). Wind energy is the cost-effective source of renewable energy generation with no emission and minimum operation and maintenance costs. It can be utilized in grid connected or islanded mode. Key factors on which the output of wind turbine depends are as follows:

1. Wind regime
2. Wind speed
3. Air density
4. Swept area of rotor
5. Tower height
6. aero-turbine performance
7. Efficiency of gearbox and generator

Figure 3. Output characteristics curve of wind turbine

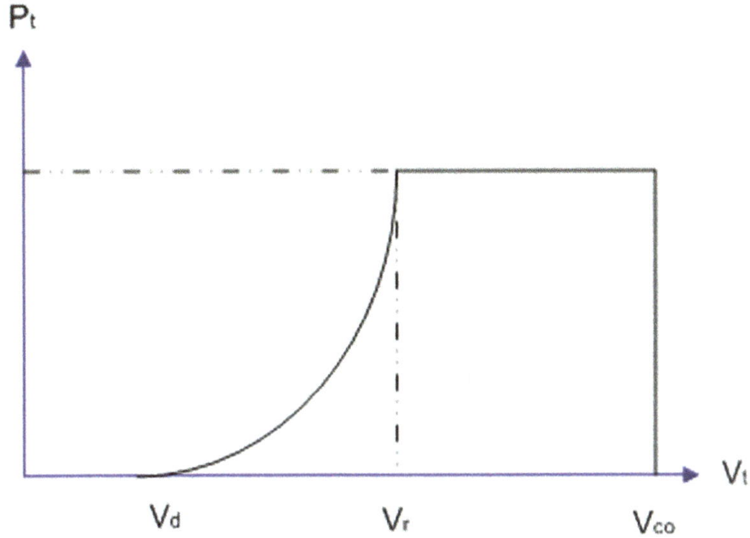

The output characteristic of wind turbine between generated power and wind speed is presented in figure 3 (Kifle et al., 2018).

As shown in the figure 3, wind energy generation is not possible if the wind velocity is less than the cut in speed. Variable energy output would be available, if the wind speed lies between cut in and rated speed. Further, constant output power would be available if the wind speed is lies between rated and cut on speed. Wind turbine would be stopped if the wind speed goes beyond the cut on speed. It is due to the violation of the mechanical safety limit of the turbine. For modelling of the wind energy system, the power curve of wind turbine will formulate in the form of polynomials as presented in equation 6 (Malheiro et al., 2015). This equation presents the energy output generated by wind turbine.

$$P_{WTG} = \begin{cases} 0 & V_t < V_{ci} \\ A + BV_t^3 & V_{ci} \le V_t \le V_r \\ P_r & V_r \le V_t \le V_{co} \\ 0 & V_t > V_{co} \end{cases} \quad (6)$$

Where,

P_{WTG} : wind turbine power output
V_t : Actual wind velocity at time t
V_{ci} : cut-in wind velocity
V_r : rated wind velocity
V_{co} : Cut-off wind velocity
P_r : rated power of the wind turbine
A and B : Parameters

$$A = \frac{P_r * V_{ci}^3}{V_{co}^3 - V_r^3} \tag{7}$$

$$B = \frac{P_r}{V_r^3 - V_{co}^3} \tag{8}$$

As the nature of wind is intermittent and random, distribution of wind velocity is adopted probability distribution function. Further, this distribution also helps to model the uncertainty of wind speed. It is clear from the statistical data that the probability distribution of wind speed follows the Weibull distribution (Aghaebrahimi and Mehdizadeh, 2011). It is a simple function, which has single-peak and two-parameter. Its probability density function and distribution are expressed by equation (9) (Marko, 2011):

$$f(V_t) = \frac{k}{c}\left(\frac{V_t}{c}\right)^{k-1} exp\left[-\left(\frac{V_t}{c}\right)^k\right] \tag{9}$$

Where,

c : Scale parameter
k : Shape parameter

Both parameters are computed by average wind velocity (μ) and the standard deviation (Ã), as presented in equations (10) and (11).

$$k = \left(\frac{\sigma}{\mu}\right)^{-1.086} \tag{10}$$

$$c = \frac{\mu}{\Gamma\left(1 + \frac{1}{k}\right)} \tag{11}$$

As from the equations (10) and (11) is clear that any changes in c and k directly affect the Weibull probability function of the wind speed, selected location wind speed data must be collected for several years.

Battery Energy Storage System

For minimizing the uncertainty and intermittency of renewable energy sources, battery energy storage system is utilized as strategic solution in the system along with renewable generators. This solution is very much helpful for enhancing the reliability of electric distribution system. The following expression can be utilized as the capacity of battery bank for a specified discharge rate, in ampere hours (Kan and Singh, 2017).

$$C_x = \frac{E_{tot}}{V_{dc}} * \frac{G_{ft}}{DOD_{max}} \tag{12}$$

Where,

C_x : Battery capacity for a specified discharge rate in ampere hours
E_{tot} : Total energy in watt hours
G_{ft} : Number of days that battery bank needs to supply
DOD_{max} : Maximum depth of discharge

DISTRIBUTION SYSTEM RELIABILITY

Distribution system reliability can be assessed with the help of reliability indices. These indices can be subdivided on the basis of duration of outages as sustained interruption indices and short interruption indices. Further, based on the energy availability and load, these indices can be subdivided into energy-based indices and load based indices. Details of these indices are not provided in this chapter as it is available in various literatures. These indices are as follows (Khan and Singh, 2017):

Sustained Interruption Indices

- System average interruption frequency index
- Transformer SAIFI
- Equivalent number of interruptions related to the installed capacity (NIEPI)
- Customer interruption
- System average interruption duration index (SAIDI)
- Transformer SAIDI
- Equivalent interruption time related to the installed capacity (TIEPI)
- Customer-minutes lost (CML)
- Customer average interruption duration index (CAIDI)
- Customer total average interruption duration index (CTAIDI)
- Customer average interruption frequency index (CAIFI)
- Average service availability index (ASAI)
- Customers experiencing multiple interruptions (CEMIn)

Indices for Short Interruption

- Momentary average interruption frequency index (MAIFI),
- Momentary average interruption event frequency index (MAIFIE),
- Customers experiencing multiple sustained interruption and momentary interruption events (CEMSMIn)

Energy-Based Indices

- Energy not supplied (ENS),
- Average energy not supplied, (AENS),
- Average customer curtailment index (ACCI)

Load-Based Indices

- Average system interruption frequency index (ASIFI)
- Average system interruption duration index (ASIDI)

Problem Formulation

The main aim of this problem is to minimize the outage cost in the electric distribution network while fulfilling the system constraints (Sreedharan et al., 2016). For finding the impacts of MG on distribution system reliability, basically two objective functions can be formulated (Fini et al., 2016; Alhelou et al., 2018; Zamani et al., 2018; Alhelou et al., 2015; Njenda et al., 2018; Haes Alhelou et al., 2018; Haes Alhelou et al., 2019). Minimization of the expected energy not supplied (EENS) is performed by the first objective function, whereas reduction of expected interruption cost (ECOST) is obtained by second objective function for electric distribution network. For finding the impact of distributed generators and battery storage system on the reliability of traditional distribution network, the above discussed two objective functions are combined. For maximizing the cost of saving in power system, renewable generation units can be incorporated in the form of MG, as denoted by equations (13) to (15). By making a comparison between objective functions with and without renewable energy generators, the aim of the selected problem can be fulfilled. The formulated objective function is as follows (Saboori et al., 2015):

$$F_g^o = \max\left(F_{WRDG}^o - F_{RDG}^o\right) \quad (13)$$

In equation 8,

$$F_{WRDG}^o = \sum_{i=1}^{n}\left(CENS*EENS_{WRDG} + ECOST_{WRDG}\right)\$/Yr \quad (14)$$

(15)
Where,

F^{o}_{WRDG} : Total outage cost without integration of renewable energy generators
F^{o}_{RDG} : Total outage cost with integration of renewable energy generators
$CENS$: cost of energy not supplied in \$/MWh

The objective function is subjected to following constraints:

1. 1. Power balance constraints:

$$\sum_{i=1}^{n} P_{TP,i}(t) + P_{Utility} = \sum_{i=1}^{n} PD_i \tag{16}$$

Where

$\sum_{i=1}^{n} P_{TP,i}(t)$ = combined power supplied by renewable energy generators and battery storage system
$P_{Utility}$ = Power supplied by utility
$\sum_{i=1}^{n} PD_i$ = total demand at load point

2. Demand constraints

$$P_{WT(t)} + P_{SPV(t)} - P_{Bout(t)} + P_{Bin(t)} = P_{RDG(t)} \tag{17}$$

$$P_{WT(t)} \geq 0,\ P_{SPV(t)} \geq 0,\ P_{Bout(t)} \geq 0,\ P_{Bin(t)} \geq 0 \tag{18}$$

Where

$P_{WT(t)}$ =wind turbine power output
$P_{SPV(t)}$ =Solar photovoltaic power output
$P_{Bout(t)}$ = outgoing power from the battery
$P_{Bin(t)}$ = incoming power in the battery
$P_{RDG(t)}$ = total power generated by renewable energy generators

3. Battery energy storage system constraints

State of charge limits

$$SOC_{min} \leq (SOC(O) + \cdot_{C}\sum_{t=1}^{t} P_{Bin(t)} - \cdot_{D}\sum_{t=1}^{t} P_{Bout(t)} \leq SOC_{max} \quad (19)$$

$$SOC_{min} = (1 - DOD)SOC_{max} \quad (20)$$

Where,

DOD = depth of discharge
$\cdot_{C}$ = charging efficiency
$\cdot_{D}$ = discharging efficiency

GENERAL CONSIDERATION

To assess the distribution network reliability, few considerations are presumed. These considerations are related to the system operation and are as follows:

1. It is considered that transmission system and distribution sub station are 100% reliable and protection system failure is generally disregarded.
2. The representation of following characteristics must be considered to integrate renewable energy generators in the reliability assessment techniques.
    a. Uncertain availability of renewable energy sources
    b. Variation of demand in chronological order
    c. Reconfiguration apparatus and protection system
    d. Chances of stand-alone operation mode
    e. Polices of prioritization for load shedding and demand transfer

By utilizing power flow solution, the adequacy of distribution network can be evaluated in terms of voltage and transfer capacity. Reliability indices that can be computed are as follows:

- Average failure rate at load point
- Outage time on annual basis
- The system indices such EENS, ECOST, SAIDI and SAIFI etc.

For eliminating capacity and voltage violation, the following corrective solutions can be incorporated in the system:

1. Components' switching: For isolating the failed sections, opening of switches and protection components is performed.
2. Load shifting between feeders: To manage load after isolating failed block, load is shifted to other feeders through connecting switches
3. Standalone mode of operation by renewable distributed generators: If the total demand is small, than it can feed with the help of distributed generators after isolating the failed section
4. Prioritization of load shedding: A priority list must be follow by the utility for load shedding purpose. First preference must be provided to the critical loads such as hospitals, airports etc. Further, if the occurred demand is higher than the renewable generation capacity, partial load must shed according to the priority policy.

CONCLUSION

Integration of renewable energy generation sources at distributed manner by MG helps distribution utility to enhance the reliability of system. Different results achieved from various reliability techniques, which are combined with renewable distributed generations are published in the cited research paper and provided that renewable distributed generation and MG improve the distribution system reliability. Further, distribution system reliability is effectively increased in islanded mode of renewable distributed generation and MG. Though, renewable energy sources have intermittent energy generation that minimized the benefits of this type of generation. Proper planning and installation of energy storage devices can minimize the disadvantages of intermittency in renewable energy generation. It will help to improve the reliability of distribution system greatly. These intermittent sources are very much important for reliability enhancement. This chapter provides an insight to the reliability enhancement of distribution system with the help of distributed generators and MG. Reliability techniques with the characteristics of renewable energy sources are discussed. Formulation of a general problem of reliability enhancement with renewable energy sources is also presented.

REFERENCES

Abbasi, A. R., & Seifi, A. R. (2015). Considering cost and reliability in electrical and thermal distribution networks reinforcement planning. *Energy*, *84*, 25–35. doi:10.1016/j.energy.2015.01.113

Adefarati, T., & Bansal, R. C. (2016). Integration of renewable distributed generators into the distribution system: A review. *IET Renewable Power Generation*, *10*(7), 873–884. doi:10.1049/iet-rpg.2015.0378

Adefarati, T., & Bansal, R. C. (2017). Reliability assessment of distribution system with the integration of renewable distributed generation. *Applied Energy*, *185*(Part 1), 158–171. doi:10.1016/j.apenergy.2016.10.087

Adefarati, T., & Bansal, R. C. (2017). The impacts of PV-wind-diesel-electric storage hybrid system on the reliability of a power system. *Energy Procedia*, *105*, 616–621. doi:10.1016/j.egypro.2017.03.364

Adegboye, B. A., & Ekundayo, K. R. (2011). Reliability assessment of 4.2 MW single shaft typhoon gas fired turbine power generation station (2003–2008). *Advanced Materials Research*, *367*, 143–150. doi:10.4028/www.scientific.net/AMR.367.143

Aghaebrahimi, M. R., & Mehdizadeh, M. (2011). A new procedure in reliability assessment of wind–diesel islanded grids. *Electric Power Components and Systems, 39*(14), 1563–76.

Ahadi, A., Kang, S. K., & Lee, J. H. (2016). A novel approach for optimal combination of wind, PV and energy storage system in diesel - free isolated communities. *Applied Energy*, *170*, 101–115. doi:10.1016/j.apenergy.2016.02.110

Al-Muhaini, M., & Heydt, G. T. (2013). Evaluating future power distribution system reliability including distributed generation. *IEEE Transactions on Power Delivery*, *28*(4), 2264–2272. doi:10.1109/TPWRD.2013.2253808

Al-Muhaini, M., Heydt, G. T., & Huynh, A. (2010). The reliability of power distribution systems as calculated using system theoretic concepts. *IEEE PES General Meeting,* Providence, RI, pp. 1-8. 10.1109/PES.2010.5588116

Alhelou, H., Hamedani-Golshan, M. E., Zamani, R., Heydarian-Forushani, E., & Siano, P. (2018). Challenges and opportunities of load frequency control in conventional, modern, and future smart power systems: A comprehensive review. *Energies*, *11*(10), 2497. doi:10.3390/en11102497

Alhelou, H. H. (2018). *Fault detection and isolation in power systems using unknown input observer. Advanced condition monitoring and fault diagnosis of electric machines* (p. 38). Hershey, PA: IGI Global.

Alhelou, H. H., Golshan, M., & Fini, M. (2018). Wind driven optimization algorithm application to load frequency control in interconnected power systems considering GRC and GDB nonlinearities. *Electric Power Components and Systems, 46*(11-12), 1223-1238.

Alhelou, H. H., & Golshan, M. E. H. (2016, May). Hierarchical plug-in EV control based on primary frequency response in interconnected smart grid. In *2016 24th Iranian Conference on Electrical Engineering (ICEE),* (pp. 561-566). IEEE. 10.1109/IranianCEE.2016.7585585

Alhelou, H. H., & Golshan, M. E. H. (2020). Decision-making-based optimal generation-side secondary-reserve scheduling and optimal LFC in deregulated interconnected power system. In *Decision making applications in modern power systems* (pp. 269–299). Academic Press. doi:10.1016/B978-0-12-816445-7.00011-6

Alhelou, H. H., Golshan, M. E. H., & Hatziargyriou, N. D. (2019). Deterministic dynamic state estimation-based optimal LFC for interconnected power systems using unknown input observer. *IEEE Transactions on Smart Grid*, 1. doi:10.1109/TSG.2019.2940199

Alhelou, H. H., Golshan, M. E. H., & Hatziargyriou, N. D. (2019). A decentralized functional observer based optimal LFC considering unknown inputs, uncertainties and cyber-attacks. *IEEE Transactions on Power Systems.*

Alhelou, H. H., Golshan, M. E. H., Zamani, R., Moghaddam, M. P., Njenda, T. C., Siano, P., & Marzband, M. (2019, June). An improved UFLS scheme based on estimated minimum frequency and power deficit. In Proceedings 2019 IEEE Milan PowerTech (pp. 1-6). IEEE. doi:10.1109/PTC.2019.8810497

Alhelou, H. H., Golshan, M. H., & Askari-Marnani, J. (2018). Robust sensor fault detection and isolation scheme for interconnected smart power systems in presence of RER and EVs using unknown input observer. *International Journal of Electrical Power & Energy Systems*, *99*, 682–694. doi:10.1016/j.ijepes.2018.02.013

Alhelou, H. H., Hamedani-Golshan, M. E., Heydarian-Forushani, E., Al-Sumaiti, A. S., & Siano, P. (2018, September). Decentralized fractional order control scheme for LFC of deregulated nonlinear power systems in presence of EVs and RER. In *2018 International Conference on Smart Energy Systems and Technologies (SEST)* (pp. 1-6). IEEE. 10.1109/SEST.2018.8495858

Alhelou, H. S. H., Golshan, M. E. H., & Fini, M. H. (2015, December). Multi agent electric vehicle control based primary frequency support for future smart micro-grid. In *Proceedings 2015 Smart Grid Conference (SGC)* (pp. 22-27).

Alkuhayli, A. A. Reliability evaluation of distribution system contained renewable distributed generations. In *Proceedings IEEE North American Power Symposium (NAPS)*, Champaign, IL. IEEE. pp. 1–6.

Anteneh, D., & Khan, B. (2019). Reliability enhancement of distribution substation by using network reconfiguration a case study at Debre Berhan Distribution Substation, *International Journal of Economy, Energy, & Environment*, *4*(2), 33.

Arifujjaman, M. D. (2015). A comprehensive power loss, efficiency, reliability and cost calculation of a 1 MW/500 kWh battery-based energy storage system for frequency regulation application. *Renewable Energy*, *74*, 158–169. doi:10.1016/j.renene.2014.07.046

Banteywalu, S., Khan, B., De Smedt, V., & Leroux, P. (2019). A novel modular radiation hardening approach applied to a synchronous buck converter. *Electronics (Basel)*, *8*(5), 513. doi:10.3390/electronics8050513

Bhandari, B., Lee, K. T., Lee, C. S., Song, C., Maskey, R. K., & Ahn, S. (2014). A novel off-grid hybrid power system comprised of solar photovoltaic, wind, and hydro energy sources. *Applied Energy*, *133*, 236–242. doi:10.1016/j.apenergy.2014.07.033

Billinton, R., & Allan, R. N. (1996). *Reliability evaluation of power systems* (2nd ed.). New York: Plenum Press. doi:10.1007/978-1-4899-1860-4

Billinton, R., & Gao, Y. (2006). Multistate wind energy conversion system models for adequacy assessment of generating systems incorporating wind energy. *IEEE Transactions on Energy Conversion*, *23*(1), 163–170. doi:10.1109/TEC.2006.882415

Billinton, R., & Gao, Y. (2008). Multistate wind energy conversion system models for adequacy assessment of generating systems incorporating wind energy. *IEEE Transactions on Energy Conversion*, *23*(1), 163–170. doi:10.1109/TEC.2006.882415

Borges, C. L. T. (2012). An overview of reliability models and methods for distribution systems with renewable energy distributed generation. *Renewable & Sustainable Energy Reviews*, *16*(6), 4008–4015. doi:10.1016/j.rser.2012.03.055

Čepin, M. (2011). Assessment of power system reliability, Springer-Verlag London, 2011.

Chowdhury, A. A., Agarwal, S. K., & Koval, D. O. (2013). Reliability modelling of distributed generation in conventional distribution systems planning and analysis. *IEEE Transactions on Industry Applications*, *39*(5), 1493–1501. doi:10.1109/TIA.2003.816554

Ding, Y., Wang, P., Goel, L., Loh, P. C., & Wu, Q. (2011). Long-term reserve expansion of power systems with high wind power penetration using universal generating function methods. *IEEE Transactions on Power Systems*, *26*(2), 766–774. doi:10.1109/TPWRS.2010.2054841

Duan, D. L., Ling, X. D., Wu, X. Y., & Zhong, B. (2015). Reconfiguration of distribution network for loss reduction and reliability improvement based on an enhanced genetic algorithm. *International Journal of Electrical Power & Energy Systems*, *64*, 88–95. doi:10.1016/j.ijepes.2014.07.036

Esau, Z., & Jayaweera, D. (2014). Reliability assessment in active distribution networks with detailed effects of PV systems. *J Mod Power Syst Clean Energy*, *2*(1), 59–68. doi:10.100740565-014-0046-2

Fini, M. H., Yousefi, G. R., & Alhelou, H. H. (2016). Comparative study on the performance of many-objective and single-objective optimisation algorithms in tuning load frequency controllers of multi-area power systems. *IET Generation, Transmission, & Distribution*, *10*(12), 2915–2923. doi:10.1049/iet-gtd.2015.1334

Gao, Y., & Billinton, R. (2009). Adequacy assessment of generating systems containing wind power considering wind speed correlation. *IET Renewable Power Generation*, *3*(2), 217–226. doi:10.1049/iet-rpg:20080036

Ghajar, R. F., & Billinton, R. (2006). Economic costs of power interruptions: A consistent model and methodology. *International Journal of Electrical Power & Energy Systems*, *28*(1), 29–35. doi:10.1016/j.ijepes.2005.09.003

Granado, P. C. D., Pang, Z., & Wallace, S. W. (2016). Synergy of smart grids and hybrid distributed generation on the value of energy storage. *Applied Energy*, *170*, 476–488. doi:10.1016/j.apenergy.2016.01.095

Gupta, K., Khan, B., & Mubeen, S. E. (2015). Available transfer capability enhancement by unified power flowcontroller. In *Proceedings IEEE International Conference on Signal Processing, Informatics, Communication, and Energy Systems (SPICES),* pp. 1, 4, 19-21.

Haes Alhelou, H., Hamedani-Golshan, M. E., Njenda, T. C., & Siano, P. (2019). A survey on power system blackout and cascading events: Research motivations and challenges. *Energies*, *12*(4), 682. doi:10.3390/en12040682

Haes Alhelou, H., Hamedani-Golshan, M. E., Njenda, T. C., & Siano, P. (2019). Wide-area measurement system-based optimal multi-stage under-frequency load-shedding in interconnected smart power systems using evolutionary computing techniques. *Applied Sciences*, *9*(3), 508. doi:10.3390/app9030508

Haes Alhelou, H., Hamedani Golshan, M. E., Njenda, T. C., & Siano, P. (2019). WAMS-based online disturbance estimation in interconnected power systems using disturbance observer. *Applied Sciences, 9*(5), 990. doi:10.3390/app9050990

Jariso, M., Khan, B., Tanwar, S., Tyagi, S., & Rishiwal, V. (2018). Hybrid energy system for upgrading the rural environment, *IEEE GlobeCom 2018.*

Jariso, M., Khan, B., Tesfaye, D., & Singh, J. (2017). Modelling and designing of stand-alone photovoltaic system (Case study: Addis Boder Health Center southwest Ethiopia. In *Proceedings IEEE International Conference on Electronics, Communication, and Aerospace Technology, ICECA 2017,* pp. 168-173, Coimbatore, India.

Khan, B., & Agnihotri, G. (2012). A novel transmission loss allocation method based on transmission usage. In *Proceedings IEEE Fifth Power India Conference*, Haryana, India. 10.1109/PowerI.2012.6479479

Khan, B., Agnihotri, G, & Gupta, G. (2013). A multipurpose matrices methodology for transmission usage, loss and reliability margin allocation in restructured environment, *Electrical & Computer Engineering: An International Journal, 2*(3).

Khan, B., Agnihotri, G., Rathore, P., Mishra, A., & Naidu, G. (2014). A cooperative game theory approach for usage and reliability margin cost allocation under contingent restructured market. *International Review of Electrical Engineering*, *9*(4), 854–862.

Khan, B., Alhelou, H. H., & Mebrahtu, F. (2019). A holistic analysis of distribution system reliability assessment methods with conventional and renewable energy sources. *AIMS Energy*, *7*(4), 413–429. doi:10.3934/energy.2019.4.413

Khan, B., & Singh, P. (2017). The current and future states of Ethiopia's energy sector and potential for green energy: A comprehensive study. *International Journal of Engineering Research in Africa, 33*, 115–139. doi:10.4028/www.scientific.net/JERA.33.115

Khan, B., & Singh, P. (2017). Optimal power flow techniques under characterization of conventional and renewable energy sources: a comprehensive analysis, *Journal of Engineering, vol. 2017,* Article ID 9539506.

Khan, B., & Singh, P. (2017). Selecting a meta-heuristic technique for smart micro-grid optimization problem: a comprehensive analysis. In IEEE Access, 5, pp. 13951-13977, 2017.

Khan, B., & Singh, P. (2019). Economic operation of smart micro-grid: a meta-heuristic approach. In Handbook of research on smart power system operation and control (pp. 330-346). Hershey, PA: IGI Global. doi:10.4018/978-1-5225-8030-0.ch014

Khan, B., & Singh, P. (2019). Economic operation of smart micro-grid: a meta-heuristic approach. In Handbook of research on smart power system operation and control (pp. 330-346). Hershey, PA: IGI Global. doi:10.4018/978-1-5225-8030-0.ch014

Khan, B., & Tanwar, S. (2019). Issues associated with microgrid integration. In Handbook of research on smart power system operation and control (pp. 252-264). Hershey, PA: IGI Global. doi:10.4018/978-1-5225-8030-0.ch010

Kifle, Y., Khan, B., & Singh, J. (2018). Designing and modeling grid connected photovoltaic system: (Case study: EEU building at Hawassa city). *International Journal of Convergence Computing*, *3*(1), 20–34. doi:10.1504/IJCONVC.2018.091113

Kifle, Y., Khan, B., & Singh, P. (2018). Assessment and enhancement of distribution system reliability by renewable energy sources and energy storage, *Journal of Green Engineering, 8*(3), 2, pp. 219-262.

Li, Y., & Zio, E. (2012). A multi-state model for the reliability assessment of distributed generation system via universal generating function. *Reliability Engineering & System Safety*, *106*, 28–36. doi:10.1016/j.ress.2012.04.008

Li, Y., & Zio, E. (2012). A multi-state model for the reliability assessment of a distributed generation system via universal generating function. *Reliability Engineering & System Safety*, *106*, 28–36. doi:10.1016/j.ress.2012.04.008

Locatelli, G., Palerma, E., & Mancini, M. (2015). Assessing the economics of large energy storage plants with an optimisation methodology. *Energy*, *83*, 15–28. doi:10.1016/j.energy.2015.01.050

Makdisie, C., Haidar, B., & Alhelou, H. H. (2018). An optimal photovoltaic conversion system for future smart grids. In Handbook of research on power and energy system optimization (pp. 601–657). Hershey, PA: IGI Global. doi:10.4018/978-1-5225-3935-3.ch018

Malheiro, A., Catrol, P. M., Lima, R. M., & Estanqueiro, A. (2015). Integrated sizing and scheduling of wind/PV/diesel/battery systems. *Renewable Energy*, *83*, 646–657. doi:10.1016/j.renene.2015.04.066

Molla, T., Khan, B., Moges, B., Alhelou, H. H., Zamani, R., & Siano, P. (2019). Integrated optimization of smart home appliances with cost-effective energy management system. *CSEE Journal of Power and Energy Systems*, *5*(2), 249–258.

Molla, T., Khan, B., & Singh, P. (2018). A comprehensive analysis of smart home energy management system optimization techniques, *Journal of Autonomous Intelligence, 1, 1*, 15–21.

Negash, K., Khan, B., Tesfaye, D., & Gayathri, M. (2017). Optimal placement of phasor measurement unit for system observability. In *Proceedings of IEEE International Conference on Computing Methodologies and Communication, (ICCMC 2017),* pp. 73-77, Erode, India.

Negash, K., Khan, B., & Yohannes, E. (2016). Artificial intelligence versus conventional mathematical techniques: A review for optimal placement of phasor measurement units [Springer]. *Journal of Technology and Economics of Smart Grids and Sustainable Energy*, *1*(1), 10. doi:10.100740866-016-0009-y

Njenda, T. C., Golshan, M. E. H., & Alhelou, H. H. (2018, November). WAMS based intelligent under frequency load shedding considering online disturbance estimation. In *2018 Smart Grid Conference (SGC)* (pp. 1-5). 10.1109/SGC.2018.8777779

Njenda, T. C., Golshan, M. E. H., & Alhelou, H. H. (2018, November). WAMS based under frequency load shedding considering minimum frequency prediction and extrapolated disturbance magnitude. In *2018 Smart Grid Conference (SGC)* (pp. 1-5).

Ogunjuyigbe, A. S. O., Ayodele, T. R., & Akinola, O. A. (2016). Optimal allocation and sizing of PV/Wind/Split-diesel/Battery hybrid energy system for minimizing life cycle cost, carbon emission and dump energy of remote residential building. *Applied Energy*, *171*, 153–171. doi:10.1016/j.apenergy.2016.03.051

Saboori, H., Hemmati, R., & Jirdehi, M. A. (2015). Reliability improvement in radial electrical distribution network by optimal planning of energy storage systems. *Energy*, *93*, 2299–2312. doi:10.1016/j.energy.2015.10.125

Shah, R., Mithulananthan, N., Bansal, R. C., & Ramachandaramurthy, V. K. (2015). A review of key power system stability challenges for large-scale PV integration. *Renewable & Sustainable Energy Reviews*, *41*, 1423–1436. doi:10.1016/j.rser.2014.09.027

Singh, P., & Khan, B. (2017). Smart microgrid energy management using a novel artificial shark optimization, Complexity, vol. 2017, Article ID 2158926. doi:10.1155/2017/2158926

Singh, P., Khan, B., Vidyarthi, A., Haes Alhelou, H., & Siano, P. (2019). Energy-aware online non-clairvoyant scheduling using speed scaling with arbitrary power function. *Applied Sciences*, *9*(7), 1467. doi:10.3390/app9071467

Souza, J., Rider, M. J., & Mantovani, J. R. S. (2015). Planning of distribution systems using mixed-integer linear programming models considering network reliability. *J. Control Autom. Electr. Syst.*, *26*(2), 170–179. doi:10.100740313-014-0165-z

Sreedharan, P., Farbes, J., Cutter, E., Woo, C. K., & Wang, J. (2016). Micro grid and renewable generation integration: University of California, San Diego. *Applied Energy*, *169*, 709–720. doi:10.1016/j.apenergy.2016.02.053

Wang, P., & Billinton, R. (2002). Reliability cost/worth assessment of distribution systems incorporating time-varying weather conditions and restoration resources. *IEEE Transactions on Power Delivery*, *17*(1), 260–265. doi:10.1109/61.974216

Yeshalem, M. T., & Khan, B. (2017). Design of an off-grid hybrid PV/wind power system for remote mobile base station: A case study. *AIMS Energy*, *5*(1), 96–112. doi:10.3934/energy.2017.1.96

Zamani, R., Golshan, M. E. H., Alhelou, H. H., & Hatziargyriou, N. (2019). A novel hybrid islanding detection method using dynamic characteristics of synchronous generator and signal processing technique. *Electric Power Systems Research*, *175*, 105911. doi:10.1016/j.epsr.2019.105911

Zamani, R., Hamedani-Golshan, M. E., Haes Alhelou, H., Siano, P., & Pota, H. R. (2018). Islanding detection of synchronous distributed generator based on the active and reactive power control loops. *Energies*, *11*(10), 2819. doi:10.3390/en11102819

Zhang, L., & Sun, M. Research on the type of load of accessing to microgrid based on reliability. In *Proceedings 2nd International conference on electrical, computer engineering and electronics*, Jinan, China. pp. 1665–70. 10.2991/icecee-15.2015.314

Zhanga, X., Bieb, Z., & Lib, G. (2011). Reliability assessment of distribution networks with distributed generations using Monte Carlo method. *Energy Procedia*, *12*, 278–286. doi:10.1016/j.egypro.2011.10.038

Zhou, H., Bhattacharya, T., Tran, D., Siew, T. S. T., & Khambadkone, A. M. (2011). Composite energy storage system involving battery and ultracapacitor with dynamic energy management in microgrid applications. *IEEE Transactions on Power Electronics*, *26*(3), 923–930. doi:10.1109/TPEL.2010.2095040

KEY TERMS AND DEFINITIONS

Microgrid: It is a small network of electricity users with a local source of supply that is usually attached to a centralized national grid but is able to function independently.

Monte Carlo Simulation Technique: It is a versatile method, which is utilized by various researchers for computing reliability of traditional distribution network as well as hybrid distribution system.

Reliability: A measure of the degree of certainty that the relay, or relay system, will perform correctly.

Chapter 9
Reliability Enhancement of Smart Distribution Network Using Reconfiguration

Degarege Anteneh
https://orcid.org/0000-0003-1260-5114
Department of Electrical and Computer Engineering, Hawassa University, Hawassa, Ethiopia

ABSTRACT

Electric power should deliver a predicable per condition for the technological, economic, and political development of any countries and it is vital for each individual. Power outage is series problem in Ethiopia at the hole of distribution network. This is due to frequent interruptions and much time service restoration. That is why most customers of Ethiopia have their day-to-day activities highly affected and they are strongly complaining to Ethiopia electric utility daily. But this power outage affected customer cost and the Ethiopian utility. In most developing countries including Ethiopia, distribution systems have received considerably less of the attention to reliability modeling and evaluation than have generating and transmitting systems. Life is directly or indirectly dependent on electric power so a utility should deliver reliable power every day for 24 hours and each year for 8,760 hours to satisfy human needs and to perform their works as much as possible with less economy.

INTRODUCTION

Deliver electrical energy to their customers is the primary function of the electric power system. This electricity is supplied as economically as possible with an acceptable degree of quality. Reliability of power system is one of the features of power quality. Economy and reliability are the two competitive features because increased reliability of supply generally requires increased capital investment. These two constraints are balanced in many ways from different countries and by different electric power utilities, although generally they are all based on various sets of criteria. A wide range of related measures or indicators can be determined using probability theory (Alkuhayli et al., 2012). A single all-purpose formula or technique does not exist. The approaches and their respective mathematical expressions

DOI: 10.4018/978-1-7998-1230-2.ch009

depend on the defined problem and determined assumptions. There are different assumptions must be made in the practical applications of probability and statistical theory. Reconfiguration of the radial distribution network is utilized for the purpose of loss minimization, reliability enhancement, voltage profile improvement and load balancing. The validity of the analysis is directly related to the validity of the model used to represent the system (Alhelou et al., 2019; Makdisie et al., 2018; Alhelou et al., 2018; Alhelou et al., 2016; Haes Alhelou et al., 2019; Njenda et al., 2018). The actual failure of the distribution system rarely complete fit the analytical descriptions used in the analysis and must be taken to ensure that significant errors are not included through oversimplification of the problem. The most important aspect of good modelling and analysis is to have a complete understanding of the engineering implications of the system. No amount of probability theory can circumvent this important engineering aspect (Singh 2014; Khan et al., 2019; Khan et al., 2018; Khan et al., 2017; Banteywalu et al., 2019; Anteneh et al., 2019; Molla et al., 2019, Molla et al., 2018, Jariso et al. 2018). There are two main categories of evaluation techniques:

- Analytical
- Simulation

Analytical techniques represent the system by a mathematical model and evaluate the measures or indicators from this model using mathematical solutions.

Simulation techniques estimate the measures or indicators by simulating the actual process and random behavior of the system. Probabilistic simulation techniques are a subset of simulation techniques that treat the problem as a series of real experiments. The input parameters of each simulated experiments are obtained using Monte Carlo selection of their values. Both categories of methods have advantages and disadvantages (Khan et al. 2014; Khan et al. 2013; Khan et al. 2012; Negash et al., 2017; Negash et al., 2016; Jariso et al., 2017; Kifle et al., 2018; Yeshalem et al., 2017; Singh et al., 2017; Gupta et al., 2015). The Monte Carlo simulation requires a large amount of computing time and is not used extensively if alternative analytical methods are available. But, if the analytical methods are too complex, the probabilistic simulations can give a good approximation of results.

Reliability means the probability of a system or component that performs their assigned task for a given period of time under the specified operating conditions stumbled upon during its anticipated life time. To achieve an acceptable level of reliability, quality and safety at an economic price, the utility have to create and improve the systems reliability continuously depending upon the requirement of the customers. Reliability assessment methods allow the evaluation of the reliability of systems.

The methods provide important information on how to improve a systems life to reduce safety risk and hazards from these main components of electrical power distribution system is a largest part of network in electrical power system. It can be defined as the part of power system that distributes power to various customers in ready-to-use form at their place of consumption. So, utilities have to ensure reliable and efficient cost-effective service, while providing service voltages and power quality within the specified range (Alhelou et al., 2019; Makdisie et al., 2018; Alhelou et al., 2018; Alhelou et al., 2016; Haes Alhelou et al., 2019; Njenda et al., 2018). The performance evaluation of the distribution sector is important in order to assess the impact of reform measures. Usually engineers try to achieve the required reliability level with minimal cost.

METHODOLOGY

The objective functions are to minimize SAIFI and SAIDI and minimize total cost (including customer interruption cost). The topology is achieved switches and protective devices can be installed in some of the feeders to allow for system reconfiguration.

Different techniques have been developed to determine the best topology for a distribution system. These include heuristic methods, linear programming, neural networks, expert systems, fuzzy logic, simulated annealing and genetic algorithms. The heuristic technique is the most commonly used due to produce fast results with good accuracy. Place and number of switches and protective devices are chosen by considering number of customers, feeder length, sensitivity of the area and economic benefits. The reliability indices SAIFI, SAIDI and cost benefits would be the main drivers for comparison of the alternatives using heuristic technique.

LITERATURE REVIEW

Electric power utilities throughout the world endeavor to meet customer demands as economically as possible at a reasonable service of reliability. In Nagy et.al (2013) to meet the customer demands, the power utility has to evolve and the distribution system have to be upgraded, operated and maintained accordingly. The economic and social effects of the loss of electric service have significant impacts on both the utility and the end user.

Reliability assessment and enhancement of electric power distribution network has been studied by different researchers using different approaches. Some of the methodologies implemented by various researches are briefly summarized below:

Mendonça and Queiroz 2012 performed a research on assessing the overall performance of Brazilian electric distribution companies. the concept of performance is focused on the service provided by companies. Traditional dimensions of quality of supply were considered. Alemu, (2012), assess the reliability assessment of Bahir Dar city distribution system on 15 kV selected four feeders. Monte Carlo simulation systems and ETAP software is used to assess and simulate the overall behavior of the distribution system reliability indices. The researcher also recommends that the improvement techniques such as reconnection of the scheme and replacing old components are economical alternatives for further improvement using additional recloser, sectionalize-switches and load break switches in cost effective locations. Levitin et at., (1995) evaluated average annual unsupplied energy and system average interruption duration index (SAIDI) for a radial distribution network with arbitrary allocation of sectionalizes and tie-lines based on the genetic algorithm. Singh (2014) utilized binary programming and genetic algorithm to identify type and location of the protective devices on a distribution feeder based on the objective of minimizing the system average interruption frequency index (SAIFI). Brown et.al (1997) utilized several intelligent algorithms to solve the problem of optimal placement of switches and protective devices in distribution networks. Teng and Lu (2008) proposed a novel algorithm for feeder automation (FA) planning. The algorithm can be used to find optimal number and locations of automated feeders and tie switches to minimize the interruption costs (CIC). Merlin and Back (1975) proposed a branch and bound type heuristic method to determine the network configuration to enhance distribution system reliability and for minimum line losses. Dorji (2009) proposed an evaluation technique for distribution reliability that uses the improved failure rate of all protective devices in a network, depending on the location of the superconducting fault current limiter. Ngaopitakkul et al. (2013) focused on the analysis of reliability with the inclusion of distributed generation (DG). C.H. Lin (2006) focused on the reliability enhancement of power distribution systems by using an algorithm to realize optimal relay setting for several test systems. Both overcurrent relays and distance relays and their coordination were considered in the optimization problem. Xuebei (2011) introduced the concept of a fluctuation environment to describe the failure rate of the distribution system components. Saumil (2007) applied both the analytical and simulation methods to evaluate and assess reliability of distribution system and risk assessment using statistical history and reports of utilities.

POWER QUALITY AND RELIABILITY

Power quality is an ambiguous term that has no constant or the same meaning for all customers and power utilities. For a customer, a power quality problem might be defined as any electric power delivering condition that causes appliances to prevent their use. Utility defined power quality problem might be perceived as noncompliance with different standards like RMS voltage or harmonics (Fini et al., 2016; Alhelou et al., 2018; Zamani et al., 2018; Alhelou et al., 2015; Njenda et al., 2018; Haes Alhelou et al., 2018; Haes Alhelou et al., 2019). Generally, power is equal to product of current and voltage at instantaneous, and it is difficult to formulate a meaningful definition for a power quality. The system, which has a good power quality indirectly tell as the system also have better voltage quality, so a distribution system has good power quality it also good in voltage quality. Utilities have no control over the current drawn by the end users and should be generally unconcerned with current waveforms. Load current can affect voltage as it interacts with system impedances, but voltage is the ultimate measure of the power quality. Both reliability and power quality are the two main and common issues of power system and also related to each other. Due to this, engineers are done different researches on the distribution system reliability and power quality enhancement. Generally, customer interruptions are considered as power quality concern since it reduced voltage magnitude to zero. Reliability is firstly concerned with the customer interruptions and a subset of power quality. Although there is general agreement that power quality includes reliability, the boundary that separates the two is not well defined. Sustained interruptions (more than a few minutes) have always been categorized as a reliability issue, but many utilities have classified momentary interruptions (less than a few minutes) as a power quality issue. The reasons are

- Momentary interruptions are the result of intentional operating practices
- Not generate a large number of customer complaints
- Difficult to measure

Reliability Evaluation

Different types of reliability indices can be utilized to evaluate of reliability to all electrical sectors i.e. generation, transmission and distribution. The basic reliability indices are as follows:

System Average Interruption Frequency Index (SAIFI)

$$SAIFI = \frac{Total\ number\ of\ interruption}{Total\ number\ of\ customerserved} \tag{1}$$

System Average Interruption Duration Index (SAIDI)

$$SAIDI = \frac{Total\ duration\ of\ all\ interrupions\ in\ customers}{total\ nomber\ of\ customer\ served} \tag{2}$$

$$SAIDI = \frac{\sum ri * Ni}{Nt}$$

Customer Average Interruption Duration Index (CAIDI)

$$CAIDI = \frac{\sum customer\ interruption\ duration}{total\ number\ of\ customer\ interruption} \tag{3}$$

$$CAIDI = \frac{SAIDI}{SAIFI}$$

Average Service Availability Index (ASAI)

$$ASAI = \frac{customer\ hours\ service\ avalilabale}{customer\ hours\ service\ demand} \tag{4}$$

$$ASAI = \frac{8760\, hours / yr - CAIDI}{8760\, hours / yr}$$

Customer Average Interruption Index (CAIFI)

$$CAIFI = \frac{Total\ number\ of\ customer\ interruption}{Number\ of\ customer\ affectted} \tag{5}$$

Average Service Unavailability Index (ASUI)

$$ASUI = 1 - ASAI \tag{6}$$

$$ASUI = \frac{SAIDI}{8760\ hours\ /\ yr}$$

Excepted Energy not Supplied Index (EENS)

$$EENS = \sum Li * Ui \tag{7}$$

Where, Li is the average connected load at load point i and Ui is average annual outage time at load point i.

Average Energy not Supplied Index (AENS)

$$AENS = \frac{Total\ energy\ not\ supplied}{Total\ number\ of\ customer\ served} \tag{8}$$

$$AENS = \frac{\sum \mathrm{Li}^*\mathrm{Ui}}{\sum \mathrm{Ni}}$$

The first five indices equations are customer-oriented indices and the last two equations are load and energy-oriented indices. These indices not only provide the information to assess the past performance of a distribution system but also predict the future system performance.

Figure 1. Optimization methodologies of distribution network reconfiguration

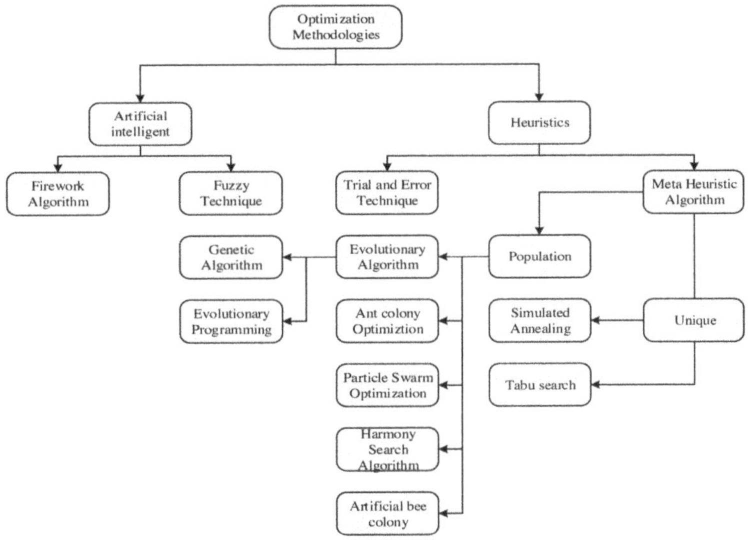

Techniques used to Improve Reliability by Network Reconfiguration

Various methods are utilized for the reconfiguration of the distribution system integrated with DGs and capacitor banks. . Reported works can be categorized into sequential and simultaneous techniques. For the former, the optimal size of the DGs needs to be determined prior to network configuration, while in the latter, optimal sizing of the DGs, capacitor and network reconfiguration are executed simultaneously (Anteneh and Khan, 2019):

Heuristic Optimization Methods

Both Heuristic and Meta-heuristic techniques are the randomized searching procedures. It is described that swarm optimization techniques and evolutionary techniques are the parts of it. So many ways have been proposed to solve the reconfiguration problem in distribution system. Baradaran and Seyed (2012) minimized the total cost of reliability of distribution network by optimal placement of fault indicators. Gustavo (2015) considered reliability as all aspects of customer interruptions, including momentary interruptions. A methodology for multi-objective optimization is proposed to minimize SAIFI, SAIDI indices and cost reduction simultaneously. Hamed et al. (2014) proposed a novel method to evaluate the proper allocation of protective devices, which is based on the sensitivity analysis. Nematollah and Rahman (2011) proposed an algorithm that is based on network portioning into groups of load buses. The line section losses between the groups of nodes are minimized. The proposed method overcomes the size restriction imposed by other reconfiguration techniques. Moshtagh and Ghasemi (2013) presented an effective application of practical distribution system automation by using genetic algorithm optimization. One location may satisfy an objective function subject to a specific reliability index. However, adding reliability index as a constraint can result in different optimum locations. Hesam (2008) presented a new algorithm for solving the distribution network reconfiguration outages. The algorithm is a simple modification from binary particle swarm optimization (BPSO) called selective particle swarm optimization

(SPSO). Selve et al. (2014) aimed to maximize the reliability of power supplied to the customers with network reconfiguration by considering reliability indexes. It selected Binary particle swarm optimization (BPSO) to determine optimal configuration of switches in the distribution network in order to maximize reliability indexes at load points.

Artificial Intelligence Methods

Now a day's various artificial intelligence-based methods such as artificial neural network (ANN), fuzzy logic, machine learning, and deep learning are utilized for the reconfiguration of the distribution network. Juhana and Mohamed (2013) discussed the application of a fuzzy logic based coordinated voltage control for the distribution network, connected with distributed generation (DG). In practical case the connection of distributed generation has created a challenge in the distribution system operators to make the voltage in the system within it acceptable levels. Xuebei (2011) proposed a new technique of evolutionary search methodology for analyzing a minimum power loss of radial configuration of distribution system, to enhance its performance of evolutionary programming by using fuzzy controlled evolutionary programming (FCEP). Jaydeep (2015) developed a method for determining the restoration plan for radial distribution system. The reliable plan among the available alternatives is determined by reliability analysis using fuzzy reasoning technique. Shahsavari et.al (2017) assessed the reliability of a distribution system using fuzzy logic for modeling of transformer and line uncertainties.

CONCLUSION

Electrical power network should provide both adequate and reliable electrical supply to its customers as much as possible with low cost. In all the distribution systems, reliability is one of the major issues. The reliability is the probability of a system or components to perform their assigned task for a given period of time under the specified operating conditions. This chapter discussed the reliability assessment and enhancement techniques by utilized reconfiguration of the distribution network. Two different types of approaches such as artificial intelligence based and heuristic based are discussed to reconfigure the distribution network.

REFERENCES

Dezaki, H. H., Abyaneh, H. A., Garmrudi, M., Mahdinia, H., & Mazlumi, K. (2014). A novel approach based on reliability sensitivity analysis to allocate protective devices. *Turkish Journal of Electrical Engineering & Computer Sciences*, *22*, 315–326. doi:10.3906/elk-1202-86

Alemu, A. (2012). Reliability assessment of Bahir Dar town distribution system, (MSc. Thesis), Bahir Dar University, 2012.

Alhelou, H., Hamedani-Golshan, M. E., Zamani, R., Heydarian-Forushani, E., & Siano, P. (2018). Challenges and opportunities of load frequency control in conventional, modern and future smart power systems: A comprehensive review. *Energies*, *11*(10), 2497. doi:10.3390/en11102497

Alhelou, H. H. (2018). *Fault detection and isolation in power systems using unknown input observer. Advanced condition monitoring and fault diagnosis of electric machines* (p. 38). Hershey, PA: IGI Global.

Alhelou, H. H., Golshan, M., & Fini, M. (2018). Wind driven optimization algorithm application to load frequency control in interconnected power systems considering GRC and GDB nonlinearities. *Electric Power Components and Systems, 46*(11-12), 1223-1238.

Alhelou, H. H., & Golshan, M. E. H. (2016, May). Hierarchical plug-in EV control based on primary frequency response in interconnected smart grid. In *2016 24th Iranian Conference on Electrical Engineering (ICEE),* (pp. 561-566). IEEE. 10.1109/IranianCEE.2016.7585585

Alhelou, H. H., & Golshan, M. E. H. (2020). Decision-making-based optimal generation-side secondary-reserve scheduling and optimal LFC in deregulated interconnected power system. In *Decision making applications in modern power systems* (pp. 269–299). Academic Press. doi:10.1016/B978-0-12-816445-7.00011-6

Alhelou, H. H., Golshan, M. E. H., & Hatziargyriou, N. D. (2019). Deterministic dynamic state estimation-based optimal LFC for interconnected power systems using unknown input observer. *IEEE Transactions on Smart Grid,* 1. doi:10.1109/TSG.2019.2940199

Alhelou, H. H., Golshan, M. E. H., & Hatziargyriou, N. D. (2019). A decentralized functional observer based optimal LFC considering unknown inputs, uncertainties, and cyber-attacks. *IEEE Transactions on Power Systems.*

Alhelou, H. H., Golshan, M. E. H., Zamani, R., Moghaddam, M. P., Njenda, T. C., Siano, P., & Marzband, M. (2019, June). An improved UFLS scheme based on estimated minimum frequency and power deficit. In 2019 IEEE Milan PowerTech (pp. 1-6). IEEE. doi:10.1109/PTC.2019.8810497

Alhelou, H. H., Golshan, M. H., & Askari-Marnani, J. (2018). Robust sensor fault detection and isolation scheme for interconnected smart power systems in presence of RER and EVs using unknown input observer. *International Journal of Electrical Power & Energy Systems, 99*, 682–694. doi:10.1016/j.ijepes.2018.02.013

Alhelou, H. H., Hamedani-Golshan, M. E., Heydarian-Forushani, E., Al-Sumaiti, A. S., & Siano, P. (2018, September). Decentralized fractional order control scheme for LFC of deregulated nonlinear power systems in presence of EVs and RER. In *2018 International Conference on Smart Energy Systems and Technologies (SEST)* (pp. 1-6). IEEE. 10.1109/SEST.2018.8495858

Alhelou, H. S. H., Golshan, M. E. H., & Fini, M. H. (2015, December). Multi agent electric vehicle control based primary frequency support for future smart micro-grid. In *2015 Smart Grid Conference (SGC)* (pp. 22-27).

Alkuhayli, A. A., Raghavan, S., & Chowdhury, B. H. (2012). Reliability evaluation of distribution systems containing renewable distributed generations, in *North American Power Symposium (NAPS),* 2012, pp. 1-6. 10.1109/NAPS.2012.6336324

Anteneh, D., & Khan, B. (2019). Reliability enhancement of distribution substation by using network reconfiguration a case study at Debre Berhan Distribution Substation. *International Journal of Economy, Energy, & Environment, 4*(2), 33.

Anteneh, D., & Khan, B. (2019). Reliability enhancement of distribution substation by using network reconfiguration a case study at Debre Berhan Distribution Substation. *International Journal of Economy, Energy, & Environment*, *4*(2), 33–40.

Banteywalu, S., Khan, B., De Smedt, V., & Leroux, P. (2019). A novel modular radiation hardening approach applied to a synchronous buck converter. *Electronics (Basel)*, *8*(5), 513. doi:10.3390/electronics8050513

Baradaran, R. H., & Seyed-Sheneva, J. (2012, March-April). Electrical distribution system reliability improvement by optimal placement of fault indicators using immune algorithm. *International Journal of Engineering Research and Applications*, *2*(2), 1383–1390.

Brown, R., Gupta, S., Christie, R., Venkata, S., & Fletcher, R. (1997). Automated primary distribution system design: Reliability and cost optimization. *IEEE Transactions on Power Delivery*, *12*(2), 1017–1022. doi:10.1109/61.584431

Chakravorty, J. (2015). Network reconfiguration of distribution system using fuzzy controlled evolutionary programming, *International Journal Of Engineering Science & Advanced Technology, 2*(2), 176–182, 2015.

de Queiroz, L. M. O., & Forrer, J. (2012). Assessing the overall performance of Brazilian electric distribution companies. George Washington University, Washington, DC.

Dorji, T. (2009). Reliability assessment of distribution systems a case study on Wang due distribution system in Bhutan, (MSc. Thesis), NTNU, Norway.

Ferreira, G. D., Bretas, A. S., & Cardoso, G. (2010, September). Optimal distribution protection design considering momentary and sustained reliability indices. In *2010 Modern Electric Power Systems* (pp. 1-8). IEEE.

Fini, M. H., Yousefi, G. R., & Alhelou, H. H. (2016). Comparative study on the performance of many-objective and single-objective optimisation algorithms in tuning load frequency controllers of multi-area power systems. *IET Generation, Transmission, & Distribution*, *10*(12), 2915–2923. doi:10.1049/iet-gtd.2015.1334

Hashim, T. J., & Mohamed, A. (2013). Fuzzy logic based coordinated voltage control for distribution network with distributed generations. *World Academy of Science, Engineering, and Technology*, 79, 24-29.

Gupta, K., Khan, B., & Mubeen, S. E. (2015). Available transfer capability enhancement by unified power flowcontroller, *IEEE International Conference on Signal Processing, Informatics, Communication and Energy Systems (SPICES),* pp. 1, 4, 19-21.

Haes Alhelou, H., Hamedani-Golshan, M. E., Njenda, T. C., & Siano, P. (2019). A survey on power system blackout and cascading events: Research motivations and challenges. *Energies*, *12*(4), 682. doi:10.3390/en12040682

Haes Alhelou, H., Hamedani-Golshan, M. E., Njenda, T. C., & Siano, P. (2019). Wide-area measurement system-based optimal multi-stage under-frequency load-shedding in interconnected smart power systems using evolutionary computing techniques. *Applied Sciences*, *9*(3), 508. doi:10.3390/app9030508

Haes Alhelou, H., Hamedani Golshan, M. E., Njenda, T. C., & Siano, P. (2019). WAMS-based online disturbance estimation in interconnected power systems using disturbance observer. *Applied Sciences*, *9*(5), 990. doi:10.3390/app9050990

Hesam, H. (2008). Distribution system protection es586b: Power system protection, University of Western Ontario.

Jariso, M., Khan, B., Tanwar, S., Tyagi, S., & Rishiwal, V. (2018). Hybrid energy system for upgrading the rural environment, *IEEE GlobeCom 2018.*

Jariso, M., Khan, B., Tesfaye, D., & Singh, J. (2017). Modelling and designing of stand-alone photovoltaic system (Case study: Addis Boder Health Center southwest Ethiopia), *IEEE International Conference on Electronics, Communication, and Aerospace Technology, ICECA 2017,* pp. 168-173, Coimbatore, India.

Khan, B., & Agnihotri, G. (2012). A novel transmission loss allocation method based on transmission usage. In *Proceedings IEEE Fifth Power India Conference*, Haryana, India. 10.1109/PowerI.2012.6479479

Khan, B., Agnihotri, G., & Gupta, G. (2013). A multipurpose matrices methodology for transmission usage, loss, and reliability margin allocation in restructured environment, *Electrical & Computer Engineering: An International Journal, 2*(3).

Khan, B., Agnihotri, G., Rathore, P., Mishra, A., & Naidu, G. (2014). A cooperative game theory approach for usage and reliability margin cost allocation under contingent restructured market. *International Review of Electrical Engineering*, *9*(4), 854–862.

Khan, B., Alhelou, H. H., & Mebrahtu, F. (2019). A holistic analysis of distribution system reliability assessment methods with conventional and renewable energy sources. *AIMS Energy*, *7*(4), 413–429. doi:10.3934/energy.2019.4.413

Khan, B., & Singh, P. (2017). The current and future states of ethiopia's energy sector and potential for green energy: A comprehensive study. *International Journal of Engineering Research in Africa*, *33*, 115–139. doi:10.4028/www.scientific.net/JERA.33.115

Khan, B., & Singh, P. (2017). Optimal power flow techniques under characterization of conventional and renewable energy sources: A comprehensive analysis, *Journal of Engineering, vol. 2017,* Article ID 9539506.

Khan, B., & Singh, P. (2017). Selecting a meta-heuristic technique for smart micro-grid optimization problem: A comprehensive analysis, In IEEE Access, 5, pp. 13951-13977.

Khan, B., & Singh, P. (2019). Economic operation of smart micro-grid: a meta-heuristic approach. In Handbook of research on smart power system operation and control (pp. 330-346). Hershey, PA: IGI Global. doi:10.4018/978-1-5225-8030-0.ch014

Khan, B., & Singh, P. (2019). Economic operation of smart micro-grid: a meta-heuristic approach. In Handbook of research on smart power system operation and control (pp. 330-346). Hershey, PA: IGI Global. doi:10.4018/978-1-5225-8030-0.ch014

Khan, B., & Tanwar, S. (2019). Issues associated with microgrid integration. In Handbook of research on smart power system operation and control (pp. 252-264). Hershey, PA: IGI Global. doi:10.4018/978-1-5225-8030-0.ch010

Kifle, Y., Khan, B., & Singh, J. (2018). Designing and modeling grid connected photovoltaic system: (Case study: EEU building at Hawassa city). *International Journal of Convergence Computing*, *3*(1), 20–34. doi:10.1504/IJCONVC.2018.091113

Kifle, Y., Khan, B., & Singh, P. (2018). Assessment and enhancement of distribution system reliability by renewable energy sources and energy storage, *Journal of Green Engineering, 8*(3), 2, pp. 219-262.

Levitin, G., Mazal-Tov, S., & Elmakis, D. (1995). Genetic algorithm for optimal sectionalizing in radial distribution systems with alternative supply. *Electric Power Systems Research*, 35(3), 149-155.

Makdisie, C., Haidar, B., & Alhelou, H. H. (2018). An optimal photovoltaic conversion system for future smart grids. In Handbook of research on power and energy system optimization (pp. 601–657). Hershey, PA: IGI Global. doi:10.4018/978-1-5225-3935-3.ch018

Merlin, A., & Back, H. (1975). Search for a minimal loss operating spanning tree configuration for an urban power distribution system. In *Proceedings of the 5th Power System Computation Conference (PSOC)*, Cambridge, pp. 1-18.

Molla, T., Khan, B., Moges, B., Alhelou, H. H., Zamani, R., & Siano, P. (2019). Integrated optimization of smart home appliances with cost-effective energy management system. *CSEE Journal of Power and Energy Systems*, *5*(2), 249–258.

Molla, T., Khan, B., & Singh, P. (2018). A comprehensive analysis of smart home energy management system optimization techniques, *Journal of Autonomous Intelligence, 1*(1), 15–21.

Moshtagh, J., & Ghasemi, S. (2013). Power restoration in distribution network using MST algorithms, *Journal of Operation and Automation in Power Engineering, 1*(1).

Nagy, S. A., Ibrahim, I. S., Ahmed, M. K., Adail, A. S., & Soliman, S. (2013). Network reconfiguration for loss reduction in electrical distribution system using genetic algorithm. *Arab Journal of Nuclear Science and Applications*, *46*(1), 78–87.

Negash, K., Khan, B., Tesfaye, D., & Gayathri, M. (2017). Optimal placement of phasor measurement unit for system observability. In *Proceedings of IEEE International Conference on Computing Methodologies and Communication, (ICCMC 2017)*, pp. 73-77, Erode, India.

Negash, K., Khan, B., & Yohannes, E. (2016). Artificial intelligence versus conventional mathematical techniques: A review for optimal placement of phasor measurement units [Springer]. *Journal of Technology and Economics of Smart Grids and Sustainable Energy*, *1*(1), 10. doi:10.100740866-016-0009-y

Nematollah, D., & Rahman, D. (2011). Optimization of recloser placement to improve reliability by genetic algorithm. *Energy and Power Engineering*, *3*(04), 508–512. doi:10.4236/epe.2011.34061

Ngaopitakkul, A., Pothisarn, C., Bunjongjit, S., Suechoey, B., Thammart, C., & Nawikavatan, A. (2013). A reliability impact and assessment of distributed generation integration to distribution system. *Energy and Power Engineering*, *5*(04), 1043–1047. doi:10.4236/epe.2013.54B199

Njenda, T. C., Golshan, M. E. H., & Alhelou, H. H. (2018, November). WAMS based intelligent under frequency load shedding considering online disturbance estimation. In *2018 Smart Grid Conference (SGC)* (pp. 1-5). 10.1109/SGC.2018.8777779

Njenda, T. C., Golshan, M. E. H., & Alhelou, H. H. (2018, November). WAMS based under frequency load shedding considering minimum frequency prediction and extrapolated disturbance magnitude. In *2018 Smart Grid Conference (SGC)* (pp. 1-5).

Saumil, A. P. (2007). *Fuzzy logic based power load balancing using fuzzy logic,* (M. E. thesis), Gujarat University, India.

Selve, V., Baskar, G., & Pradheeba, R. (2014, March). Distribution feeder reconfiguration for reliability evaluation by using binary particle swarm optimization. *International Journal of Innovative Research in Science, Engineering, and Technology*, *3*(Special Issue), 3.

Shahsavari, A., Sadeghi-Mobarakeh, A., Stewart, E. M., Cortez, E., Alvarez, L., Megala, F., & Mohsenian-Rad, H. (2017). Distribution grid reliability versus regulation market efficiency: An analysis based on micro-PMU data. IEEE Transactions on Smart Grid, 8(6), 2916-2925.

Singh, P., & Khan, B. (2017) Smart microgrid energy management using a novel artificial shark optimization, Complexity, vol. 2017, Article ID 2158926. doi:10.1155/2017/2158926

Singh, P., Khan, B., Vidyarthi, A., Haes Alhelou, H., & Siano, P. (2019). Energy-aware online non-clairvoyant scheduling using speed scaling with arbitrary power function. *Applied Sciences*, *9*(7), 1467. doi:10.3390/app9071467

Singh, U. (2014). *Radial distribution system reconfiguration for loss minimization Using exhaustive search method.* India: Thapar University.

Teng, J. H., & Lu, C. N. (2008). Value-based distribution feeder automation planning. *International Journal of Electrical Power & Energy Systems*, *28*(3), 186–194. doi:10.1016/j.ijepes.2005.11.008

Yeshalem, M. T., & Khan, B. (2017). Design of an off-grid hybrid PV/wind power system for remote mobile base station: A case study. *AIMS Energy*, *5*(1), 96–112. doi:10.3934/energy.2017.1.96

Yu, X. (2011). *Distribution system reliability enhancement,* (MSc thesis), Elec. and Comp. Eng., Georgia Institute of Technology.

Zamani, R., Golshan, M. E. H., Alhelou, H. H., & Hatziargyriou, N. (2019). A novel hybrid islanding detection method using dynamic characteristics of synchronous generator and signal processing technique. *Electric Power Systems Research*, *175*, 105911. doi:10.1016/j.epsr.2019.105911

Zamani, R., Hamedani-Golshan, M. E., Haes Alhelou, H., Siano, P., & Pota, H. R. (2018). Islanding detection of synchronous distributed generator based on the active and reactive power control loops. *Energies*, *11*(10), 2819. doi:10.3390/en11102819

KEY TERMS AND DEFINITIONS

Power Quality Issue: A power quality problem might be defined as any electric power delivering condition that causes appliances to prevent their use.

Protective Relay: It is an intelligent electronic devices (IEDs) which receive measured signals from the secondary side of CTs and VTs and detect whether the protected unit is in a stressed condition (based on their type and configuration) or not. A trip signal is sent by protective relays to the circuit breakers to disconnect the faulty components from power system if necessary.

Reliability: A measure of the degree of certainty that the relay, or relay system, will perform correctly.

Chapter 10

Impacts of Distributed Generations on Power System:
Transmission, Distribution, Power Quality, and Power Stability

Kamlesh Kumar
Mehran University of Engineering and Technology, Jamshoro, Pakistan

Mahesh Kumar
Mehran University of Engineering and Technology, Jamshoro, Pakistan

ABSTRACT

With increasing population and urbanization, the demand of electricity also increases day by day; to fulfill this demand, clean and environment-friendly distributed generations are being installed, but these have some issues in power section. With the integration of DG load curve is levelized, feeder voltage is improved; loading effect on the transformer and branches is reduced, and provides electricity with no pollution. This chapter investigates impacts of DGs to the power system; distributed generation means to generate electric power near the power consumption point. Power quality and reliability can be enhanced by the interconnection of distribution generation to an existing distribution system. However, there are so many effects of distributed generation e.g. changing of load losses, increasing of short circuit levels, voltage transient, congestions in the system branches, power quality, and reliability and network protection issues such as false tripping, nuisance tripping, unintentional islanding, neutral shifting is mainly affected.

DOI: 10.4018/978-1-7998-1230-2.ch010

INTRODUCTION

Mostly centralized generation based electric power system is present, in which electric power is send from the generation point to the consumers located at very farthest places, which results in huge power loss and un-regularity in system voltage, also huge amount of money is required for transmission system and their maintenance as well. Hence, now a day's decentralized generation technique is being used and more and more electric power is being generated by using the renewable energy source (solar, wind and micro-hydro) to avoid these losses and to complete the fast-increasing electrical energy requirement. Distributed generation is done from renewable energy resources and cogeneration. Distributed generation (DG) is generally less than 10MW in size, interconnected with distribution feeder, substation and at consumer level. Distributed generation has some positive points over centralized electric power system e.g. rather free location in network area, according to the availability of resources size of distributed generation is planned, grid becomes more reliable, and also there is no considerable impact of greenhouse effect of this technology. Distributed generation system results in low transmission and distribution losses, service quality is enhanced and voltage regulation is also improved. But, on the other hand, it may create many more problems of power system efficiency and their reliability, over voltage and safety issues, there may be occurrence of transients because of connection and disconnection of generators according to the requirement of power or also during their operation and level of short circuit is increased (Lucian I.D, M. A. 2014). The generation of electricity at consumer end (distributed generation) come into practice because of its low cost, enhancement of efficiency, and clean energy from solar panel, Wind turbine, tidal power plants, Magneto hydro dynamic (MHD) generation, and fuel cell etc. (Alhelou et al., 2019; Makdisie et al., 2018; Alhelou et al., 2018; Alhelou et al., 2016; Haes Alhelou et al., 2019; Njenda et al., 2018).

BACKGROUND

In recent years, due to low investment, environment friendly and high flexibility, distribution generations are becoming an important part of smart gird. But, with the addition of distribution generations, steady state and dynamic characteristics of distribution networks may change; due to uncertainty of output, power flow in distribution network is bidirectional, which creates uncertainty in operating conditions of distribution networks. With the integration of distributed generation load curve is levelized, feeder voltage is improved; loading effect on the transformer and branches is reduced, and provide electricity without any pollution. Along with that, there occur an improvement in overall power system e.g. reduction of losses, transmission and distribution lines investment is saved, avoiding of uncertainty of fuel prices etc. There are several reasons of development of the Distributed Generation such as market liberalization, small and local power plants offer electric power at low prices because of having no huge transmission losses, demand for reliable, uninterrupted power and process heat generated is employed for useful purposes (Khan et al., 2019; Khan et al., 2018; Khan et al., 2017; Banteywalu et al., 2019; Anteneh et al., 2019; Molla et al., 2019, Molla et al., 2018, Jariso et al., 2018). Therefore, local power becomes economically competitive. In distributed generations, power plants types and the method of electric power being sent to the consumption point are different than that of centralized generations. In distributed generations, power can be used at the place where it's generated, for a lower number of distributed generations, plants operating system might be independent of each other and if the numbers of DGs are more than

Table 1. Different categories of DGs according to size (Philip P. Barker, R. W. 2000)

Technology	Power module available size
Micro Hydro	25KW-1MW
Small Hydro	1-100MW
Wind Turbine	200W-3MW
PV array	20W-100KW
Biomass gasification	100KW-20MW
Ocean energy	100KW-5MW
Geothermal energy	5-100MW

it can be controlled by system's control center. New developing technologies of energy generation are smaller in size as compare to conventional energy source of centralized grid system, so these are connected with the distribution grids. The impacts of distributed generation on the existing centralized system is according to its scale or size for power output and the technology used for the generation of electric power (Fini et al., 2016; Alhelou et al., 2018; Zamani et al., 2018; Alhelou et al., 2015; Njenda et al., 2018; Haes Alhelou et al., 2018; Haes Alhelou et al., 2019). There also occur increasing in speed of synchronous generator during fault and reactive power also increased, which results in decreasing in terminal voltage and synchronous torque decrease and rotor speed become more faster, when the asynchronous generator are connected far away from synchronous generator it means that is weak coupling of asynchronous generator with synchronous generator. When the power electronics technologies are used with the distributed generation, then the maximum rotor speed deviation of synchronous generator decreased, because whenever fault occur there is a disconnection of distributed generator with the help of power electronic switching circuits, terminal voltage is decreased and rotor speed of synchronous generator is decreased and also huge amount of energy is lost (J. G. Slootweg, W. K., n.d.). The present radial distribution system has major power losses because of their high resistance to reactance ratio. So, the installation of distributed generation-based distribution system is more beneficial in order to decrease power losses, improvement in voltage profile, and increase security of the system (Kumar et al. 2017). The reliability, controllability, and stability of the power system are mainly dependent on the quality and design scheme of protection, that's why protection of power system is an important consideration for the continuity and flexibility of power system (Adly Girgis, S. B. n.d.). Distributed generation based on the renewable energy resources (solar and wind) results in generation of intermittent nature (Khan et al., 2014; Khan et al., 2013; Khan et al., 2012; Negash et al., 2017; Negash et al., 2016; Jariso et al., 2017; Kifle et al., 2018; Yeshalem et al., 2017; Singh et al., 2017; Gupta et al., 2015). Because of variation of season and time, the amount of heat and wind speed changes that cause variation in the electric power output. These variations in generated output electric power results in so many obstacles i.e. rise and dip in voltage, oscillations in power, power losses and voltage stability, that's why proper sizing and placement of renewable distributed generations becomes more beneficial for the power system (Kumar et al., 2017). Different categories of DGs according to size are mentioned in Table 1.

Traditional concept of electrical power generation and new concept adopted for electric power generation are presented in Figure 1 and Figure 2.

Figure 1. Traditional concept of electrical energy supply)

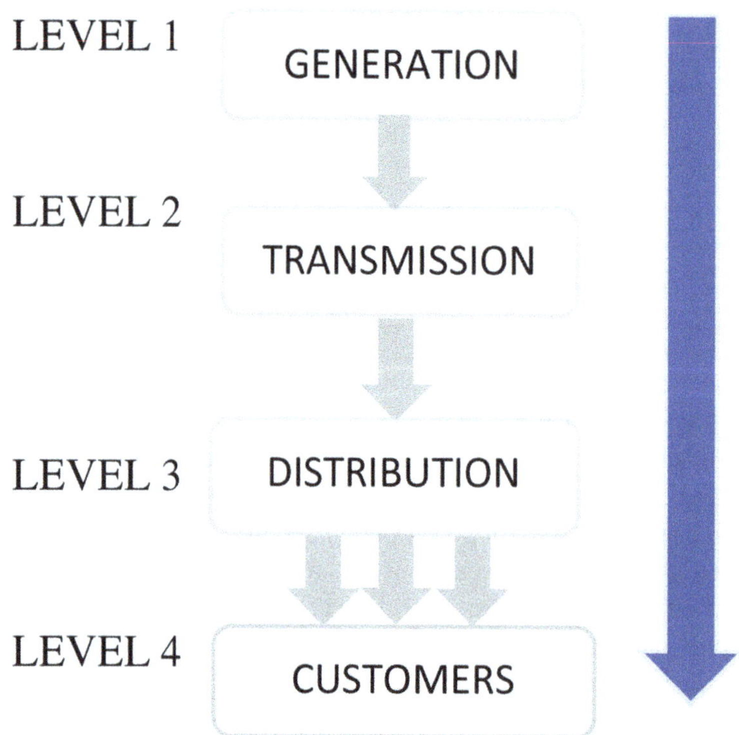

DISTRIBUTED GENERATION TECHNOLOGIES

Distributed generations can be classified into two types from power conversion point of view. Its output may be alternating or direct current, hence for changing that voltage to nominal voltage and frequency inverter are used after the generation points. Following are mostly available distributed generation technologies used now days:

Photovoltaic Systems

In photovoltaic system, when solar cell made up of semiconductor materials (p type and n type) are exposed to sun light, they convert the photon energy into electrical energy as shown in Figure 3. There is no any emission, but they require large spaces and the initial cost, according to the sun light the variable DC output voltage is produced that is afterwards converted into alternating current for the requirement of supply for ac loads. Residential solar system consists of solar panel, battery for storing of energy, charge controller, inverter and load as given in Figure 4.

Figure 2. New concept of industrial electrical energy (Philip P. Barker, R. W. 2000)

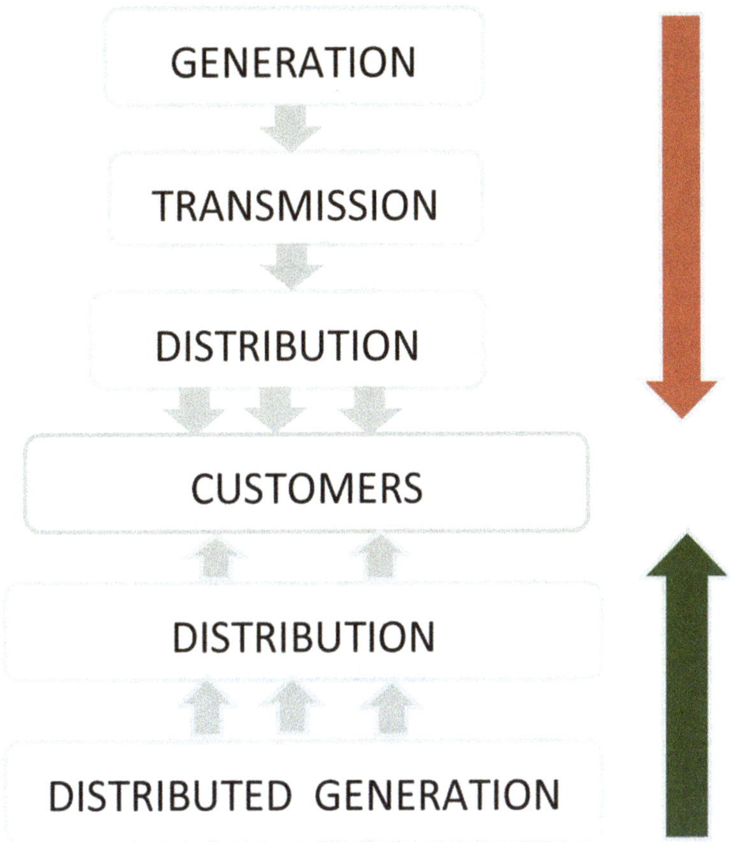

Figure 3. Photovoltaic solar power panel

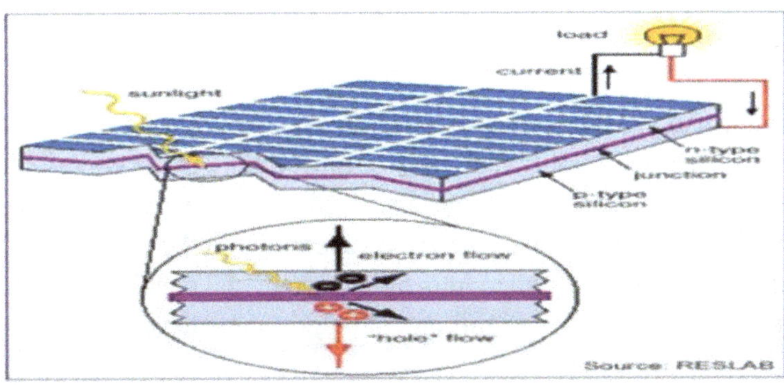

Figure 4. Residential solar system
(Ingenieros, A. M. 2011)

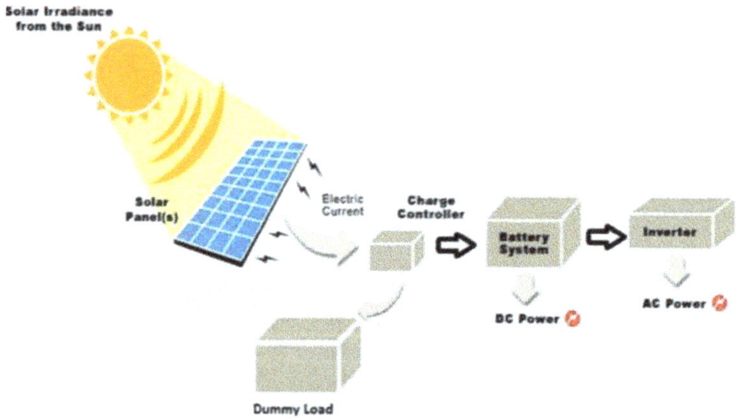

Figure 5. Working Diagram of Wind Turbine Power Plant (tutorvista. (n.d.).

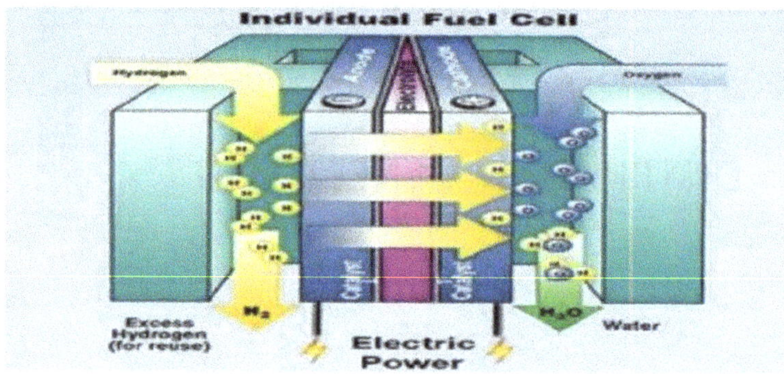

Wind Power Plant

The working diagram of wind turbine power is given in Figure 5, which shows that the conversion of wind energy into electricity consists of two processes. Firstly, the kinetic energy of the wind extracted by the rotor is acting as mechanical torque in the shaft, and then generation system converts this torque into electrical energy. This system generates electricity according to the wind speed, hence that variable generated AC voltage is initially transformed into direct current and again into alternating with the help of inverters; while directly connected wind turbine to the grid must have fixed speed.

Fuel Cells

The operation of fuel cells is same as of battery, which charges by the mixture of hydrogen and fuel. DC voltage, output is produced by the chemical reaction of hydrogen and oxygen, then for the delivery to grid that DC voltage is converted into AC with the help of inverters. Reliability of this technology is

Figure 6. Schematic diagram of Fuel Cell (Technologies Inc. February 15, 2011).

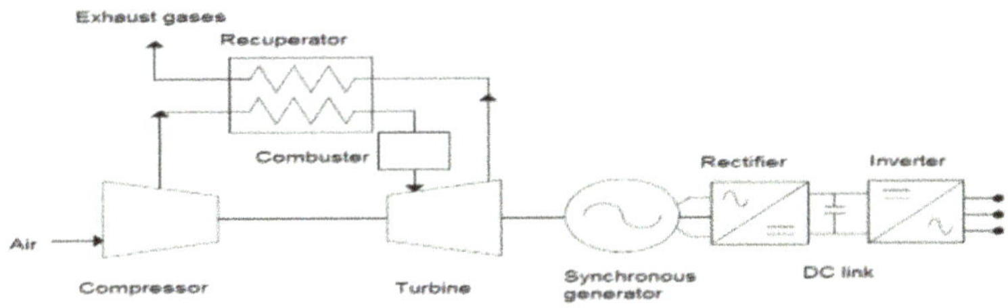

Figure 7. Schematic diagram of micro turbine (Gonzalez-Longatt, F. M. 2008).

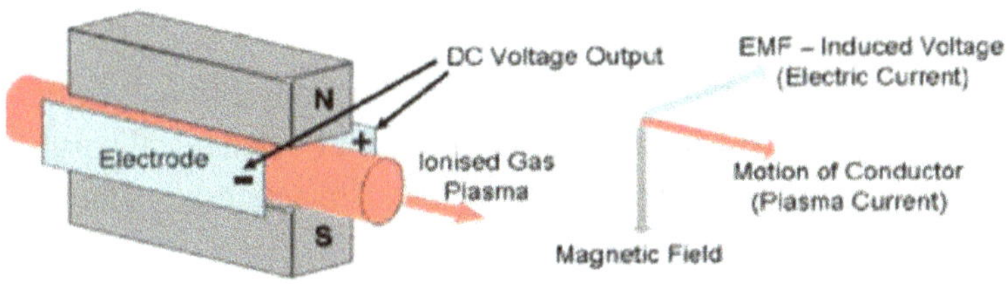

high because no moving part is used; hence no noise is created. Schematic diagram of the fuel cell is given in Figure 6.

Micro-Turbines

In micro-turbines thermal energy is converted into mechanical with flow of gas. The gas is mixed with air in combustion chamber, which rotate and impulse the generator and compressor. At one similar shaft, generator, compressor, and turbine are mounted, and for the requirement of standard frequency and voltage, the produced power from this technology is converted into direct current and then back again into alternating current as shown in Figure 7. The electricity generated by this technique is clean and having good efficiency, but this system requires higher maintenance.

MHD Generation

In a conventional generator, rotor conductor is made to rotate in the magnetic field. When rotor conductor cuts the magnetic fields, it induces an emf in the conductor. Figure 8 presents the working diagram of MHD generator, which shows that, there is only difference conducting fluid (ionized gas). Whenever this ionized gas is made to flow, it cuts the magnetic field which in turns induces an emf in the conducting fluid. This induced EMF appears across the electrode plates. Following are the advantages of MHD generation,

Figure 8. Working diagram of MHD generator

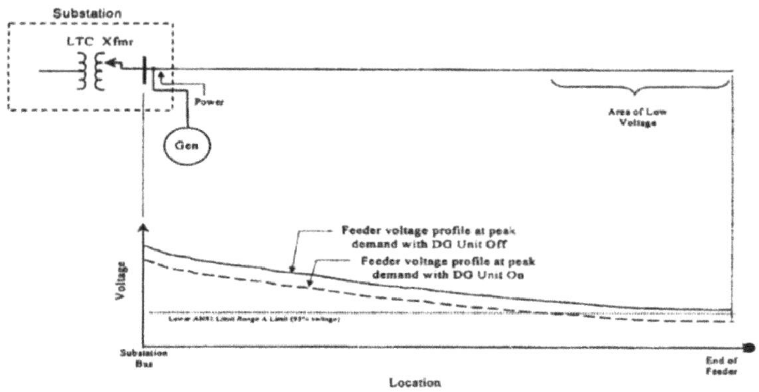

- The running cost of MHD generators is much lower than conventional generators.
- In this technology there is no any movement or presence of mechanical part to move, that result in reduction of losses to zero and become as more dependable energy resource.
- Highest efficiency

Bio-Mass

The electricity or electric power generated from biological materials is known as biomass energy; according to the requirement for heating and electric power this energy is generated. Biomass has following characteristics:

- It is a renewable energy source.
- It protects the clean water supplies.
- It helps in reducing of acid rain and smog.
- It reduces the greenhouse effect.

Tidal Power Plant

Due to greater pull from the moon, tide Become higher, causing water to rotate the Turbine. While in low tide period, water comes out again turning turbine. Thereby generating electricity from tides, but can't be generated at high amount, it's only the alternate way to generate electricity.

IMPACTS OF DISTRIBUTED GENERATIONS

The introduction of DG to the existing radial and centralized designed systems can significantly impact (positive or negative) to the voltage as well as protection devices. The impact is assessed by comparing the results obtained in two conditions i.e. system with and without DG. The main problem is discussed below:

Figure 9.Voltage profile with and without Distributed Generation (Philip P. Barker, R. W. 2000).

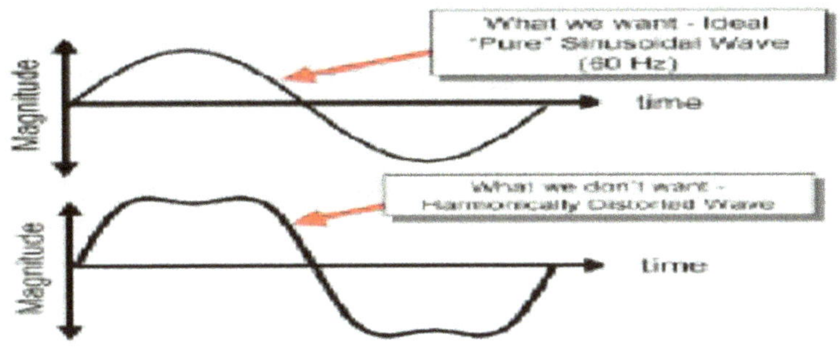

Figure 10. Comparison of pure sinusoidal and distorted waveform (Khan, U. N. 2008)

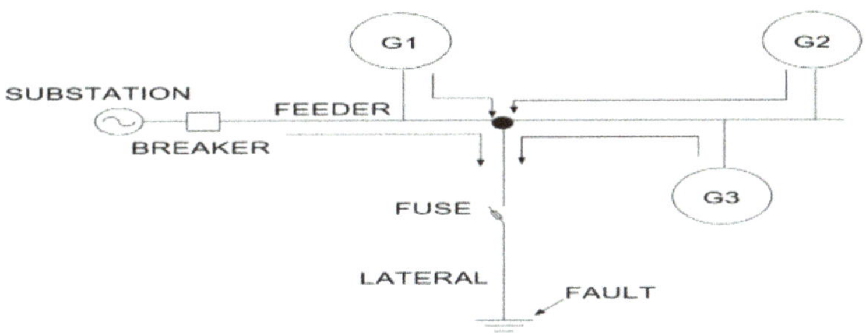

Losses

Feeder losses are highly affected by the decentralized techniques, which are avoided by proper selection of the installation place. Proper location of DGs is basically same to that of banks of capacitors in order to eliminate the losses. Only dissimilarity is that proper location of DG units is suitable for the apparent power components contribution, but capacitors bank gives the reactive power. Load flow analysis software is used for the investigation of the suitable location of DG for reducing the losses. For instance, high losses occurring on the feeder are easily removed by using small capacity DGs, because thermal characteristics of overhead lines and cables must to be taken in consideration at the time of selection of DGs units. Mostly, losses are lower for DGs and loads are near (Vu Van Thong, J. D. 2004).

Voltage Regulation

At substations, voltage regulation is done by load tap changing transformers (LTC) and on the feeder or along the line voltage is regulated by shunt capacitor method. The introduction of DG may cause changes in feeder's voltage because of variation in magnitude and direction of power components. Distributed generation may have positive or negative impacts on voltage regulation (Philip P. Barker, R. W. 2000). Suppose a distribution transformer interconnected with distributed generation at the point where the voltage on the primary of distribution transformer is above ANSI upper level, then voltage at some of

Table 2. Hormonic current injection requirements for distributed generators per IEEE 519-1992 (Philip P. Barker, R. W. 2000)

Order of harmonics	Allowed Level relative to fundamental (odd harmonics)
<11th	4%
< 11th to 17th	2%
< 17th to 23rd	1.5%
< 23rd to 35th	0.6%
35th or greater	0.3%
Total Harmonic Distortion	5%

*Even harmonics are limited to the 25 of odd values.

Figure 11. Fault contributions due to DG units 1, 2, & 3

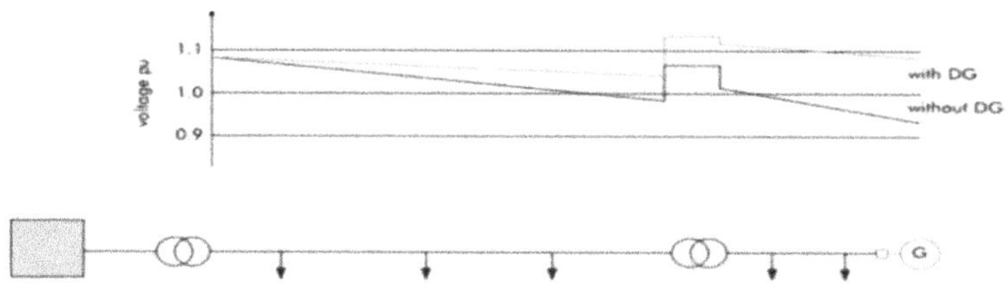

the costumers may be raised and used to damage their equipments of daily life usage. (Philip P. Barker, R. W. 2000)

One can be observed from the figure 9, that the introduction of distributed generation results in a reduction of voltage for the feeder. One can use voltage regulator for avoiding of that variation in voltage, but as the voltage regulator itself consume electric power, hence again voltage is reduced, and less voltage is present at the end of feeder. It means the output of the DGs installation is not so much good as expected. This problem can be solved by the addition of distributed generation units to the upstream side of the regulator. By DGs, a small transformer shared with several loads may raise the secondary side voltage. It also causes overvoltage because of injection of active and reactive power. For smaller DGs there is small effect but, the higher DGs may affect more to the regulation (Philip P. Barker, R. W. 2000).

Harmonics

A not pure sinusoidal wave is called harmonic wave. Mostly harmonics occur due to electronic devices are always present in the power system as given in Figure 10. Mostly converters, inverters, variable frequency drives, switching circuits, and arc furnaces are the hormonic producer at huge amount.

There may be presence of harmonics because of synchronous generator at power generating station, also present in control circuits made up of semiconductor devices, e.g. inverters based on silicon controlled rectifier are the producer of harmonics in huge amount. The cleaner output or pure sinusoidal waveform can be obtained by using inverters by the new technology (Khan, U. N. 2008). Harmonics are also produced by rotating generators (Philip P. Barker, R. W. 2000). According to IEEE 519-1992,

Figure 12. Voltage Rise Due to Distributed Generation

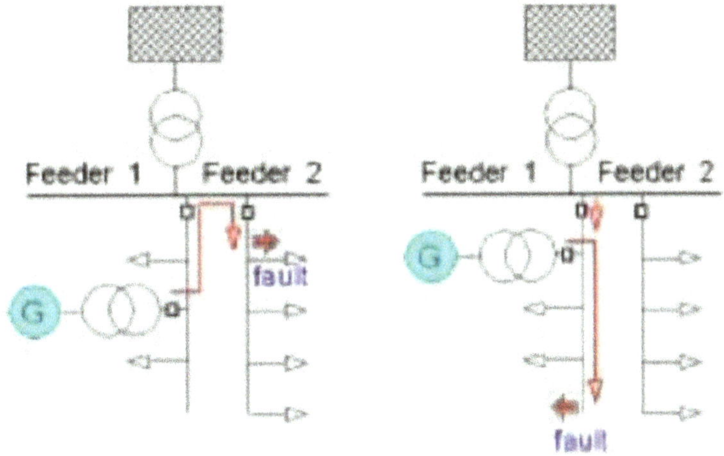

hormonic current injection requirements for distributed generators are given in Table 2. Winding pitch of 2/1 is considered as best winding because they produce least third harmonic. Normally, it's observed that distributed generation produce lesser hormonic impacts with respect to other impacts on the power system (Philip P. Barker, R. W. 2000). Usually these problems occur due to resonance with capacitor banks and the inability of harmonic sensitive equipment's.

Short Circuit Levels of the Network

The DGs may increase the fault current of system and can affects the reliability and safety of distribution system. There may not be a much effect of single or smaller DG units, but the larger DG unit's affect very much to the protection components i.e. relays, circuit breaker and relay.

It is shown from figure 11 that in the normal condition a fuse lateral on a feeder operating safely, except fault conditions (Philip P. Barker, R. W. 2000).

The fault current is observed to be decreased, when the location of distributed generation is inside the utility substation and the fault, which result in a less tripping and other problems (Khan, U. N. 2008). Short circuit levels are depending on the nature of the DG. In first few cycles' induction generator and synchronous generator contribute equally, after that synchronous generator is highest contributing DG to the faults. Very little fault current is provided by Inverter interfaced DG type (Martin-Arnedo, J. A. 2009).

Change of Short Circuit Capacity

The short circuit capacity level is increased with the installation of distributed generators in distribution networks. Sometimes high short circuit capacity is required to indicate a problem.

Voltage Level of Integration and Interconnection

In the present centralized distribution network, there is variation of voltage occur as the distance changes from the generator or transformer. These voltage drops are in the tolerance range under normal condition and easy to predict, but the addition of distributed generations may cause current to be flows in both directions and the voltage will be increased as shown in Figure12.

Power Quality

Different distributed generations creates different power quality issues due to their characteristics. Now a day's power electronics devices and modern controls introduced more power quality problems. The integration of distributed generation with existing radial system results in uncertainty in output of DGs, also uncertainty occur in the operation of networks because of bidirectional flow of power and with the variation of voltage and frequency are challenges for quality of power and overall system (Shiyu Liu, Y. Y. n.d.).

Power Security

Whenever there occur a short circuit fault, then distributed generation as well as power system both supply fault current according to network configuration, grid impedances, size and location of distributed generation, that results in increased level of fault current. If the magnitude of that fault current is greater than the capacity of circuit breaker present in the feeder; circuit will not be able to isolate that fault, which may result in destruction of overall power system. At substation, fault current decrease on feeder because of fault current supplying by DGs, which may result in desensitization of relay to detect fault. (Carmen L.T. Borges, D. F. 2006, Shah Arifur Rahman, B. D. n.d.).

False Tripping Of Feeders

It is well known that integration of distributed generation to the centralized power system may contribute to fault current in healthy part, which results in increasing fault current above pick up level of over current relays connected with healthy feeder. Hence these protective devices trip and make the healthy part to be isolated without clearance of fault in faulted feeder (Shah Arifur Rahman, B. D. n.d).

Impact of DG on Feeder Protection

For the continuous supply and protection purpose relays, circuit breaker and fuses are placed along with the feeders and distribution network. Designing of this protection equipment is done by taking some characteristics into consideration, keeping in mind that entire network straight can't be protected directly. Generally, in complex systems various protective devices are used for the protection of maximum distance and to design the protection scheme of a network (Khan, U. N. 2008).

When a non-faulted component is removed that fault is called mal-trip, and fail-to-trip fault occurs if faulted component is not removed. In mal-trip ("Sympathetic tripping") unrequired device trips on detection of fault before required, device instead of the other. This type of tripping results in reduction of reliability of network because of isolating of healthy part of network (M. Häger, F. S. 2006). In other

Figure 13. Mal-trip and fail to trip (M. Häger, F. S. 2006).

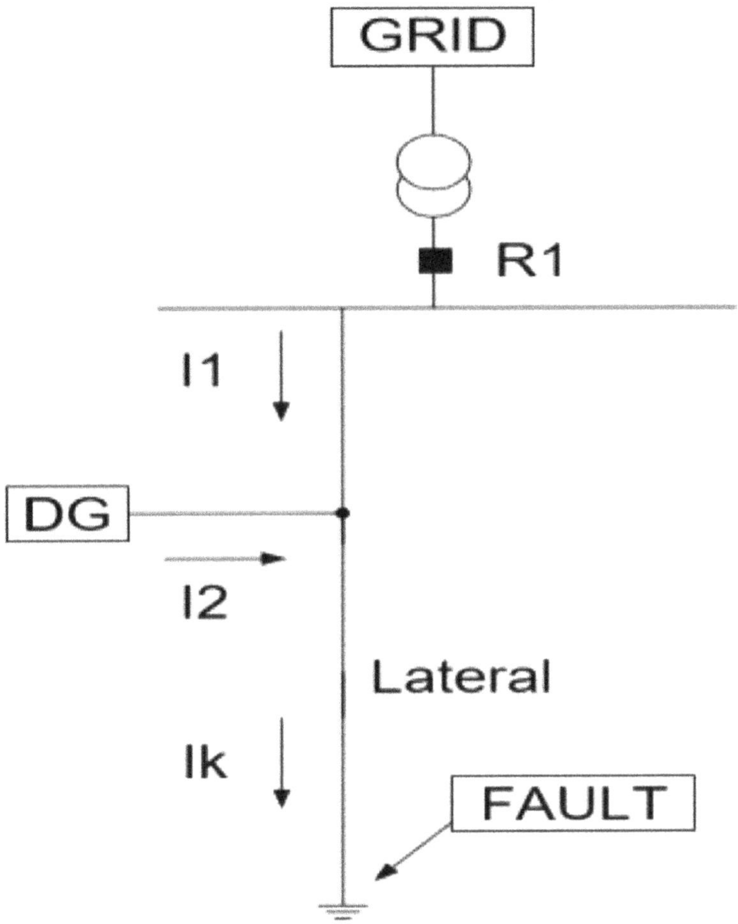

type, fault current originated from DG is below the setting point for the over current protection. Hence, there occurs no any effect on the protection system.

In the given figure13, the relay and circuit breaker at feeder 2, must have to trip on fault condition in order to remove the damaged part from safe part, and breaker/relay of the feeder 1 are used for backup tripping. But, in actual on the occurrence of fault, relay of feeder 1 is tripped first, that cause an excessive current to the DG, which is sensed by the relay of feeder 2 as a fault current and then it trips.

Reduction of Reach of Protective Devices

"Reduction of reach", might be occurring due to DGs interpretation, the over current relays may perform well, when many smaller units connected to the distribution network. As shown in Figure14. It is observed that fault current depends on both size of DGs and distance as well (Thomas Gallery, L. M. n.d.).

Neutral Shifting

When single line to ground fault occurs in the distribution system, circuit breaker is isolated that cause neural shifting; for the integration of distributed generation to existing power system if delta or ungrounded transformer is used, then neutral shifting results in a overvoltage to the feeder except faulty one, which will damage the equipment of costumer and also safety hazards. (Shah Arifur Rahman, B. D. n.d).

Automatic Reclosing

If during the open interval of recloser, distributed generation units are continuously operating then there may not be extinguish of fault arc of the recloser and temporary fault converted into permanent fault and auto reclosing operation is affected (Shah Arifur Rahman, B. D. n.d).

Islanding

If such portion of utility that is not connected or is isolated (because of upstream of circuit breaker or automatic sectionalizing switch) from main power system is continuously energized through distributed energy, then it is called islanding. Islanding is only possible if the generator is capable to sustain the load of separated or isolated part. It is not necessary that system should be capable of islanding each part, because it may cause the power quality and safety problem (Philip P. Barker, R. W. 2000).

CONCLUSION

The critical impacts that occur on the protection and different configuration of distribution system, with the introduction of DG are being discussed in this chapter. It is concluded that with the integration of the distributed generation to existing power system (which was not designed with the consideration of addition of distributed generation), there may arise many problems of power quality, system reliability decreased, efficiency also decreased, voltage regulation, harmonics, security and safety issues, while with the addition of distributed generation to well-designed distribution system results in the improvement of power system. Different technologies have different impacts on distribution system; can be classified according to the generation technologies and connection types with main utility. Power electronic converters are used for synchronous or induction generators. Induction generators are not contributing as Synchronous generators contribute as source for faults. Due to the limited rating of power electronic converters, these cannot hold high over-currents. The fault that arises in power system is being analyzed in given chapter and some conclusion was taken as follow:

- Risk of fault is increased due to introducing of DG into the distribution system.
- Radial power flow characteristics of system are reduced with the penetration of a DG.
- There will be no coordination, if the downstream devices feel the same fault current to that of upstream devices; because these are only able to analyze their own fault current.
- The value of the fault current decreases, when distance between the DG and the fault location is increased.

- Protective devices become less sensitive on the presence of DG, which cause the short circuit current to be to be decreased.
- Transient stability is not affected by the asynchronous generators based distributed generation as much as by the synchronous generator based DG.

ACKNOWLEDGMENT

The author is thanking the Mehran University of Engineering and Technology Jamshoro for providing the necessary facilities for doing this research.

REFERENCES

Adly Girgis, S. B. (n.d.). Effect of distributed generation on protective device coordination in distribution system. IEEE, (pp. 115-119).

Alhelou, H., Hamedani-Golshan, M. E., Zamani, R., Heydarian-Forushani, E., & Siano, P. (2018). Challenges and opportunities of load frequency control in conventional, modern and future smart power systems: A comprehensive review. *Energies, 11*(10), 2497. doi:10.3390/en11102497

Alhelou, H. H. (2018). *Fault detection and isolation in power systems using unknown input observer. In Advanced condition monitoring and fault diagnosis of electric machines* (p. 38). Hershey, PA: IGI Global.

Alhelou, H. H., Golshan, M., & Fini, M. (2018). Wind driven optimization algorithm application to load frequency control in interconnected power systems considering GRC and GDB nonlinearities. *Electric Power Components and Systems, 46*(11-12), 1223-1238.

Alhelou, H. H., & Golshan, M. E. H. (2016, May). Hierarchical plug-in EV control based on primary frequency response in interconnected smart grid. In *2016 24th Iranian Conference on Electrical Engineering (ICEE),* (pp. 561-566). IEEE. 10.1109/IranianCEE.2016.7585585

Alhelou, H. H., & Golshan, M. E. H. (2020). Decision-making-based optimal generation-side secondary-reserve scheduling and optimal LFC in deregulated interconnected power system. In *Decision making applications in modern power systems* (pp. 269–299). Academic Press. doi:10.1016/B978-0-12-816445-7.00011-6

Alhelou, H. H., Golshan, M. E. H., & Hatziargyriou, N. D. (2019). Deterministic dynamic state estimation-based optimal LFC for interconnected power systems using unknown input observer. *IEEE Transactions on Smart Grid*, 1. doi:10.1109/TSG.2019.2940199

Alhelou, H. H., Golshan, M. E. H., & Hatziargyriou, N. D. (2019). A decentralized functional observer based optimal LFC considering unknown inputs, uncertainties and cyber-attacks. *IEEE Transactions on Power Systems.*

Alhelou, H. H., Golshan, M. E. H., Zamani, R., Moghaddam, M. P., Njenda, T. C., Siano, P., & Marzband, M. (2019, June). An improved UFLS scheme based on estimated minimum frequency and power deficit. In 2019 IEEE Milan PowerTech (pp. 1-6). IEEE. doi:10.1109/PTC.2019.8810497

Alhelou, H. H., Golshan, M. H., & Askari-Marnani, J. (2018). Robust sensor fault detection and isolation scheme for interconnected smart power systems in presence of RER and EVs using unknown input observer. *International Journal of Electrical Power & Energy Systems*, *99*, 682–694. doi:10.1016/j.ijepes.2018.02.013

Alhelou, H. H., Hamedani-Golshan, M. E., Heydarian-Forushani, E., Al-Sumaiti, A. S., & Siano, P. (2018, September). Decentralized fractional order control scheme for LFC of deregulated nonlinear power systems in presence of EVs and RER. In *2018 International Conference on Smart Energy Systems and Technologies (SEST)* (pp. 1-6). IEEE. 10.1109/SEST.2018.8495858

Alhelou, H. S. H., Golshan, M. E. H., & Fini, M. H. (2015, December). Multi agent electric vehicle control based primary frequency support for future smart micro-grid. In *2015 Smart Grid Conference (SGC)* (pp. 22-27).

Anteneh, D., & Khan, B. (2019). Reliability enhancement of distribution substation by using network reconfiguration a case study at Debre Berhan Distribution Substation. *International Journal of Economy, Energy, & Environment*, *4*(2), 33.

Banteywalu, S., Khan, B., De Smedt, V., & Leroux, P. (2019). A novel modular radiation hardening approach applied to a synchronous buck converter. *Electronics (Basel)*, *8*(5), 513. doi:10.3390/electronics8050513

Barker, P. P., & De Mello, R. W. (2000, July). Determining the impact of distributed generation on power systems. I. Radial distribution systems. In *2000 Power Engineering Society Summer Meeting* (Cat. No. 00CH37134) (Vol. 3, pp. 1645-1656). IEEE.

Carmen, L. T., & Borges, D. F. (2006). Optimal distributed generation allocation for reliability, losses, and voltage improvement. *Electrical Power and Energy Systems,* 413-420.

Chan, T. F. (2001). Synchronous machines. Encyclopedia of Life Support Systems, Polythecnic University, Hong Kong. Energy or Technologies, Inc. (n.d.). Retrieved from http://www.energyor.com/energyor/fuel_cell.cfm

Fini, M. H., Yousefi, G. R., & Alhelou, H. H. (2016). Comparative study on the performance of many-objective and single-objective optimisation algorithms in tuning load frequency controllers of multi-area power systems. *IET Generation, Transmission & Distribution*, *10*(12), 2915–2923. doi:10.1049/iet-gtd.2015.1334

Gonzalez-Longatt, F. M. (2008, Junio). Impacto de la Generaci6n Distribuida en el Comportamiento de los Sistemas de Potencia. Universidad Central de Venezuela.

Gupta, K., Khan, B., & Mubeen, S. E. (2015). Available transfer capability enhancement by unified power flowcontroller, *IEEE International Conference on Signal Processing, Informatics, Communication, and Energy Systems (SPICES),* pp. 1, 4, 19-21.

Haes Alhelou, H., Hamedani-Golshan, M. E., Njenda, T. C., & Siano, P. (2019). A survey on power system blackout and cascading events: Research motivations and challenges. *Energies*, *12*(4), 682. doi:10.3390/en12040682

Haes Alhelou, H., Hamedani-Golshan, M. E., Njenda, T. C., & Siano, P. (2019). Wide-area measurement system-based optimal multi-stage under-frequency load-shedding in interconnected smart power systems using evolutionary computing techniques. *Applied Sciences*, *9*(3), 508. doi:10.3390/app9030508

Haes Alhelou, H., Hamedani Golshan, M. E., Njenda, T. C., & Siano, P. (2019). WAMS-based online disturbance estimation in interconnected power systems using disturbance observer. *Applied Sciences*, *9*(5), 990. doi:10.3390/app9050990

Häger, M. F. S. (2006). The impact of distributed energy resources on distribution-system protection. Ludvika, Sweden.

IngenierosA. M. (2011). grupoglobalgreen. Retrieved from www.grupoglobalgreen.es

Jariso, M., Khan, B., Tanwar, S., Tyagi, S., & Rishiwal, V. (2018). Hybrid energy system for upgrading the rural environment, *IEEE GlobeCom 2018*.

Jariso, M., Khan, B., Tesfaye, D., & Singh, J. (2017). Modelling and designing of stand-alone photovoltaic system (Case study: Addis Boder Health Center southwest Ethiopia). In *Proceedings IEEE International Conference on Electronics, Communication, and Aerospace Technology, ICECA 2017*, pp. 168-173, Coimbatore, India.

Khan, B., & Agnihotri, G. (2012). A novel transmission loss allocation method based on transmission usage. In *Proceedings IEEE Fifth Power India Conference*, Haryana, India. 10.1109/PowerI.2012.6479479

Khan, B., Agnihotri, G., & Gupta, G. (2013). A multipurpose matrices methodology for transmission usage, loss, and reliability margin allocation in restructured environment, *Electrical & Computer Engineering: An International Journal, 2*(3).

Khan, B., Agnihotri, G., Rathore, P., Mishra, A., & Naidu, G. (2014). A cooperative game theory approach for usage and reliability margin cost allocation under contingent restructured market. *International Review of Electrical Engineering*, *9*(4), 854–862.

Khan, B., Alhelou, H. H., & Mebrahtu, F. (2019). A holistic analysis of distribution system reliability assessment methods with conventional and renewable energy sources. *AIMS Energy*, *7*(4), 413–429. doi:10.3934/energy.2019.4.413

Khan, B., & Singh, P. (2017). The current and future states of Ethiopia's energy sector and potential for green energy: A comprehensive study. *International Journal of Engineering Research in Africa*, *33*, 115–139. doi:10.4028/www.scientific.net/JERA.33.115

Khan, B., & Singh, P. (2017). Optimal power flow techniques under characterization of conventional and renewable energy sources: A comprehensive analysis, *Journal of Engineering, vol. 2017,* Article ID 9539506.

Khan, B., & Singh, P. (2017). Selecting a meta-heuristic technique for smart micro-grid optimization problem: A comprehensive analysis. In IEEE Access, 5, pp. 13951-13977.

Khan, B., & Singh, P. (2019). Economic operation of smart micro-grid: a meta-heuristic approach. In Handbook of research on smart power system operation and control (pp. 330-346). Hershey, PA: IGI Global. doi:10.4018/978-1-5225-8030-0.ch014

Khan, B., & Singh, P. (2019). Economic operation of smart micro-grid: A meta-heuristic approach. In Handbook of research on smart power system operation and control (pp. 330-346). Hershey, PA: IGI Global. doi:10.4018/978-1-5225-8030-0.ch014

Khan, B., & Tanwar, S. (2019). Issues associated with microgrid integration. In Handbook of research on smart power system operation and control (pp. 252-264). Hershey, PA: IGI Global. doi:10.4018/978-1-5225-8030-0.ch010

Kifle, Y., Khan, B., & Singh, J. (2018). Designing and modeling grid connected photovoltaic system: (Case study: EEU building at Hawassa city). *International Journal of Convergence Computing*, *3*(1), 20–34. doi:10.1504/IJCONVC.2018.091113

Kifle, Y., Khan, B., & Singh, P. (2018). Assessment and enhancement of distribution system reliability by renewable energy sources and energy storage, *Journal of Green Engineering, 8*(3), 2, pp. 219-262.

Kumar, M., Nallagownden, P., & Elamvazuthi, I. (2017). Optimal placement and sizing of renewable distributed generations and capacitor banks into radial distribution systems. *Energies*, 10, 1–25.

Shiyu Liu, Y. Y. (n.d.). *Impact of distributed generation on electric power distribution system reliability*. IEEE. China.

Lucian, I. D. M. A. (2014). Effects of distributed generation on electric power system. *Science Direct,* 681-686.

Mahat, P. (2010, September). *Control and operation of islanded distribution system.* (Doctoral dissertation, Aalborg University, Denmark).

Mahesh, K., Nallagownden, P., & Elamvazuthi, I. (2017). Advanced pareto front non-dominated sorting multi-objective particle swarm optimization for optimal placement and sizing of distributed generation. *Energies*, 9, 1–23.

Makdisie, C., Haidar, B., & Alhelou, H. H. (2018). An optimal photovoltaic conversion system for future smart grids. In Handbook of research on power and energy system optimization (pp. 601–657). Hershey, PA: IGI Global. doi:10.4018/978-1-5225-3935-3.ch018

Martin-Arnedo, J. A. (2009, October). Impacts of distributed generation on protection and power quality. In *IEEE Power & Energy Society General Meeting.*

Molla, T., Khan, B., Moges, B., Alhelou, H. H., Zamani, R., & Siano, P. (2019). Integrated optimization of smart home appliances with cost-effective energy management system. *CSEE Journal of Power and Energy Systems*, *5*(2), 249–258.

Molla, T., Khan, B., & Singh, P. (2018). A comprehensive analysis of smart home energy management system optimization techniques, *Journal of Autonomous Intelligence, 1*(1), 15–21.

Negash, K., Khan, B., Tesfaye, D., & Gayathri, M. (2017). Optimal placement of phasor measurement unit for system observability. In *Proceedings of IEEE International Conference on Computing Methodologies and Communication, (ICCMC 2017),* pp. 73-77, Erode, India.

Negash, K., Khan, B., & Yohannes, E. (2016). Artificial intelligence versus conventional mathematical techniques: A review for optimal placement of phasor measurement units [Springer]. *Journal of Technology and Economics of Smart Grids and Sustainable Energy, 1*(1), 10. doi:10.100740866-016-0009-y

Njenda, T. C., Golshan, M. E. H., & Alhelou, H. H. (2018, November). WAMS based intelligent under frequency load shedding considering online disturbance estimation. In *2018 Smart Grid Conference (SGC)* (pp. 1-5). 10.1109/SGC.2018.8777779

Njenda, T. C., Golshan, M. E. H., & Alhelou, H. H. (2018, November). WAMS based under frequency load shedding considering minimum frequency prediction and extrapolated disturbance magnitude. In *2018 Smart Grid Conference (SGC)* (pp. 1-5).

Ochoa, L. F., Padilha-Feltrin, A., & Harrison, G. (2006). Evaluating distributed generation impacts with a multi objective index. *IEEE Transactions on Power Delivery, 21*(3), 1452–1458. doi:10.1109/TPWRD.2005.860262

Shah Arifur Rahman, B. D. (n.d.). *Impact of distributed generation on power system protection.* IEEE, (pp. 1-9). Pakistan.

Singh, P., & Khan, B. (2017). Smart microgrid energy management using a novel artificial shark optimization, Complexity, vol. 2017, Article ID 2158926. doi:10.1155/2017/2158926

Singh, P., Khan, B., Vidyarthi, A., Haes Alhelou, H., & Siano, P. (2019). Energy-aware online non-clairvoyant scheduling using speed scaling with arbitrary power function. *Applied Sciences, 9*(7), 1467. doi:10.3390/app9071467

Singh, P., & Khan, B. (2017). Smart microgrid energy management using a novel artificial shark optimization, Complexity, vol. 2017, Article ID 2158926.

Slootweg, J. G., & Kling, W. L. (n.d.). *Impacts of distributed generation on power system transient stability.* IEEE, (pp. 862-867).

Thomas Gallery. L. M. (n.d.). Impact of distributed generation on distribution network protection. ESBI Engineering & Facility Management, Ireland.

tutorvista. (n.d.). Retrieved from http://www.tutorvista.com/physics/wind-electric-power

Vu Van Thong, J. D. (2004, January). Interconnection of distributed generators and their influences on power system. Katholieke Univeristeit Leuven, Belgium.

Yeshalem, M. T., & Khan, B. (2017). Design of an off-grid hybrid PV/wind power system for remote mobile base station: A case study. *AIMS Energy, 5*(1), 96–112. doi:10.3934/energy.2017.1.96

Zamani, R., Golshan, M. E. H., Alhelou, H. H., & Hatziargyriou, N. (2019). A novel hybrid islanding detection method using dynamic characteristics of synchronous generator and signal processing technique. *Electric Power Systems Research, 175*, 105911. doi:10.1016/j.epsr.2019.105911

Zamani, R., Hamedani-Golshan, M. E., Haes Alhelou, H., Siano, P., & Pota, H. R. (2018). Islanding detection of synchronous distributed generator based on the active and reactive power control loops. *Energies*, *11*(10), 2819. doi:10.3390/en11102819

KEY TERMS AND DEFINITIONS

Distributed Generation: It is done from renewable energy resources and cogeneration. Distributed generation (DG) is generally less than 10MW in size, interconnected with distribution feeder, substation and at consumer level.

Harmonics: A not pure sinusoidal wave is called harmonic wave.

Protective Relay: It is an intelligent electronic devices (IEDs) which receive measured signals from the secondary side of CTs and VTs and detect whether the protected unit is in a stressed condition (based on their type and configuration) or not. A trip signal is sent by protective relays to the circuit breakers to disconnect the faulty components from power system if necessary.

Reliability: A measure of the degree of certainty that the relay, or relay system, will perform correctly.

Chapter 11
Smart Home Energy Management System

Tesfahun Molla
Hawassa University, Hawassa, Ethiopia

ABSTRACT

With the development of smart grid technology, residents can schedule their power consumption pattern in their home to minimize electricity expense, reducing peak-to-average ratio (PAR) and peak load demand. The two-way flow of information between electric utilities and consumers in smart grid opened new areas of applications. In this chapter, the general architectures of the home energy management systems (HEMS) are introduced in a home area network (HAN) based on the smart grid scenario. Efficient scheduling methods for home power usage are discussed. The energy management controller (EMC) receives the demand response (DR) information indicating the Time-of use electricity price (TOUP) through the home gateway (HG). With the DR signal, the EMC achieves an optimal power scheduling scheme that can be delivered to each electric appliance by the HG.

INTRODUCTION

Due to the ever-growing demand of electrical energy, the percentage of renewable energy generation is greatly raised along with the increased energy prices. Various environmental constraints also posed a limit on energy generation from conventional energy sources (Khomami and Javidi, 2013). All these challenges motivated power industry to shift their focus on smart demand side management techniques. Electricity usage report in United States suggests that at least 30% of electric power is wasted from 72% of energy consumed by residential and commercial users (Geng and Lu, 2016). In Europe, as target by 2020, 20% share of renewable energy production and 20% of energy efficiency have been fixed to be met. The advancement of information and communication technologies (ICTs) increased the demand of reliable and quality power supply (Zhao et al., 2013; Khan et al., 2019; Khan et al., 2018; Khan et al., 2017; Banteywalu et al., 2019; Anteneh et al., 2019; Molla et al., 2019, Molla et al., 2018, Jariso et al. 2018). These ICTs are important part of smart grid system that transfers information from one system to another one. This information transfer is very helpful to control and coordinate various smart

DOI: 10.4018/978-1-7998-1230-2.ch011

grid technologies to respond immediately under varying demand conditions. Smart grids can use local renewable energies like wind and solar energy to solve environmental problems, increase reliability of equipment and system and reduce the costs of infrastructure (Dogaheh and Dogaheh, 2017). Further to control the demand from consumer premises, these ICT systems are very helpful. These are the integral part of smart home appliances energy management system, which is one aspect of smart grid (Mohammadi et al., 2013).

Today, the concept of micro-grid is being used for the purpose of helping the environment through using several renewable and available resources such as wind and solar energy, along with other energy generators such as micro-turbines and fuel cells to reduce generation costs and reduce environmental pollutions. The resources are not only interrelated, but also the interaction continues in higher levels such as distribution networks (Alhelou et al., 2019; Makdisie et al., 2018; Alhelou et al., 2018; Alhelou et al., 2016; Haes Alhelou et al., 2019; Njenda et al., 2018). Restriction of fossil fuels and increasing growth of demand for energy, enhanced living standards, global warming and environmental problems have led to increasing advancements in technology and use of modern and cleaner energies. The idea of smart grid was begun with the idea of Advanced Metering Infrastructure (AMI) to develop demand side management, increase energy efficiency and a self-healing electric grid, so that it could improve reliability and respond to natural disasters or deliberate sabotage (Dogaheh and Dogaheh, 2017). Advancement of smart grid system allows consumers to reduce their energy consumption through proper scheduling of different appliances (Khan et al., 2014; Khan et al., 2013; Khan et al., 2012; Negash et al., 2017; Negash et al., 2016; Jariso et al., 2017; Kifle et al., 2018; Yeshalem et al., 2017; Singh et al., 2017; Gupta et al., 2015). This is possible with the information of different electricity pricing techniques such as real time electricity pricing (RTEP) and time of use pricing (TOUP). The demand side management (DSM) techniques are very old in power system that incorporates DR techniques with load shifting, energy efficiency and conservation program. Shifting of consumers' load from peak hours to off peak hours is the main function of DR techniques. For that purpose incentives are offered to consumers.

DR programs transfer customers' load during periods of high-demand to off-peak periods by offering them incentives and can reduce critical peak demand or daily peak demand (Qayyum et al. 2015). In smart grid environment there are different DR pricing schemes. The most common pricing schemes include real-time pricing (RTP), time of use pricing (TOUP) and critical peak pricing (CPP). Within the concept of demand-side energy management, residential energy management recently evokes increasing desire from the research community. The traditional grid has demand response programs for only large-scale consumers such as commercial buildings and industrial plants. However, for the residential consumers it does not have a similar mechanism due to two main reasons (Fini et al., 2016; Alhelou et al., 2018; Zamani et al., 2018; Alhelou et al., 2015; Njenda et al., 2018; Haes Alhelou et al., 2018; Haes Alhelou et al., 2019). First, it has been difficult to handle a large number of residential customers without efficient automation tools, sensors and communication. The second reason is, compared with their implementation cost; the impact of demand response programs has been considered to be relatively small. However, in smart grid system, smart appliances, low-cost sensors and efficient communication system set a stage for efficient energy management technique with the interaction between the utility grid, the users and electrical devices.

Background of Smart Grid

Traditional power grid is insufficient to meet the electricity demand of modern society. In order to meet these challenges the concept of smart grid was introduced. Smart grid increases efficiency of the power, storage capacity, grid sustainability and customer engagement (Hassan et al. 2018). The traditional power grid is known to be service provider and centrally controlled and networked system. It allows the transmission of power in one way from the generating station to the consumers. In developed countries to satisfy the fluctuating demand of the consumers, the electricity suppliers supply sufficient power and the consumers utilize without consumption limit. When the power demand and fluctuation of consumption increases abruptly, it could be problematic to operate without using the feedback technology. For this reason, it is important to change the power grid into a much smarter level. There is no common definition for a smart grid. Generally, smart grid can be defined as; it is a modern electric power grid that is sustainable and technically superior to that of traditional power grid. A smart grid is an autonomous electricity environment which delivers electricity in a smarter and much controlled way from the point of generating station into the consumers. It is integrated with power electronic devices, sensors, actuators and communication devices. The consumers are known to be the integral part of a smart grid because they can affect the pattern of electricity consumption behavior based on the information received and the incentives provided by the electric utility. Smart grid allows two-way communications between customers and electricity utility companies; smart appliances, smart meters and renewable energy sources (RES) helps to achieve reliable, cost effective and intelligent control of the Utility power grid. Figure 1 presented the overview of Smart grid system.

Smart grid has four common features:

- An indoor home-in-display (HID) and feedback interface to display the information, controlling and schedule the appliances in remote way using tablets or PCs.
- A smart meter for monitoring the energy consumed by the customer and the energy sell back to the grid when there is DG connected to the grid.
- A distribution network for bidirectional communication between the utility and the consumer.
- An interface in the utility side

Smart Building

A smart building is an automated type of building, which monitors the electrical operations of the building such as lighting, heating, ventilation, air conditioning, internal and external security and other related systems automatically. In order to collect data and monitor the entire system and services based on the desire of the users, the building uses different sensors, microchips and actuators. The owners, facility managers and operators are benefited from this type of infrastructure due to maximization of reliability and performance, optimized energy usage and reduction of the environmental impact on the building. Traditional buildings are not integrated with automation technologies, sensors and communication system. It used as shelters and provide temperature control and safety with the same efficiency level for years. But, the oldest traditional structures and newer buildings have been converted in to smart buildings, are changing constantly. Smart buildings connected to highly adaptable and intelligent software environment makes the way of living easy, increase comfort of occupants, increase security and have

Figure 1. Overview of smart grid (Hassan et al. 2018)

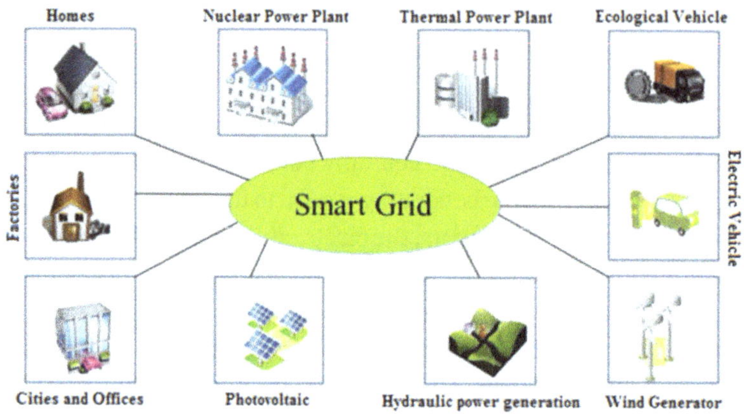

low impact on environment than buildings which are not connected. There are different smart buildings around the world. Those include educational facilities, hospitals, stadiums and others.

According to the Navigant research it was estimated that the market share of smart building technology will generate $8.5 billion revenue in 2020 up from $4.7 billion in 2016 which was growing with a compound annual growth rate of 15.9% over the forecasted period (SB, nd). Making buildings smart begins by interconnecting the core systems like lighting systems, heating systems, water pumps, fire alarms etc. with sensors and efficient advanced control system. In this thesis, the main focus is limited to lowering cost of energy consumption of appliances without highly affecting comfort level of occupants, reduction of electricity billing and regulating the peak load demand while maintaining the horizontal load distribution within a day by using demand side management (DSM) technique based on the demand response signal received from the electric utility company.

Levels of Energy Management under Smart Grid Communication Network (SGCN)

There are different levels of communication network in smart grid which helps to effectively monitor energy management system and transfer information between electric utility company and consumers. Those networks include Local area network (LAN), Neighborhood Area network (NAN), Wide Area Network (WAN).

Consumers under Home Area Network (HAN)

HAN is a type of local area network, which is located at buildings and facilitates the communication or exchange of information between EMC and home appliances. This enables the EMC to identify each schedulable home appliances based on its' address and efficiently monitor their operation within a day. A feedback signal which shows the electric pricing data and billing cost are displayed via home in display. The electricity consumption data are sent to the utility grid via smart meter after receiving the information from each appliance. Therefore, HAN is a backbone of the communication between schedulable appliances at home and the smart meter.

Aggregator Under Neighborhood Area Network (NAN)

The exchange of information between utility companies' WANs and smart metering communications on customer premises are facilitated by neighborhood area network (NAN). It covers from the start point of smart meters on customer side which is the most important element known as heart of smart grid revolution up to Utilities' WAN. The smart meter records real time or nearly real time data. The net energy which is the difference between energy consumed from utility grid and energy exported to utility grid recorded by smart meters is sent to Utility Company. Similarly, a feedback signal like real time energy pricing data and other important information sent to the customer from utility company via smart meter or home in display (HID) to view the information received. NAN acts as an aggregator between the smart meter domain and WAN domain to facilitate the two-way communication or exchange of information. NAN is the most crucial element in smart grid communication system which transfers a huge amount of data and distributing control signals among electric utility companies and an abundant volume of devices installed at customers' premises.

Electric Power Utility Under Wide Area Network (WAN)

A WAN is used as an aggregator of data from multiple NANs and sends it to the private network owned by the electric utility company. It covers long-haul distances from NAN to a control center. WAN applications, including wide-area control, wide-area monitoring and protection, have been identified as the next-generation solution to achieve power system operation, power system planning and power system protection in the smart grid Wide-area monitoring, protection and control applications which provides higher data resolution and shorter response time than classical supervisory control and data acquisition (SCADA) and energy management (EMS) systems (Kuzlu et al., 2014).

Demand Response

Demand response is a technique in which end-users are involved in demand side management of electrical power consumption based on electrical pricing signal received from electric power utility company. When end-users are creating awareness about their energy consumption, they tend to manage their total usage. The demand response program in traditional grid system mostly applied for large scale consumers such as commercial buildings and industrial plants; however, such mechanism does not exist for residential users mainly due to two reasons. The first reason is it is too difficult to handle a large number of home appliances without using automation technology, sensors and communication system. The second reason is, compared to their implementation cost the impact of demand response program is relatively small (Qayyum et al. 2015). Studies have suggested that by employing automated energy management strategies, it is possible to let the users to participate in demand response program and control their load consumption pattern based on the pricing information. In order to implement this strategy, different techniques can be adopted, for example by using internet and controlling the energy consumption via software installed on PC. Demand response can be categorized into two. The first demand response program is price-based demand response such as time-of-use (TOU) pricing, critical peak pricing (CPP) and real-time pricing (RTP), which provides time-varying rates of electricity consumption at different periods of time. This encourages customers to use less electricity when the price of electricity is high and they tend to shift their loads to non-peak hours when electricity pricing is low. The second type of demand response programs are incentive-based demand response program in which the utility company pays the customers who are participating in the program to reduce their loads in the time requested by the utility company or program sponsor, which is triggered by reliability problem in the grid system or highest electricity prices.

Demand response techniques uses smart appliances, sensors, actuators, smart meters and two-way communication. So, it is considered to be part of smart grid technologies.

Demand Side Management

In demand side management (DSM), customers will contribute to:

- Reduction of peak load in peak-hours of the day and reduction of pollutant power plants by using renewable energy sources, thus minimizing operational and maintenance cost.
- Integration into the main distribution grids of micro generators (MG) and distributed generators (DG) which are mostly exposed to generation volatility.
- Maximizing stability of the main distribution network and, thus minimizing interruption problems of power supply, grid maintenance and operational cost.

DSM consists of all activities that manage time of energy consumption. One of the most important goals of DSM is to reduce the peak load. The most frequently used techniques in demand side management are the following (Qayyum et al., 2015). Figure 2 presented the Demand side management techniques.

1. **Peak Clipping:** It refers to demand reduction or load cutting during heavy load periods of the day using a method of direct load control (DLC) of customer appliances by the utility via remotely controllable switches. The duration for which peak occurs is reduced by methods like distributed generation and shutting down the consumers equipment.
2. **Valley Filling:** refers when the cost of production is lower compared to peak hours, it promotes the consumption of energy during off-peak hours. So, low demand periods are filled by building off-peak capacities. There are various incentives like discounts that motivate consumers to change their habits of energy consumption.
3. **Strategic Conservation:** refers to the reduction of seasonal energy consumption mainly by minimizing the energy wastage, to increase the efficiency of energy consumption. In this method, load shape optimization is achieved by applying demand reduction methods on customer premises.
4. **Strategic Load Growth:** It monitors the maximum consumption of electrical energy in seasonal pattern. This technique optimizes the significant daily demand response when the demand exceeds beyond valley filling technique. The method is based on maximizing the market share of loads which is supported by a combination of energy conversion technique and energy storage systems (ESS) or distributed energy resources (DG).
5. **Load Shifting:** It shifts the energy usage from the peak period to off-peak periods during which the consumption of energy is lower on recurring basis. This is often done by using the energy stored during off-peak periods and using it during peak hours. It may also be supported by using distributed generation.

Figure 2. Demand side management techniques (Gaur et al. 2017)

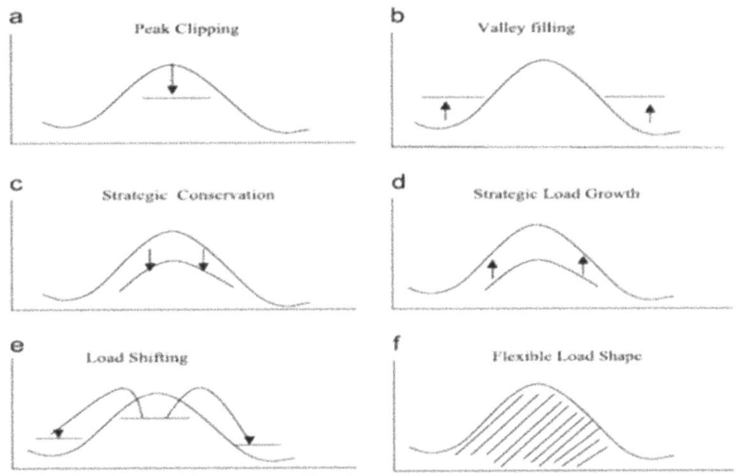

6. **Flexible Modeling:** It consists of actions, following an integrated plan between the consumer and concessionaire, according to the need of moment. Through installation of the load-limiting devices, it limits the energy use of a consumer during certain moments without affecting the actual security conditions.

Time Based Pricing Programs

Time based pricing programs also known as time-based rate programs are prices incurred on customers that vary on different times of the days. In traditional grid system it consisting of small commercial and residential customers and those customers are subjected to two types of pricing schemes known as flat rates and tiered rates. In flat rates, customers are charged with the same rate for their consumption during a given period of time (e.g., during one month or 30 days of billing cycle). In tiered rates, certain blocks of consumption on a given period of time (e.g., first 100 kWh vs. next 100 kWh) are charged based on different pricing (this might be season to season, day to day or hour to hour). Time based pricing is generally grouped into the following types (TBRP, nd):

- **Time of Use Pricing (TOU):** This type of pricing is applied to a certain broad blocks of hours. For example in a typical summer week day on-peak 6 hours; off-peak hours in other hours in summer days of the month.
- **Real Time Pricing (RTP):** The pricing rates are applied to consumption on an hourly basis.
- **Variable Peak Pricing (VPP):** This is the combination of time of use and real time pricing.
- **Critical Peak Pricing (CPP):** when utility companies anticipate or consider a high wholesale market prices or emergency conditions in power system, for a specified period of time they may call critical events (e.g., on a hot summer weekday 4 p.m. to 6 p.m.), during these time, the electric price may be substantially raised.

Figure 3. Overview of home energy management system

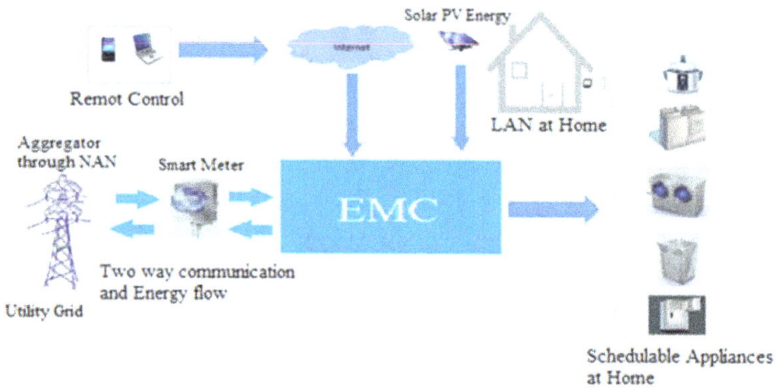

Except TOU pricing, the pricing is not known with certainty ahead of time. So, all the electricity pricing listed excluding TOU pricing scheme are known as *Dynamic pricing*.

The Structure of Home Energy Management System in Smart Grid

It is assumed that a smart home consists of various appliances having adequate and proper communication interfaces in order to facilitate the exchange of information with the EMC. It is also assumed that each of the appliances communicate with the EMC and can also communicate with each other in the most sophisticated and effective scheduling. Figure 3 presents the structure of home energy management system with different data communication domains, those domains include the internet domain, home area network (HAN) domain and the smart meter domain (AMI).

The internet domain enables consumers to monitor and control their power consumption profile, the scheduling of their appliance power consumption and etc. through an in-home display (IHD) which consists of computer, tablet or smart phone. The smart meter domain includes a large number of interconnected smart meters called “Automatic metering infrastructure (AMI)” which is installed and monitored by utility companies in order to transmit load information and DR signals between the smart homes and power market. Based on the pricing signals received 24 hours day a head, the home energy management controller (EMC) schedules the electricity usage pattern for the next day.

DEMAND SIDE MANAGEMENT AND OPTIMIZATION OF RESIDENTIAL DEMAND

Load Type Categorization

In energy management system of residential building, scheduling of different loads can be achieved by specifying the type of load to be scheduled and characteristics of the given load. Those characteristics include the operation duration and the average energy consumption of each appliance. Generally, Home appliances are classified into three major groups as follows (Ruan et al. 2014).

1. **Base LINE LOADS (NON-SHIFT ABLe):** Those are type of appliances which operate immediately when the residents are requested. Such appliance includes Light, Television, air conditioner, computer, etc. the energy supplied to such appliances are considered to be non-schedulable.
2. **Uninterruptible Flexible Loads:** Those types of home appliances are operated continuously until the completion of the task. The starting and ending time of such appliances can be set flexibly. Such appliances include washing machine, Refrigerator, dish washer and Electric Oven.
3. **Interruptible Flexible Loads:** It refers to the appliances which are allowed to operates continuously and can be interrupted when the user desire to shut down in the specified interval of time. Such appliances include clothes dryer, pool pump and floor cleaning robot.

In the home area network (HAN) domain the Electric vehicles (EV) and Energy storage system (ESS) may be available and are integrated into the system. The EV can be used as a load during charging mode or grid to vehicle (G2V) modes of operation and it can also be used as energy storage system during discharging mode or vehicle to grid (V2G) modes of operation. To maximize the comfort level of the user, baseline loads are left uncontrollable in-home energy management system problem. It is also assumed that all the home appliances operate at their rated power during their working periods of time and all constraints needs to be met.

Mathematical Model of HEMS in Smart Grid

The total power consumption of household appliances can be calculated as:

$$\sum_{t=1}^{N} P_a^t \,.\, \tau .\, X_a^t = E_a . \tag{1}$$

Appliance operation duration;

$$\mathrm{t} = [\alpha_a \; \beta_a \;] \tag{2}$$

Length of operation time for an appliance for non-interruptible appliances;

$$\sum_{t=\alpha_a}^{\beta_a} X_a^t = 1 \tag{3}$$

Length of operation time for an appliance for interruptible appliances;

$$\sum_{t=1}^{N} \text{Ä} X_a^t = L_a \tag{4}$$

Where, X_a^t represents the on and off status of an appliance n. $X_a^t = 1$ represents the appliance is on and $X_a^t = 0$ represents the appliance is off. P_a^t represents power consumption value of household appliances and E_a is the maximum power consumption limit of appliances in kWh; [α_a β_a] is the boundary limit for the operation of an appliance which is set by the user. L_a is the length of operation time for an appliance *a* (hour). T is the resolution time in minutes. The priority of operation for each household appliance is set by the user to switched on or off at the same time. This helps to maximize user satisfaction. Assuming that, the objective functions are minimizing users' cost of electricity consumption and minimization of customer's dissatisfaction value due to scheduling;

1. 1. Minimizing users' cost of electricity consumption:

$$_{\min}\mathrm{C} = \sum_{n=1}^{N} \tau . C_{grid} . P_{w,grid} \tag{5}$$

$$\mathrm{C}_{\mathrm{gird}} = \begin{cases} C_{TOU} & P_{w,grid} \geq 0 \\ C_{Sold} & P_{w,grid} \leq 0 \end{cases} \tag{6}$$

Where C_{grid} the grid electricity cost and $P_{w,grid}$ is the power supplied by the grid at time slot t (kW), C_{TOU} is time of use electricity pricing ($ /kWh) and C_{Sold} is the cost of electricity sold back to the grid.

2. Minimization of customer's dissatisfaction value due to scheduling

One of the objectives of this paper is to minimize customer's discomfort level while minimizing the cost of electricity and reducing the peak load. In order to model and quantify user dissatisfaction, a delay time rate function is introduced.

$$\text{minimize} \sum a \epsilon \mathrm{A}\, f_{sn} \tag{7}$$

Where, f_{sn} is the discomfort associated with a shiftable appliance which is obtained by defining the delay time rate (DTR) of shift able home appliances as shown in equation (8) (Zhao et al., 2013):

$$DTR = \left(\frac{t_a - \alpha_a}{\beta_a - la - \alpha_a} \right) \forall\, a \in \mathrm{A} \tag{8}$$

Where, α_a and β_a are the start and end time limits to finish operation of appliance which is set by the user. 'l_a' is length of operation duration for a given appliance and t_a is the actual starting time of a given appliance. If the appliance starts operation later than the actual operation starting time (t_a), the DTR would increase. The Value of DTR for an appliance varies between 0 and 1. Assume $t_a \in [\alpha_a, \beta_a]$ and LOT is l_a, if the appliance started its' operation at time slot α_a the DTR could be 0. This is because, there is no any delay time to start operation. If the appliance start its' operation at time slot $\beta_a - la$, the value of DTR could be 1. Further a delay parameter $\mathfrak{z} > 1$ may be introduced to equate f_{sn} as $\mathfrak{z}^{DTR}$. Thus, discomfort associated with shift able appliance would be as follows:

$$f_{sn} = \sum \wp^{DTR} \tag{9}$$

By satisfying energy and timing constraints the above objective functions would be solved easily.

CONCLUSION

Demand side management is the most effective technique used in optimization of appliance load scheduling in smart grid environment to minimize cost of electricity billing and peak to average ratio with minimum peak load and nearly optimum horizontal load distribution within 24 hours of the day. Therefore, this chapter discussed the various DSM techniques with the formulation of optimization problem for Residential Demand response.

REFERENCES

Alhelou, H., Hamedani-Golshan, M. E., Zamani, R., Heydarian-Forushani, E., & Siano, P. (2018). Challenges and opportunities of load frequency control in conventional, modern and future smart power systems: A comprehensive review. *Energies, 11*(10), 2497. doi:10.3390/en11102497

Alhelou, H. H. (2018). *Fault detection and isolation in power systems using unknown input observer. In Advanced condition monitoring and fault diagnosis of electric machines* (p. 38). Hershey, PA: IGI Global.

Alhelou, H. H., Golshan, M., & Fini, M. (2018). Wind driven optimization algorithm application to load frequency control in interconnected power systems considering GRC and GDB nonlinearities. *Electric Power Components and Systems, 46*(11-12), 1223-1238.

Alhelou, H. H., & Golshan, M. E. H. (2016, May). Hierarchical plug-in EV control based on primary frequency response in interconnected smart grid. In *2016 24th Iranian Conference on Electrical Engineering (ICEE),* (pp. 561-566). IEEE. 10.1109/IranianCEE.2016.7585585

Alhelou, H. H., & Golshan, M. E. H. (2020). Decision-making-based optimal generation-side secondary-reserve scheduling and optimal LFC in deregulated interconnected power system. In *Decision making applications in modern power systems* (pp. 269–299). Academic Press. doi:10.1016/B978-0-12-816445-7.00011-6

Alhelou, H. H., Golshan, M. E. H., & Hatziargyriou, N. D. (2019). Deterministic dynamic state estimation-based optimal LFC for interconnected power systems using unknown input observer. *IEEE Transactions on Smart Grid*, 1. doi:10.1109/TSG.2019.2940199

Alhelou, H. H., Golshan, M. E. H., & Hatziargyriou, N. D. (2019). A decentralized functional observer based optimal lfc considering unknown inputs, uncertainties and cyber-attacks. *IEEE Transactions on Power Systems*.

Alhelou, H. H., Golshan, M. E. H., Zamani, R., Moghaddam, M. P., Njenda, T. C., Siano, P., & Marzband, M. (2019, June). An improved UFLS scheme based on estimated minimum frequency and power deficit. In 2019 IEEE Milan PowerTech (pp. 1-6). IEEE. doi:10.1109/PTC.2019.8810497

Alhelou, H. H., Golshan, M. H., & Askari-Marnani, J. (2018). Robust sensor fault detection and isolation scheme for interconnected smart power systems in presence of RER and EVs using unknown input observer. *International Journal of Electrical Power & Energy Systems*, *99*, 682–694. doi:10.1016/j.ijepes.2018.02.013

Alhelou, H. H., Hamedani-Golshan, M. E., Heydarian-Forushani, E., Al-Sumaiti, A. S., & Siano, P. (2018, September). Decentralized fractional order control scheme for LFC of deregulated nonlinear power systems in presence of EVs and RER. In *2018 International Conference on Smart Energy Systems and Technologies (SEST)* (pp. 1-6). IEEE. 10.1109/SEST.2018.8495858

Alhelou, H. S. H., Golshan, M. E. H., & Fini, M. H. (2015, December). Multi agent electric vehicle control based primary frequency support for future smart micro-grid. In *2015 Smart Grid Conference (SGC)* (pp. 22-27).

Anteneh, D., & Khan, B. (2019). Reliability enhancement of distribution substation by using network reconfiguration a case study at Debre Berhan Distribution Substation. *International Journal of Economy, Energy, & Environment*, *4*(2), 33.

Banteywalu, S., Khan, B., De Smedt, V., & Leroux, P. (2019). A novel modular radiation hardening approach applied to a synchronous buck converter. *Electronics (Basel)*, *8*(5), 513. doi:10.3390/electronics8050513

Dogaheh, Z., & Dogaheh, M. (2017). Optimizing the performance of smart grids in relation with residential energy centers equipping with solar power units (PV), *Sindhological Studies, 2017 (1),* 14-22.

Fini, M. H., Yousefi, G. R., & Alhelou, H. H. (2016). Comparative study on the performance of many-objective and single-objective optimisation algorithms in tuning load frequency controllers of multi-area power systems. *IET Generation, Transmission, & Distribution*, *10*(12), 2915–2923. doi:10.1049/iet-gtd.2015.1334

Gaur, G., Mehta, N., Khanna, R., & Kaur, S. (2017). Demand side management in a smart grid environment. In *Proceedings IEEE International Conference on Smart Grid and Smart Cities*, PEC University of Technology, Chandigarh, India, 2017. 10.1109/ICSGSC.2017.8038581

Geng, L., & Lu, Z. (2016). A novel model for home energy management system based on internet of things. *IEEE International Conference on Power and Renewable Energy*, 2016.

Gupta, K., Khan, B., & Mubeen, S. E. (2015). Available transfer capability enhancement by unified power flow controller. In *Proceedings IEEE International Conference on Signal Processing, Informatics, Communication and Energy Systems (SPICES),* 2015, pp. 1, 4, 19-21.

Haes Alhelou, H., Hamedani-Golshan, M. E., Njenda, T. C., & Siano, P. (2019). A survey on power system blackout and cascading events: Research motivations and challenges. *Energies*, *12*(4), 682. doi:10.3390/en12040682

Haes Alhelou, H., Hamedani-Golshan, M. E., Njenda, T. C., & Siano, P. (2019). Wide-area measurement system-based optimal multi-stage under-frequency load-shedding in interconnected smart power systems using evolutionary computing techniques. *Applied Sciences*, *9*(3), 508. doi:10.3390/app9030508

Haes Alhelou, H., Hamedani Golshan, M. E., Njenda, T. C., & Siano, P. (2019). WAMS-based online disturbance estimation in interconnected power systems using disturbance observer. *Applied Sciences*, *9*(5), 990. doi:10.3390/app9050990

Jariso, M., Khan, B., Tanwar, S., Tyagi, S., & Rishiwal, V. (2018). Hybrid energy system for upgrading the rural environment, IEEE GlobeCom 2018.

Jariso, M., Khan, B., Tesfaye, D., & Singh, J. (2017). Modelling and designing of stand-alone photovoltaic system (Case study: Addis Boder Health Center southwest Ethiopia). In *Proceedings IEEE International Conference on Electronics, Communication, and Aerospace Technology, ICECA 2017*, pp. 168-173, Coimbatore, India.

Khan, B., & Agnihotri, G. (2012). A novel transmission loss allocation method based on transmission usage. In *Proceedings IEEE Fifth Power India Conference*, Haryana, India. 10.1109/PowerI.2012.6479479

Khan, B., Agnihotri, G., & Gupta, G. (2013). A multipurpose matrices methodology for transmission usage, loss and reliability margin allocation in restructured environment, *Electrical & Computer Engineering: An International Journal, 2*(3).

Khan, B., Agnihotri, G., Rathore, P., Mishra, A., & Naidu, G. (2014). A cooperative game theory approach for usage and reliability margin cost allocation under contingent restructured market. *International Review of Electrical Engineering*, *9*(4), 854–862.

Khan, B., Alhelou, H. H., & Mebrahtu, F. (2019). A holistic analysis of distribution system reliability assessment methods with conventional and renewable energy sources. *AIMS Energy*, *7*(4), 413–429. doi:10.3934/energy.2019.4.413

Khan, B., & Singh, P. (2017). The current and future states of Ethiopia's energy sector and potential for green energy: A comprehensive study. *International Journal of Engineering Research in Africa*, *33*, 115–139. doi:10.4028/www.scientific.net/JERA.33.115

Khan, B., & Singh, P. (2017). Optimal power flow techniques under characterization of conventional and renewable energy sources: a comprehensive analysis, *Journal of Engineering, vol. 2017,* Article ID 9539506.

Khan, B., & Singh, P. (2017). Selecting a meta-heuristic technique for smart micro-grid optimization problem: A comprehensive analysis. In IEEE Access, 5, pp. 13951-13977.

Khan, B., & Singh, P. (2019). Economic operation of smart micro-grid: A meta-heuristic approach. In Handbook of research on smart power system operation and control (pp. 330-346). Hershey, PA: IGI Global. doi:10.4018/978-1-5225-8030-0.ch014

Khan, B., & Singh, P. (2019). Economic operation of smart micro-grid: A meta-heuristic approach. In Handbook of research on smart power system operation and control (pp. 330-346). Hershey, PA: IGI Global. doi:10.4018/978-1-5225-8030-0.ch014

Khan, B., & Tanwar, S. (2019). Issues associated with microgrid integration. In Handbook of research on smart power system operation and control (pp. 252-264). Hershey, PA: IGI Global. doi:10.4018/978-1-5225-8030-0.ch010

Khomami, H., & Javidi, M. (2013). An efficient home energy management system for automated residential demand response. In *13th IEEE International Conference on Environment and Electrical Engineering (EEEIC).*

Kifle, Y., Khan, B., & Singh, J. (2018). Designing and modeling grid connected photovoltaic system: (Case study: EEU building at Hawassa city). *International Journal of Convergence Computing*, *3*(1), 20–34. doi:10.1504/IJCONVC.2018.091113

Kifle, Y., Khan, B., & Singh, P. (2018). Assessment and enhancement of distribution system reliability by renewable energy sources and energy storage, *Journal of Green Engineering, 8*(3), 2, pp. 219-262.

Kuzlu, M., Pipattanasomporn, M., & Rahman, S. (2014). Communication network requirements for major smart grid applications in HAN, NAN, and WAN. *Computer Networks*, *67*, 74–88. doi:10.1016/j.comnet.2014.03.029

Makdisie, C., Haidar, B., & Alhelou, H. H. (2018). An optimal photovoltaic conversion system for future smart grids. In Handbook of research on power and energy system optimization (pp. 601–657). Hershey, PA: IGI Global. doi:10.4018/978-1-5225-3935-3.ch018

Mohammadi, S., Momtazpour, M., & Sanaei, E. (2013). Optimization-based home energy management in the presence of solar energy and storage. In *Proceedings 21st Iranian Conference on Electrical Engineering (ICEE)*, Mashhad, pp. 1-6. 10.1109/IranianCEE.2013.6599649

Molla, T., Khan, B., Moges, B., Alhelou, H. H., Zamani, R., & Siano, P. (2019). Integrated optimization of smart home appliances with cost-effective energy management system. *CSEE Journal of Power and Energy Systems*, *5*(2), 249–258.

Molla, T., Khan, B., & Singh, P. (2018). A comprehensive analysis of smart home energy management system optimization techniques, *Journal of Autonomous Intelligence, 1, 1*, 15–21.

Negash, K., Khan, B., Tesfaye, D., & Gayathri, M. (2017). Optimal placement of phasor measurement unit for system observability. In *Proceedings of IEEE International Conference on Computing Methodologies and Communication, (ICCMC 2017),* pp. 73-77, Erode, India.

Negash, K., Khan, B., & Yohannes, E. (2016). Artificial intelligence versus conventional mathematical techniques: A review for optimal placement of phasor measurement units [Springer]. *Journal of Technology and Economics of Smart Grids and Sustainable Energy*, *1*(1), 10. doi:10.100740866-016-0009-y

Njenda, T. C., Golshan, M. E. H., & Alhelou, H. H. (2018, November). WAMS based intelligent under frequency load shedding considering online disturbance estimation. In *2018 Smart Grid Conference (SGC)* (pp. 1-5). 10.1109/SGC.2018.8777779

Njenda, T. C., Golshan, M. E. H., & Alhelou, H. H. (2018, November). WAMS based under frequency load shedding considering minimum frequency prediction and extrapolated disturbance magnitude. In *2018 Smart Grid Conference (SGC)* (pp. 1-5).

Qayyum, F. A., Naeem, M., Khwaja, A. S., Anpalagan, A., Guan, L., & Venkatesh, B. (2015). Appliance scheduling optimization in smart home networks, Special section on smart grids, IEEE. doi:10.1109/ACCESS.2015.2496117

Ruan, B., Yang, Q., Fang, X., & Yan, W. (2014). Demand response under real-time pricing for domestic energy system with DGs. In *Proceedings International Conference on Power System Technology (POWERCON 2014)*, Chengdu, 10.1109/POWERCON.2014.6993601

Singh, P., & Khan, B. (2017). Smart microgrid energy management using a novel artificial shark optimization, Complexity, vol. 2017, Article ID 2158926. doi:10.1155/2017/2158926

Singh, P., Khan, B., Vidyarthi, A., Haes Alhelou, H., & Siano, P. (2019). Energy-aware online non-clairvoyant scheduling using speed scaling with arbitrary power function. *Applied Sciences*, *9*(7), 1467. doi:10.3390/app9071467

Smart Building (SB). Retrieved from https://www.rcrwireless.com/20160725/business/smart-building-tag31-tag99

Time based rate programs (TBRP). (n.d.). Retrieved from https://www.smartgrid.gov/recovery_act/time_based_rate_programs.html

ul Hassan, C. A., Khan, M. S., Ghafar, A., Aimal, S., Asif, S., & Javaid, N. (2018). Energy optimization in smart grid using grey wolf optimization algorithm and bacterial foraging algorithm, Advances in Intelligent Networking and Collaborative Systems, Springer International Publishing, AG 2018, Islamabad, Pakistan.

Yeshalem, M. T., & Khan, B. (2017). Design of an off-grid hybrid PV/wind power system for remote mobile base station: A case study. *AIMS Energy*, *5*(1), 96–112. doi:10.3934/energy.2017.1.96

Zamani, R., Golshan, M. E. H., Alhelou, H. H., & Hatziargyriou, N. (2019). A novel hybrid islanding detection method using dynamic characteristics of synchronous generator and signal processing technique. *Electric Power Systems Research*, *175*, 105911. doi:10.1016/j.epsr.2019.105911

Zamani, R., Hamedani-Golshan, M. E., Haes Alhelou, H., Siano, P., & Pota, H. R. (2018). Islanding detection of synchronous distributed generator based on the active and reactive power control loops. *Energies*, *11*(10), 2819. doi:10.3390/en11102819

Zhao, Z., Lee, W., Shin, Y., & Song, K. (2013, September). An optimal power scheduling method for demand response in home energy management system. *IEEE Transactions on Smart Grid*, *4*(3).

KEY TERMS AND DEFINITIONS

Demand Side Management (DSM): These techniques are very old in power system that incorporates DR techniques with load shifting, energy efficiency and conservation program. Shifting of consumers' load from peak hours to off peak hours is the main function of DR techniques.

Smart Building: It is an automated type of building, which monitors the electrical operations of the building such as lighting, heating, ventilation, air conditioning, internal and external security and other related systems automatically.

Smart Grid: It can be defined as; it is a modern electric power grid that is sustainable and technically superior to that of traditional power grid. A smart grid is an autonomous electricity environment which delivers electricity in a smarter and much controlled way from the point of generating station in to the consumers.

Chapter 12
Generation Extension Arrangement in Power Engineering Networks Using Chaotic Grasshopper Optimization Algorithm:
Concepts, Solutions, and Management

Suriya Ponnambalam
Annamalai University, India

Subramanian Srikrishna
https://orcid.org/0000-0002-9888-5382
Annamalai University, India

Ganesan Sivarajan
Government College of Engineering, Salem, India

Abirami Manoharan
Government College of Engineering, Srirangam, India

ABSTRACT

Electric utilities over the domain affected with ecological issues associated with standard fossil fuel-established plants are examining more within the potentiality of interposing energy sources type of plants into the system as an alternative. Integration of Demand Side Management (DSM) and Supply Side Management (SSM) is required in a rational power system planning that implies concurrent deliberation of both qualitative and quantitative problems like costs, fuel mix, and reliability of power supply. This chapter examines the economic and environmental ability of power supplies initiation into an existing peak deficit power system, incorporating both DSM and SSM plans. The Generation Expansion Planning (GEP) study is carried out in the power system for the period of 24 years planning horizon.

DOI: 10.4018/978-1-7998-1230-2.ch012

INTRODUCTION

Demand for electric power carries on rising every day, therefore an acceptable strategic planning for the hike of the power system is critical in future. There is a prerequisite to fabricate new assets in the sort of generation plants and transmission lines while thinking about the impacts of these new assets on power system operations and long-haul motion of the economy. Therefore, in modern years, there has been a vast inflation in the consumption of electrical energy. Hence, the GEP has been viewed as a vital task for a long time. GEP can be performed for short, medium, and long-term periods (Alhelou et al., 2019; Makdisie et al., 2018; Alhelou et al., 2018; Alhelou et al., 2016; Haes Alhelou et al., 2019; Njenda et al., 2018).

The decision power system planners make are where to locate new investment, the capacity of new investment, the timing of new investment and what type of technology should be installed. One fourth most important task to be done before planning is performed load forecasting. The electric load forecasts for the complete planning period; planning constraints assure that the supply of generation will be sufficient for the period of the planning horizon (Fini et al., 2016; Alhelou et al., 2018; Zamani et al., 2018; Alhelou et al., 2015; Njenda et al., 2018; Haes Alhelou et al., 2018; Haes Alhelou et al., 2019).

The major difference between the electricity and other commodities is that electricity cannot be gathered in a large quantity economically; therefore electricity is consumed at the time of its production. Almost each attempt, the power system operator has made is to meet system demand with generation on a minute to minute basis. Generally, demand has the deficient adaptability to react to an irregularity amongst generation and demand, so changing generation following the demand difference turns the major way to keep demand/supply in adjusts.

GENERATION EXPANSION PLANNING - RECENT REPORTS

Forecasting involves the predictions of the future based on the analysis of trends of past and present, comprising three noteworthy parts, for example, input factors (present and past information), forecasting/evaluating techniques (investigation of trend) and output variables (future predictions), forecast the load demand employing statistical and yield Computational Intelligence (CI) approach are conceded in Debnath and Mourshed, (2018) that shows precision of CI strategies for forecasting was better than that of statistical ones. The energy planning issue is analyzed by Ervural et al., (2018) by developed a multi-objective assessment, creating model by means of fuzzy constraints to achieve various goals simultaneously. Abdin and Zio, (2018) develop an incorporated operational flexibility evaluation scheme that is based on consistent horizon reduction methods driven by an explicit optimization objective to dodge predispositions that can emerge from specially appointed approximations and embraces reasonable measurements to quantitative survey the adaptability level of the obtained plans. GEP model that joins operational flexibility depicted in Hua et al., (2018), Unit Commitment (UC) problem is solved through a convex relaxation. This convex relaxation is a primal definition of the Lagrangian dual problem of UC. Karunanithi et al., (2017) addressed to study the economic and environmental influence of Renewable Energy Sources (RES) introduction into an existing peak deficit power system, in the state of Tamil Nadu, India, using the Long-Range Energy Alternative Planning system (LEAP) an energy-economic model, integrating both demand and supply side management strategies.

Least cost GEP with solar based and wind power plant using a DE algorithm with the cost of adjusting customary base load frameworks is drawing the convergence of power system organizers over the world is presented in (Rajesh et al. 2016; Khan et al., 2019; Khan et al., 2018; Khan et al., 2017; Banteywalu et al., 2019; Anteneh et al., 2019; Molla et al., 2019, Molla et al., 2018, Jariso et al. 2018). A column generation path for solving GEP through huge renewable energy diffusion is granted in Flores-Quiroz *et al,* (2016). Hemmati et al., (2016) presents a combined planning for Energy Storage Systems (ESSs) scheduling and generation expansion in the meantime, such substantial scale ESSs are furnished with diminishing ecological contamination and diminish GEP cost. A detailed study of electric power GEP allowing for uncertainty of climate fluctuations has been organized in Li et al., (2016) because, climate fluctuations has been a serious global concern as it has long-term effects on human systems.

GEP issue through load uncertainty is formulated based on joint Chance-Constrained Programming (CCP) and is worked out by coordinating sensitivity toward iterative algorithms introduced in Manickavasagam et al., (2015), these algorithms build up the parts of the system and its reaction to load variations. Uncertainties of load as well as wind power accessibility are designed by Park and Baldick (2015), the Gaussian copula method is used to correlate random variables under texas wind regime and sample scenarios are produced. Tri-objective linear programming method Chen et al., (2015) will be applied to a case study of the power GEP system in china and is solved by Dinkelbach's Algorithm (DA). GEP problem is determined for Tamil Nadu, an Indian state intended for the duration of 30 years from 2012 to 2041.

PROBLEM STATEMENT

The power system expansion planning problem includes generation expansion, transmission system expansion and generation expansion with DR resources. The optimal power system expansion planning problem attempts to minimize the investment cost and the reliability cost while meeting the system demand subjected to various operational and technical constraints. The power system expansion planning problem involves finding the least cost generating unit and transmission lines to meet the load demand.

The retirement of existing generating unit is included in generation expansion planning problem, also managing the balance between supply and demand by introducing load flexibility instead of only altering generation levels, at almost all operational timescales. So, the proposed model is designed as the classical expansion planning problem with DR resources that minimizes peak load demand in the peak hours of the daytime.

SUPPLY SIDE MANAGEMENT STRATEGY IN GEP

SSM indicates to actions interpreted to assure the generation, transmission and distribution of energy are carried efficiently. This has become peculiarly significant for the deregulation of the electric power industry in several countries, where the effective usage of obtainable power supply becomes necessary to abide aggressive and is presented in (supply side management module 13). SSM is used primarily with reference to electricity but it can also be applied to actions concerning the supply of other energy resources such as fossil fuels and renewable. Utility companies may look at means of modifying their load profile to allow their least efficient generating equipment to be used as little as possible (compared with high efficiency equipment that should be used to the maximum) (Fini et al., 2016; Alhelou et al., 2018; Zamani et al., 2018; Alhelou et al., 2015; Njenda et al., 2018; Haes Alhelou et al., 2018; Haes Alhelou et al., 2019). They may improve maintenance and control of existing equipment, or upgrade equipment with state-of-the-art technologies. Energy users will generally concentrate their efforts on Demand Side Management (DSM) methods however some will regard the supply side as well. For instance, they may look at on site generation alternatives together with cogeneration (or) consider varying to different fuel sources such as solar, wind, bio-fuels and natural gas.

Retirements of Existing Generating Units

When a generation unit stretches to its lifetime, most forth times, it enhances in vogue and inefficient. The modelling of the retirement of specific old incompetent generating units is considered in this thesis. Then, once a generation unit attains its lifetime a new investment is admitted to be made at the similar place.

This evaluation is important because of old units are being operated at the higher cost and a diminished practical capacity, which makes unreliability of the framework to be more serious. In recent years, the devaluation in greenhouse gases and the consumer welfare have attained more importance where the solution that only decreases the total planning costs may not be suitable for real power frameworks.

IMPACT OF DEMAND SIDE MANAGEMENT IN GEP

Various reasons are put forward for promoting or undertaking DSM. For example, DSM may be aimed at addressing the following issues (University of Warwick, REEEP, 2005):

- **Cost Reduction:** Many DSM and energy efficiency efforts have been imported in the circumstance of unified resource planning moreover intended on decreasing total costs of meeting power demand;
- **Ecological and Social Enhancement:** Energy proficiency along with DSM may be chased to attain ecological as well as social destinations by decreasing energy consumption, lead to decreased greenhouse gas discharges;
- **Reliability and Network Problems:** Ameliorative and/or avoiding issues in the electric power network by decreasing demand in manners that retain system reliability in the instant term and above the longer term remit the necessity for network extension;
- **Enhanced Markets:** Short-term responses to electric power market conditions i.e. demand response, especially by decreasing load through intervals of high market expenses induced by decreased generation or system capacity.

An energy customer may have many reasons for selecting a certain DSM activity. Generally these would be economic, environmental, marketing or regulatory. The aforementioned statements are uttered in somewhat distinct technique (Satish Saini, 2004), where it is debated that the advantages of DSM to users, firms, utilities and society can be realized through:

- Reductions in customer energy bills;
- Reductions in the need for new power plant, transmission and distribution networks;
- Stimulation of cost-effective improvement;
- Formation of long-term works owing to novel innovations and technologies;
- Enhances in the contention of local industries;
- Diminution in air contamination;
- Decreased enslavement on alien power sources;
- Diminutions in crest energy costs for electricity.

A new feature (Satish Saini, 2004) is that of improved energy defense over a decreased dependence on overseas energy sources. While the susceptibility to the instability of global energy markets may not be the regard of an individual utility, industry or commercial company, at the national level, diminished dependency on energy significances can have vital security of energy supply implications. For example, the dependency of many countries on oil, which as a primary resource is concentrated in only a relatively few countries, is creating geo-political tensions.

Expansion by way of DR Resources

The traditional GEP and TEP issue is to establish the size, location, and time taken to introduce generation and transmission advances over a planning horizon to delight the forecasted future load demand. Nonetheless, in the context of smart grids, the accessible, smart grid technologies will impact the extension planning issue. Normally, the impacts of smart grid technologies, mostly involve a decrease in peak load, reduction in operational cost, improvement of investment efficiency and environmental concerns.

Keeping in mind the end goal is to accomplish technical performance of the new advances in the expansion planning, the modelling of the issues, which are connected with various technologies and factors need to be analyzed and contemplated. A smart grid ordinarily specifies to a class of innovation utilized by people to pass power utility deliverance structures into modern power frameworks utilizing PC–based computerization and isolated control. The number of applications that can be used on the smart grid increases, once the data communications technology organized is increasing as fast as inventive companies. Benefits comprise improved cyber-security, handling sources of electricity and even integrating.

CHALLENGES IN GEP

The economic production and the associated development in economic activities increase the demand for electricity. Electricity is used in all kinds of human actions, such as industrial production, domestic/residential affair, agricultural activities, transportation, lighting and heating. There is no appropriate method to store the electric energy in large quantities; hence it is mandatory to keep up a continuous and a nearly spontaneous balance between the production and the consumption of electricity.

Some added generation will be kept as reserve to satisfy the deviations in demand. If the supply system is inadequate to meet up the demand, load shedding is unavoidable. The increased installed capacity can reduce the power shortage. The overinvestment and the high operating costs may lead to increase energy cost and may reflect in the bill paid by the consumer. In contrast, underinvestment and low generation margins will companion to the inaccessibility of energy and the poor reliability to consumers. The identification and the examination of energy improvement and issues of energy policy options such as consumption, distribution and planning are most important.

PROBLEM FORMULATION OF GENERATION EXPANSION PLANNING

The GERP problem can be stated by minimizing the present value of the total cost of the combination of generating units under the constraints. Depending on the variation of forecasted demand, the combination of generating units has to be altered to meet the power balance between generation and demand of a power system. The problem is detailed over a long-term horizon and the projected model is hence created based on a year index. The single objective model is used for undecided parameters of the problem while fulfilling the system constraints.

Formulation of GEP in Conjunction with Supply Side Management

Mathematically, solving an economical GEP problem is equivalent in finding a set of optimal choice vectors that diminish total cost subjected to various constraints over a planning horizon. The single objective GEP with unit retirement issue is formulated as follows.

To minimize the sum of yearly discounted costs over a planning horizon is the main objective which is given by Equation (1) and subjected to constraints.

$$minTC = \sum_{t=1}^{T}\left[Inv(M_t) + Sal(M_t) + OM(X_t) + Out(X_t)\right] \tag{1}$$

where

$$X_t = X_{t-1} + M_t - R_t \quad (t = 1,2,\ldots T)$$

The yearly discounted cost is composed of investment costs Equation (2), salvage cost Equation (3), expected O&M cost Equation (4) and outage cost Equation (5). The investment cost is assumed to appear in the beginning of the year, while the maintenance cost is occurring in the middle of the year and is calculated by the comparable energy function method moreover salvage cost is pretended to appear at the final step of the planning horizon. For compelling, demonstrating of the investment cost in the planning horizon, it is re-established with the related yearly proportional incentive as stated in Equation (2). The salvage cost is usually positive appropriate to the value of retired facilities.

$$Inv(M_t) = (1+r)^{-2t} \sum_{i=1}^{N}(CI_i * M_{t,i}) \times r(1+r)^{LT_i} \Big/ \left((1+r)^{LT_i} - 1\right) \tag{2}$$

$$Sal(M_t) = (1+r)^{-2(T+1)} \sum_{i=1}^{N} [CI_i * \delta_i^{2(T-t+1)} * M_{t,i}] \times f_{t,i} \tag{3}$$

where

$$f_{t,i} = (w^e_{(t-1),i} - w^e_{t,i})$$

$$w^e_{(t+1),i} \leq w^e_{t,i}$$

$$w^n_{(t+1),i} \geq w^n_{t,i}$$

$$OM(X_t) = \sum_{g'=0}^{1} \left[(1+r)^{-(2.5+t'+g')} \times \sum_{i=1}^{N} ((w^e_{t,i} * X_{t,i} * FC_i) + VC_i * EES_{t,i}) \right] \tag{4}$$

where

$$t' = 2(t-1)\, and\, T' = 2*T - t'$$

$$Out(X_t) = OUC \times \sum_{g'=0}^{1} \left((1+r)^{-(2.5+t'+g')} \right) \tag{5}$$

where

$$OUC = EENS_t \times EENS\, cost$$

$$EENS = \sum_{\{j|U_j > R_j\}} (U_j - R_j) P(U_j) Ot_j \tag{6}$$

Formulation of GEP with Demand Side Management

The objective of the presented optimization model is to focus on the total cost consisting of operation and maintenance, investment, generation, reliability and environmental cost. The GEP issue via DR resources is comparable in finding an arrangement of optimal decision vectors over a planning horizon and is formulated as follows:

$$Total\ cost = Inv\_cost + O\_M\ cost + Gen\_cost + Reli\_cost + Env\_cost \tag{7}$$

$$s.t\ \sum_{j=1}^{J} b_{tj} + \sum_{i=1}^{I} e_{ti} \geq L_{td} + l_t$$

$$0 \leq x_{ti} \leq x_{ti_{\max}}$$
$$0 \leq e_{ti} \leq x_{ti}$$
$$0 \leq b_{tj} \leq x_{tj_{\max}}$$

where

$$Inv.cost = \sum_{t=1}^{T} (1+a)^{-t} \sum_{i=1}^{I} a_{ti} x_{ti} \tag{8}$$

$$O\_m.cost = \sum_{t=1}^{T} (1+a)^{-t} \sum_{i=1}^{I} m_{ti} x_{ti} + \sum_{t=1}^{T} (1+a)^{-t} \sum_{j=1}^{J} g_{tj} q_{tj} \tag{9}$$

$$Gen\_cost = \sum_{t=1}^{T} (1+a)^{-t} \sum_{i=1}^{I} e_{ti} f_{ti} + \sum_{t=1}^{T} (1+a)^{-t} \sum_{j=1}^{J} b_{tj} d_{tj} \tag{10}$$

$$Reli\_cost = \sum_{t=1}^{T} (1+a)^{-t} lol_t * Clol_t \tag{11}$$

Considering the impact of DR, the added generation capacity in the planning period can be decreased because of its participation, so the generation cost equation is changed from Equations (10)-(12).

$$Gen\_cost = \sum_{t=1}^{T} (1+a)^{-t} \sum_{i=1}^{I} e_{ti} f_{ti} + \sum_{t=1}^{T} (1+a)^{-t} \sum_{j=1}^{J} \left(b_{tj} d_{tj} - p_t b_{tj} h_t \right) \tag{12}$$

$$Env\_cost = \sum_{t=1}^{T} \left(\sum_{j=1}^{J} b_{tj} B_{tj} + \sum_{i=1}^{I} e_{ti} E_{ti} \right) \tag{13}$$

GEP Constraints

In GEP model, seven types of constraints are presented that consists of reserve margin, fuel mix ratio, reliability criterion, upper construction limit, load demand, emission and fuel demand constraints.

Reserve Margin

The cumulative capacity of recently added unit has to meet the minimum as well as maximum reserve margin.

$$(1+Res_{min})\times L_t \leq \sum_{i=1}^{N} X_{t,i} \leq (1+Res_{max})\times L_t \tag{14}$$

Fuel Mix Ratio

Liquefied Natural Gas (LNG), oil, nuclear and coal are the various types of generating units considered in GEP. The chosen units together with the existing units of all types must meet the fuel mixture ratio.

$$FMR_{\min}^{j} \leq X_{t,j} \Big/ \sum_{i=1}^{N} X_{t,i} \leq FMR_{\max}^{j} \quad j=1,2,\ldots N \tag{15}$$

Reliability Criterion

This constraint establishes that chosen unit together with the existing units must meet reliability criterion on loss of load probability.

$$LOLP(X_t) \leq \varepsilon \tag{16}$$

Upper Construction Limit

The maximum tolerable number of new generating units to be committed during period t is reflected in this constraint.

$$0 \leq M_t \leq M_{\max,t} \tag{17}$$

Demand

The power demand constraint establish that the expected power demand is less than or equal to the cumulative capacity of the added unit at stage *t*.

$$\sum_{i=1}^{N} X_{t,i} \geq L_t \tag{18}$$

Emission Constraints

This constraint reflects the emission rate the pollutant *s* emitted by each unit of the type *i* should be less than or equal to the maximum allowance for emitting pollutant *s* in year *t*.

$$\sum_{i=1}^{N} X_{t,i} \times Em_{i,s} \leq q_{t,s} \tag{19}$$

Fuel Demand Constraints

This constraint establishes that fuel demand for each unit of the type *i* in the period *t* should be lower than the imported fuel.

$$\sum_{i=1}^{N} CCF_i \times EES_{t,i} - y_{t,i} \leq 0 \tag{20}$$

SUPPLY AND DEMAND SIDE MANAGEMENT OF GEP USING CGOA

In the typical GOA, there is no need to remain linearly decrements, however in the CGOA a chaotic variable differing C_1 and C_2 may be more advantageous for the search, which may also make the method converge to the optimum value with a fast speed. The exploitation and exploration of the whole swarm around the goal is balanced by the parameter C_1 balances using Equation (21). The values produced by the chaotic map as exposed in Gokhale and Kale, **(2016)** should recline in the interval [0, 1] as required by the GOA.

$$y_{qd} = C_1(t)\left(\sum_{\substack{m=1,\\ m\neq q}}^{Q} C_2(t)\left((UB_d - LB_d)/2\right)\times S\left(\left|y_{md} - y_{qd}\right|\right)\times\frac{\left(y_{md} - y_{qd}\right)}{dist\left(y_m, y_q\right)}\right) + y_d^* \tag{21}$$

Here, the value of C_1 is recouped by the values generated by chaotic maps in order to offer chaotic performance for the C_1 parameter. $C_1(t)$ in (21) performs the value produced from the chaotic map in the tth iteration as a substitute of linear decrement values by novel GOA presented in Saremi et al., **(2017)**. The parameter C_2 reduces the repulsion zone, attraction zone and comfort zone among grasshoppers which linearly reduces the space in order to conduct the grasshoppers to either exploit or explore.

Moreover, combination of above-mentioned parameters (C_1 and C_2) shows the subsequent explanation may help how the projected chaotic algorithms are theoretically effectual.

- The chaotic C_1 assist GOA to restore the places of grasshoppers chaotically that develops exploration.
- The chaotic C_2 allows GOA to alter the attraction zone, repulsion zone and comfort zone superior than GOA because there might be dissimilar values for C_2 parameter.
- Different chaotic maps for C_1 and C_2 give different exploitation and exploration samples for the GOA algorithm. Since, chaotic maps illustrate chaotic performance, a generation of GOA might offer prominence to either exploration or exploration.
- In provision for local optima entrap problem, the chaotic parameters lead GOA to exit from it.
- These chaotic parameters aid GOA to chaotically exploit the region, in case of finding an efficient part of search space.

The parameter C_1 is accountable for dropping the movements of grasshoppers just about the target which means that C_1 basically poises the exploitation and exploration in the area of the goal for the intact swarm. On the other hand, the parameter C_2 linearly reduces the space in order to direct the grasshoppers to locate the optimal solution in the search space.

An allied point to judge is that the adaptive parameter C_1 contributes to the diminution in repulsion or attraction forces among grasshoppers relative to the number of iterations, whereas C_2 decreases the search exposure around the target as the number of iteration raises. In order to poise the exploitation and exploration, C_1 is reduced proportionally with the rise in the number of iterations.

This approach allowed GOA to perform effective exploitation toward the later stages of the optimization. Similarly, the values of C_2 are decreased in order to reduce the comfort zone with respect to rise in the iteration count. Both parameters (C_1 and C_2) are considered as a single parameter and it is adapted using circle map by Equation (22):

$$x_{k+1} = x_k + b - \left(\frac{C}{2\pi}\right)\sin(2\pi x_k) \tag{22}$$

Figure 1. Flow chart for generation expansion planning

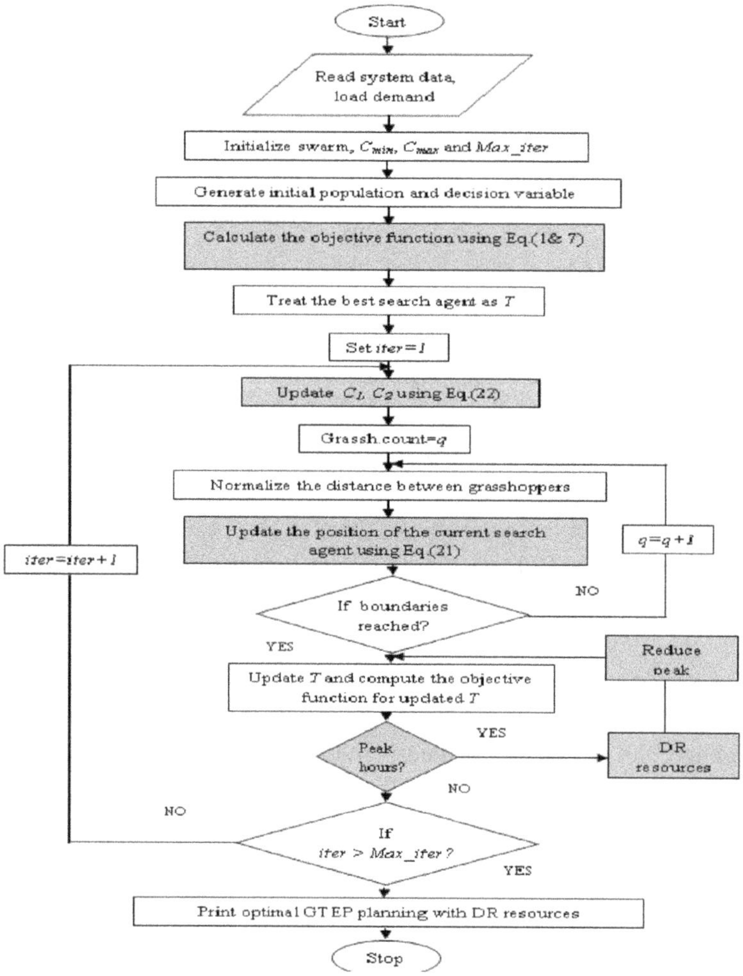

This equation will generate chaotic numbers between (0, 1) by using *C*=0.5 and *b*=0.2.

The step-by-step procedure of implementing CGOA for GEP problem for long-term planning is summarized as follows.

Step1: Read the test system data and load demand for study period.

Step 2: Initialize swarm and CGOA parameters such as C_{min}, C_{max} and maximum number of iterations (*Max-iter*).

Step 3: According to the swarm and number of decision variables initial population are generated randomly.

Step 4: Compute the total cost for each search agents using Equation (1&7) and subjected to constraints.

Step 5: Treat the best search agent as *T*.

Step 6: Iteration=iteration+1.

Step 7: Update C_1 and C_2 for current iteration using Equation (22).

Step 8: Repeat the following steps for all search agent.

Step 9: Adjust the space among grasshoppers.

Step 10: Renew the location of the present search agent by Equation (21).

Step 11: Bring the current search agent back if it goes outside the boundaries.

Step 12: Update *T* if there is a better solution.

Step 13: Repeat the procedure from step 6 until the maximum number of iteration is reached.

The flow chart for generation and transmission expansion planning with DR resources is shown in Figure 1.

SOLUTIONS AND RECOMMENDATIONS

The CGOA is developed in MATLAB 7.9 platform and is executed on personal computer configured with Intel core i3 processor 2.20 GHz and 2 GB random access memory. The test system with five types of candidate options and twelve types of existing units is considered for GEP. The load demand for 6-year planning horizon and additional data for GEP are captured from Kannan et al., (2009). 5000 MW is the forecasted peak demand at stage zero. The test system uses various types of units (depending upon the type of fuel used: Oil, LNG, CCGT, Nuclear and Coal) as candidate sets. 20% and 40% are the reserve margin of lower and upper bounds respectively, 0.1, 0.1, 0.15, 0.2 and 0.2 are the salvage factor (δ) for oil, LNG, coal, PWR and PHWR respectively, 0.05 $/KWh is the EENS cost and discounted rate is fixed to 8.5%.

Table 1. Optimal expansion and retirement planning using CGOA

Stage	No. of units selected					Existing units capacity (MW)	Newly added capacity (MW)
	Oil (200MW)	LNG C/C (450 MW)	CCGT (500 MW)	Nuc. (PWR) (1000 MW)	Coal (700 MW)		
1	1	1	**2**	**2**	0	5000	3650
2	1	0	0	1	1	8650	1900
3	0	1	1	0	0	10550	950
4	0	0	1	1	1	11500	2200
5	0	0	0	0	1	13700	700
6	1	2	0	0	0	14400	1100
7	2	1	0	1	0	15500	1850
8	1	0	2	0	0	17350	1200
9	1	0	0	0	1	18550	900
10	0	0	3	1	1	19450	3200
11	1	3	0	0	0	22650	1550
12	1	0	1	0	2	24200	2100

Table 2. Best solution obtained for investment and outage cost optimized individually

Cost ($)	Objective	
	Min. outage cost (NSGA-II)	Min. outage cost including unit retirement (CGOA)
Investment cost	1.26242e10	1.22871e10
Outage cost	3.23075e6	2.01954e6
Total cost	1.26275e+010	1.22891e+010
Computation Time (s)	800	700

It is assumed that a 7-year planning horizon is taken into account in the simulation procedure. The discount rate is locate to be 10%, the lower and upper bounds of LOLP are measured as 0.001 and 0.01, the price of wind energy production index is 10 $/MWh, the EENS cost is place at 50 $/MWh, and the forced outage rate for various technologies is ranged from 2% to 10%. Moreover, the limits of yearly utilization factor for different technologies like nuclear, oil, coal, CCGT and wind are assumed as 80%~90%, 40%~70%, 60%~80%, 50%~70% and 20%~40% respectively.

Figure 2. Capacity Addition of 24-Year Planning Horizon Including Unit Retirement

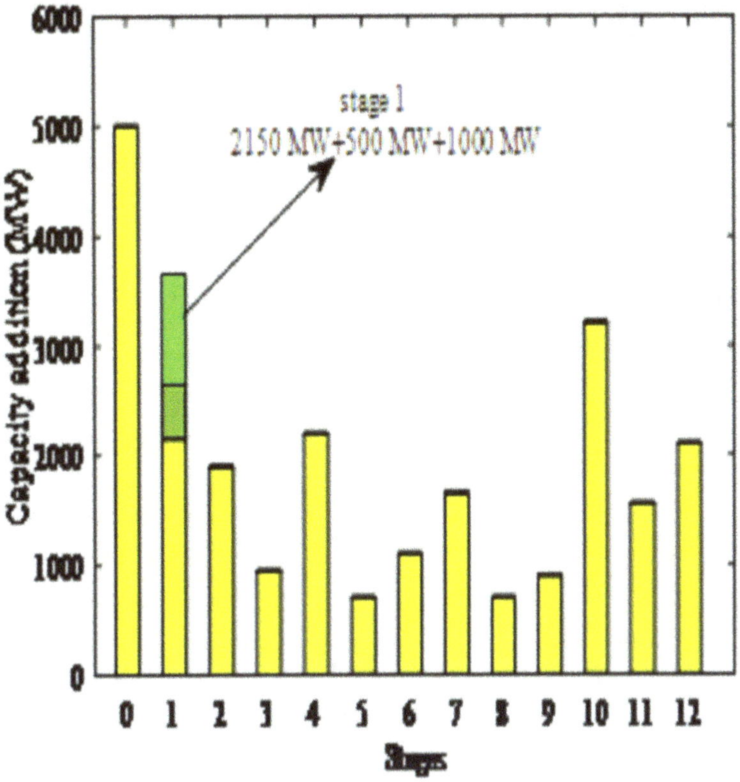

Co-summation of GEP Including Supply Side Management

In this case, the binary variables of the existing units are agreed to 1 in the first year to avoid their retirement in the starting point of the study. The abridged scenarios satisfied in the previous study are also used here as the preliminary set. The salvage cost associated with retired units is assumed to be zero (Khan et al. 2014; Khan et al. 2013; Khan et al. 2012; Negash et al., 2017; Negash et al., 2016; Jariso et al. 2017; Kifle et al. 2018; Yeshalem et al. 2017; Singh et al. 2019; Gupta et al., 2015). The existing generating units in pre-expansion state consist of fifteen generating units. It is assumed that coal#1 (2*250 MW), coal#2 (500 MW) and coal#3 (500 MW), units are retired in 24-year planning horizon. So, the existing unit capacity is reduced from 5000 MW to 3500 MW.

To compensate the retirement of the generating units, one CCGT (500 MW) and one Nuclear(PWR) (1000 MW) are additional generating unit in the post expansion condition also shown in Table 1, this clearly indicate the number of newly added unit for 24-year planning horizon including unit retirement using GOA. From the Table 1 one CCGT (500 MW) unit and one Nuclear(PWR) (1000 MW) unit is added in stage-1. Because of retirement, capacity in stage-1 is increased from 2150 MW to (2150MW +500MW+ 1000MW) 3650 MW. Results for NSGA-II and GOA in terms of investment cost, outage cost, and total costs are given in Table 2. The outage cost and total cost for GEP excluding unit retirement using GOA is 2.14940e6 and 1.28254e+010 respectively.

Figure 3. Comparison between GOA and NSGA-II

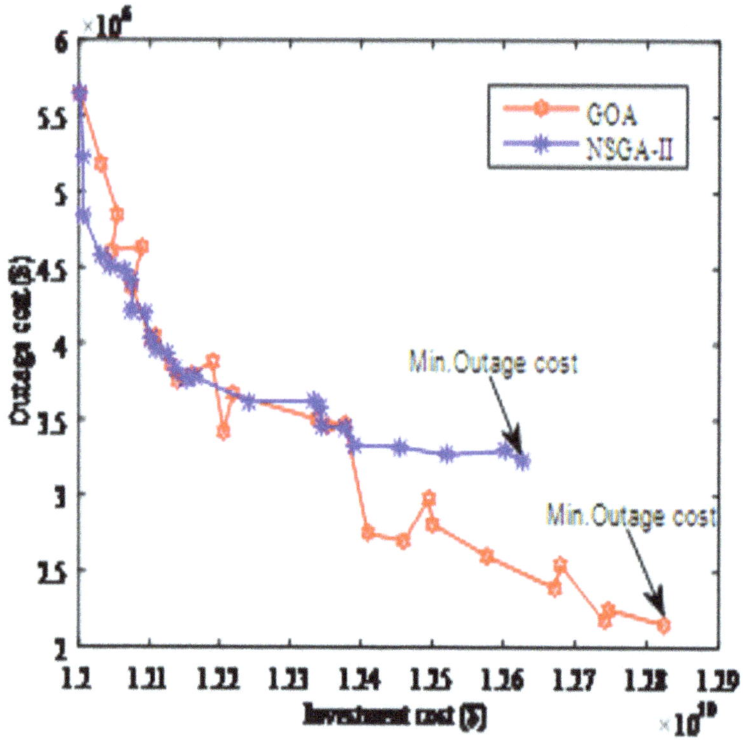

Table 3. Optimal generation expansion planning without DR using CGOA

Unit type / Time period	Nuclear	Oil	Coal	CCGT	Wind	CCS	Total added capacity (MW)	Total cost (M$)
Y_1	3	1	0	1	12	0	2052	10020.7
Y_2	2	0	1	4	10	3	3265	6979.31
Y_3	5	1	0	5	7	2	3972	6275.86
Y_4	1	3	1	2	9	4	2814	6048.28
Y_5	4	0	0	5	10	1	3695	5737.93
Y_6	1	4	3	3	7	2	3007	5613.79

Let assume the salvage costs linked with retired units are zero. Capacity addition, for 24-year planning horizon is shown in Figure 2, it is observed that by allowing the retirement option in GEP, two 200MW oil units retire in the 1st year, while an additional new 450MW unit is invested in the 2nd year, these decisions improve the cost function of GEP.

Table 4. Optimal generation expansion planning with DR using CGOA

Unit type/ Time period	Nuclear	Oil	Coal	CCGT	Wind	CCS	Total added capacity (MW)	Total cost (M$)
Y_1	2	1	3	0	12	2	1921	8262.15
Y_2	2	0	1	1	13	3	2297	5782.85
Y_3	5	1	0	4	8	2	3658	5535.19
Y_4	1	3	2	1	9	4	2464	5102.57
Y_5	4	0	3	2	10	1	2645	4504.33
Y_6	1	4	1	1	11	2	2411	4422.71

Figure 4. Weekly peak load reduction for GEP using CGOA

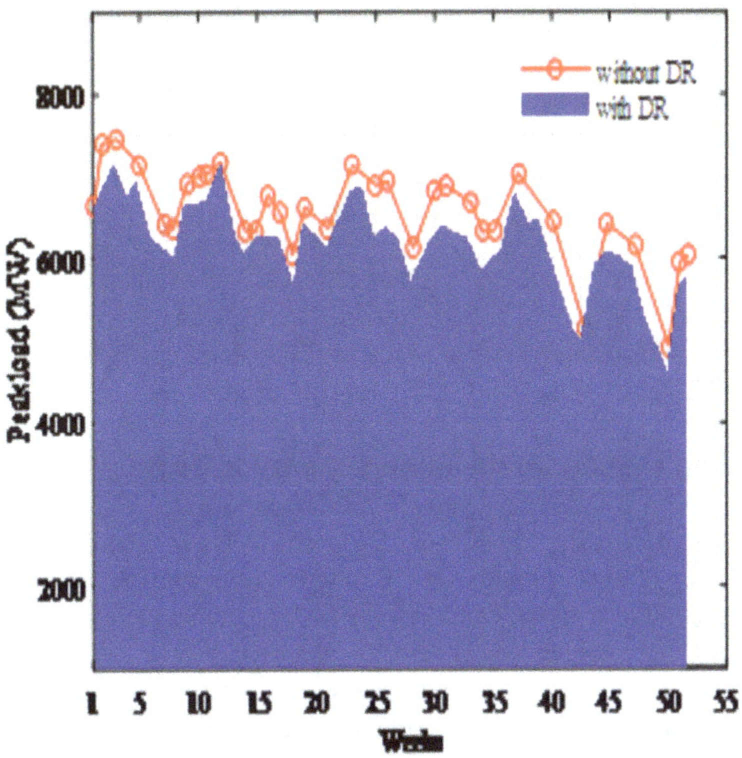

In these simulations salvage cost is assumed to be zero, the retirement choices of various existing generating units are neither through their substantial contributions in the system unreliable nor their costly operations (including fuel and O&M costs). While considering 24-years planning, capacity addition in the stage-1 is increased from 2150 MW to (2150MW +500MW+ 1000MW) 3650 MW.

Figure 5. Total Cost for GEP using CGOA

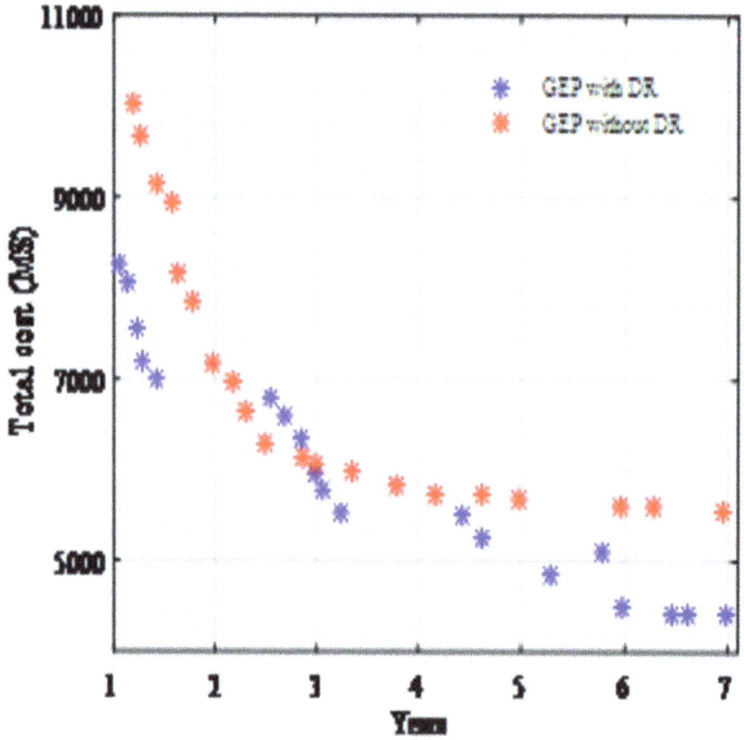

Comparison between GOA and NSGA-II is shown in Figure 3. From the comparison between NSGA-II and GOA it is obvious that outage cost is decreased and a viable solution is reached. The outage cost and total cost using GOA is 2.14940e6 and 1.28254e+010 respectively.

Corollary Including Demand Side Management for GEP

Table 3 exhibits that the generation technologies by means of lower economic cost will be preferred over the planning horizon if the decision-maker is decided on achieving the lowest-cost plan along with the eco-friendly generation technologies will be specified greater deliberation in the lowest emission planning scheme, when the uncertain load demand is calculated.

In addition, the total system cost, including the utility investment and operational costs without DR is 10020 M$. Considering the collision of the uncertain DR resources on the proposed generation expansion planning, the optimal robust expansion plans are offered in Table 4.

With the consideration of saving economic cost in the planning, as many CCGT generation units as feasible are accommodated in the investment plan. The results in Table 4 shows total system cost for GEP with DR concerning the uncertain DR resources is 8262 M$, which demonstrates that the DR resources has been the potential to keep generation capacity investment.

Simulation results for GEP without and with DR is explained in following studies, theoretically DR has the ability of not only dropping the peak load to relieve up the forces of generation expansion, but also helping a feasible regulation resource in the power system to support the wind power integration capacity. The outcomes of the numerical study establish these characteristics.

The results of the weekly peak load decline of GEP using CGOA are established in Figure 4 and peak demands are 7452 MW, 7168 MW for without and with DR respectively. Compared with GEP without DR, the projected model can attain a higher peak load reduction, since it organizes the optimal DR solution, while attaining the optimal investment and operation plans.

In other words, the proposed model is a valuable tool and also helping system operators create assessments on when and how much DR resources should be utilized. The harmonized use of DR is significant in terms of ensuring the economical, safe and reliable operations of a power system with power saturation. In addition Figure 4 shows the weekly peak load comparison between GEP without and with DR respectively.

Figure 5 shows the total cost comparison between GEP without DR and with DR correspondingly via CGOA. From Figure 5, the total cost for GEP with DR is much lower, as compared with GEP without DR.

FUTURE RESEARCH DIRECTIONS

This work can be further extended to examine various other pertaining issues such as-

- Consideration of the volatility of different market parameters such as fuel prices and the electricity market price. In the present work sensitive analysis has been carried out to examine their effects, comprehensive analysis is developed in the future.
- The uncertainty associated with price and demand variations in the long-term may also is considered in detail and appropriate stochastic modeling framework need to be develop to examine their effects.

CONCLUSION

With increasing demand for energy worldwide and the resources being limited or becoming ever more expensive, it is important (and usually cost effective) to improve the efficiency of energy supply. In turn, this usually means a benefit to the energy consumer in terms of lower energy prices. Improved efficiency on the supply side will also make a valuable contribution to reducing the impact of energy use on the environment. While demand-side improvements are certainly important, supply-side options also need to be identified, evaluated and implemented where the economics justify.DSM in its various forms is an important tool for enabling a more efficient use of the energy resources available to a country. DSM employed to electric power systems can palliate electrical system crises, decrease swoon and improve system reliability, abridge need on costly imports, decrease energy prices, afford relief to the power grid and generation plants, defer investments in generation, transmission and distribution networks and contribute to lower environmental emissions. Similar benefits can be achieved from DSM when applied to the use of other types of energy. Thus DSM can offer significant economic and environmental benefits.

REFERENCES

Abdin, I. F., & Zio, E. (2018). An integrated framework for operational flexibility assessment in multi-period power system planning with renewable energy production. *Applied Energy*, *222*, 898–914. doi:10.1016/j.apenergy.2018.04.009

Alhelou, H., Hamedani-Golshan, M. E., Zamani, R., Heydarian-Forushani, E., & Siano, P. (2018). Challenges and opportunities of load frequency control in conventional, modern and future smart power systems: A comprehensive review. *Energies*, *11*(10), 2497. doi:10.3390/en11102497

Alhelou, H. H. (2018). *Fault detection and isolation in power systems using unknown input observer. In Advanced condition monitoring and fault diagnosis of electric machines* (p. 38). Hershey, PA: IGI Global.

Alhelou, H. H., Golshan, M., & Fini, M. (2018). Wind driven optimization algorithm application to load frequency control in interconnected power systems considering GRC and GDB nonlinearities. *Electric Power Components and Systems, 46*(11-12), 1223-1238.

Alhelou, H. H., & Golshan, M. E. H. (2016, May). Hierarchical plug-in EV control based on primary frequency response in interconnected smart grid. In *2016 24th Iranian Conference on Electrical Engineering (ICEE),* (pp. 561-566). IEEE. 10.1109/IranianCEE.2016.7585585

Alhelou, H. H., & Golshan, M. E. H. (2020). Decision-making-based optimal generation-side secondary-reserve scheduling and optimal LFC in deregulated interconnected power system. In *Decision making applications in modern power systems* (pp. 269–299). Academic Press. doi:10.1016/B978-0-12-816445-7.00011-6

Alhelou, H. H., Golshan, M. E. H., & Hatziargyriou, N. D. (2019). Deterministic dynamic state estimation-based optimal LFC for interconnected power systems using unknown input observer. *IEEE Transactions on Smart Grid*, 1. doi:10.1109/TSG.2019.2940199

Alhelou, H. H., Golshan, M. E. H., & Hatziargyriou, N. D. (2019). A decentralized functional observer based optimal LFC considering unknown inputs, uncertainties and cyber-attacks. *IEEE Transactions on Power Systems.*

Alhelou, H. H., Golshan, M. E. H., Zamani, R., Moghaddam, M. P., Njenda, T. C., Siano, P., & Marzband, M. (2019, June). An improved UFLS scheme based on estimated minimum frequency and power deficit. In 2019 IEEE Milan PowerTech (pp. 1-6). IEEE. doi:10.1109/PTC.2019.8810497

Alhelou, H. H., Golshan, M. H., & Askari-Marnani, J. (2018). Robust sensor fault detection and isolation scheme for interconnected smart power systems in presence of RER and EVs using unknown input observer. *International Journal of Electrical Power & Energy Systems*, *99*, 682–694. doi:10.1016/j.ijepes.2018.02.013

Alhelou, H. H., Hamedani-Golshan, M. E., Heydarian-Forushani, E., Al-Sumaiti, A. S., & Siano, P. (2018, September). Decentralized fractional order control scheme for LFC of deregulated nonlinear power systems in presence of EVs and RER. In *2018 International Conference on Smart Energy Systems and Technologies (SEST)* (pp. 1-6). IEEE. 10.1109/SEST.2018.8495858

Alhelou, H. S. H., Golshan, M. E. H., & Fini, M. H. (2015, December). Multi agent electric vehicle control based primary frequency support for future smart micro-grid. In *2015 Smart Grid Conference (SGC)* (pp. 22-27).

Anteneh, D., & Khan, B. (2019). Reliability enhancement of distribution substation by using network reconfiguration a case study at Debre Berhan Distribution Substation. *International Journal of Economy, Energy, & Environment*, *4*(2), 33.

Banteywalu, S., Khan, B., De Smedt, V., & Leroux, P. (2019). A novel modular radiation hardening approach applied to a synchronous buck converter. *Electronics (Basel)*, *8*(5), 513. doi:10.3390/electronics8050513

Chen, F., Huang, G., & Fan, Y. (2015). A linearization and parameterization approach to tri-objective linear programming problems for power generation expansion planning. *Energy*, *87*, 240–250. doi:10.1016/j.energy.2015.04.104

Debnath, K. B., & Mourshed, M. (2018). Forecasting methods in energy planning models. *Renewable & Sustainable Energy Reviews*, *88*, 297–325. doi:10.1016/j.rser.2018.02.002

Energy efficiency and demand side management, Course Module, University of Warwick, REEEP, April, 2005.

Ervural, B. C., Evren, R., & Delen, D. (2018). A multi-objective decision-making approach for sustainable energy investment planning. *Renewable Energy*, *126*, 387–402. doi:10.1016/j.renene.2018.03.051

Fini, M. H., Yousefi, G. R., & Alhelou, H. H. (2016). Comparative study on the performance of many-objective and single-objective optimisation algorithms in tuning load frequency controllers of multi-area power systems. *IET Generation, Transmission, & Distribution*, *10*(12), 2915–2923. doi:10.1049/iet-gtd.2015.1334

Flores-Quiroz, A., Palma-Behnke, R., Zakeri, G., & Moreno, R. (2016). A column generation approach for solving generation expansion planning problems with high renewable energy penetration. *Electric Power Systems Research*, *136*, 232–241. doi:10.1016/j.epsr.2016.02.011

Gokhale, S. S., & Kale, V. S. (2016). An application of a tent map initiated chaotic firefly algorithm for optimal overcurrent relay coordination. *International Journal of Electrical Power & Energy Systems*, *78*, 336–342. doi:10.1016/j.ijepes.2015.11.087

Gupta, K., Khan, B., & Mubeen, S. E. (2015). Available transfer capability enhancement by unified power flow controller. In *Proceedings IEEE International Conference on Signal Processing, Informatics, Communication, and Energy Systems (SPICES)*, pp. 1, 4, 19-21.

Haes Alhelou, H., Hamedani-Golshan, M. E., Njenda, T. C., & Siano, P. (2019). A survey on power system blackout and cascading events: Research motivations and challenges. *Energies*, *12*(4), 682. doi:10.3390/en12040682

Haes Alhelou, H., Hamedani-Golshan, M. E., Njenda, T. C., & Siano, P. (2019). Wide-area measurement system-based optimal multi-stage under-frequency load-shedding in interconnected smart power systems using evolutionary computing techniques. *Applied Sciences*, *9*(3), 508. doi:10.3390/app9030508

Haes Alhelou, H., Hamedani Golshan, M. E., Njenda, T. C., & Siano, P. (2019). WAMS-based online disturbance estimation in interconnected power systems using disturbance observer. *Applied Sciences*, *9*(5), 990. doi:10.3390/app9050990

Hemmati, R., Saboori, H., & Jirdehi, M. A. (2016). Multistage generation expansion planning incorporating large scale energy storage systems and environmental pollution. *Renewable Energy*, *97*, 636–645. doi:10.1016/j.renene.2016.06.020

Hua, B., Baldick, R., & Wang, J. (2018). Representing operational flexibility in generation expansion planning through convex relaxation of unit commitment. *IEEE Transactions on Power Systems*, *33*(2), 2272–2281. doi:10.1109/TPWRS.2017.2735026

Jariso, M., Khan, B., Tanwar, S., Tyagi, S., & Rishiwal, V. (2018). Hybrid energy system for upgrading the rural environment, IEEE GlobeCom 2018.

Jariso, M., Khan, B., Tesfaye, D., & Singh, J. (2017). Modelling and designing of stand-alone photovoltaic system (Case study: Addis Boder Health Center southwest Ethiopia). In *Proceedings IEEE International Conference on Electronics, Communication, and Aerospace Technology, ICECA 2017*, pp. 168-173, Coimbatore, India.

Kannan, S., Baskar, S., McCalley, J. D., & Murugan, P. (2009). Application of NSGA-II algorithm to generation expansion planning. *IEEE Transactions on Power Systems*, *24*(1), 454–461. doi:10.1109/TPWRS.2008.2004737

Karunanithi, K., Saravanan, S., Prabakar, B. R., Kannan, S., & Thangaraj, C. (2017). Integration of demand and supply side management strategies in generation expansion planning. *Renewable & Sustainable Energy Reviews*, *73*, 966–982. doi:10.1016/j.rser.2017.01.017

Khan, B., & Agnihotri, G. (2012). A novel transmission loss allocation method based on transmission usage. In *Proceedings IEEE Fifth Power India Conference*, Haryana, India. 10.1109/PowerI.2012.6479479

Khan, B., Agnihotri, G, & Gupta, G. (2013). A multipurpose matrices methodology for transmission usage, loss and reliability margin allocation in restructured environment, *Electrical & Computer Engineering: An International Journal, 2*(3).

Khan, B., Agnihotri, G., Rathore, P., Mishra, A., & Naidu, G. (2014). A cooperative game theory approach for usage and reliability margin cost allocation under contingent restructured market. *International Review of Electrical Engineering*, *9*(4), 854–862.

Khan, B., Alhelou, H. H., & Mebrahtu, F. (2019). A holistic analysis of distribution system reliability assessment methods with conventional and renewable energy sources. *AIMS Energy*, *7*(4), 413–429. doi:10.3934/energy.2019.4.413

Khan, B., & Singh, P. (2017). The current and future states of Ethiopia's energy sector and potential for green energy: A comprehensive study. *International Journal of Engineering Research in Africa*, *33*, 115–139. doi:10.4028/www.scientific.net/JERA.33.115

Khan, B., & Singh, P. (2017). Optimal power flow techniques under characterization of conventional and renewable energy sources: a comprehensive analysis, *Journal of Engineering, vol. 2017,* Article ID 9539506.

Khan, B., & Singh, P. (2017). Selecting a meta-heuristic technique for smart micro-grid optimization problem: A comprehensive analysis. In IEEE Access, 5, pp. 13951-13977.

Khan, B., & Singh, P. (2019). Economic operation of smart micro-grid: a meta-heuristic approach. In Handbook of research on smart power system operation and control (pp. 330-346). Hershey, PA: IGI Global. doi:10.4018/978-1-5225-8030-0.ch014

Khan, B., & Tanwar, S. (2019). Issues associated with microgrid integration. In Handbook of research on smart power system operation and control (pp. 252-264). Hershey, PA: IGI Global. doi:10.4018/978-1-5225-8030-0.ch010

Kifle, Y., Khan, B., & Singh, J. (2018). Designing and modeling grid connected photovoltaic system: (Case study: EEU building at Hawassa city). *International Journal of Convergence Computing*, *3*(1), 20–34. doi:10.1504/IJCONVC.2018.091113

Kifle, Y., Khan, B., & Singh, P. (2018). Assessment and enhancement of distribution system reliability by renewable energy sources and energy storage, *Journal of Green Engineering, 8*(3), 2, pp. 219-262.

Li, S., Coit, D. W., & Felder, F. (2016). Stochastic optimization for electric power generation expansion planning with discrete climate change scenarios. *Electric Power Systems Research, 140*, 401–412. doi:10.1016/j.epsr.2016.05.037

Makdisie, C., Haidar, B., & Alhelou, H. H. (2018). An optimal photovoltaic conversion system for future smart grids. In Handbook of research on power and energy system optimization (pp. 601–657). Hershey, PA: IGI Global. doi:10.4018/978-1-5225-3935-3.ch018

Manickavasagam, M., Anjos, M. F., & Rosehart, W. D. (2015). Sensitivity-based chance- constrained generation expansion planning. *Electric Power Systems Research, 127*, 32–40. doi:10.1016/j.epsr.2015.05.011

Molla, T., Khan, B., Moges, B., Alhelou, H. H., Zamani, R., & Siano, P. (2019). Integrated optimization of smart home appliances with cost-effective energy management system. *CSEE Journal of Power and Energy Systems*, *5*(2), 249–258.

Molla, T., Khan, B., & Singh, P. (2018). A comprehensive analysis of smart home energy management system optimization techniques, *Journal of Autonomous Intelligence, 1*(1), 15–21.

Negash, K., Khan, B., Tesfaye, D., & Gayathri, M. (2017). Optimal placement of phasor measurement unit for system observability. In *Proceedings of IEEE International Conference on Computing Methodologies and Communication, (ICCMC 2017),* pp. 73-77, Erode, India.

Negash, K., Khan, B., & Yohannes, E. (2016). Artificial intelligence versus conventional mathematical techniques: A review for optimal placement of phasor measurement units [Springer]. *Journal of Technology and Economics of Smart Grids and Sustainable Energy*, *1*(1), 10. doi:10.100740866-016-0009-y

Njenda, T. C., Golshan, M. E. H., & Alhelou, H. H. (2018, November). WAMS based intelligent under frequency load shedding considering online disturbance estimation. In *2018 Smart Grid Conference (SGC)* (pp. 1-5). 10.1109/SGC.2018.8777779

Njenda, T. C., Golshan, M. E. H., & Alhelou, H. H. (2018, November). WAMS based under frequency load shedding considering minimum frequency prediction and extrapolated disturbance magnitude. In *2018 Smart Grid Conference (SGC)* (pp. 1-5).

Park, H., & Baldick, R. (2015). Stochastic generation capacity expansion planning reducing greenhouse gas emissions. *IEEE Transactions on Power Systems*, *30*(2), 1–9. doi:10.1109/TPWRS.2014.2386872

Rajesh, K., Bhuvanesh, A., Kannan, S., & Thangaraj, C. (2016). Least cost generation expansion planning with solar power plant using differential evolution algorithm. *Renewable Energy*, *85*, 677–686. doi:10.1016/j.renene.2015.07.026

Saini, S. (2004). Conservation vs. generation-The significance of demand side management (DSM), its tools and techniques. *Energy Management,* 52-54.

Saremi, S., Mirjalili, S., & Lewis, A. (2017). Grasshopper optimization algorithm: Theory and applications. *Advances in Engineering Software*, *105*, 30–47. doi:10.1016/j.advengsoft.2017.01.004

Singh, P., & Khan, B. (2017). Smart microgrid energy management using a novel artificial shark optimization, Complexity, vol. 2017, Article ID 2158926. doi:10.1155/2017/2158926

Singh, P., Khan, B., Vidyarthi, A., Haes Alhelou, H., & Siano, P. (2019). Energy-aware online non-clairvoyant scheduling using speed scaling with arbitrary power function. *Applied Sciences*, *9*(7), 1467. doi:10.3390/app9071467

Sustainable energy regulation and policymaking training manual-demand side management module 14.

Sustainable energy regulation and policymaking training manual-supply side management module 13.

Yeshalem, M. T., & Khan, B. (2017). Design of an off-grid hybrid PV/wind power system for remote mobile base station: A case study. *AIMS Energy*, *5*(1), 96–112. doi:10.3934/energy.2017.1.96

Zamani, R., Golshan, M. E. H., Alhelou, H. H., & Hatziargyriou, N. (2019). A novel hybrid islanding detection method using dynamic characteristics of synchronous generator and signal processing technique. *Electric Power Systems Research*, *175*, 105911. doi:10.1016/j.epsr.2019.105911

Zamani, R., Hamedani-Golshan, M. E., Haes Alhelou, H., Siano, P., & Pota, H. R. (2018). Islanding detection of synchronous distributed generator based on the active and reactive power control loops. *Energies*, *11*(10), 2819. doi:10.3390/en11102819

KEY TERMS AND DEFINITION

Demand-Side Management: (**DSM**): It is the modification of consumer demand for energy through various methods such as financial incentives and behavioral change through education.

Generation Expansion Planning: It is finding an optimal solution for the planning problem in which the installation of new generation units satisfies both technical and financial limits.

Supply-Side Management (SSM): It refers to actions taken to ensure the generation; transmission and distribution of energy are conducted efficiently.

Chapter 13
Micro-Grid Planning and Resilience Within Bulk System Planning and Operation

Tefaye Belay
Mettu University, Ethiopia

ABSTRACT

Micro grid is widely used in real worlds for advanced forecasting and demand response of renewable energy source, grid integration, and operations. Micro grid consists of conventional and nonconventional energy source such as wind energy, solar energy, biomass energy, hydro power, diesel power, fuel cell, geothermal power, thermal power, etc. Micro grid is a combination of AC power and DC power such as wind, solar, fuel cell, biomass, and Hydro power, which is mostly used in micro grids. Grid can be operated by grid connected mode or islanding modes. Micro grid is classified into traditional micro grids and smart micro grids.

INTRODUCTION

Smart grid is the future of power grid in the real world with regards to powers systems planning, operation, design and control (Akshata et al. 2014, SGETP, 2008). Microgrid is also the feature of smart grid. It consists of renewable and non-renewable energy source such as hydro power, solar power, wind power, geothermal, thermal power, biomass, fuel cell, gas storage and many more (Dohn, et al. 2011, IEEE, 2009). Micro grid includes two main systems in it i.e. distributed generator (DG) and distribution storage (DS) (Gonen et al. 2013). Micro grid can be operates either grid connected mode and islanding mode (Mehrizi-Sani et al. 2010). In grid connected mode, micro grid is connected to the utility grid and the power for customer is sufficiently available as compare to islanded mode in which it is disconnected from the grid. This switching between micro grid and main grid is performed to connect and disconnect the micro grid from the main grid for various purposes (Lidula et al.2011, Mehrizi-Sani et al. 2010). Generally micro grid switching for grid connected mode or islanding mode is performed by circuit breaker or protective relay and when circuit breaker or relay is closed it is called grid connected

DOI: 10.4018/978-1-7998-1230-2.ch013

Figure 1. Conventional Power system (Nikmehr et al. 2015, Wu D et al. 2016, Mahmoodi et al. 2015, Hina Fathima et al. 2015)

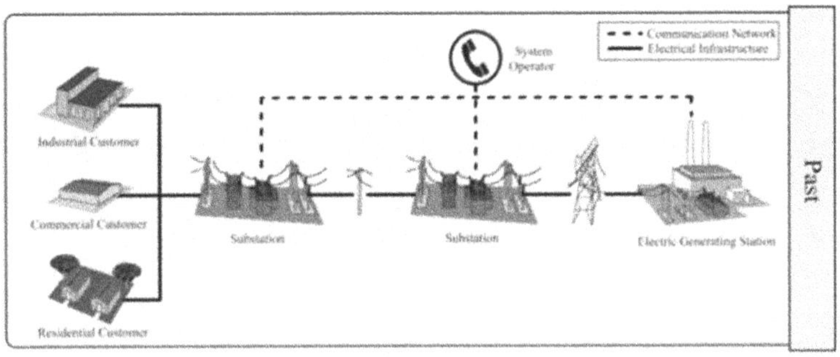

mode, when it is open it is called islanded mode (Alhelou et al., 2019; Makdisie et al., 2018; Alhelou et al., 2018; Alhelou et al., 2016; Haes Alhelou et al., 2019; Njenda et al., 2018). This connection and disconnection is performed at point of common coupling (PCC) (Che et al. 2014). Switch of micro grid can be made in and out by the transmission systems (Kuo et al. 2011). Micro grid is consists of AC power as well as DC power, when say AC power, it is from main grid and when say DC power, it is from solar. In the grid connected mode, Micro grid can supply sufficient power while in islanded mode the power sometimes reached to lower level as per the availability of the resources (Akshata et al. 2014, Mehrizi-Sani et al. 2010). Micro grid can be used to reduce the global warming (Khan et al., 2019; Khan et al., 2018; Khan et al., 2017; Banteywalu et al., 2019; Anteneh et al., 2019; Molla et al., 2019, Molla et al., 2018, Jariso et al. 2018).

EVOLUTION OF POWER SYSTEMS

There are three power evolutions in power systems.

Conventional Power Systems

As shown in figure 1, conventional power systems are traditional power system model, which consists of Generator, transformer, transmission line and distribution load. Load can be consists of industrial, residential and commercial customers. In conventional power system, power losses are high due to traditional power system component. In this power systems, there is no automatic or intelligence transmission and distribution control (Nikmehr et al. 2015, Wu D et al. 2016).

Present Power System

As shown in figure 2, power systems is modern power systems model in power systems, which consists of generator, transformer, transmission line and distribution load. Load can be consists of industrial, residential and commercial loads. In the present power systems, power losses are low due to automatic power system component. As compare to traditional power system, present power systems has automatic transmission and distribution control (Nikmehr et al. 2015, Wu D et al. 2016, Mahmoodi et al. 2015,

Figure 2. Present Power system (Nikmehr et al. 2015, Wu D et al. 2016, Mahmoodi et al. 2015, Hina Fathima et al. 2015)

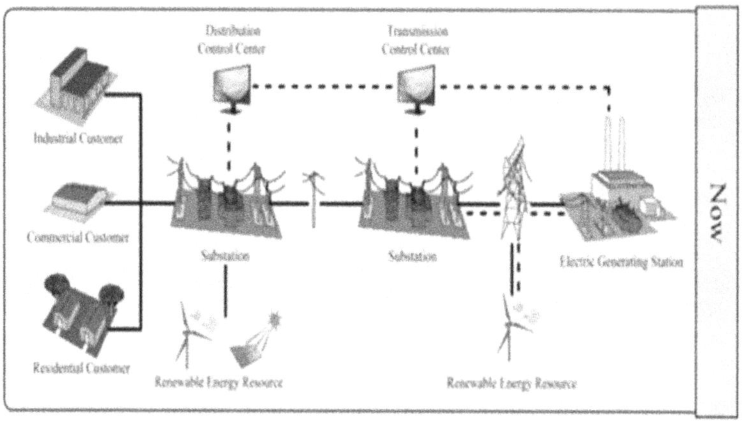

Hina Fathima et al. 2015; Fini et al., 2016; Alhelou et al., 2018; Zamani et al., 2018; Alhelou et al., 2015; Njenda et al., 2018; Haes Alhelou et al., 2018; Haes Alhelou et al., 2019).

Future Power System

As shown in figure 3, Future power system is the latest power systems model in which generator, transformer, transmission line and distribution load is presented. Load can be consists of industrial, residential and commercial loads. In future power system, power losses are low due to advanced intelligence power systems component. As compare to present power system, future power system has advanced artificial intelligence based smart transmission control, service provider and distribution control (Nikmehr et al. 2015, Wu D et al. 2016, Mahmoodi et al. 2015, Hina Fathima et al. 2015).

CONCEPT OF MICRO GRID

Definition of Micro Grid

Micro grid is the small power plant (Energy source), which is the combination of renewable energy and conventional energy source. Micro grid can be implemented on small area or small community. Micro grid is operated either in grid connected mode or islanded mode. In most of cases micro grid is consists of renewable energy source such as wind, solar, geothermal, biomass and hydropower. Hydropower plant can act as main grid for micro grids, by which small grid is connected.

Component of Micro Grid

Micro grid is consists of solar energy and wind energy in most of the cases but it may includes fuel cell, biomass, diesel and many more resources. The old micro grid also connected to diesel, thermal and other conventional energy sources.

Figure 3. Future Power system (Nikmehr et al. 2015, Wu D et al. 2016, Mahmoodi et al. 2015, Hina Fathima et al. 2015)

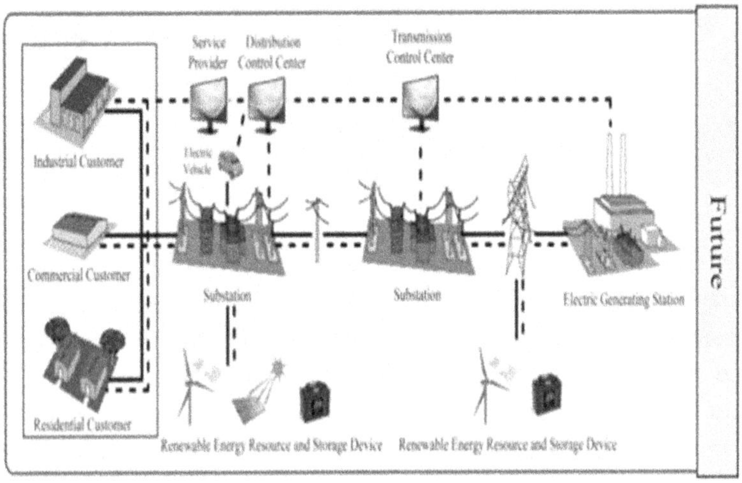

Wind Energy

Wind energy is the energy, which is produced by using speed of wind. Kinetic energy of wind is the input for wind power plants. It means wind is used to rotate the wind turbines. Wind turbines are connected to generators in order to produce electricity. Generator converts mechanical energy into electrical energy, so wind turbine produced mechanical energy and generator converted this energy into electrical energy in wind power plants. Wind power plant contains wind turbine, blade, pole, and generator (Hajipour et al. 2015; Khan et al. 2014; Khan et al. 2013; Khan et al. 2012; Negash et al., 2017; Negash et al., 2016; Jariso et al. 2017; Kifle et al. 2018; Yeshalem et al. 2017; Singh et al. 2019; Gupta et al., 2015). There are two types of wind turbine normally utilized in wind power plants. These are horizontal axis wind turbine (HAWT) and vertical axis wind turbine (VAWT). Horizontal axis wind turbine is perpendicular to gear box and it's mostly used in power plants. Vertical axis wind turbine is above the of gear box and are the conventional wind turbines, which produced low power compare to horizontal axis wind turbines (Bove et al. 2012, Shokrzadeh et al. 2015). Total power from the wind is equal to the half of the product of the air density, area cover by blade and cubic of input speed.

$$\mathcal{P} = \frac{1}{2}\rho' \, \mathbb{V}^3$$

Where P is power produce by wind power, ρ is the Air density, A is the Area of blades and V is the wind speed. As shown in figure 4 the operation of wind power plant is as follows.

Figure 4. Block diagram of wind energy generation plant

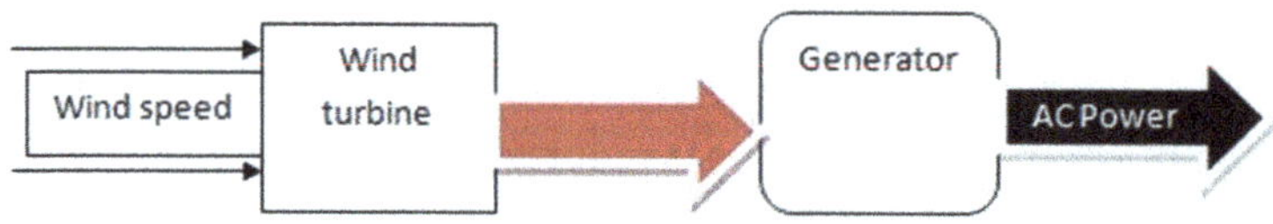

Solar Energy

Solar energy is the energy available from the sun light and harnessed by using cell, module and array. These cells converted the sunlight into DC current. This means solar photovoltaic generation plant consists of cell, module and array. Cells are the basic element of the solar generation plant, which together develop modules and modules develop arrays. There are different types of solar power plant based on their solar energy conversion techniques such as, solar thermal and photo voltaic (PV) plants. Solar thermal is not used in micro grid, because energy from it is used for boiler not for electricity. PV cell directly produced electricity and it's suitable for micro grid (Kuo et al. 2001). As shown in figure 5 the operation PV cell is as follows:

Fuel Cell

Fuel cell is the important element of modern micro grid, which utilized oxygen and hydrogen gases to produce electricity, heat and liquid. The initial cost of fuel cell is very high but the running cost is low as compare to wind, solar, hydro, and geothermal. This element of micro grid is very much utilized in modern micro grids (Dragicevi et al., 2016, Saboori et al., 2015). As shown in figure 6, the block diagram of fuel cell based generating plant is as follows:

Bulk Power System or Utility Grid

Bulk power system means utility grid, which produce huge power when compare to micro grid. It acts as main grid for other grid integrations. Bulk system consists of generator, transformer, transmission line and loads. Generator generates electricity from turbine of hydro or thermal power plants and transformer is used to step up and down the voltage levels, transmission line is carries power from generation to Load. Load utilized the electrical energy and it's classifies as residential, commercial and industrial load. In bulk system many protective devices such as relay, circuit breaker and isolator are available. Figure 7 presented the block diagram of bulk power system:

Figure 5 block diagram of solar power.

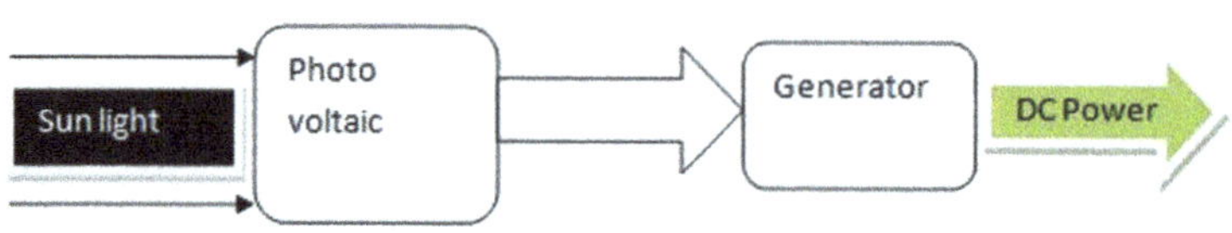

Figure 6. block diagram of fuel cell based generating plant

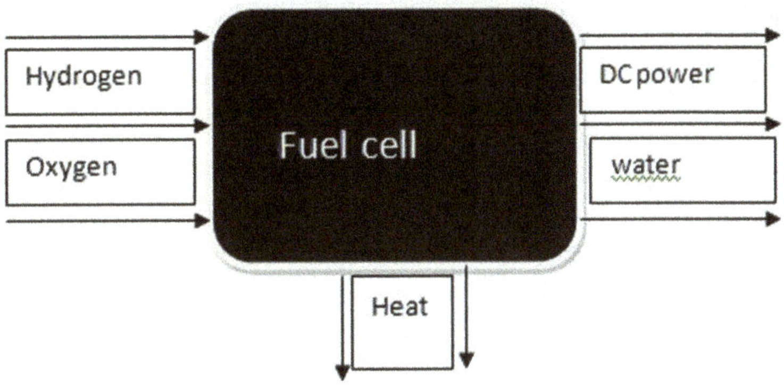

Battery

Battery is the storage device which is used to store the electrical energy. It is utilized in micro grid in order to store energy generated from intermittent renewable energy sources such as wind and solar. Battery the basic necessity for the micro grid to store generated power in modern micro grids (Bahmani-Firouzi B, 2014).

Converter

Converter is the device used to convert power from AC to DC, DC to AC, AC to AC or DC to DC. Converters generally utilized in micro grids are chopper, inverter, and rectifier. Converter is acts as translator of power in power systems.

Types of Micro Grid

Micro grid is classified into two categories based on the component and technology used:

Traditional Micro Grid

Traditional micro grid is the oldest micro grid, which used conventional elements of micro grids such as battery storage, power converter, and diesel power. Power loss in traditional micro grid is high when compare to smart micro grids. Generally traditional micro grid has one direction flow of electricity and information. It isn't used by service provider and other services (Mohsenian-Rad et al., 2012, ETPSmart-Grids et al., 2008).

Figure 7. Block diagram of bulk systems.

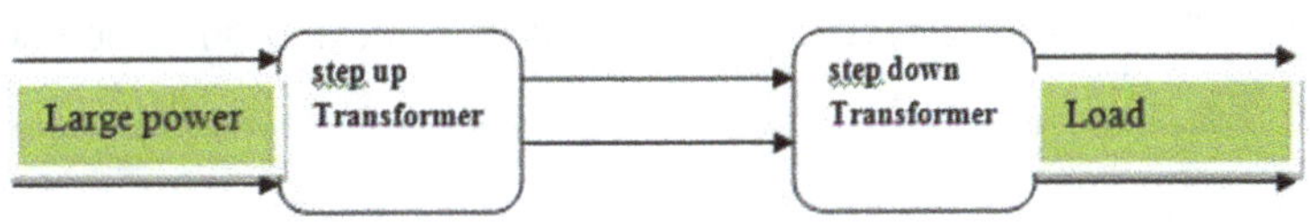

Table 1. Comparison of Traditional Power Grids and Smart Micro grids (Akshata et al. 2014)

Performance Measure	Traditional Power Grids	Smart Micro grids
Cost	High	Low
Fuel Efficiency	Low efficiency is 30% to 50%.	High from 70%–90%
Emissions	Large	small
Security	Major	minor
Construction Constraints	Difficult	Easy.
Reliability	Low	High

Smart Micro Grid

Smart micro grid is modern micro grid, which used advanced elements of micro grids such as advanced energy storage devices, advanced power electronic devices-based power converters, and artificial intelligence based switches and protective relay, and modern communication infrastructure. Power losses in the smart micro grid are very low as compare to traditional micro grids. Generally, smart micro grid has multi direction flow of electricity and information. It's used by service provider and other services (LaCommare et al., 2004, Lu, X., Bahramirad et al., 2015, Mohsenian-Rad et al., 2012, Wang et al., 2015b). In smart micro grid, most of the power plants are based on renewable energy sources such as wind energy, solar energy, and biomass energy.

Features of Micro Grid

Micro grid has energy sources which include renewable and conventional energy sources operated in parallel with main grid (bulk system) and distributed generators. The main advantages of micro grid is produce green energy, reduce air pollution or greenhouse gases (GHG). The main aim of micro grid is to produce reliable and secure power to commercial, residential and industrial customers.

Table 1 presented the Comparison of Traditional Power Grids and Smart Micro grids:

OPERATION OF MICRO GRID

Micro Grid Operation

The main aim of micro grid is to produce power in the real world. It has the advantage of green energy consumption, low carbon emission and less tariff as compare to conventional power plant. Micro grid can be operated in grid connected mode or islanded mode. In grid connected mode, micro grid is connected to the main grid. In islanded mode, micro grid is disconnected from main grid. The operation of micro grid is different for solar, wind and bulk power systems (Dragicevi et al. 2016, Bove et al. 2012, Shokrzadeh et al. 2015, Kuo, et al. 2001). The operation of solar and wind power plants is shown in figures 8 and 9 respectively:

Figure 8. Solar power operations with main grid

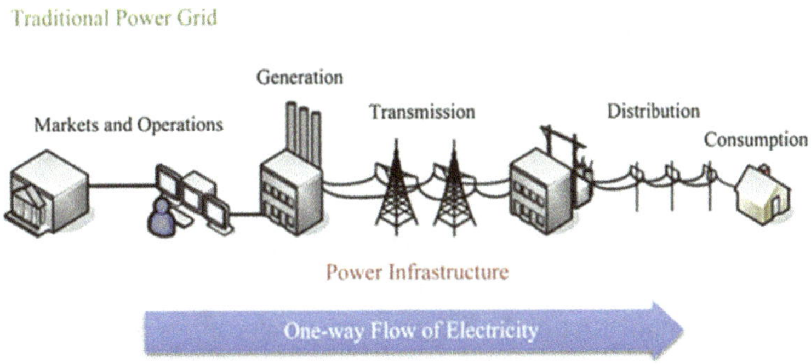

The power is flow from bulk power generation is connected to transmission line by using step up transformer. This transformer is used to step up voltage from bulk power generation and then it is connected to step down transformer and renewable energy source (wind and solar PV Energy) and the loads. Industrial load is generally supplied by bulk generating stations. Commercial and residential loads can be supplied by the micro grid as shown in the figure 10. Various protective devices are utilized to supply different demands in order to keep the systems in normal conditions (Dragicevi et al. 2016, Mohsenian-Rad et al. 2012, Shahidehpour et al. 2013). Figure 10 presented the operation of the micro grids connected with the main grid:

Operational Differences Between Micro Grids and the Main Grid

Generally the operation of micro grid and main grid is different (Gonen et al. 2013, Guerrero et al. 2011, IEEE et al. 2009, Mehrizi-Sani et al. 2010). These differences are as follows:

- Micro grid is implemented on the small area but main grid is developed on large area

Figure 9. Wind power operation with main grid.

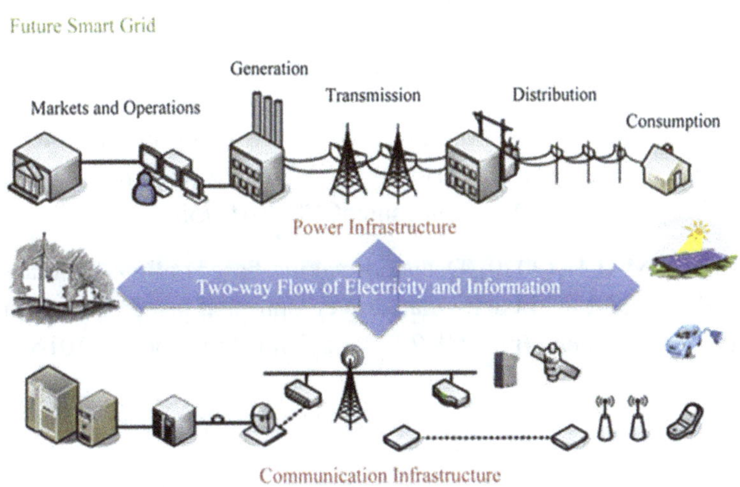

Figure 10. Block diagram of Micro grid system

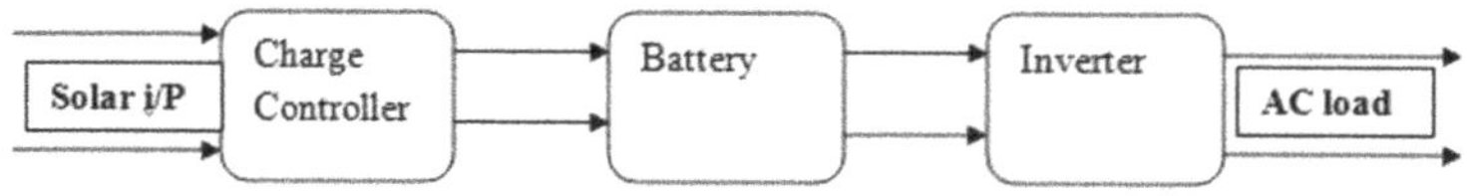

- Micro grid produced low amount of power while main grid produces large amount of power.
- Micro grid necessarily consists of energy storage while in the main grid is it not
- Micro grid required low construction as well as operation cost as compare to main grid.

CONCLUSION

Micro grid is the small power plant (Energy source), which is the combination of renewable and conventional energy sources. This chapter discussed the various types of power plants such as conventional, modern and smart power plants. Further, it discussed the micro grid, different types of micro grid as traditional and modern one. The operation of micro grid in the grid connected mode is also discussed in this chapter.

REFERENCES

Akshata, A. S., Harshada, V. H., Madhuri, S., & Mandasmita, R. K. (2012). Seminar report on microgrid.

Alhelou, H., Hamedani-Golshan, M. E., Zamani, R., Heydarian-Forushani, E., & Siano, P. (2018). Challenges and opportunities of load frequency control in conventional, modern and future smart power systems: A comprehensive review. *Energies, 11*(10), 2497. doi:10.3390/en11102497

Alhelou, H. H. (2018). *Fault detection and isolation in power systems using unknown input observer. In Advanced condition monitoring and fault diagnosis of electric machines* (p. 38). Hershey, PA: IGI Global.

Alhelou, H. H., Golshan, M., & Fini, M. (2018). Wind driven optimization algorithm application to load frequency control in interconnected power systems considering GRC and GDB nonlinearities. *Electric Power Components and Systems, 46*(11-12), 1223-1238.

Alhelou, H. H., & Golshan, M. E. H. (2016, May). Hierarchical plug-in EV control based on primary frequency response in interconnected smart grid. In *2016 24th Iranian Conference on Electrical Engineering (ICEE),* (pp. 561-566). IEEE. 10.1109/IranianCEE.2016.7585585

Alhelou, H. H., & Golshan, M. E. H. (2020). Decision-making-based optimal generation-side secondary-reserve scheduling and optimal LFC in deregulated interconnected power system. In *Decision making applications in modern power systems* (pp. 269–299). Academic Press. doi:10.1016/B978-0-12-816445-7.00011-6

Alhelou, H. H., Golshan, M. E. H., & Hatziargyriou, N. D. (2019). Deterministic dynamic state estimation-based optimal LFC for interconnected power systems using unknown input observer. *IEEE Transactions on Smart Grid*, 1. doi:10.1109/TSG.2019.2940199

Alhelou, H. H., Golshan, M. E. H., & Hatziargyriou, N. D. (2019). A decentralized functional observer based optimal LFC considering unknown inputs, uncertainties, and cyber-attacks. *IEEE Transactions on Power Systems*.

Alhelou, H. H., Golshan, M. E. H., Zamani, R., Moghaddam, M. P., Njenda, T. C., Siano, P., & Marzband, M. (2019, June). An improved UFLS scheme based on estimated minimum frequency and power deficit. In 2019 IEEE Milan PowerTech (pp. 1-6). IEEE. doi:10.1109/PTC.2019.8810497

Alhelou, H. H., Golshan, M. H., & Askari-Marnani, J. (2018). Robust sensor fault detection and isolation scheme for interconnected smart power systems in presence of RER and EVs using unknown input observer. *International Journal of Electrical Power & Energy Systems*, *99*, 682–694. doi:10.1016/j.ijepes.2018.02.013

Alhelou, H. H., Hamedani-Golshan, M. E., Heydarian-Forushani, E., Al-Sumaiti, A. S., & Siano, P. (2018, September). Decentralized fractional order control scheme for LFC of deregulated nonlinear power systems in presence of EVs and RER. In *2018 International Conference on Smart Energy Systems and Technologies (SEST)* (pp. 1-6). IEEE. 10.1109/SEST.2018.8495858

Alhelou, H. S. H., Golshan, M. E. H., & Fini, M. H. (2015, December). Multi agent electric vehicle control based primary frequency support for future smart micro-grid. In *2015 Smart Grid Conference (SGC)* (pp. 22-27).

Anteneh, D., & Khan, B. (2019). Reliability enhancement of distribution substation by using network reconfiguration a case study at Debre Berhan Distribution Substation. *International Journal of Economy, Energy, & Environment*, *4*(2), 33.

Bahmani-Firouzi, B., & Azizipanah-Abarghooee, R. (2014). Optimal sizing of battery energy storage for micro-grid operation management using a new improved bat algorithm. In *Electric Power Energy Systems, 56,* 42-54.

Banteywalu, S., Khan, B., De Smedt, V., & Leroux, P. (2019). A novel modular radiation hardening approach applied to a synchronous buck converter. *Electronics (Basel)*, *8*(5), 513. doi:10.3390/electronics8050513

Bove, R., Bucher, M., & Ferretti, F. (2012). Integrating large shares of wind energy in macro economical cost-effective way. *Energy, 43,* 438-47.

Che, L., Khodayar, M., & Shahidehpour, M. (2014). Only connect: Microgrids for distribution system restoration. *IEEE Power & Energy Magazine*, *12*(1), 70–81. doi:10.1109/MPE.2013.2286317

Dohn, R. L. (2011). The business case for micro grids. White paper: The new face of energy modernization.

Dragicevi, T., Lu, X., Vasquez, J. C., & Guerrero, J. M. (2016a). DC micro grids part I: A review of control strategies and stabilization techniques. *IEEE Transactions on Power Electronics*, *31*(7), 4876–4891.

Dragicevi, T., Lu, X., Vasquez, J. C., & Guerrero, J. M. (2016b). DC micro grids part II: A review of power architectures, applications, and standardization issues. *IEEE Transactions on Power Electronics, 31*(5), 3528–3549. doi:10.1109/TPEL.2015.2464277

Fini, M. H., Yousefi, G. R., & Alhelou, H. H. (2016). Comparative study on the performance of many-objective and single-objective optimisation algorithms in tuning load frequency controllers of multi-area power systems. *IET Generation, Transmission, & Distribution, 10*(12), 2915–2923. doi:10.1049/iet-gtd.2015.1334

Gonen, T. (2013). Electric power distribution engineering (3rd ed.). Boca Raton, FL: CRC Press.

Guerrero, J. M., Vasquez, J. C., Matas, J., de Vicuña, L. G., & Castilla, M. (2011). Hierarchical control of droop-controlled AC and DC microgrids: A general approach toward standardization. *IEEE Transactions on Industrial Electronics, 58*(1), 158–172. doi:10.1109/TIE.2010.2066534

Gupta, K., Khan, B., & Mubeen, S. E. (2015). Available transfer capability enhancement by unified power flow controller. In *Proceedings IEEE International Conference on Signal Processing, Informatics, Communication, and Energy Systems (SPICES),* pp. 1, 4, 19-21.

Haes Alhelou, H., Hamedani-Golshan, M. E., Njenda, T. C., & Siano, P. (2019). A survey on power system blackout and cascading events: Research motivations and challenges. *Energies, 12*(4), 682. doi:10.3390/en12040682

Haes Alhelou, H., Hamedani-Golshan, M. E., Njenda, T. C., & Siano, P. (2019). Wide-area measurement system-based optimal multi-stage under-frequency load-shedding in interconnected smart power systems using evolutionary computing techniques. *Applied Sciences, 9*(3), 508. doi:10.3390/app9030508

Haes Alhelou, H., Hamedani Golshan, M. E., Njenda, T. C., & Siano, P. (2019). WAMS-based online disturbance estimation in interconnected power systems using disturbance observer. *Applied Sciences, 9*(5), 990. doi:10.3390/app9050990

Hajipour, E., Bozorg, M., & Fotuhi-Firuzabad, M. (2015). Stochastic capacity expansion planning of remote micro grids with wind farms and energy storage. *IEEE Trans. Sust. Energy, 6*(2), 491–498. doi:10.1109/TSTE.2014.2376356

Hina Fathima, A., & Palanisamy, K. (2015). Optimization in microgrids with hybrid energy systems – a review. *Renewable & Sustainable Energy Reviews, 45*, 431–446. doi:10.1016/j.rser.2015.01.059

IEEE. (2009). IEEE application guide for IEEE Std 1547(TM). IEEE Standard for Interconnecting Distributed Resources with Electric Power Systems. IEEE Std 1547.

Jariso, M., Khan, B., Tanwar, S., Tyagi, S., & Rishiwal, V. (2018). Hybrid energy system for upgrading the rural environment, IEEE GlobeCom 2018.

Jariso, M., Khan, B., Tesfaye, D., & Singh, J. (2017). Modelling and designing of stand-alone photovoltaic system (Case study: Addis Boder Health Center southwest Ethiopia). In *Proceedings IEEE International Conference on Electronics, Communication, and Aerospace Technology, ICECA 2017,* pp. 168-173, Coimbatore, India.

Khan, B., & Agnihotri, G. (2012). A novel transmission loss allocation method based on transmission usage. In *Proceedings IEEE Fifth Power India Conference*, Haryana, India. 10.1109/PowerI.2012.6479479

Khan, B., Agnihotri, G, Gupta, G. (2013). A multipurpose matrices methodology for transmission usage, loss and reliability margin allocation in restructured environment, *Electrical & Computer Engineering: An International Journal, 2*(3).

Khan, B., Agnihotri, G., Rathore, P., Mishra, A., & Naidu, G. (2014). A cooperative game theory approach for usage and reliability margin cost allocation under contingent restructured market. *International Review of Electrical Engineering*, *9*(4), 854–862.

Khan, B., Alhelou, H. H., & Mebrahtu, F. (2019). A holistic analysis of distribution system reliability assessment methods with conventional and renewable energy sources. *AIMS Energy*, *7*(4), 413–429. doi:10.3934/energy.2019.4.413

Khan, B., & Singh, P. (2017). The current and future states of Ethiopia's energy sector and potential for green energy: A comprehensive study. *International Journal of Engineering Research in Africa, 33*, 115–139. doi:10.4028/www.scientific.net/JERA.33.115

Khan, B., & Singh, P. (2017). Optimal power flow techniques under characterization of conventional and renewable energy sources: A comprehensive analysis, *Journal of Engineering, vol. 2017,* Article ID 9539506.

Khan, B., & Singh, P. (2017). Selecting a meta-heuristic technique for smart micro-grid optimization problem: A comprehensive analysis. In IEEE Access, 5, pp. 13951-13977.

Khan, B., & Singh, P. (2019). Economic operation of smart micro-grid: a meta-heuristic approach. In Handbook of research on smart power system operation and control (pp. 330-346). Hershey, PA: IGI Global. doi:10.4018/978-1-5225-8030-0.ch014

Khan, B., & Singh, P. (2019). Economic operation of smart micro-grid: a meta-heuristic approach. In Handbook of research on smart power system operation and control (pp. 330-346). Hershey, PA: IGI Global. doi:10.4018/978-1-5225-8030-0.ch014

Khan, B., & Tanwar, S. (2019). Issues associated with microgrid integration. In Handbook of research on smart power system operation and control (pp. 252-264). Hershey, PA: IGI Global. doi:10.4018/978-1-5225-8030-0.ch010

Kifle, Y., Khan, B., & Singh, J. (2018). Designing and modeling grid connected photovoltaic system: (Case study: EEU building at Hawassa city). *International Journal of Convergence Computing*, *3*(1), 20–34. doi:10.1504/IJCONVC.2018.091113

Kifle, Y., Khan, B., & Singh, P. (2018). Assessment and enhancement of distribution system reliability by renewable energy sources and energy storage, *Journal of Green Engineering, 8*(3), 2, pp. 219-262.

Kuo, Y. C., Liang, T. J., & Chen, J. F. (2001). Novel maximum-power-point-tracking controller for photovoltaic energy conversion system. *IEEE Transactions on Industrial Electronics*, *48*(3), 594–601. doi:10.1109/41.925586

LaCommare, K., & Eto, J. H. (2004). *Understanding the cost of power interruptions to U. S. electricity consumers, technical report*. Lawrence Berkeley National Laboratory. doi:10.2172/834270

Lidula, N. W. A., & Rajapakse, A. D. (2011). Microgrids research: A review of experimental microgrids and test systems. *Renewable & Sustainable Energy Reviews*, *15*(1), 186–202. doi:10.1016/j.rser.2010.09.041

Lu, X., Bahramirad, S., Wang, J., & Chen, C. (2015). Bronzeville community micro grids: A reliable, resilient, and sustainable solution for integrated energy management with distribution systems. *The Electricity Journal*, *28*(10), 29–42. doi:10.1016/j.tej.2015.11.009

Mahmoodi, M., Shamsi, P., & Fahimi, B. (2015). Economic dispatch of a hybrid microgrid with distributed energy storage. *IEEE Transactions on Smart Grid*, *6*(6), 2607–2614. doi:10.1109/TSG.2014.2384031

Makdisie, C., Haidar, B., & Alhelou, H. H. (2018). An optimal photovoltaic conversion system for future smart grids. In Handbook of research on power and energy system optimization (pp. 601–657). Hershey, PA: IGI Global. doi:10.4018/978-1-5225-3935-3.ch018

Mehrizi-Sani, A., & Iravani, R. (2010). Potential-function based control of a microgrid in islanded and grid-connected modes. *IEEE Transactions on Power Systems*, *25*(4), 1883–1891. doi:10.1109/TPWRS.2010.2045773

Mohsenian-Rad, H. (2012). *Introduction to smart grid*. Texas Tech University.

Molla, T., Khan, B., Moges, B., Alhelou, H. H., Zamani, R., & Siano, P. (2019). Integrated optimization of smart home appliances with cost-effective energy management system. *CSEE Journal of Power and Energy Systems*, *5*(2), 249–258.

Molla, T., Khan, B., & Singh, P. (2018). A comprehensive analysis of smart home energy management system optimization techniques, *Journal of Autonomous Intelligence, 1*(1), 15–21.

Negash, K., Khan, B., Tesfaye, D., & Gayathri, M. (2017). Optimal placement of phasor measurement unit for system observability. In *Proceedings of IEEE International Conference on Computing Methodologies and Communication, (ICCMC 2017),* pp. 73-77, Erode, India.

Negash, K., Khan, B., & Yohannes, E. (2016). Artificial intelligence versus conventional mathematical techniques: A review for optimal placement of phasor measurement units [Springer]. *Journal of Technology and Economics of Smart Grids and Sustainable Energy*, *1*(1), 10. doi:10.100740866-016-0009-y

Nikmehr, N., & Najafi Ravadanegh, S. (2015). Optimal power dispatch of multi-microgrids at future smart distribution grids. *IEEE Transactions on Smart Grid*, *6*(4), 1648–1657. doi:10.1109/TSG.2015.2396992

Njenda, T. C., Golshan, M. E. H., & Alhelou, H. H. (2018, November). WAMS based intelligent under frequency load shedding considering online disturbance estimation. In *2018 Smart Grid Conference (SGC)* (pp. 1-5). 10.1109/SGC.2018.8777779

Njenda, T. C., Golshan, M. E. H., & Alhelou, H. H. (2018, November). WAMS based under frequency load shedding considering minimum frequency prediction and extrapolated disturbance magnitude. In *2018 Smart Grid Conference (SGC)* (pp. 1-5).

Olken, M. (2009). Beyond the gridlock. *IEEE Power & Energy Magazine*, *7*(2), 4–6. doi:10.1109/MPE.2009.931830

Saboori, H., Hemmati, R., & Jirdehi, M. A. (2015). Reliability improvement in radial electrical distribution network by optimal planning of energy storage systems. *Energy, 93,* 2299-2312.

Shahidehpour, M., & Khodayar, M. (2013). Cutting campus energy costs with hierarchical control: The economical and reliable operation of a micro grid. *IEEE Electr. Mag.*, *1*(1), 40–56. doi:10.1109/MELE.2013.2273994

Shokrzadeh, S., Jozani, M. J., Bibeau, E., & Molinski, T. (2015). A statistical algorithm for predicting the energy storage capacity for baseload wind power generation in the future electric grids. *Energy, 89,* 793-802.

Singh, P., & Khan, B. (2017). Smart microgrid energy management using a novel artificial shark optimization. Complexity, vol. 2017, Article ID 2158926, doi:10.1155/2017/2158926

Singh, P., Khan, B., Vidyarthi, A., Haes Alhelou, H., & Siano, P. (2019). Energy-aware online non-clairvoyant scheduling using speed scaling with arbitrary power function. *Applied Sciences*, *9*(7), 1467. doi:10.3390/app9071467

The smart grids European technology platform (SGETP). Retrieved from http://www.smartgrids.eu/ETP SmartGrids.

Wang, Z., Chen, B., Wang, J., Begovic, M., & Chen, C. (2015b). Coordinated energy management of networked microgrids in distribution systems. *IEEE Transactions on Smart Grid*, *6*(1), 45–53. doi:10.1109/TSG.2014.2329846

Wu, D., Lesieutre, B., Ramanathan, P., & Kakunoori, B. (2016). Preserving privacy of AC optimal power flow models in multi-party electric grids. *IEEE Transactions on Smart Grid*, 1–11.

Yeshalem, M. T., & Khan, B. (2017). Design of an off-grid hybrid PV/wind power system for remote mobile base station: A case study. *AIMS Energy*, *5*(1), 96–112. doi:10.3934/energy.2017.1.96

Zamani, R., Golshan, M. E. H., Alhelou, H. H., & Hatziargyriou, N. (2019). A novel hybrid islanding detection method using dynamic characteristics of synchronous generator and signal processing technique. *Electric Power Systems Research*, *175*, 105911. doi:10.1016/j.epsr.2019.105911

Zamani, R., Hamedani-Golshan, M. E., Haes Alhelou, H., Siano, P., & Pota, H. R. (2018). Islanding detection of synchronous distributed generator based on the active and reactive power control loops. *Energies*, *11*(10), 2819. doi:10.3390/en11102819

KEY TERMS AND DEFINITIONS

Fuel Cell: A fuel cell is an electrochemical cell that converts the chemical energy of a fuel (often hydrogen and an oxidizing agent (often oxygen) into electricity through a pair of redox reactions.

Protective relay: It is an intelligent electronic devices (IEDs) which receive measured signals from the secondary side of CTs and VTs and detect whether the protected unit is in a stressed condition (based on their type and configuration) or not. A trip signal is sent by protective relays to the circuit breakers to disconnect the faulty components from power system if necessary.

Reliability: A measure of the degree of certainty that the relay, or relay system, will perform correctly.

Smart Grid: an electricity supply network that uses digital communications technology to detect and react to local changes in usage.

APPENDIX

List of Abbreviations

AC Alternate current
DC Direct Current
DER Distributed Energy Resources
DG Distributed generators
DS Distributed Storage
GHG Greenhouse gas
HAWT Horizontal axis wind turbine
MPPT Maximum Power Point Tracking
PCC Point of common coupling
PV Photovoltaic
RES Renewable energy sources
VAWT Vertical axis wind turbine

Chapter 14
Management of Electrical Maintenance of University Buildings Using Deterioration Models

Bassel Alhassan
Tishreen University, Syria

ABSTRACT

Buildings maintenance has received increasing international attention in various fields of scientific research. As a result, the maintenance of buildings has changed from the preventive to the predictive approach. This is done through an evaluation model to support and assist the management of the facility in selecting alternatives and making appropriate decisions in maintenance according to building status and maintenance budget. This chapter investigated the reasons for the electrical maintenance of the university buildings and the degree of importance of each element of electrical maintenance through the design of a questionnaire in which the electrical components were divided into elements and then each element was linked to all maintenance items that related to it. At the end of the research, mathematical models were developed, and they help to forecast the electrical maintenance items and distribution of the maintenance budget, and to verify the validity of these models. The models have been applied to study the case of dorm buildings in Tishreen University.

INTRODUCTION

Maintenance is defined as a continuous process and a permanent activity to preserve the buildings and the keeping of the equipment in the best case for normal use. The maintenance includes the entire building with its various parts (structural, health, mechanical, electrical, and ventilation) (Bin Akasah, 2007).

Maintenance can also be defined as the protection of the buildings from damage in order to perform its function fully as long as possible with the repair of defects in order to preserve the building and to reduce the costs resulting from negligence or deferral by sharing all technical and administrative means.

DOI: 10.4018/978-1-7998-1230-2.ch014

These include preparation, planning and implementation of the maintenance process as well as follow-up and control during and after implementation (Al-Hassan; jrad, 2015). Maintenance is the set of technical and executive works that aim at keeping a building items in the state in which it can perform its required functions. Maintenance of any component or material must not be less than the necessary minimum limit. There have been many views that the maintenance process is complex and requires high costs (Bin Akasah, 2007), but it leads to a longer life of the facility and the keeping of cost if implemented according to the correct schedule (Arditi; Nawakorawit, 1999).

To shift the maintenance work from the approach that depends on the repair of the failure when it occurs, which leads to the accumulation of faults to the predictive approach, which helps management to anticipate the failures that can occur in the construction and forecast the budget required for maintenance. To achieve that, modeling method was proposed (Khan et al., 2019; Khan et al., 2018; Khan et al., 2017; Banteywalu et al., 2019; Anteneh et al., 2019; Molla et al., 2019, Molla et al., 2018, Jariso et al. 2018). The mathematical model is a simplified version of a complex thing and is used to solve problems and make predictions and as a basis for related ideas (Olagunju,2012). The mathematical model can be described as a process of simplification of complex life issues, which is expressed as mathematical equation. The predictive model can be described as the construction of an evaluation model, helping to evaluate different alternatives.

There has been a lot of research on deterioration models. One of these researches (Abdelaty, 2012) aims to set up deterioration models for all important components of the metro to determine the time between two periodic maintenance and maintenance budget. The researcher set priorities for the maintenance of the subway components through the impact of the component on the service level in the metro. In another research in the Niger state of Nigeria (Olagunju, 2012), deterioration models have been developed for residential buildings, these models based on the basic components of the building. To determine the most significant components of the building, the researcher put twenty five variables expressing the different components of the building and conducting a questionnaire. The researcher found that eight variables expressed by 94% of the status or condition of the building. Another study (Frangopol, 2003), aims to develop a reliable deterioration model to predict the maintenance work and optimize maintenance cost. The researcher worked on a mathematical model to optimize the cost of maintenance of infrastructure under uncertainty, where the researcher studied several facilities that need maintenance work with a focus on bridges.

THE AIM AND IMPORTANCE OF RESEARCH

In order to preserve our real estate wealth and due to the fact that it is impossible to build a building that does not require maintenance and in the light of the lack of maintenance manual in our country Syria. The aim of this research is to develop deterioration models of electrical maintenance of university buildings, which help the administration to quickly evaluate the construction situation in terms of the electrical component, as well as identify the main reasons for the electrical maintenance and the expected electrical maintenance work for the next year, arrange these works and implement the most urgent items according to the budget set for the maintenance work (Khan et al. 2014; Khan et al. 2013; Khan et al. 2012; Negash et al., 2017; Negash et al., 2016; Jariso et al. 2017; Kifle et al. 2018; Yeshalem et al. 2017; Singh et al. 2017; Gupta et al., 2015)

RESEARCH METHODS AND MATERIALS

In this chapter, author used a descriptive approach. A questionnaire on maintenance was designed. The data obtained were then subjected to several methods of analysis using business intelligence by power bi application and using fuzzy logic. Before the questionnaire was designed, the building was divided as it came (Alhelou et al., 2019; Makdisie et al., 2018; Alhelou et al., 2018; Alhelou et al., 2016; Haes Alhelou et al., 2019; Njenda et al., 2018):

FUNCTIONAL DIVISION OF THE BUILDING

Due to the large number of electrical elements in the construction, and for the ease of control of these elements, author divided the construction, faults and electrical faults that can be happened according to two methods:

1. Division of electrical components
2. Spatial division of building
   a. The division of the electrical components:

The electrical component is functionally divided into the set of basic elements. Each of these elements will be linked to all faults that can occur in this element. This division is shown in Figure (1):

Thus, author has 3 electrical elements associated with 20 defects that can occur in these elements. Author have taken the advantage of the French maintenance manual for residential buildings to get these defects (Bornand, 1991) after making the necessary adjustments to these items to fit the reality and nature of university buildings in our country through the distribution of an experimental questionnaire on engineers with experience in the field of electrical maintenance of buildings until author has reached the final version of the questionnaire, and this division will help us to develop deterioration models for electrical maintenance of university buildings.

Spatial Division of the Building

The structure within the questionnaire was divided spatially into 5 main sections showing in the figure (2). In the questionnaire, the defects were distributed according to their location.

THE QUESTIONNAIRE

The deterioration models for the electrical components of the university buildings were based on the mathematical equations, each equation consists of two parts. The first is a constant coefficient of weight associated with each item of maintenance and it remains constant in all the buildings. The second is variable and relates to the status and condition of the item or component (component condition). It depends on the evaluation of the engineer supervising the maintenance works.

Figure 1. Elements and maintenance items in the electrical component

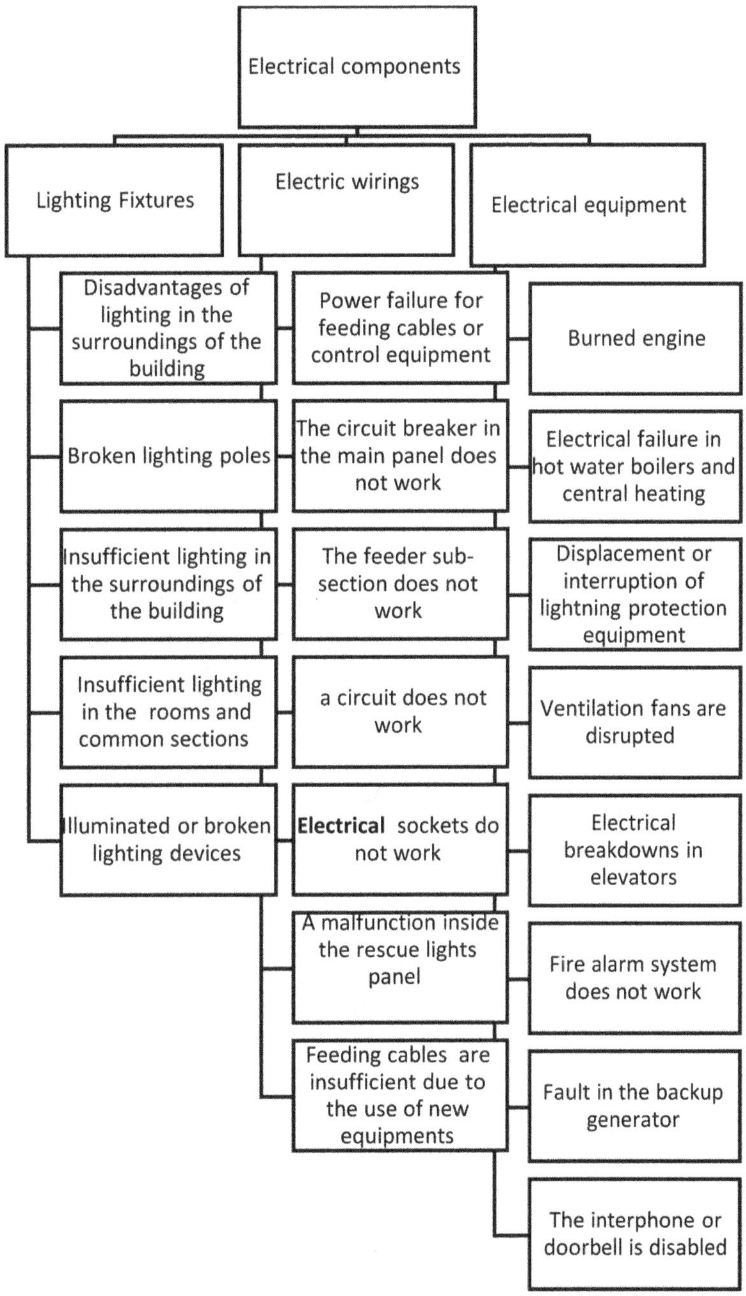

To obtain the first part of the deterioration model, which is the constant or the weighting coefficient, a questionnaire was drawn up, which includes all the electrical maintenance items described in Figure 1, which author called defect and author linked each item of maintenance with a set of criteria. criteria that reflects the importance of the item for maintenance work, whether preventive maintenance or repair in case of failure, which have been observed in the previous studies are: Probability of occurrence, risk to

Figure 2. Spatial division of construction

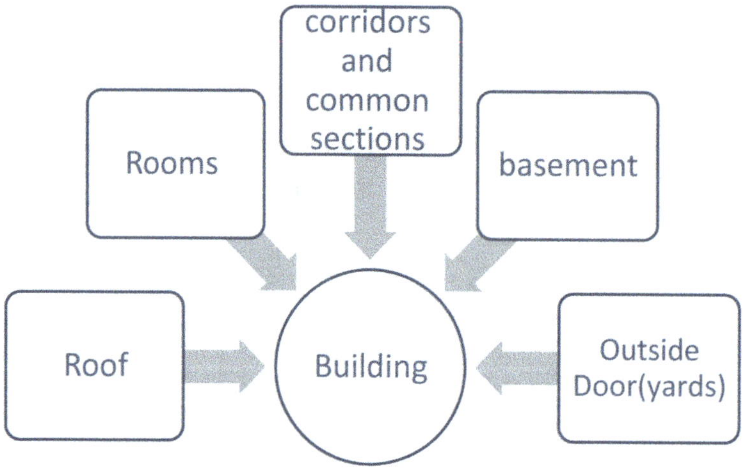

the building, time of maintenance, cost of maintenance, its impact on the level of service in construction, its impact on the environment, its impact on the security and safety of the enterprise. After returning to the experience, author found that the first 5 criteria (probability of occurrence, risk to the building, time of maintenance, cost of maintenance, service level) are the most appropriate for the nature of the facilities under study (university buildings). The first phase was the development of electrical maintenance items, by distributing a set of experimental questionnaires to engineers with experience in the field of electrical maintenance to obtain maintenance items (defects) that are more appropriate to the reality and nature of university buildings in Syria. After making the necessary adjustments, author obtained the final version of the questionnaire. After that, 112 questionnaires were distributed to the engineers working in the maintenance field in various government departments in Lattakia Governorate, from that author got 95 completed questionnaire. After obtaining all the questionnaires, these questionnaires were processed using business intelligence through the application of power bi, the statistical analysis of the results of the questionnaire is given below (Fini et al., 2016; Alhelou et al., 2018; Zamani et al., 2018; Alhelou et al., 2015; Njenda et al., 2018; Haes Alhelou et al., 2018; Haes Alhelou et al., 2019) .

Figure 3. deterioration model equation

Figure 4. Input and Output in the fuzzy logic

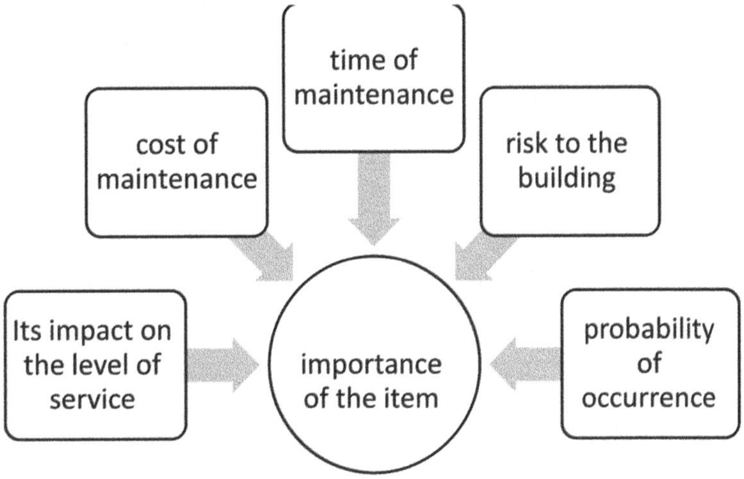

SURVEY ANALYSIS

Importance Degree

After processing the questionnaires, author obtained about the maintenance and calculating averages, each watch or item of maintenance has been linked to a set of percentages that express:

1. Probability of occurrence
2. Risk to the building
3. Time of maintenance
4. Cost of maintenance
5. Its impact on the level of service

To obtain the constant or coefficient of weighting that falls into the deterioration model for each maintenance item, author combined these values with each other. For values in this integration, author used the fuzzy logic that allows the use of ambiguous and uncertain concepts.

By using fuzzy tech 6.06 application, fuzzy logic, author get degree of importance for each item in electrical component as shown in the next table:

Element-Level

The electrical component was divided into 3 elements and author will analyze the data related to these elements obtained from the survey results. It should be noted that the lower the percentage of the item, the greater indication of this item:

Table 1. importance degree for electrical component items

maintenance item name	importance degree
Disadvantages of lighting in the surroundings of the building	77.5
Broken lighting poles	81
Insufficient lighting in the surroundings of the building	81
Insufficient lighting in the rooms and common sections	83
Illuminated or broken lighting devices	83
Power failure for feeding cables or control equipment	55.5
The circuit breaker in the main panel does not work	62
The subsection circuit breaker does not work	67
a circuit does not work	81
Power sockets do not know	83
A malfunction inside the rescue lights panel	83
Feeding cables are insufficient due to the use of new equipment's	64.5
Burned engine	52
Electrical failure in hot water boilers and central heating	62.5
Displacement or interruption of lightning protection equipment	83
Ventilation fans are disrupted	62
Electrical breakdowns in elevators	50
Fire alarm system does not work	68.5
Fault in the backup generator	45.5
The interphone or doorbell is disabled	83

Probability of Occurrence

Figure 5. Electrical component elements distribution according to the probability of occurrence

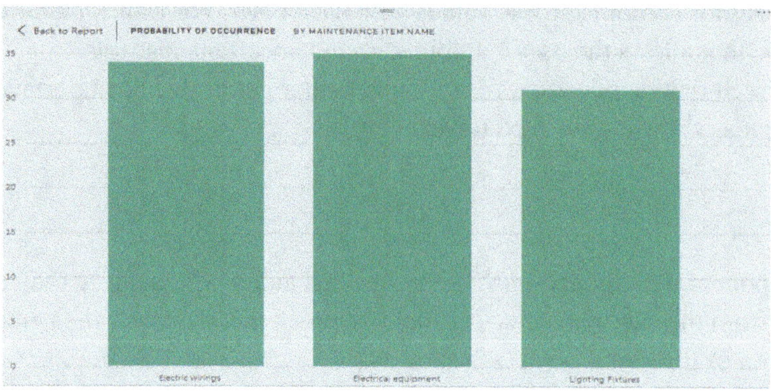

Risk to the Building

Figure 6. Electrical component elements distribution according to the risk to the building

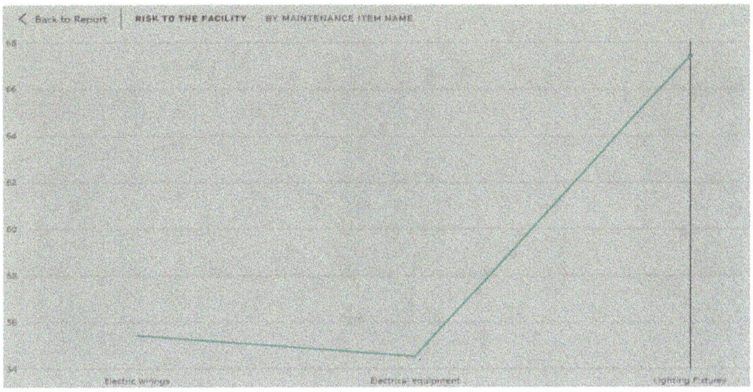

Time of Maintenance

Figure 7. Electrical component elements distribution according to the time of maintenance

Cost of Maintenance

Figure 8. Electrical component elements distribution according to the cost of maintenance

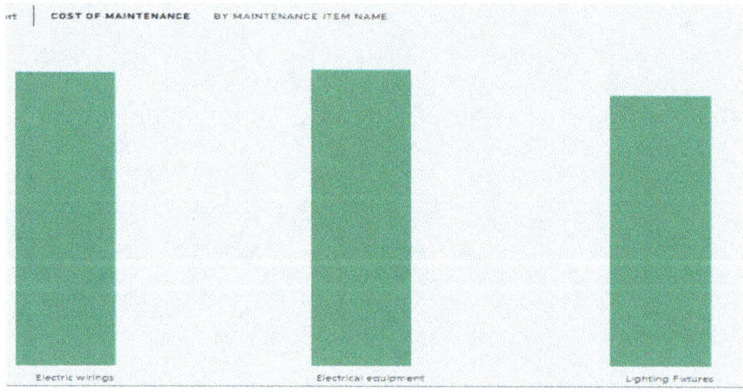

Impact on the Level of Service

Figure 9. Electrical component elements distribution according to the impact on the level of service

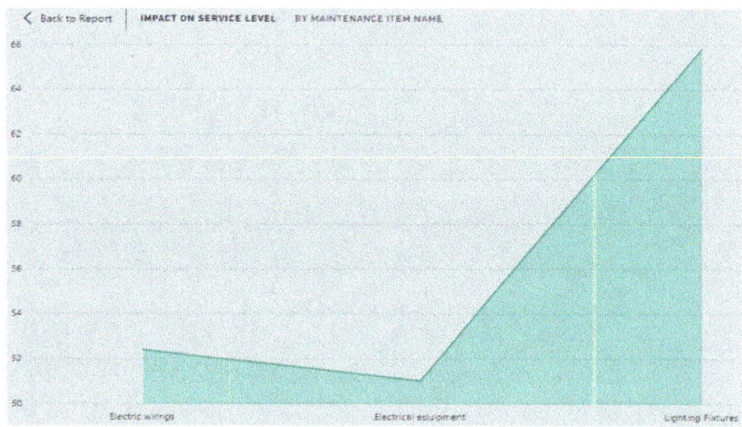

ITEM-LEVE

Probability of Occurrence

Figure 10. Electrical component items distribution according to the probability of occurrence

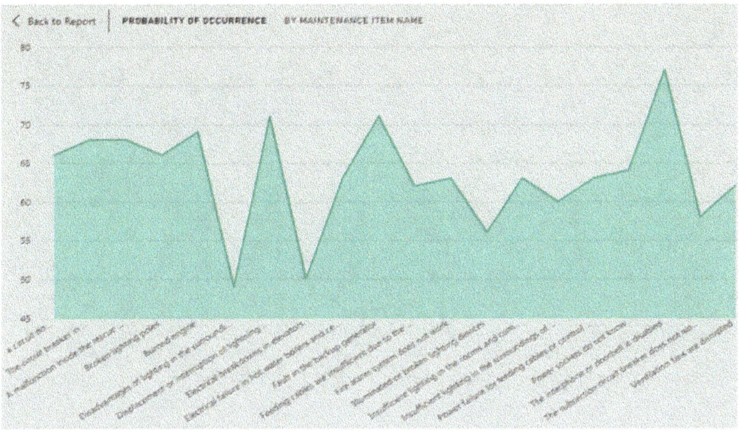

Risk to the Building

Figure 11. Electrical component items distribution according to the risk to the building

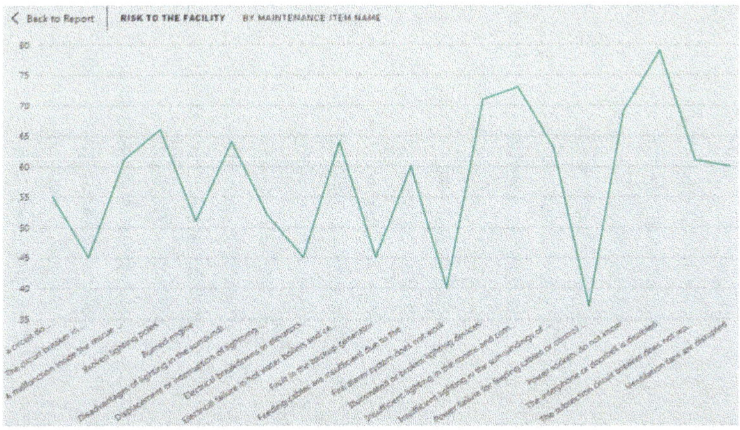

Time of Maintenance

Figure 12. Electrical component items distribution according to the time of maintenance

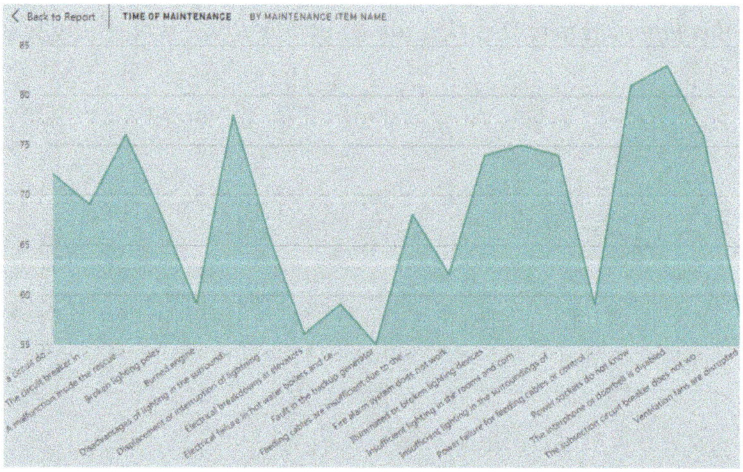

Cost of Maintenance

Figure 13. Electrical component items distribution according to the cost of maintenance

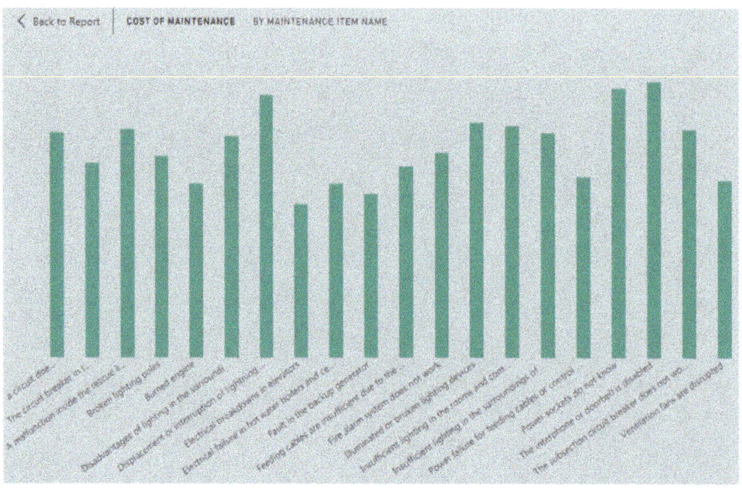

Impact on the Level of Service

Figure 14. Electrical component items distribution according to the impact on the level of service

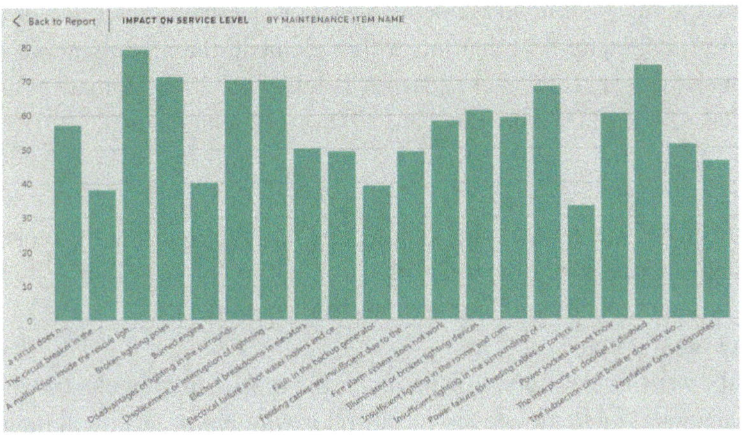

DISCUSSION OF THE RESULTS OF THE STATISTICAL ANALYSIS OF ELECTRICAL COMPONENTS

Lighting Fixtures Element:

Element of lighting fixtures is the most important electrical component of the possibility of the occurrence of faults in it as the lighting devices are the most number of electrical elements used by the occupants of construction and be exposed to breakage or combustion due to weather conditions, but these elements are the least electrical elements of the risk to the building and the cost and time of maintenance are relatively small if compared to the rest of the electrical elements, as well as the impact on the level of service is low. The items of maintenance included in the Lighting Fixtures element:

1. Disadvantages of lighting in the surroundings of the building: where this item was the most common items of electrical maintenance is likely to occur, the presence of lighting outside the building makes them more likely to crash or breakage due to weather conditions.
2. Broken lighting poles: This item can occur because of accident or result of deliberate sabotage and is treated by replacing broken columns.
3. Insufficient lighting in the surroundings of the building: This is due to poor electricity supply or poor lighting equipment.
4. Insufficient lighting in the rooms and common sections.
5. Illuminated or broken lighting devices: because of an accident or because of the end of service life and shall be replaced.

Electrical Equipment Element

This element was the least electrical element likely to be faulted. However, it is the most dangerous electrical element of the building in the event of a failure in it, as well as the most time and cost consuming elements and the most common element impact on the level of service in the building due to the importance of electrical equipment that fall within element. the main items of maintenance in the electrical equipment element are (Fini et al., 2016; Alhelou et al., 2018; Zamani et al., 2018; Alhelou et al., 2015; Njenda et al., 2018; Haes Alhelou et al., 2018; Haes Alhelou et al., 2019):

1. Burned engine:
2. Electrical failure in hot water boilers and central heating: This item was the most common electrical maintenance items that needed time to fix it.
3. Displacement or interruption of lightning protection equipment.
4. Ventilation fans are disrupted.
5. Electrical breakdowns in elevators: this item was the most common electrical maintenance items cost during maintenance and the second most important item of electrical maintenance.
6. Fire alarm system does not work
7. Fault in the backup generator: This item is the most important electrical maintenance items because of the importance of the backup generator.
8. The interphone or doorbell is disabled: This item was the least important electrical maintenance items likely to occur as well as the least common items in terms of maintenance time and cost where the interferon bell is present in many government buildings, but its use is very little, and in some cases, it is never used.

ELECTRIC WIRINGS ELEMENT

Main Maintenance Items Included in this Element

1. Power failure for feeding cables or control equipment: This item was the most dangerous item of electrical maintenance to the building in case it occurs, Where the failure may mean the absence of electrical power for the complete building or important equipment in the building, such as the elevator.
2. The circuit breaker in the main panel does not work.
3. The feeder sub-section does not work.
4. A circuit does not work.
5. Electrical sockets do not work
6. A malfunction inside the rescue lights panel: This item was the least important of the electrical maintenance items affecting the service level, but the risk to the establishment is relatively large in the event of a failure where the function of the rescue light is to provide lighting in case of power failure in the event of fire or disaster in construction.
7. Feeding cables are insufficient due to the use of new equipments.

Table 2. Distribution of the evaluation grades for the engineer supervising maintenance

grade	Description of the component condition
1	Failure of the component as the component does not work with the possibility of structural or environmental damage or damage to security and safety at the building.
2	The component works but is frequently disrupted with clear signs of deterioration of the component's status and potential damage to the building
3	The component is medium-state as it works but needs monitoring and needs preventive maintenance
4	The component works well with small and superficial faults that do not affect component work and maintenance can be delayed
5	The component works as if it were new and there are no signs of a malfunction with a new exterior

Table 3. Deterioration model for electrical component elements

Element name	Deterioration model
Electrical equipment	$Y1 = 0.52X1 + 0.63X2 + 0.83X3 + 0.62X4 + 0.5X5 + 0.69X6 + 0.46X7 + 0.83X8$
Electric wirings	$Y2 = 0.56X1 + 0.62X2 + 0.67X3 + 0.81X4 + 0.83X5 + 0.83X6 + 0.65X7$
Lighting Fixtures	$Y3 = 0.78X1 + 0.81X2 + 0.81X3 + 0.83X4 + 0.83X5$

Due to the presence of electric panels and most of the electrical wiring in the basement, therefore the electrical breakdown in the basement is the most serious electrical failures in the building as well as the repair takes time and cost significantly compared to other sections, While the electrical faults in the rooms and their fixtures, which are mainly related to lighting devices less faults in term of the cost of repair, as well as electrical faults that occur in the rooms are at less risk than the faults that occur in the rest of the sections.

DETERIORATION MODELS

To benefit from the results of the statistical study, deterioration models were developed to predict electrical maintenance work. This predictive model is used for the control and rapid evaluation of the buildings. This is achieved by dividing the electrical component into elements, studying each element and writing the deterioration model, and then integrating these elements to obtain a deterioration model of the electrical component for the building completely and use this model in order to predict the state of the electrical component of the next year. All this done through a logical scientific method rather than relying on the personal factors. In this way author can maintain the continuity of the construction and the continuation of the electrical network to perform its work perfectly and using a small operating cost. After processing the questionnaire author linked each maintenance item with a number that reflects the importance of this item and its impact on the performance in the building and this number will enter the equation of deterioration model as a constant or coefficient of weighting is smaller or equal to one.

The status or condition of the element is the second component of the deterioration model, through the evaluation of the engineer supervising the maintenance work of this component and to obtain this evaluation scientifically, author relied on a paper approved by (Australasian Association of Higher Education Facilities Officers, 2000).

Deterioration Models for Electrical Components

Deterioration model for Lighting Fixtures element:

$$Y_3 = m_1X_1 + m_2X_2 + m_3X_3 + m_4X_4 + m_5X_5$$

$$Y_3 = 0.78X_1 + 0.81X_2 + 0.81X_3 + 0.83X_4 + 0.83X_5$$

$m_1, m_2, m_3, m_4 and\, m_5$ are the weighting coefficients of the maintenance items associated with the Lighting Fixtures element where back to the construction division. Author note the existence of the five maintenance items associated with the Lighting Fixtures element. These items are disadvantages of the lighting in the surroundings of the building, broken lighting poles, insufficient lighting in the surroundings of the building, insufficient lighting in the rooms and common sections, illuminated or broken lighting devices.

X_1, X_2, X_3, X_4, X_5, are the evaluation of the maintenance engineer to determine the defects associated with the Lighting Fixtures element. Each evaluation is multiplied by the associated weighting factor. After the compensation of the evaluation of the engineer responsible for maintenance in the deterioration model of the Lighting Fixtures element author obtained the deterioration model value of the Lighting Fixtures element.

The previous equation is valid for a deterioration model for the Lighting Fixtures element in the different types of buildings, after compensating the maintenance engineer's assessments in the previous equation; author obtained a value that reflects the condition of the Lighting Fixtures element in the facility in terms of its need for maintenance. In a similar way author got the deterioration models of the other elements in the electrical component, shown in the following table:

Thus, author obtained 3 deterioration models for the electrical components that represent the elements. Using these equations, one can obtain a deterioration model that expresses electrical components where the number of variables is equal to the number of elements. In deterioration model for electrical component author considered Y_1, Y_2, and Y_3 values are constants or weight coefficients. L_1, L_2 and L_3 are the evaluators' assessment of the state of these elements according to the maintenance engineer:

$$Z_2 = y_1L_1 + y_2L_2 + y_3L_3$$

Knowing the value of the relapse model for the electrical component gives an idea of an improvement or decline in the electrical component and helps to evaluate the maintenance performance performed in a previous year.

Table 4. Deterioration model value for dorm buildings in Tishreen University

Element name	Evaluation of engineers supervising maintenance Maintenance item- level								Deterioration Model Value
	X1	X2	X3	X4	X5	X6	X7	X8	Y%
Electrical equipment	1	3	1	1	1	2	1	1	17.58%
Electric wirings	3	3	3	5	3	3	3		47.23%
Lighting Fixtures	3	2	4	3	4				52.04%
Deterioration model for electrical component $Z1 = 0.1758L1 + 0.4723L2 + 0.5204L3$									

CASE STUDY

To ascertain the validity and reliability of the deterioration models built in the previous section, it is applied to the dorm building in Tishreen University, where author obtained the evaluation of engineers supervising maintenance in the dorm buildings for various electrical components of the building and compensate the value of these assessments in previous deterioration models. In this way, author obtained deterioration models for all electrical components of dorm building housing, where author obtained the deterioration model value of Lighting Fixtures element in the following manner:

$$Y_3 = 0.78X_1 + 0.81X_2 + 0.81X_3 + 0.83X_4 + 0.83X_5$$

$$Y_3 = 0.78*3 + 0.81*2 + 0.81*4 + 0.83*3 + 0.83*4 = 13.01$$

This is the degree author obtained from 25, where author have five values for x each evaluated from 5 degrees and therefore the highest result author can get is twenty-five and to make this result percentage must be multiplied by 4 author get the value of the deterioration model for Lighting Fixtures:

$$Y_3 = 13.01 \backslash 25 = 52.04\%$$

In a similar way author got deterioration models to other elements in the electrical component shown in the following table:

DISCUSSION OF DETERIORATION MODELS FOR ELECTRICAL ELEMENTS IN DORM BUILDINGS

As shown in Table (2-4), the values of deterioration models have shown that there is an urgent need for maintenance of the electrical equipment element in the dorm building, which was the value of the model (17.58%). The most urgent equipment for maintenance is the generator in case of power failure due to its great importance and its need for permanent preventive maintenance as well as the elevator, which was suffering from permanent breakdowns and water pumping engines with electrical breakdowns in water generation boilers and complete disrupted fans of ventilation and air conditioning in all the dorm buildings and the displacement or interruption of lightning protection equipment. The Electric wirings element was relatively good, with the model value (47.23%), noting the possibility of power failure from the feeder cables, control equipment and the need to provide more feeder cables for the new equipment. The Lighting Fixtures element was relatively good, with a deterioration model (52.04%) with a break in the building's lighting poles and insufficient lighting in the surroundings of the building.

CONCLUSION

Given that the community will expect high performance levels of university buildings, so the level of maintenance carried out is of great importance to the life of the buildings so maintenance will remain an important part of the work of the construction industry.

In this research author tried to highlight the subject of electrical for university buildings maintenance and the main findings of the research:

1. Divide electrical components to elements.
2. Determine the main reasons for electrical maintenance and connect these reasons with the electrical component elements.
3. Get deterioration models by combining two important elements: the condition of the component and the second element is the component importance to the building.
4. Application of deterioration models on dorm buildings.
5. Find the most important elements in the construction that need to be carried out for maintenance according to the budget available for maintenance.

The deterioration models help giving senior management a comprehensive view of the electrical component in the building and a rapid assessment of the status of this component. In addition, allocate the budget for maintenance according to the priority of the elements.

Thus, author have obtained a scientific tool to assist in evaluating the status of the electrical component in the building and the most important maintenance items and maintenance budget and identify the most important resources necessary for the implementation of maintenance.

REFERENCES

Abdel Aty, A. (2013). *Maintaining subways infrastructures using BIM.* (Master Thesis, Cairo University).

Abdel Aty, A. (2012). Framework for monitoring and maintenance subways stations using building information modeling. Construction Research Congress, 2320-2328.

Akasah. B. (2007) Generic process model for maintenance management of school buildings, Malaysia University.

Alhassan, B., & Jrad, F. (2015). *Enhancing maintenance management for governmental buildings using building information modelling (BIM),* (Master Thesis, Teshreen University, Latakia).

Alhelou, H., Hamedani-Golshan, M. E., Zamani, R., Heydarian-Forushani, E., & Siano, P. (2018). Challenges and opportunities of load frequency control in conventional, modern and future smart power systems: A comprehensive review. *Energies, 11*(10), 2497. doi:10.3390/en11102497

Alhelou, H. H. (2018). *Fault detection and isolation in power systems using unknown input observer. In Advanced condition monitoring and fault diagnosis of electric machines* (p. 38). Hershey, PA: IGI Global.

Alhelou, H. H., Golshan, M., & Fini, M. (2018). Wind driven optimization algorithm application to load frequency control in interconnected power systems considering GRC and GDB nonlinearities. *Electric Power Components and Systems, 46*(11-12), 1223-1238.

Alhelou, H. H., & Golshan, M. E. H. (2016, May). Hierarchical plug-in EV control based on primary frequency response in interconnected smart grid. In *2016 24th Iranian Conference on Electrical Engineering (ICEE),* (pp. 561-566). IEEE. 10.1109/IranianCEE.2016.7585585

Alhelou, H. H., & Golshan, M. E. H. (2020). Decision-making-based optimal generation-side secondary-reserve scheduling and optimal LFC in deregulated interconnected power system. In *Decision making applications in modern power systems* (pp. 269–299). Academic Press. doi:10.1016/B978-0-12-816445-7.00011-6

Alhelou, H. H., Golshan, M. E. H., & Hatziargyriou, N. D. (2019). Deterministic dynamic state estimation-based optimal LFC for interconnected power systems using unknown input observer. *IEEE Transactions on Smart Grid,* 1. doi:10.1109/TSG.2019.2940199

Alhelou, H. H., Golshan, M. E. H., & Hatziargyriou, N. D. (2019). A decentralized functional observer based optimal LFC considering unknown inputs, uncertainties, and cyber-attacks. *IEEE Transactions on Power Systems.*

Alhelou, H. H., Golshan, M. E. H., Zamani, R., Moghaddam, M. P., Njenda, T. C., Siano, P., & Marzband, M. (2019, June). An improved UFLS scheme based on estimated minimum frequency and power deficit. In 2019 IEEE Milan PowerTech (pp. 1-6). IEEE. doi:10.1109/PTC.2019.8810497

Alhelou, H. H., Golshan, M. H., & Askari-Marnani, J. (2018). Robust sensor fault detection and isolation scheme for interconnected smart power systems in presence of RER and EVs using unknown input observer. *International Journal of Electrical Power & Energy Systems, 99,* 682–694. doi:10.1016/j.ijepes.2018.02.013

Alhelou, H. H., Hamedani-Golshan, M. E., Heydarian-Forushani, E., Al-Sumaiti, A. S., & Siano, P. (2018, September). Decentralized fractional order control scheme for LFC of deregulated nonlinear power systems in presence of EVs and RER. In *2018 International Conference on Smart Energy Systems and Technologies (SEST)* (pp. 1-6). IEEE. 10.1109/SEST.2018.8495858

Alhelou, H. S. H., Golshan, M. E. H., & Fini, M. H. (2015, December). Multi agent electric vehicle control based primary frequency support for future smart micro-grid. In *2015 Smart Grid Conference (SGC)* (pp. 22-27). IEEE.

Anteneh, D., & Khan, B. (2019). Reliability enhancement of distribution substation by using network reconfiguration a case study at Debre Berhan Distribution Substation. *International Journal of Economy, Energy & Environment*, *4*(2), 33.

Arditi, N. (1999). Designing buildings for maintenance: Designers' perspective, Technical Papers, 117–132.

Banteywalu, S., Khan, B., De Smedt, V., & Leroux, P. (2019). A novel modular radiation hardening approach applied to a synchronous buck converter. *Electronics (Basel)*, *8*(5), 513. doi:10.3390/electronics8050513

Bin Akasah. (2007). *Generic process model for maintenance management of school buildings*. Malaysia University.

Bornand, E. (1991) Le guide de maintenance, Berne.

Fini, M. H., Yousefi, G. R., & Alhelou, H. H. (2016). Comparative study on the performance of many-objective and single-objective optimisation algorithms in tuning load frequency controllers of multi-area power systems. *IET Generation, Transmission, & Distribution*, *10*(12), 2915–2923. doi:10.1049/iet-gtd.2015.1334

Frangopol, D. (2003). Reliability deterioration and lifetime maintenance cost optimization. University of Colorado, Boulder.

Gupta, K., Khan, B., & Mubeen, S. E. (2015). Available transfer capability enhancement by unified power flowcontroller. In *Proceedings IEEE International Conference on Signal Processing, Informatics, Communication, and Energy Systems (SPICES)*, pp. 1, 4, 19-21.

Haddad, *R., & Hassan, B.* (2001). *Develop a methodology to analyze the factors affecting maintenance costs in government hospitals,* Tishreen University Magazine.

Haes Alhelou, H., Hamedani-Golshan, M. E., Njenda, T. C., & Siano, P. (2019). A survey on power system blackout and cascading events: Research motivations and challenges. *Energies*, *12*(4), 682. doi:10.3390/en12040682

Haes Alhelou, H., Hamedani-Golshan, M. E., Njenda, T. C., & Siano, P. (2019). Wide-area measurement system-based optimal multi-stage under-frequency load-shedding in interconnected smart power systems using evolutionary computing techniques. *Applied Sciences*, *9*(3), 508. doi:10.3390/app9030508

Haes Alhelou, H., Hamedani Golshan, M. E., Njenda, T. C., & Siano, P. (2019). WAMS-based online disturbance estimation in interconnected power systems using disturbance observer. *Applied Sciences*, *9*(5), 990. doi:10.3390/app9050990

Jariso, M., Khan, B., Tanwar, S., Tyagi, S., & Rishiwal, V. (2018). Hybrid energy system for upgrading the rural environment, IEEE GlobeCom 2018.

Jariso, M., Khan, B., Tesfaye, D., & Singh, J. (2017). Modelling and designing of stand-alone photovoltaic system (Case study: Addis Boder Health Center southwest Ethiopia). In *Proceedings IEEE International conference on Electronics, communication, and aerospace Technology, ICECA 2017,* pp. 168-173, Coimbatore, India.

Khan, B., & Agnihotri, G. (2012) Novel transmission loss allocation method based on transmission usage. In *Proceedings IEEE Fifth Power India Conference*, Haryana, India. 10.1109/PowerI.2012.6479479

Khan, B., Agnihotri, G., & Gupta, G. (2013). A multipurpose matrices methodology for transmission usage, loss and reliability margin allocation in restructured environment, *Electrical & Computer Engineering: An International Journal, 2*(3).

Khan, B., Agnihotri, G., Rathore, P., Mishra, A., & Naidu, G. (2014). A cooperative game theory approach for usage and reliability margin cost allocation under contingent restructured market. *International Review of Electrical Engineering*, *9*(4), 854–862.

Khan, B., Alhelou, H. H., & Mebrahtu, F. (2019). A holistic analysis of distribution system reliability assessment methods with conventional and renewable energy sources. *AIMS Energy*, *7*(4), 413–429. doi:10.3934/energy.2019.4.413

Khan, B., & Singh, P. (2017). The current and future states of Ethiopia's energy sector and potential for green energy: A comprehensive study. *International Journal of Engineering Research in Africa*, *33*, 115–139. doi:10.4028/www.scientific.net/JERA.33.115

Khan, B., & Singh, P. (2017). Optimal power flow techniques under characterization of conventional and renewable energy sources: A comprehensive analysis, *Journal of Engineering, vol. 2017,* Article ID 9539506.

Khan, B., & Singh, P. (2017). Selecting a meta-heuristic technique for smart micro-grid optimization problem: a comprehensive analysis, in IEEE Access, 5, pp. 13951-13977, 2017.

Khan, B., & Singh, P. (2019). Economic operation of smart micro-grid: a meta-heuristic approach. In Handbook of research on smart power system operation and control (pp. 330-346). Hershey, PA: IGI Global. doi:10.4018/978-1-5225-8030-0.ch014

Khan, B., & Singh, P. (2019). Economic operation of smart micro-grid: a meta-heuristic approach. In Handbook of research on smart power system operation and control (pp. 330-346). Hershey, PA: IGI Global. doi:10.4018/978-1-5225-8030-0.ch014

Khan, B., & Tanwar, S. (2019). Issues associated with microgrid integration. In Handbook of research on smart power system operation and control (pp. 252-264). Hershey, PA: IGI Global. doi:10.4018/978-1-5225-8030-0.ch010

Kifle, Y., Khan, B., & Singh, J. (2018). Designing and modeling grid connected photovoltaic system: (Case study: EEU building at Hawassa city). *International Journal of Convergence Computing*, *3*(1), 20–34. doi:10.1504/IJCONVC.2018.091113

Kifle, Y., Khan, B., & Singh, P. (2018). Assessment and enhancement of distribution system reliability by renewable energy sources and energy storage, *Journal of Green Engineering, 8*(3), 2, pp. 219-262.

Makdisie, C., Haidar, B., & Alhelou, H. H. (2018). An optimal photovoltaic conversion system for future smart grids. In Handbook of research on power and energy system optimization (pp. 601–657). Hershey, PA: IGI Global. doi:10.4018/978-1-5225-3935-3.ch018

Molla, T., Khan, B., Moges, B., Alhelou, H. H., Zamani, R., & Siano, P. (2019). Integrated optimization of smart home appliances with cost-effective energy management system. *CSEE Journal of Power and Energy Systems*, *5*(2), 249–258.

Molla, T., Khan, B., & Singh, P. (2018). A comprehensive analysis of smart home energy management system optimization techniques, *Journal of Autonomous Intelligence, 1(1)*, 15–21.

Negash, K., Khan, B., Tesfaye, D., & Gayathri, M. (2017). Optimal placement of phasor measurement unit for system observability. In *Proceedings of IEEE International Conference on Computing Methodologies and Communication, (ICCMC 2017),* pp. 73-77, Erode, India.

Negash, K., Khan, B., & Yohannes, E. (2016). Artificial intelligence versus conventional mathematical techniques: A review for optimal placement of phasor measurement units [Springer]. *Journal of Technology and Economics of Smart Grids and Sustainable Energy*, *1*(1), 10. doi:10.100740866-016-0009-y

Njenda, T. C., Golshan, M. E. H., & Alhelou, H. H. (2018, November). WAMS based intelligent under frequency load shedding considering online disturbance estimation. In *2018 Smart Grid Conference (SGC)* (pp. 1-5). 10.1109/SGC.2018.8777779

Njenda, T. C., Golshan, M. E. H., & Alhelou, H. H. (2018, November). WAMS based under frequency load shedding considering minimum frequency prediction and extrapolated disturbance magnitude. In *2018 Smart Grid Conference (SGC)* (pp. 1-5).

Olagunju, R. E. (2012). Predictive modelling for sustainable residential building maintenance in developing countries: A Nigeria case. *Interdisciplinary Journal of Contemporary Research in Business*, *4*(6), 1273–1283.

Olagunju, R. E. (2012). Sustainability of residential buildings in Nigeria: An appraisal of the factors that influence maintenance of residential buildings 'standards. *Civil and Environmental Research*, *2*(4), 20–29.

Singh, P., & Khan, B. (2017). Smart microgrid energy management using a novel artificial shark optimization, Complexity, vol. 2017, Article ID 2158926. doi:10.1155/2017/2158926

Singh, P., Khan, B., Vidyarthi, A., Haes Alhelou, H., & Siano, P. (2019). Energy-aware online non-clairvoyant scheduling using speed scaling with arbitrary power function. *Applied Sciences*, *9*(7), 1467. doi:10.3390/app9071467

Yeshalem, M. T., & Khan, B. (2017). Design of an off-grid hybrid PV/wind power system for remote mobile base station: A case study. *AIMS Energy*, *5*(1), 96–112. doi:10.3934/energy.2017.1.96

Zamani, R., Golshan, M. E. H., Alhelou, H. H., & Hatziargyriou, N. (2019). A novel hybrid islanding detection method using dynamic characteristics of synchronous generator and signal processing technique. *Electric Power Systems Research*, *175*, 105911. doi:10.1016/j.epsr.2019.105911

Zamani, R., Hamedani-Golshan, M. E., Haes Alhelou, H., Siano, P., & Pota, H. R. (2018). Islanding detection of synchronous distributed generator based on the active and reactive power control loops. *Energies*, *11*(10), 2819. doi:10.3390/en11102819

KEY TERMS AND DEFINITION

Deterioration Models: This predictive model is used for the control and rapid evaluation of the buildings.

Lighting Fixtures Element: Element of lighting fixtures is the most important electrical component of the possibility of the occurrence of faults in it as the lighting devices are the most number of electrical elements used by the occupants of construction and be exposed to breakage or combustion due to weather conditions.

Maintenance: It is defined as a continuous process and a permanent activity to preserve the buildings and the keeping of the equipment in the best case for normal use.

Chapter 15
State Estimation of Active Distribution Networks

Razan Al Rhia
Tishreen University, Syria

Haithm Daghrour
Tishreen University, Syria

ABSTRACT

Monitoring and controlling the electrical distribution system for real time is becoming very important to improve its operating performance after transition to active distribution systems. So, many sensors are needed to monitor all parts in the systems. But if sensors are installed at all buses, investment cost becomes huge. To reduce the number of sensors, state estimation approach can be used to estimate the voltage of buses, which do not have sensors. State Estimation (SE) algorithms are broadly classified into Static State Estimator (SSE) and Dynamic State Estimator (DSE). This chapter classifies most algorithms used in active distribution networks, also State estimation in unbalanced distribution systems, and Role of PMU in Distribution System State Estimation.

INTRODUCTION

Distribution systems are the link between the transmission system and the end-users. While electricity travels in transmission lines at a high voltage, end-users consume electricity at a low voltage, figure 1 illustrate typical transmission and distribution structure with the corresponding voltage levels.

Distribution networks have been affected by the changes in technology, changes in the values of society, changes in economy and in environment. Examples of changes are

- The liberalization of the energy market
- The increase of distributed generation DG, mainly renewable energy sources
- New Distributed Energy Resources (DER) such as Electric Vehicle

DOI: 10.4018/978-1-7998-1230-2.ch015

Figure 1. Typical transmission and distribution structure with the corresponding voltage levels

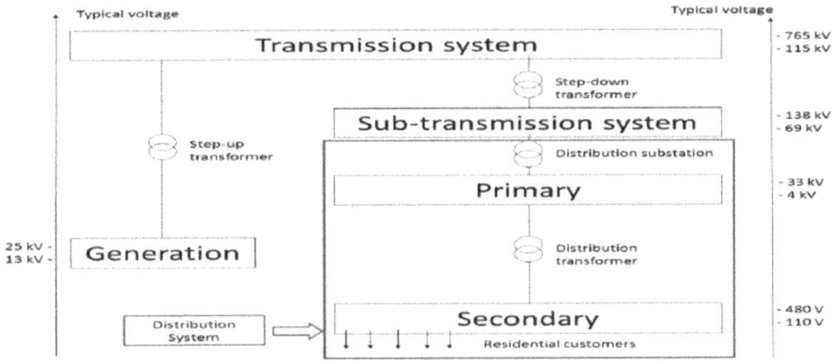

Thus, system security, operation safety, environmental protection, power quality, cost of supply and energy efficiency need to be examined in new ways in response to changing requirements in a liberalized market environment. Technologies should also demonstrate reliability, sustainability and cost effectiveness.

The evolution of electricity grids is referred to smart grid. Which can be defined According to the European Technology Platform of Smart Grids, as "an electricity network that can intelligently integrate the actions of all users connected to it – generators, consumers and those that assume both roles – in order to efficiently deliver sustainable, economic and secure electricity supplies". A smart grid employs innovative products and services together with intelligent monitoring, control, communication and self-healing technology.

The appearance of smart grid has contributed of the transformation from the passive distribution network into a very complex and active network, where passive networks designed to accept bulk power from transmission system and distribute to customers, real time control problem resolved at planning stage, and the principle "fit and forget" will cause to have no control over DER and limiting capacity of DER to be absorbed by the existing networks (Hatziargyriou, 2014; Alhelou et al., 2019; Makdisie et al., 2018; Alhelou et al., 2018; Alhelou et al., 2016; Haes Alhelou et al., 2019; Njenda et al., 2018).

It is worth noting that power systems have always been smart, especially at the transmission network level. However, the distribution network is now testing a development that needs more intelligence for:

- Facilitate the integration of highly distributed generation based on renewable energy sources (RESs)
- Enable local energy demand management, interacting with end-users through smart measurement systems.
- Utilizing technologies already applied in transmission grids such as dynamic control techniques to offer high level of safety, quality and reliability of energy.

In summary, the transition from passive distribution networks into active one, meaning that control and decision-making have been distributed, and power flow has become bi-directional (Khan et al., 2019; Kifle et al. 2018; Khan et al., 2018; Khan et al., 2017; Anteneh et al., 2019; Molla et al., 2019, Molla et al., 2018, Jariso et al. 2018).

The main function of an active distribution network is to link the efficiency of power generation with consumer demands, allowing the best way to work in real time. The assessment of power flow, voltage control and protection requires cost-competitive technologies and new communications systems as well as information and communication technology (ICT) playing a key role (Hatziargyriou, 2014).

Every large complex system requires a monitoring system to provide the information necessary to ensure reliable operation. In power systems, the state of the power system is being identified by the bus voltage and phase angles. Real time accurate estimates of the states enable the calculation of power flow between the branches, the power injections at the buses and other important parameters. The method of estimating the state is called **state estimation** (**SE**).

The advantage of the state estimator is that it provides a means to monitor all the parameters of the power system, with fewer real time measurements from the field. The state estimator, therefore, becomes the key component of the monitoring system used in power grids, and state estimation forms the core of every Energy Management System (EMS) (Francy, 2013).

An Overview of State Estimation

State Estimation was first introduced to power systems by Fred Schweppes in 1970 (Shweppe & Rom,1969). Nowadays it represents an important function in the management and control of the operation of electric transmission networks all over the world. The state estimator obtains the system state using the SCADA, PMU and pseudo measurements in the system in addition to topology information. After the state is known, various functions of EMS like contingency analysis, security analysis, optimal power flow and other functions can be carried out as shown in Figure (2).Therefore, SE is the backbone function of Transmission system EMS.

State estimation in distribution system was not required due to its passive nature with unidirectional power flows since there was no active generation at this level. After transition to smart grids and active distribution networks, the shape of power in distribution system is changing due to bidirectional power flows and distributed generation. This establishes a need for bringing Distribution System into the operation circle of monitoring and control, which makes the role of Distribution System State Estimation (DSSE) more significant. Research on DSSE began near 1990.

However, due to the different properties between the distribution grid and the transmission grid, state estimators in the transmission grid cannot be implemented on the distribution grid directly. Specifically, the main differences between the distribution grid and the transmission grid are in the following ways (kumar et al,2015; Ahmad et al.,2018):

Construction: Most distribution networks have a radial construction (whereas transmission systems are more meshed).

High R/X Ratios: Due to low voltage levels and comparatively shorter lengths, Distribution System have higher R/X ratios than the Transmission Systems. Therefore, Distribution Systems cannot be modeled and analyzed as Transmission Systems due to the fact that the assumptions made for these networks are not true for Distribution networks. Iterative algorithms that use the Newton-Raphson simply do not converge for networks with higher R/X ratios. Moreover, for such networks, DC approximation and decoupled power-flow solutions also become invalid.

Figure 2. Role of SE in EMS/SCADA

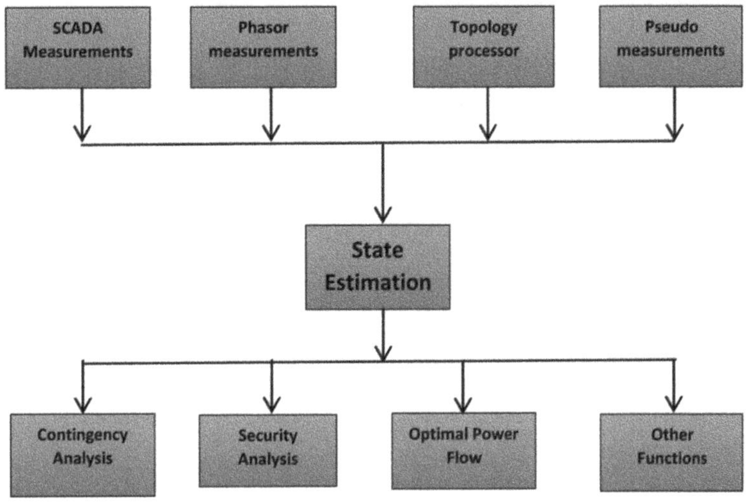

Redundancy: For technical and economic reasons the number of measurement points in distribution networks is much lower than in transmission networks and are not enough for the observability analysis required in the estimation process. Systems are under-determined, rather than over-determined. The conventional SE assumes the system to be over-determined by having redundant measurements but DSs are under-determined. Various attempts have been made to solve this measurement scarcity problem in DSs by generating pseudo measurements using load data.

Measurement Types: Most of the available input data at the distribution level are measurements (or pseudo measurements) of power or current injections. Direct measurements of voltages and power flows are rare.

Scale and Complexity: Distribution systems are diverse (e.g. networks in rural areas are very different from those in urban areas) and have very large numbers of components.

This means that the methods developed for DSSE (distributed or multiarea DSSE techniques) need to be scalable to achieve all sort of complexities, have a relatively low computational burden, and be applicable across a range of different network types. to achieve all sort of complexities.

Phase Imbalances: Conventional SE techniques assume that the network is a balanced system. However, distribution systems can have significant phase imbalances, requiring the use of full three-phase system models.

Complex Measurement Functions: The measurements available at feeders are current and power injections. Direct voltage and power measurements are rarely available, which complicates the measurements functions. Recently PMUs or μPMUs have been researched for Distribution Systems which provide direct measurements of voltage and current phasors thereby eliminating the nonlinearity of measurement functions (Khan et al., 2014; Khan et al., 2013; Khan et al., 2012; Negash et al., 2017; Negash et al., 2016; Jariso et al., 2017; Kifle et al., 2018; Yeshalem et al., 2017; Singh et al., 2019).

The main differences are summarized in the following table 1 (Blaauwbroek,2018).

Table 1. Comparison of transmission and distribution system state estimation

	Transmission systems	Distribution systems
Number of nodes	10-100	100-10000
Branch characteristics	Mainly inductive	Mainly resistive
Topology characteristics	Meshed	Radial
Loading balance	Three-phase balanced	Unbalanced or single-phase
Loading aggregation level	Highly aggregated	Non-aggregated
Loading stochasticity	Low-stochastic	High-stochastic
Measurement availability	Redundant	Limited, smart meters
Nature of data	Non-personal	Partly personal

What is State Estimation?

State estimation (SE) is a technique used to find unknown values of the state variables based on some imperfect measurements. In power systems, state estimation is considered as a set of techniques able to provide a quantitative snapshot of the state of an entire network on the basis of partial or incomplete set of data.

To obtain information of the system and high measurements redundancy, large number of meters should be installed. To better utilize the redundant data, a state estimation method is developed based on statistic skills which replaces the traditional power flow calculations due to its high accuracy by considering the noise from smart meters (zhang, 2018; Fini et al., 2016; Alhelou et al., 2018; Zamani et al., 2018; Alhelou et al., 2015; Njenda et al., 2018; Haes Alhelou et al., 2018; Haes Alhelou et al., 2019).

The objective of SE is to estimate the complex bus voltage phasors (the states) at every bus in a given power system. This is accomplished by processing the measurements of various line flows, bus injections, voltage and line current magnitudes as well as the information about the status of the circuit breakers, switches, transformer taps and the parameters of the transmission lines, transformers, and shunt capacitors and reactors. Measurement types, locations, and accuracies vary depending on the individual power system.

The estimator is designed to produce the best estimation of the system by minimizing the cost which is defined by the difference between the estimated value and the actual measurement after considering the noise in the actual measurement. After processing these raw data, the most probable state of the system is obtained and sent to the EMS, which utilizes this information for further power system applications such as a contingency analysis, optimal power flow, etc. (Zhang, 2018).

State estimation has been traditionally a key function in the transmission systems and the resultant estimates are used for many functions in the operation and planning throughout the entire system, including power flow optimization and bad data analysis (Huang et al., 2012).

The state estimators typically include the following functions (Abur, 2004), as illustrated in figure 3.

Figure 3. Relation between state estimation and energy management functions

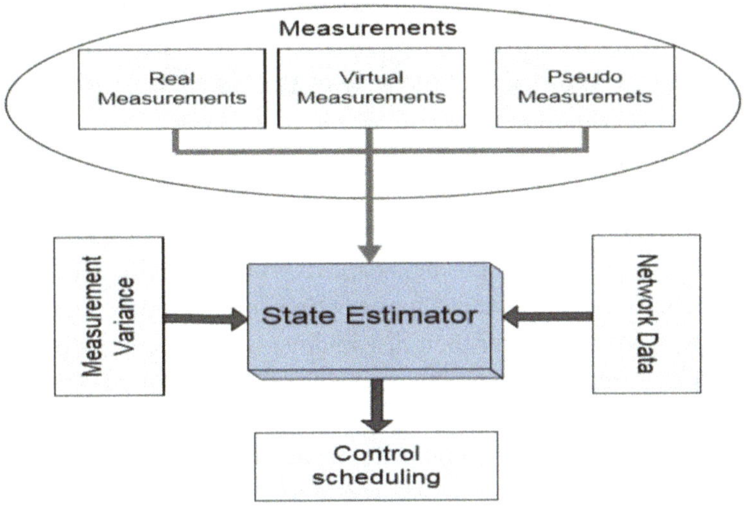

- **Topology Processor:** Gathers status data about the circuit breakers and switches, and configures the one-line diagram of the system.
- **Observability Analysis:** Determines if a state estimation solution for the entire system can be obtained using the available set of measurements. Identifies the unobservable branches, and the observable islands in the system if any exist.
- **State Estimation Solution:** Determines the optimal estimate for the system state, which is composed of complex bus voltages in the entire power system, based on the network model and the gathered measurements from the system. Also provides the best estimates for all the line Hows, loads, transformer taps, and generator outputs.
- **Bad Data Processing:** Detects the existence of gross errors in the measurement set. Identifies and eliminates bad measurements provided that there is enough redundancy in the measurement configuration.
- **Parameter and Structural Error Processing:** Estimates various network parameters, such as transmission line model parameters, tap changing transformer parameters, shunt capacitor or reactor parameters. Detects structural errors in the network configuration and identifies the erroneous breaker status provided that there is enough measurement redundancy.

Why do we Need State Estimation in Active Distribution System?

The operation of a power network requires accurate monitoring of its state: bus voltages, line currents, consumption, and generation, and this is the function of state estimation as mentioned above (Picallo et al., 2018).

Typically distribution networks used to have a simple radial structure with a single source bus, the point of common coupling (PCC), connected to the grid and injecting power, and therefore SE was not as necessary as in transmission networks (Picallo et al.,2018). But now after transition to smart grids and active distribution systems, with Increasing number of DGs like PV-panels, plug-in electric vehicles, batteries, etc. in the distribution networks which resulted bidirectional power flow that make the load profile becoming more volatile (primadianto et al., 2016), and complicates the operation of distribution systems and will require enhanced distribution automation (DA). The system operators will therefore need more timely and reliable knowledge of the state of the system to properly monitor, control and economically dispatch power (Huang et al., 2012).

State estimation in active distribution systems is necessary for the following reasons:

- In order to facilitate smart grid features such as demand response (DR) and two-way power flow, and to enable real-time system monitoring, adaptive protection with network restoration, energy loss minimization, and voltage / reactive power optimization, utility companies will need to have more timely and accurate models for their distribution systems. This calls for SE at the distribution level, which places more stringent requirements on SE algorithms (primadianto et al., 2016; (Huang et al., 2012b).
- New regulations and market pricing competition may require utility companies to share more information and monitor the grid over large geographical areas. This calls for distributed control, and hence, distributed SE to facilitate interconnection-wide coordinated monitoring (Huang et al., 2012b).
- These devices are the traditional options of voltage and VAR control: Switchable shunt capacitor/reactor, tapped line voltage regulators, on load tap changer (OLTC) in primary substation, and now they are required to operate more actively for providing voltage control in coordination with the fast power electronics based reactive power available from the DGs. The control of these devices needs to be coordinated in distribution substation control for optimal operation. So, the operational visibility of the network is required. This is provided by the SE (Nanchian, 2015).
- For near real-time monitoring of the power grid advanced measurement technologies are used in active distribution system like PMUs. Typically, a PMU takes 30 measurements/s, thereby offering the possibility of a much more timely view of the power system dynamics than conventional measurements. More importantly, all PMU measurements are synchronized, as they are time stamped by the global positioning system's (GPS's) universal clock. However, PMUs with their higher measurement frequency put enormous strain on the communication and data processing infrastructure of the grid. This drives the need for resource-efficient, event-triggered SE solutions that employ on-demand (event-triggered) sensing, estimation, and communication(Huang et al., 2012b).

However, as the electric power grid becomes smarter, more DA will be needed and SE at the distribution level will become more important. The control mechanism in the distribution system will most likely be distributed and active in nature, so will be the corresponding SE functions (Huang et al., 2012b).

The use of appropriate state estimation will bring the below advantages:

Figure 4. Diagram of a classic distribution management system(DMS)

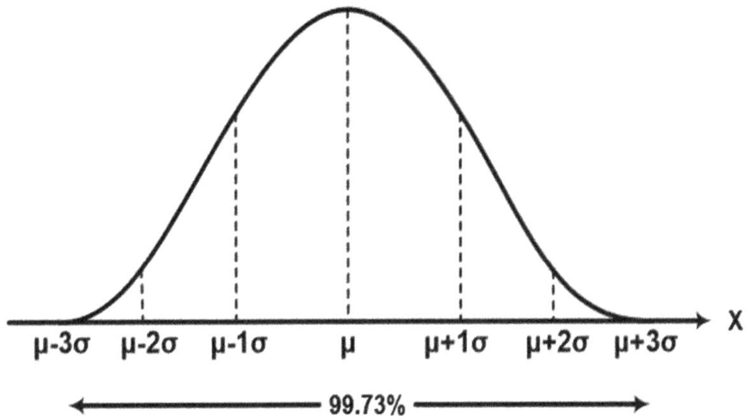

1. **Peak Shaving:** The use of smart network can affect both the generation and consumptions sides which a direct result is peak load shaving.
2. **Fossil Fuel Reduction:** By reducing the loss of energy in the feeders as well as possibility of altering the topology through some useful techniques such as reconfiguration will result in notable fossil fuel reduction.
3. **Interruption Reduction:** Reducing the outage time, the number of interruptions and unwanted shut downs.
4. **Reducing Investment Costs:** The main reason for new investments in the current electric grids is the load growth which is mainly caused at peak load hours. By the use of idea of smart grid and intelligent management techniques, the load growth can be controlled which will reduce the investment costs.
5. **Reducing Switching Cost:** Decreasing costs associated with the switching of remote subscribers (Khorshidi et al., 2016).

Input to the Distribution System State Estimation:

DSSE consists of all the inputs to the DSSE block and it consists of following components as shown in figure 4:

Measurement

Most commonly used measurement types are described as follows (liu,2017):

SCADA Measurements:

The SCADA system usually provides data on the substation level including

- active and reactive power, current magnitude on each distribution feeder
- voltage magnitude at the substation bus
- branch current (magnitude) from the feeder.

For the errors of the measurements are usually taken as follows:

- 1% for the magnitude of the voltage
- 3% for the power flow and current magnitude measurements

AMI Measurements

The AMI system consists of smart meters installed in the user side. Smart meters collect detailed real-time data on energy consumption such as active power, reactive power, apparent power, power factor, voltage magnitude etc. and deliver the raw data through two-way communication networks to the data management system, where raw data is processed and then provided for advanced application.

The measurement error is suggested to be within 10%.

PMU Measurements

The main output of the PMU installed in the distribution system is the phasor information of the sampled waveforms. Based on the basic measurement, other measurements can be computed and derived.

It is indicated that the PMU can measure (1) voltage, branch current and power flow; or (2) voltage, load current and power according to the installation position. In general, it can be assumed that the PMUs can provide the following measurements:

- Voltage (magnitude and angle)
- Branch current (magnitude and angle)
- Load current (magnitude and angle)
- Active and reactive power
- Active and reactive power flow.

For the measurement error, the synchrophasor measurement standard allows a 1% vectorial error in phasor estimation, which is translated into a maximum deviation of 1% in amplitude estimation or a maximum deviation of 1crad for phase angle.

Figure 5. Normal distribution curve

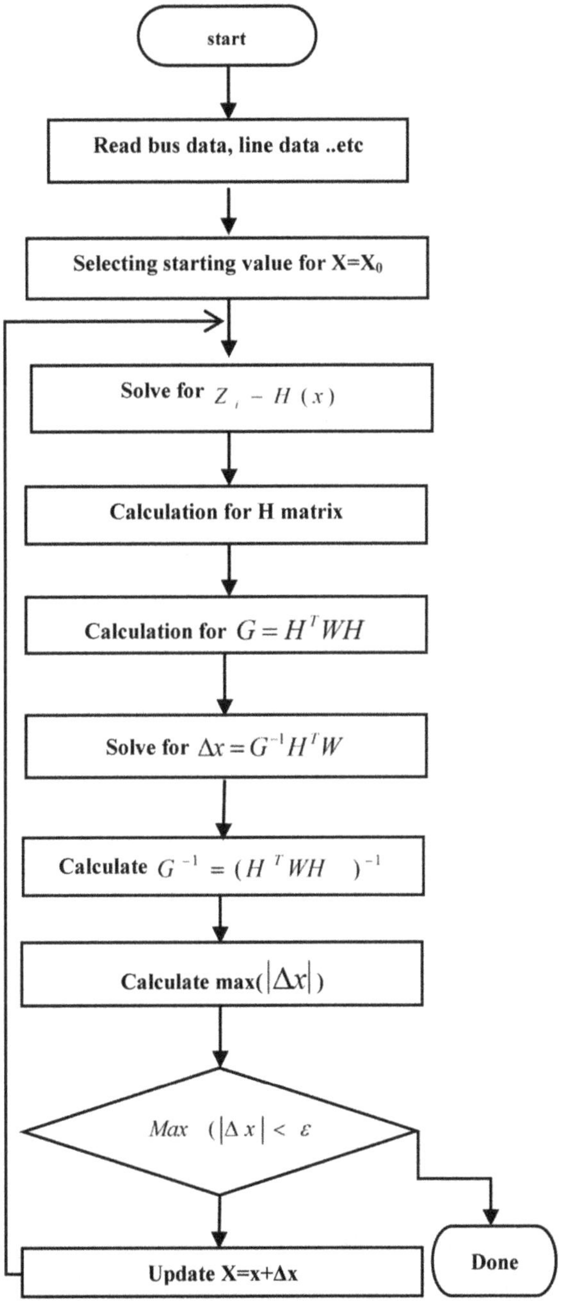

Pseudo Measurements

These measurements are used in the case of absence of any real measurement of loads, which are highly distributed and diverse. The measurements values are obtained on historical data to improve measurement redundancy in poorly monitored areas. Forecasted loads at load buses, and scheduled generation at generator buses are examples of this type of measurements.

The measurement error of the pseudo measurement, of course, depends on the load modelling/forecasting algorithm and the availability of the historical data. Sometimes, a large error may occur when there are no adequate historical data or loads change greatly and may reach to 50%.

Virtual Measurements

These are the zero power injections at the buses which have no loads and no generations.

Measurement Noise (Variance)

Measurement noise is an important input to SE algorithm. It determines the accuracy of the measurements for accurate estimation of the system states.

Measurement variance can be obtained from the equation (1). The standard deviation is used to provide the weight of each measurement as all the measurements are modelled assumed to have a normal distribution.

As shown in figure (5) $\pm 3\sigma$ deviation covers around 99.73% area of Gaussian curve.

Hence, for a measurement z given % of error about mean μz, the standard deviation of error can be obtained from equation (1).

Network Data

The modelling of various network components of DSSE is a challenging task. For instance, the presence of unsymmetrical-network components and unbalanced-load on unbalanced distribution system makes it essential to consider the exact model of the system components (three-phase model). It includes information about line model, transformer model, Switch model, Load model (it is illustrated in (Nanchian,2015).

Literature Review on DSSE

Fred Schweppe first formulated the state estimation as Weighted Least Squares (WLS) which estimate the most likely state of the system by minimizing the non-linear objective function which sums the weighted squares of the error or difference between the estimated values and the actual values. WLS is the most popular technique so far and the major difference among researchers is either choosing the nodal voltage or line current as the variable. An analysis of the DSSE algorithms shows that node voltage or branch current based estimates provide similar accuracy (Primadianto & Lu,2016b). An efficient branch-current DSSE was proposed in (Pau et al.,2013) that allows synchronized PMU measurements to be included in the estimate. A three-phase SE algorithm was developed in ((Baran & Kelley,1994; Li,1996). A linear state estimator was proposed in (Haughton& Heydt,2012) for full three phase unbalanced distribution system. The concept of minimum additional measurements for improving accuracy of the SE in WLS framework have proposed in (Roytelman& Shahidehpour,1993; Meliopoulos & Zhang,1996), Since the quality and adequacy of measurements are less than required.

Baran & Kelley (1995) expressed the current function only in one phase to divide the problem into three problems. This technique has been used in weakly-meshed networks.

Lin and Tang (1996) proposed a new fast decoupled state estimator with equality constraints to increase the speed of convergence.

Celik and Liu (1999) have applied a Gauss-Seidel load flow algorithm for state estimation in radial distribution networks. Deng et al.(2002) used complex node voltage as variable parameter and decomposed the whole system's WLS problem into a series of WLS subproblems dealing with only one branch state estimation.

Ghosh and others (1997b) have proposed a different method for state estimation in distribution network which applies a probabilistic extension of the radial load flow algorithm.

There are also algorithms, which are called Forecasting Aided State Estimation (FASE), estimating the state of the system by considering both the past state of the system and the real-time measurements. One of the most commonly used FASE is the Kalman filter (Kalman,1960). The Kalman filter (KF) applied to nonlinear systems like the power system is in the form of extended Kalman filter (EKF), which linearizes all nonlinear transformations (Qi et al.,2016).

Gelagaev (2008) made a comparison between the WLS and the Extended Kalman Filter (EKF) methods and proposed a first analysis of the relative importance of process and measurement covariance matrices.

Julier et al. (2016) proposed the unscented Kalman filter (UKF) as a derivative-free alternative to EKF in the framework of state estimation. Thus, the UKF turns out to be an improvement of the EKF SE as it is much easier to implement than the EKF, because there is no need for the linearization and the derivation of Jacobian matrices.

Valverde and Terzija (2011) showed the advantage of the unscented Kalman filter (UKF) over the EKF and WLS methods.

The researchers start to apply the ensemble Kalman filter for the dynamic state estimation, that both EKF and UKF have some inconveniences (Zhou et al.,2014; Ray& Subudhi,2012).

Methods used for dynamic SE have been also proposed in (Durgaprasad& Thakur,1998; Sinha& Mondal,1999; Shih& Huang,2002; Huang et al.,2007; Ghahremani& Kamwa,2011; de la O Serna et al.,2011; Zhang et al.,2011).

Distribution System State Estimation Algorithms:

SE algorithms are broadly classified into Static State Estimator (SSE) and Dynamic State Estimator (DSE)

Static State Estimators assumes the power system to be quasi-static, which implies the system state stays constant between the two consecutive updates of the state vector. Therefore, to improve real time monitoring of the grid the updates need to be performed within shorter intervals.

Dynamic State Estimation (DSE) is a step closer to real time monitoring when compared to SSE. There are two parts to the DSE: Prediction and Filtering. The prediction of the state variables involves the modeling of the power system behavior. The prediction is calculated based on a mathematical model which takes into account the nonlinearities of the measurement functions unlike in SSE where a linear model is developed for the ease of computation (Francy,2013).

Based on the timing and evolution of the estimates, state variables chosen, and the treatment of load and bad data, the existing DSSE methodologies can be classified into different paradigms (Primadianto& Lu, 2017).

WLS-Based Static DSSE

Weighted least square (WLS) is the most common algorithm, the formulation of the problem based on WLS mathematically represents the relationship between system states and measurements. Many of the earlier research papers on DSSE focused on adapting conventional WLS techniques to distribution networks (Lu et al., 1995; Roytelman & Shahidehpour,1993; Meliopoulos& Zhang,1996; Lin & Teng,1996).

The analog quantities pass through transducers and analog-to-digital converters and the digital outputs are then telemetered to the energy control center over various communication links. The data acquired always contains inaccuracies which are unavoidable since physical measurements cannot be entirely free of random errors or noise.

1) 1) Measurements have certain Error associated with the true value of measured quantity

$$[Z] = [Z_{true}] + [e] \tag{2}$$

$$[Z] = [h(x)[x] + [e] \tag{3}$$

$$Z = \begin{bmatrix} Z_1 \\ Z_2 \\ \cdot \\ \cdot \\ \cdot \\ Z_m \end{bmatrix} = \begin{bmatrix} h1(x1, x2, x3, \ldots, xn) \\ h2(x1, x2, x3, \ldots, xn) \\ \cdot \\ \cdot \\ \cdot \\ hm(x1, x2, x3, \ldots, xn) \end{bmatrix} + \begin{bmatrix} e1 \\ e2 \\ \cdot \\ \cdot \\ \cdot \\ em \end{bmatrix} \tag{4}$$

Where,
Z is measurement vector
Z true is true value of measured quantity
(x) is the non linear function expressing measurements in terms of State
x is the system state vector
e is the measurement error
m is the number of measurements
n is the number of state variables

2) 2) Error between actual measurements Z and the true (unknown) value of measured quantities:

$$[\mathbf{e}] = [\mathbf{Z}] - \{\ [\mathbf{H}][\mathbf{x}]\ \} = [\mathbf{Z}] - [\mathbf{Z\_true}\ ] \quad (5)$$

Regarding the general statistical properties of the measurement errors, the following assumptions are made:

- The measurement error e_i is assumed to have a normal distribution with zero mean and known standard deviation σ_i, i.e. $E(e_i)=0$;
- The measurement errors are assumed to be independent, i.e $E(e_i\ e_j)=0$.

Hence, the covariance matrix of the measurement errors R is diagonal

$$R = Cov(e) = E[e.e^T] = diag\left\{\sigma_1^2, \sigma_2^2, \ldots, \sigma_m^2\right\}$$

The standard deviation σ_i of measurement i is set to reflect the expected accuracy of the corresponding meter used.

3) 3) True values of state cannot be determine so we calculate estimates of states:

$$\mathbf{e} = [\mathbf{Z}] - [\mathbf{H}] * [\mathbf{x}] \quad (6)$$

x^ is the estimated value of state variables
e^ is the difference between actual measurements and their estimated values

4) Because of the errors or noise the true values of physical quantities are never known and the best possible estimates have to be calculated of the unknown quantities. a criterion must be selected for calculating the estimates $x^\wedge$ from which error $e^\wedge$ and $Z^\wedge$ are to be computed. If algebraic sum of the errors is minimized, then positive and negative errors would offset one another. So Sum of Squares of Error is minimized.

To ensure that measurements from meters of known greater accuracy are treated more favorably than less accurate measurements, each term in the sum in squares is multiplied by an appropriate weighting factor W to give the objective function:

$$f = \sum_{i=1}^{n} w_i e_i^2 \tag{7}$$

5) Selecting the best estimates of the state variables as those values which cause objective function f to take on its minimum value.
6) 6) For minimizing f, the estimates $x^{\wedge}$ are those values of $x^{\wedge}$, which satisfy the equation,

$$\left[\frac{\partial(\hat{e})}{\partial x}\right][w][\hat{e}] = 0 \tag{8}$$

7) 7) Substituting $\hat{e} = Z - H\hat{x}$ in equation (8)

$$[\frac{\partial(z - H\hat{x})}{\partial x}][w][z - H\hat{x}] = 0 \tag{9}$$

8) 8) Solving for $x^{\wedge}$

$$\Delta\hat{x} = (H^T WH)^{-1} H^T wz \tag{10}$$

The iterative approach is used to solve for the state

$$x^{k+1} = x^k + \Delta\hat{x} \tag{11}$$

Where, H is Jacobian Matrix, ∂x, ∂h
h is Measurement Function
x are the state variables
k is iteration index
G= H^T W H is gain matrix
xk is value of State vector at kth itteration
x k+1 is the value of State vector
***Δ*x^** is update vector of the state vector

Figure 6. Flowchart for WLS state estimation

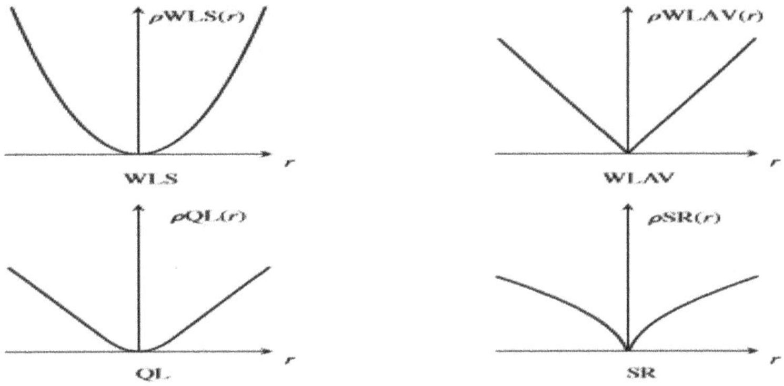

$$H = \begin{bmatrix} 0 & \frac{\partial v_{mag}}{\partial v} \\ \frac{\partial p_{inj}}{\partial \delta} & \frac{\partial p_{inj}}{\partial v} \\ \frac{\partial Q_{inj}}{\partial \delta} & \frac{\partial Q_{inj}}{\partial v} \\ \frac{\partial p_{ij}}{\partial \delta} & \frac{\partial p_{ij}}{\partial v} \\ \frac{\partial Q_{ij}}{\partial \delta} & \frac{\partial Q_{ij}}{\partial v} \end{bmatrix} \quad (12)$$

Where,
Vmag is the voltage magnitude.
Pinj and Qinj are the real and reactive power injections respectively.
Pij and Qij are the real and reactive power flow from bus i to j respectively.
Resulting: The WLS state estimator solution is based on solving the following optimization problem:

$$\begin{aligned} \min_{m} J(x) &= \sum_{1}^{m} w_i (z_i - h_i(x))^2 \\ &= [z - h(x)]^T W [z - h(x)] \end{aligned} \quad (13)$$

The flowchart of WLS state estimation is described in figure (6).

There are two main classes of WLS state estimators used in distribution systems: state estimators based on node voltage state variables and estimators based on a state expressed in terms of branch-currents. Both can be formulated in polar and rectangular coordinates. These methods have been extended for three-phase analysis (Baran et al.,1995).

Static DSSE methods typically fall into the following categories:

Node Voltage Based DSSE Methods

State variables in this case are node voltages, which be in polar forms:

$$x = [|V| \quad \theta] \tag{14}$$

where θ and v correspond to the phase-angle and the node voltage magnitude of every node, respectively. In polar forms, the entries of measurement and gain matrix in the normal equation must be recalculated at each iteration

In rectangular form state variable is:

$$x = [V_{real} \quad V_{imag}] \tag{15}$$

By adopting node voltage in rectangular form as the state variables, the Jacobian terms of the converted bus current injection measurements become constant (Primadianto& Lu, 2017).

If only traditional measurements (node voltage magnitudes, branch-current magnitudes, power injections and power flows) are available, the slack bus is usually adopted as reference bus and the state vector is x = [θ2...θn,v1...vn], with θ1 conventionally set equal to zero. In this way, all the voltage phase angles represent phase differences expressed with reference to the slack bus. Instead, if synchronized measurements (voltage and current phasors provided by PMUs) are available, voltage phase-angles can be expressed in their absolute values (with reference to the UTC time), removing the need of using a reference bus. Then, the state vector can be extended to include also the slack bus voltage phase-angle.

Voltage based state estimation methods attempt to reconstruct the voltage phasors by computing the maximum likelihood estimate based on sensor measurements using a WLS approach (Pau et al,2013). More details can be found at (Baran&Kelley,1994 ; Lu et al.,1995; Haughton&Heydt,2013; Deng et al.,2002).

2) 2) **Branch Current Based DSSE Methods:** This approach is by far the most popular method tested. Feeder branch currents in rectangular form are chosen as state variables (Primadianto& Lu,2017).

$$x = [I_{branch\_real} \; I_{branch\_imag}] \tag{16}$$

In particular when using Branch Current _DSSE, power measurements are translated into equivalent current ones to ensure that all Jacobian matrix elements are constants. Then, the estimation of the state x in the Branch Current-DSSE is obtained by an iterative three steps algorithm. Each iteration consists of:

- Definition/update of the equivalent measurements;
- Branch-currents estimation applying a WLS approach;
- Network voltages state computation through a forward-sweep calculation, descending the network graph starting from the slack bus voltage.

Figure 7. Examples of objective functions

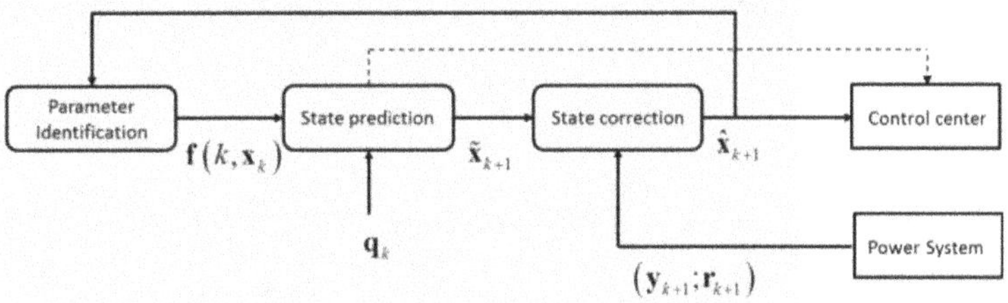

In case of only traditional measurements, the state vector is extended by adding the slack bus voltage magnitude, while the slack bus phase-angle is chosen as reference and conventionally set equal to zero. Similarly to the voltage state estimators, in case of PMU measurements there is no need of using a reference bus, thus the vector state is modified by considering the real and imaginary parts of the slack bus voltage (Pau et al.,2013). Some of Literatures that studied Branch current in DSSE are (Baran & Kelley,1995; lin et al., 2001; Teng, 2002; Baran et al., 2009; Wang & Schulz, 2004).

Load Estimation for DSSE

In DSSE, the number of telemetered devices that can provide system measurements is often very limited, and not sufficient to allow observability of the entire network, or bad data identification. In many cases, DSSE relies on pseudo-measurements of the demands at each load point in the network, based on historical data or load forecasts, which have significantly lower accuracy than actual measurements. Load estimation techniques and their application to DSSE are discussed in (Ghosh et al., 1997; Singh et al., 2010; Baran & Kelley, 1994)

Robust DSSE Methods

The estimator is called robust if the estimated states are unaffected by large deviations in measured data. The robust estimator is remaining neutral regardless of the existence of bad data (Nanchian, 2015).

The robust estimation was initially introduced by Huber (1992). Later Mili et al. (1994) applied this theory in concept of power system state estimation.

A robust estimator generally is one approach to cope with bad data in measurements to use other objective functions which gradually suppress the effect of bad measurements during the iterative estimation step by reducing the weights assigned to the suspected bad data points. When the estimated states remain unaffected by major deviations in a limited number of redundant measurements, the corresponding estimator can be considered statistically robust. Hence, there is no need to perform bad data elimination after estimation (Primadianto & Lu, 2017; Uatrongjit, 2018).

Consider the objective function:

$$\phi_r(x) = \sum_{k=1}^{M} \rho(r_k) \tag{17}$$

Where r_k is the residual of the *k*th measurement. Some examples of ρ are the WLAV, QL and square root(SR) functions which are described by:

$$\rho WLAV(r) = \left| r/\sigma \right| \tag{18}$$

$$\rho QR(r) = \begin{cases} (r/\sigma)^2; \left| r \right| \le a\sigma \\ 2a \ \left| r/\sigma \right| - a^2); otherwise \end{cases} \tag{19}$$

$$\rho SR(r) = \begin{cases} (r/\sigma)^2; \left| r \right| \le a\sigma \\ 2a^2(\sqrt{a \left| r/\sigma \right|} - 3); otherwise \end{cases} \tag{20}$$

Where r and σ denote residual and its standard deviation, respectively. It can be noticed that (17) also includes the case of WLS if we define $\rho WlS(r)= r^2/\sigma^2$. The shapes of these objective functions are shown in figure (7). Compared with WLS, other objective functions tend to have less emphasis on a large residual value r.

Dynamic DSSE Methods

Dynamic State Estimation (DSE), also known as Forecasting-Aided SE (FASE) or alternative SE, is a recursive estimation method based on several measurement snapshots in a time sequence (Della Giustina et al., 2014). DSE with a relative simplified model for tracking state vectors was also introduced early in 1970 by Debs and Larson. DSE can predict state vectors one-time step ahead based on a priori knowledge and that can be corrected with the subsequent measurement sets. Depending on the techniques, estimated DSE variables can be either static state variables, e.g., bus voltage magnitudes and phase angles, or dynamic state variables, e.g., speed variables of generators (Nguyen et al., 2013).

A dynamic model that monitors system operating conditions more completely than static models can be represented by a process equation (21) and a measurement equation (22) as follows:

$$X_{k+1} = f(k, x_k) + q_k \tag{21}$$

$$y_k = h(k, x_k) + r_k \tag{22}$$

Figure 8. Dynamic state estimation process

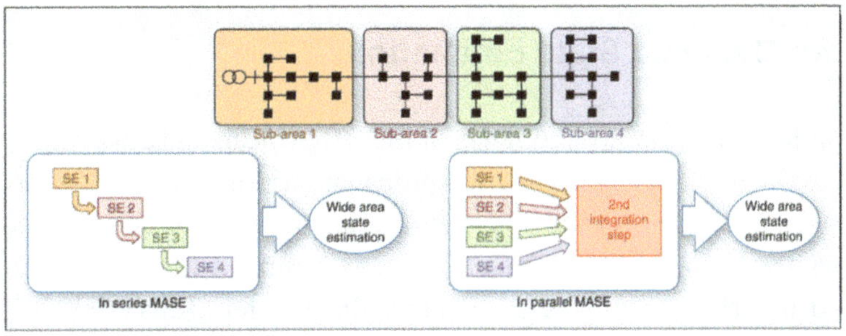

where k is the time sample; x_k is the state vector; q_k represents modelling uncertainties corresponding to a white Gaussian noise with zero mean and covariance matrix Q_k; y_k is the measurement vector; h is a set of nonlinear load-flow functions for the current network configuration; and r_k is a Gaussian error vector with zero mean and diagonal covariance matrix R_k.

In this dynamic state-space model, equation (21) can be interpreted as the memory of the system state time evolution, and equation (22) is considered its refreshment. Such memory will be responsible for the forecasting capability of the model. Depending on the availability of measurements, the model can be adequate or parsimonious. The basic idea of a state estimation function is to determine the most likely system state vector x for either the static steady-state or dynamic state of the system:

$$x = [\theta_i, V_i, P_{inject}, \varpi_i,, \delta_i]^T \tag{23}$$

Based on the quantities measured and acquired by remote terminal units (RTUs), presented as:

$$y = [V, P_{flow}, Q_{flow}, P_{inject}, Q_{inject}]^T \tag{24}$$

Static state estimation delivers only a snapshot of the present state vector (find X_k at a certain time *k* by using the set of measurement data received at the same instant of time k), the aims of the dynamic state estimation is to offer not only time-varying behavior of the system but also expect the future state of the system.

In general, DSE is achieved by implementing three steps, i.e., parameter identification, state prediction (forecasting), and state correction (filtering), as illustrated in Figure (9). In the first step, the parameters of the transition function $f(k,x_k)$ are identified. This will be associated with the processing noise q_k to predict a step-ahead state variable vector $\tilde{x}_{k+1}$. Through the measurement function $h(k, \tilde{X}_{k+1})$ and measurement data (y_{k+1}; r_k,), the state vector is corrected as $\hat{X}_{k+1}$, which will be a fine-grained estimation for use at the control center. While SSE based on WLS provides only a snapshot of the current state vector, i.e., x_k at a certain time k, DSE aims not only to provide time-varying solutions but also to predict the system state one step ahead.

DYNAMIC STATE ESTIMATION METHODS

Kalman Filters for Dynamic State Estimation

Kalman filter (KF) utilizes the first two moments of the state (mean and covariance) in its update rule . It consists of two steps: in prediction step, the filter propagates the estimate from last time step to current time step; in update step, the filter updates the estimate using collected measurements. DSE has been implemented by extended Kalman filter (EKF), unscented Kalman filter (UKF), square-root unscented Kalman filter, extended particle filter, and ensemble Kalman filter (Taha et al.,2015) DSE techniques are described in detail in (simon,2006).

The extended Kalman filter and unscented Kalman filter are sufficient techniques in the estimation of dynamic state vector element under transient condition.

Extended Kalman Filter

One of the greatest used techniques in power system dynamic state estimation is Extended Kalman Filter (EKF) that takes into its accounts both the entering measurements and the expected state to get an optimal estimate of the system state. Kalman filter is appropriate for real-time estimation since the estimation is finished for any instantaneous time. EKF is a discrete Kalman filter adapting to nonlinear system estimation through linearization.

For a nonlinear dynamic system defined by differential algebraic functions (DAEs) in (25) and further in discrete form in (26), the aim of EKF is to reduce the covariance of the incompatibility between the estimated states and the states.

$$\begin{aligned} \frac{dy}{dx} &= f(x, y, u, w) \\ 0 &= g_c(x, y, u, v) \end{aligned} \tag{25}$$

Where the vector represents the state variables, the vector represents the algebraic variables, is the vector of input variables, and are processing noise and measurement noise. The subscript "c" represents the continuous form. The discrete form of (21,22) is shown below:

$$\begin{aligned} X_k &= X_{k-1} + f_c(X_{k-1}, u_{k-1}, w_{k-1})\Delta t \\ &= (X_{k-1}, u_{k-1}, w_{k-1}) \\ y_k &= h(X_k, u_k, v_k \end{aligned} \tag{26}$$

The EKF problem can be resolved in a two-step

1) 1) Step 1 -predicted state

$$\hat{X}_k^- = f(\hat{X}_{k-1}, h(\hat{X}_{k-1}, u_{k-1}), u_{k-1}, 0)$$
$$P_k^- = A_k P_{k-1} A_k^T + W_k Q_{k-1} W_k^T \tag{27}$$

2) 2) Step 2 –corrected state

$$K_k = P_k^- H_k^T (H_k P_k^- H_k^T + V_k R_k V_k^T)$$
$$\hat{X}_k = \hat{X}_k^- + K_k (Z_k - h(\hat{X}_k^-, u_k, 0)) \tag{28}$$
$$P_k = (I - K_k H_k) P_k^-$$

where "–" means a priori state, A_k and W_k are the Jacobians process at step k, P_k is a covariance matrix of the state estimation error and is also named gain factor matrix, and Q_k is the process noise covariance at step k. H_k and V_k are the measurement Jacobians at step k, and R_k is the measurement noise covariance at step k.

$$A = \frac{\partial f}{\partial x}, H = \frac{\partial h}{\partial x}, = \frac{\partial f}{\partial w}, V = \frac{\partial h}{\partial v} \tag{29}$$

Unscented Kalman Filter

The UKF is a proficient discrete-time recursive filter capable of solving estimation issues which are having the formula shown below

$$X_k = f(X_{k-1}, k-1) + q_{k-1} \tag{30}$$

$$y_k = h(X_k, k) + r_k \tag{31}$$

Where X is the state vector, y is the measurement vector, while q_{k-1} and r_k are, respectively, the system and measurement Gaussian noises with zero mean and uncorrelated covariance matrices Q and R, respectively. Functions f and h are non-linear functions demonstrating the system and measurements models in terms of the state variables and other system inputs.

The UKF merges the classical Kalman filter idea and the UT . The prime benefit of the UKF over the EKF is the fact that (30) and (31) were not approximated, so (30) and (31) were not linearized; which averts missing of higher order information and therefore enhances the features of the estimator. Moreover, not any Jacobian or Hessian matrices are required, which decreases the CPU necessities and submits computational improvements over the EKF. The UKF will provide higher precision than the EKF by way of higher order relations of the non-linear model functions are taken into consideration. The UKF involves three main steps which are demonstrated below:

Step 1: sigma points calculation

It is required to generate sets of 2n+1 sigma points, by means of the state vector x at time k -1 and the equivalent covariance matrix P_{k-1}

$$X_{k-1} = [x_{k-1} \cdots x_{k-1}] + \sqrt{c}[0 \quad \sqrt{P_{k-1}} \quad -\sqrt{P_{k-1}}] \tag{32}$$

Where $c = n + \lambda$ and $\lambda = \alpha^2(n+k) - n$, λ is a scaling parameter with the propagated constant $(10^{-4} \le \alpha \le 1)$ which proposed for Gaussian distribution, the scaling parameter k can be taken k=3 or 0.In fact, The purpose of used parameter k is to decrease the high order error of approximation. For the determination of the estimation initialization (i.e. when k = 0), the initial state vector also the initial covariance matrix must be clear beforehand according to a priori familiarity of the system.

Step 2: Kalman filter state prediction

Calculate, one by one, the series of sigma points determined in phase 1 through the state update function

$$\hat{X}_k^i = f(X_{k-1}^i, k-1) \qquad i = 0, \ldots\ldots, 2n \tag{33}$$

Where $X_{k-1}{}^i$ is the $(i + 1)$th column of matrix X_{k-1} and the resultant $\hat{X}_k$ is a $n \times (2n + 1)$ matrix holding the generated sigma points. Then, calculate the expected state mean vector x_k^- and the expected covariance matrix P_k^- as follows

$$x_k^- = \sum_{i=0}^{2n} W_i^m \hat{X}_k^i \tag{34}$$

$$P_k^- = \sum_{i=0}^{2n} W_i^c [(\hat{X}_k^i - X_k^-)(\hat{X}_k^i - X_k^-)^T] + Q_{k-1} \tag{35}$$

The weight factors in expression (34) and (35) is evaluated by the following Formulae

$$W_0^m = \frac{\lambda}{n+\lambda}$$
$$W_0^c = \frac{\lambda}{n+\lambda} + (1 - \alpha^2 + \beta) \tag{36}$$

$$W_i^m = W_i^c = \frac{1}{2(n+\lambda)} \tag{37}$$

Where the variable β can be taken a value of two (β=2, the optimum value of variable β for Gaussain distribution function is two).

Figure 9. In series and in parallel Multi-Area state estimation applied to MV distribution network

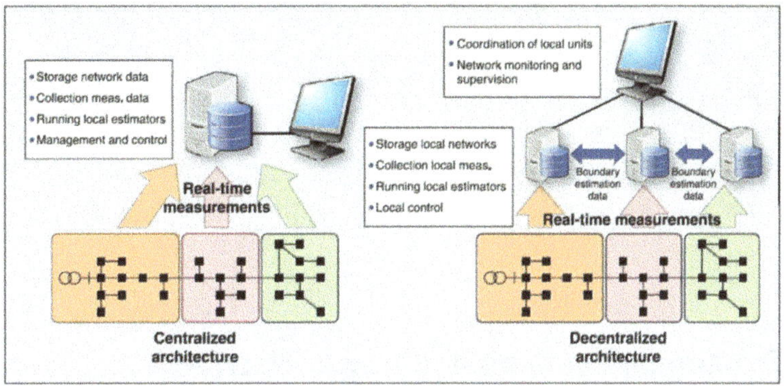

Step 3: Kalman filter state correction

Compute the sigma points matching to the expected state mean vector and the covariance matrix

$$X_k^- = [x_k^- \dots x_k^-] + \sqrt{c}[0 \quad \sqrt{P_k^-} \quad -\sqrt{P_k^-}] \tag{38}$$

Propagate the new sigma points obtained from equation (37) through the measurement update function

$$Y_k^- = h(X_k^-, k \tag{39}$$

The mean of the propagate new points is computed from the following expression

$$\mu_k = \sum_{i=0}^{2n} W_i^m Y_k^{-i} \tag{40}$$

Calculate the measurement covariance matrix. In addition, the cross-covariance of the state and measurement:

$$S_k = \sum_{i=0}^{2n} W_i^c [(Y_k^{-i} - \mu_k)(Y_k^{-i} - \mu_k)^T] + R_k \tag{41}$$

$$C_k = \sum_{i-0}^{2n} W_i^c \left[\left(X_k^{-i} - x_k^-\right)\left(Y_k^{-i} - \mu_k\right)^T \right] \tag{42}$$

Then, the filter's gain matrix is obtained as the following

$$K_K = C_k S_k^{-1} \tag{43}$$

The correction state vector mean and its corresponding covariance are evaluated by the following expression

$$x_k = x_k^- + K_K[y_k - \mu_k] \tag{44}$$

$$P_k = P_k^- - K_K S_k K_k^T \tag{45}$$

Observers for Dynamical Systems and Power Networks

In contrast with KF and its diverse derivatives, dynamic observers have also been thoroughly developed and investigated for different classes of systems. However, in comparison with KF-based techniques, observers have not been fully utilized in DSE for power systems. In other words, the status quo of DSE is based on KF and its variants. The basic principle governing most of the designed observers is the utilization of the plant's outputs and inputs to generate real-time estimates of the plant states and reconstruction of unknown inputs and sensor faults. The cornerstone of most observers is the innovation function sometimes a simple gain matrix designed to nullify the effect of unknown inputs and faults. Linear and nonlinear functional observers, sliding-mode observers, unknown input observers, and observers for fault detection and isolation are all examples of observers developed for different classes of systems, under different assumptions. With high sampling rates of PMUs — that can be considered as sensors — nonlinear dynamic observers inherently possess the computational, technical and theoretical capabilities to perform excellent estimation of the grid's physical status. In fact, recently designed observers have the potential to eclipse many KF-based methods in terms of (a) convergence guarantees, (b) estimation error bounds and (c) tolerance to unknown inputs, attack vectors and parametric uncertainty. Observers are typically used for deterministic systems, whereas KFs are used for stochastic dynamical systems (Taha et al.,2015).

Distributed DSSE Methods

Distribution grids consist of numerous regional substations, feeders, and nodes, so it requires a high computing time for the whole network SE (Primadianto& Lu, 2017), this led to conceive a scalable approach that leads to distributed state estimation (which known as Multi-Area State Estimation (MASE)) methods and to hierarchical bottom-up estimation methods (Della Giustina et al.,2014).

Distributed DSSE divides the distribution network into several sub-areas according to geographical, topological, and measurement points and solves the problem in local estimators (Primadianto& Lu, 2017).

Distributed DSSE can execute several estimators in series or parallel. Figure (10) shows two possible options for developing MASE in distribution systems. Peculiarities of the two approaches are (Della Giustina et al.,2014):

In Series: The distribution system is partitioned in several areas. SE is performed sequentially and each area exploits results of the neighboring upstream areas as input for its estimation process. In this way, dimensions of each problem are reduced, but overall execution times can be still high.

In Parallel: The distribution system is partitioned in several areas, and each one has its own estimator running simultaneously in parallel with the ones of the other areas. In this way execution time is reduced, making the SE compliant with real-time applications requiring high reporting rates.

practical implementation to use multi-area approaches is affected by:

- Partitioning criterion due to the low number of real measurements is a challenging problem. In general, partitioning can be performed mainly on the basis of topological or technical criteria.
- Communication infrastructure needed for data transmission and the coordination stage.

From this point of view, a key role is played by the architecture. Also in this case, two opposite options can be considered:

A Centralized Scheme: All of the areas communicate and send data to a central master. in the control center, where Computing centers are physically located, SE is run and eventually made to work in parallel. With this architecture, all of the following functions of are concentrated on the central unit:

- Collection of the measurements
- Storage of network data
- Integration of the sub-areas estimation results
- Management and control of the network

Such a solution may allow a more straightforward management of SE and control functions, but it may imply also high communication exchange costs and storage requirements.

A Decentralized Scheme: All of the areas have a dedicated computing center where all the processing tasks and control functions are carried out. Each area collects local measurements, keeps local network data and exchanges data with neighboring areas to integrate and improve estimation results. Final state estimates can then be communicated to the central master, which has only monitoring and supervisory tasks. the advantages of this scheme are reduction of communication and storage requirements but the disadvantage is the cost of having a dedicated center in each sub-area.

Figure 11 is a schematic overview of the architectural options.

State Estimation in Unbalanced Distribution Systems

SE algorithms for distribution networks should be designed taking into account the specific features of such networks. One of the important features is a high level of the phase unbalance due to unsymmetrical loads. Moreover, some distribution grids have two-phase and single-phase branches. Consequently, distribution networks cannot be treated in the context of a single-phase model of the state estimation problem. So three phase model is used and the applied state estimation will be three phase based on WLS in its methodologies a branch current and node voltages which were described above. Where branch current method are preferred on node voltages to represent the system state x. This will be useful to decouple the Jacobian matrix H on a per-phase basis, allowing conventional SE methods to be applied to distribution systems which are unbalanced, or have single phase or two-phase lateral feeders. A number of papers have built on this approach in order to develop robust and accurate three-phase DSSE techniques (Thukaram et al., 2000; Teng, 2002).

The objective function will be as described above (see eq.13).

The main difference between the three-phase and single-phase SE is the set of functions h_i. In the three-phase SE it is formulated using power flow equations, taking into account the relationship between the network phases. The real and reactive power injections for the three-phase SE can be expressed as follows (Chusovitin et al., 2016):

$$P_i^{ph} = V_i^{ph} \sum_{l=1}^{3} \sum_{j=1}^{n} V_j^l [G_{ij}^{ph,l} \cos(\delta_i^{ph} - \delta_j^l) + B_{ij}^{ph,l} \sin(\delta_i^{ph} - \delta_j^l)] \tag{46}$$

$$Q_i^{ph} = V_i^{ph} \sum_{l=1}^{3} \sum_{j=1}^{n} V_j^l [G_{ij}^{ph,l} \sin(\delta_i^{ph} - \delta_j^l) - B_{ij}^{ph,l} \cos(\delta_i^{ph} - \delta_j^l)] \tag{47}$$

Where

G+jB is the system admittance matrix,

n is number of buses

and l is the number of phases that can be 1, 2 or 3 phase.

The branch real power flow P_{ij}^{ph} and reactive power flow Q_{ij}^{ph} equations from bus i to bus j for phase ph can be:

$$\begin{aligned} P_{ij}^{ph} &= V_i^{ph} \sum_{l=1}^{3} V_i^l [G_{ij}^{ph,l} \cos(\delta_i^{ph} - \delta_i^l) + B_{ij}^{ph,l} \sin(\delta_i^{ph} - \delta_i^l)] \\ &- V_i^{ph} \sum_{l=1}^{3} V_j^l [G_{ij}^{ph,l} \cos(\delta_i^{ph} - \delta_j^l) + B_{ij}^{ph,l} \sin(\delta_i^{ph} - \delta_j^l)] \end{aligned} \tag{48}$$

$$\begin{aligned} Q_{ij}^{ph} &= -V_i^{ph} \sum_{l=1}^{3} V_i^l [G_{ij}^{ph,l} \sin(\delta_i^{ph} - \delta_i^l) - B_{ij}^{ph,l} \cos(\delta_i^{ph} - \delta_i^l)] \\ &- V_i^{ph} \sum_{l=1}^{3} V_j^l [G_{ij}^{ph,l} \sin(\delta_i^{ph} - \delta_j^l) - B_{ij}^{ph,l} \cos(\delta_i^{ph} - \delta_j^l)] \end{aligned} \tag{49}$$

In three phase system the state variable vector x is typically defined as the voltage angles and magnitudes at each node and each phase (Gao& Yu,2017).

$$x = [\delta_1^{ph} \ldots \delta_i^{ph} \ldots \delta_n^{ph} V_1^{ph} \ldots V_i^{ph} \ldots V_n^{ph}]^T \qquad (50)$$

Where the three-phase voltage magnitude and voltage angle at bus i are respectively:

$$V_i^{ph} = \begin{bmatrix} V_i^a \\ V_i^b \\ V_i^c \end{bmatrix}, \delta_i^{ph} = \begin{bmatrix} \delta_i^a \\ \delta_i^b \\ \delta_i^c \end{bmatrix}$$

The covariance matrix of the errors are given by:

$$R = Cov(e) = E(ee^T) = diag(\sigma_1^2, \ldots \sigma_i^2, \ldots \sigma_m^2) \text{ where } \sigma_i = [\sigma_i^a \sigma_i^b \sigma_i^c]$$

Where $\sigma_i^2$is the variance of the ith measurement error. the three-phase modelling of various components of the network such as line, transformers and switches. Are illustrated in (Majumdar & Pal, 2016). In addition to measurement functions associated with measurement devices, there are equality and inequality constraints to the objective function as following:

$$\begin{aligned} J &= [z - h(x)]^T W [z - h(x)] \\ \textit{subject to}&: \\ &c_{eq}(x) = 0 \\ &c_{ineq}(x) \le 0 \end{aligned} \qquad (51)$$

Where: $c_{eq}(x)$ vector of zero injection measurements.
$c_{ineq}(x)$ vector of inequality constraints.

A. A. Equality constraints c_{eq} (x): The equality constraints are the set of equations corresponding to virtual measurements.

$$\begin{aligned} 0 &= P_{Gi}^{ph} - P_{Di}^{ph} - P_i^{ph} \\ 0 &= Q_{Gi}^{ph} - Q_{Di}^{ph} - Q_i^{ph} \end{aligned} \qquad (52)$$

Where P_{Gi}^{ph} and Q_{Gi}^{ph} are the real and reactive power injected at bus i respectively,
P_{Di}^{ph} and Q_{Di}^{ph} are the load demand at the same bus.
n is number of buses
l is the number of phases which can be 1, 2 or 3 phase.

B. Inequality constraints c_{ineq} (x) These are the set of constraints of state variables that represent the system operational and security limits, such as setting upper and lower limits for control variables. The constraints are as follows:

- Bus voltage - Voltage magnitudes at each bus in the network:

$$V^{ph}_{\min,i} \leq V^{ph}_{i} \leq V^{ph}_{\max,i}$$

- Bus angle - The bus angle at each bus in the network:

$$-\delta^{ph}_{\min,i} \leq \delta^{ph}_{i} \leq \delta^{ph}_{\max,i}$$

The algorithm to solve the state estimation problem stated in (19) is presented. To solve this nonconvex optimization problem, the Lagrange multiplier theorem is invoked in geo or by WLS as in (Majumdar & Pal, 2016).

Role of PMU in Distribution System State Estimation

The introduction of PMUs has increased the accuracy of SE algorithms, by measuring not only magnitudes, but also phase angle differences between voltages phasors at different nodes. PMUs measure either nodal voltage phasors or branch current phasors.

Unfortunately, for economic and technical constraints, a PMU cannot be used in each node of the network for observability. Thus, the best solution is to combine traditional and PMU measurements. The inclusion of PMUs in state estimation creates two critical problems:

- How to combine traditional and PMU measurements; which can be handled by combined the PMU data with those resulting from traditional measurements in a single stage, Or by using a two-stage scheme to improve the state estimated from traditional measurement equipment, e.g. by applying a post-processing step based on PMU phasor data.
- How to place the PMUs in the network in order to maximize observability and performances: The optimal PMU Placement (OPP) problem is a constrained optimization problem that aims at minimizing the number of devices while ensuring observability and given uncertainty boundaries. Alternatively, we can maximize the estimation accuracy when a maximum number of PMUs is available. This problem can been tackled in several different ways, e.g. using genetic algorithms, particle swarm optimization (PSO), integer programming or binary search.

General measurement model described by equation (20):

$$Z = h(x) = \begin{bmatrix} p(x) \\ Q(x) \\ V(x) \\ \theta(x) \end{bmatrix} + e \qquad or\, Z = h(x) = \begin{bmatrix} p(x) \\ Q(x) \\ I(x) \\ \beta(x) \end{bmatrix} + e \tag{53}$$

Where I(x), β(x) are the magnitudes and the phases of the line currents, other elements are illustrated above.

State estimation based WLS generally be used. it is well known that WLS method suffers from possible of convergence problems due to numerical instabilities, which may occur when R^{-1} is ill-conditioned or when virtual measurements (e.g. zero-power injection) exist. In such cases, the standard WLS methods can be modified by using the QR Chorlesky decomposition or the Lagrange multiplier techniques. (Barchi,2015)

Intelligent Optimization Algorithms

Due to nonlinear characteristics of some equipments in active distribution systems, such as voltage-control equipments, static var compensators (SVCs), and DGs, the objective of the optimization problem (represented in equation 12) is no longer differentiable and continuous. Thus, the conventional Distribution SE methods, iterative convergence method, and sensitivity analysis could not effectively solve such kind of non-differentiable and discrete problems. However, a successful solution can be obtained by using various intelligent optimization algorithms such as:

- Particle Swarm optimization (PSO) (Naka et al., 2003; Niknam et al., 2009; Ramesh et al., 2012)
- ant colony algorithm (Niknam et al., 2005)
- honey-bee mating optimization (Niknam, 2008; Arefi et al., 2009)
- genetic algorithm (Souza et al., 2005; (Hossam-Eldin et al., 2009)
- artificial neural networks (Hippert et al., 2001; Manitsas et al., 2012)
- fuzzy logic (Shabani et al.,1996)

CONCLUSION

This chapter summarized the most issues related to state estimation in active distribution systems, and focused only the state estimator without other functions like bad data and observability. This chapter illustrated the differences between state estimation in transmission and distribution systems, the necessity of state estimation in active distribution systems, and the classification of state estimation algorithms in its forms static and dynamic. State estimation in unbalanced system and the role of PMU in DSSE are described briefly. At last due to nonlinear characteristics of some equipments in active distribution systems Intelligent Optimization Algorithms were used.

REFERENCES

Ahmad, F., Rasool, A., Ozsoy, E., Sekar, R., Sabanovic, A., & Elitaş, M. (2018). Distribution system state estimation-A step towards smart grid. *Renewable & Sustainable Energy Reviews*, *81*, 2659–2671. doi:10.1016/j.rser.2017.06.071

Alhelou, H., Hamedani-Golshan, M. E., Zamani, R., Heydarian-Forushani, E., & Siano, P. (2018). Challenges and opportunities of load frequency control in conventional, modern and future smart power systems: A comprehensive review. *Energies*, *11*(10), 2497. doi:10.3390/en11102497

Alhelou, H. H. (2018). *Fault detection and isolation in power systems using unknown input observer. In Advanced condition monitoring and fault diagnosis of electric machines* (p. 38). Hershey, PA: IGI Global.

Alhelou, H. H., Golshan, M., & Fini, M. (2018). Wind driven optimization algorithm application to load frequency control in interconnected power systems considering GRC and GDB nonlinearities. *Electric Power Components and Systems, 46*(11-12), 1223-1238.

Alhelou, H. H., & Golshan, M. E. H. (2016, May). Hierarchical plug-in EV control based on primary frequency response in interconnected smart grid. In *2016 24th Iranian Conference on Electrical Engineering (ICEE),* (pp. 561-566). IEEE. 10.1109/IranianCEE.2016.7585585

Alhelou, H. H., & Golshan, M. E. H. (2020). Decision-making-based optimal generation-side secondary-reserve scheduling and optimal LFC in deregulated interconnected power system. In *Decision making applications in modern power systems* (pp. 269–299). Academic Press. doi:10.1016/B978-0-12-816445-7.00011-6

Alhelou, H. H., Golshan, M. E. H., & Hatziargyriou, N. D. (2019). Deterministic dynamic state estimation-based optimal LFC for interconnected power systems using unknown input observer. *IEEE Transactions on Smart Grid*, 1. doi:10.1109/TSG.2019.2940199

Alhelou, H. H., Golshan, M. E. H., & Hatziargyriou, N. D. (2019). A decentralized functional observer based optimal LFC considering unknown inputs, uncertainties and cyber-attacks. *IEEE Transactions on Power Systems*.

Alhelou, H. H., Golshan, M. E. H., Zamani, R., Moghaddam, M. P., Njenda, T. C., Siano, P., & Marzband, M. (2019, June). An improved UFLS scheme based on estimated minimum frequency and power deficit. In 2019 IEEE Milan PowerTech (pp. 1-6). IEEE. doi:10.1109/PTC.2019.8810497

Alhelou, H. H., Golshan, M. H., & Askari-Marnani, J. (2018). Robust sensor fault detection and isolation scheme for interconnected smart power systems in presence of RER and EVs using unknown input observer. *International Journal of Electrical Power & Energy Systems*, *99*, 682–694. doi:10.1016/j.ijepes.2018.02.013

Alhelou, H. H., Hamedani-Golshan, M. E., Heydarian-Forushani, E., Al-Sumaiti, A. S., & Siano, P. (2018, September). Decentralized fractional order control scheme for LFC of deregulated nonlinear power systems in presence of EVs and RER. In *Proceedings 2018 International Conference on Smart Energy Systems and Technologies (SEST)* (pp. 1-6). IEEE. 10.1109/SEST.2018.8495858

Alhelou, H. S. H., Golshan, M. E. H., & Fini, M. H. (2015, December). Multi agent electric vehicle control based primary frequency support for future smart micro-grid. In *2015 Smart Grid Conference (SGC)* (pp. 22-27).

Anteneh, D., & Khan, B. (2019). *Reliability enhancement of distribution substation by using network reconfiguration a case study at Debre Berhan Distribution Substation. International Journal of Economy, Energy, & Environment*, *4*(2), 33.

Arefi, A., Haghifam, M. R., Fathi, S. H., Niknam, T., & Olamaei, J. (2009, June). A novel algorithm based on honey bee mating optimization for distribution harmonic state estimation including distributed generators. In 2009 IEEE Bucharest PowerTech (pp. 1-7). IEEE. doi:10.1109/PTC.2009.5282101

Baran, M. E., Jung, J., & McDermott, T. E. (2009, July). Including voltage measurements in branch current state estimation for distribution systems. In 2009 IEEE Power & Energy Society General Meeting (pp. 1-5). IEEE. doi:10.1109/PES.2009.5275479

Baran, M. E., & Kelley, A. W. (1994). State estimation for real-time monitoring of distribution systems. *IEEE Transactions on Power Systems*, *9*(3), 1601–1609. doi:10.1109/59.336098

Baran, M. E., & Kelley, A. W. (1995). A branch-current-based state estimation method for distribution systems. *IEEE Transactions on Power Systems*, *10*(1), 483–491. doi:10.1109/59.373974

Barchi, G. (2015). *Algorithms and performance analysis for synchrophasor and grid state estimation* (Doctoral dissertation, University of Trento).

Blaauwbroek, N. (2018). Flexibility in active distribution network management: A new horizon in monitoring and control for grid supportive demand side management.

Celik, M. K., & Liu, W. H. (1999, January). A practical distribution state calculation algorithm. In *IEEE Power Engineering Society. 1999 Winter Meeting (Cat. No. 99CH36233)* (Vol. 1, pp. 442-447). IEEE. 10.1109/PESW.1999.747496

Chusovitin, P., Polyakov, I., & Pazderin, A. (2016, November). Three-phase state estimation model for distribution grids. In *2016 IEEE International Conference on the Science of Electrical Engineering (ICSEE)* (pp. 1-5). IEEE. 10.1109/ICSEE.2016.7806063

de la O Serna, J. A., & Rodriguez-Maldonado, J. (2011). Instantaneous oscillating phasor estimates with Taylor \$^ K \$-Kalman filters. *IEEE Transactions on Power Systems*, *26*(4), 2336–2344. doi:10.1109/TPWRS.2011.2157539

Della Giustina, D., Pau, M., Pegoraro, P. A., Ponci, F., & Sulis, S. (2014). Electrical distribution system state estimation: Measurement issues and challenges. *IEEE Instrumentation & Measurement Magazine*, *17*(6), 36–42. doi:10.1109/MIM.2014.6968929

Deng, Y., He, Y., & Zhang, B. (2002). A branch-estimation-based state estimation method for radial distribution systems. *IEEE Transactions on Power Delivery*, *17*(4), 1057–1062. doi:10.1109/TPWRD.2002.803800

Durgaprasad, G., & Thakur, S. S. (1998). Robust dynamic state estimation of power systems based on M-estimation and realistic modeling of system dynamics. *IEEE Transactions on Power Systems*, *13*(4), 1331–1336. doi:10.1109/59.736273

Fini, M. H., Yousefi, G. R., & Alhelou, H. H. (2016). Comparative study on the performance of many-objective and single-objective optimisation algorithms in tuning load frequency controllers of multi-area power systems. *IET Generation, Transmission, & Distribution*, *10*(12), 2915–2923. doi:10.1049/iet-gtd.2015.1334

Francy, R. (2013). *Enhanced power system state estimation techniques for the incorporation of variable energy* (Doctoral dissertation, Masdar Institute of Science and Technology).

Gao, Y., & Yu, N. (2017, April). State estimation for unbalanced electric power distribution systems using AMI data. In 2017 IEEE Power & Energy Society Innovative Smart Grid Technologies Conference (ISGT) (pp. 1-5). IEEE. doi:10.1109/ISGT.2017.8085999

Gelagaev, R., Vermeyen, P., & Driesen, J. (2008, September). State estimation in distribution grids. In *2008 13th International Conference on Harmonics and Quality of Power* (pp. 1-6). IEEE. 10.1109/ICHQP.2008.4668803

Ghahremani, E., & Kamwa, I. (2011). Dynamic state estimation in power system by applying the extended Kalman filter with unknown inputs to phasor measurements. *IEEE Transactions on Power Systems*, *26*(4), 2556–2566. doi:10.1109/TPWRS.2011.2145396

Ghosh, A. K., Lubkeman, D. L., Downey, M. J., & Jones, R. H. (1997). b). Distribution circuit state estimation using a probabilistic approach. *IEEE Transactions on Power Systems*, *12*(1), 45–51. doi:10.1109/59.574922

Ghosh, A. K., Lubkeman, D. L., & Jones, R. H. (1997). Load modeling for distribution circuit state estimation. *IEEE Transactions on Power Delivery*, *12*(2), 999–1005. doi:10.1109/61.584427

Gomez-Exposito, A., & Abur, A. (2004). Power system state estimation: theory and implementation. Boca Raton, FL: CRC Press.

Haes Alhelou, H., Hamedani-Golshan, M. E., Njenda, T. C., & Siano, P. (2019). A survey on power system blackout and cascading events: Research motivations and challenges. *Energies*, *12*(4), 682. doi:10.3390/en12040682

Haes Alhelou, H., Hamedani-Golshan, M. E., Njenda, T. C., & Siano, P. (2019). Wide-area measurement system-based optimal multi-stage under-frequency load-shedding in interconnected smart power systems using evolutionary computing techniques. *Applied Sciences*, *9*(3), 508. doi:10.3390/app9030508

Haes Alhelou, H., Hamedani Golshan, M. E., Njenda, T. C., & Siano, P. (2019). WAMS-based online disturbance estimation in interconnected power systems using disturbance observer. *Applied Sciences*, *9*(5), 990. doi:10.3390/app9050990

Hatziargyriou, N. (Ed.). (2014). *Microgrids: Architectures and control*. John Wiley & Sons.

Haughton, D. A., & Heydt, G. T. (2012). A linear state estimation formulation for smart distribution systems. *IEEE Transactions on Power Systems*, *28*(2), 1187–1195. doi:10.1109/TPWRS.2012.2212921

Haughton, D. A., & Heydt, G. T. (2013). A linear state estimation formulation for smart distribution systems. *IEEE Transactions on Power Systems*, *28*(2), 1187–1195. doi:10.1109/TPWRS.2012.2212921

Hippert, H. S., Pedreira, C. E., & Souza, R. C. (2001). Neural networks for short-term load forecasting: A review and evaluation. *IEEE Transactions on Power Systems*, *16*(1), 44–55. doi:10.1109/59.910780

Hossam-Eldin, A. A., Abdallah, E. N., & El-Nozahy, M. S. (2009, July). A modified genetic based technique for solving the power system state estimation problem. In *International Conference on World Academy of Science, Engineering and Technology* (pp. 311-320).

Huang, J., Gupta, V., & Huang, Y. F. (2012, March). Electric grid state estimators for distribution systems with microgrids. In *2012 46th Annual Conference on Information Sciences and Systems (CISS)* (pp. 1-6). IEEE. 10.1109/CISS.2012.6310739

Huang, Y. F., Werner, S., Huang, J., Kashyap, N., & Gupta, V. (2012b). State estimation in electric power grids: Meeting new challenges presented by the requirements of the future grid. *IEEE Signal Processing Magazine*, *29*(5), 33–43. doi:10.1109/MSP.2012.2187037

Huang, Z., Schneider, K., & Nieplocha, J. (2007, December). Feasibility studies of applying Kalman filter techniques to power system dynamic state estimation. In *2007 International Power Engineering Conference (IPEC 2007)* (pp. 376-382). IEEE.

Huber, P. J. (1992). Robust estimation of a location parameter. In *Breakthroughs in statistics* (pp. 492–518). New York, NY: Springer. doi:10.1007/978-1-4612-4380-9_35

Jariso, M., Khan, B., Tanwar, S., Tyagi, S., & Rishiwal, V. (2018). Hybrid energy system for upgrading the rural environment, IEEE GlobeCom 2018.

Jariso, M., Khan, B., Tesfaye, D., & Singh, J. (2017). Modelling and designing of stand-alone photovoltaic system (Case study: Addis Boder Health Center southwest Ethiopia). In *Proceedings IEEE International Conference on Electronics, Communication, and Aerospace Technology, ICECA 2017,* pp. 168-173, Coimbatore, India.

Kalman, R. E. (1960). A new approach to linear filtering and prediction problems. *Journal of Basic Engineering*, *82*(1), 35–45. doi:10.1115/1.3662552

Khan, B., & Agnihotri, G. (2012). A novel transmission loss allocation method based on transmission usage. In *Proceedings IEEE Fifth Power India Conference*, Haryana, India. 10.1109/PowerI.2012.6479479

Khan, B., Agnihotri, G., & Gupta, G. (2013). A multipurpose matrices methodology for transmission usage, loss, and reliability margin allocation in restructured environment, *Electrical & Computer Engineering: An International Journal, 2*(3).

Khan, B., Agnihotri, G., Rathore, P., Mishra, A., & Naidu, G. (2014). A cooperative game theory approach for usage and reliability margin cost allocation under contingent restructured market. *International Review of Electrical Engineering*, *9*(4), 854–862.

Khan, B., Alhelou, H. H., & Mebrahtu, F. (2019). A holistic analysis of distribution system reliability assessment methods with conventional and renewable energy sources. *AIMS Energy*, *7*(4), 413–429. doi:10.3934/energy.2019.4.413

Khan, B., & Singh, P. (2017). The current and future states of Ethiopia's energy sector and potential for green energy: A comprehensive study. *International Journal of Engineering Research in Africa, 33*, 115–139. doi:10.4028/www.scientific.net/JERA.33.115

Khan, B., & Singh, P. (2017). Optimal power flow techniques under characterization of conventional and renewable energy sources: A comprehensive analysis, *Journal of Engineering, vol. 2017,* Article ID 9539506.

Khan, B., & Singh, P. (2017). Selecting a meta-heuristic technique for smart micro-grid optimization problem: a comprehensive analysis, in IEEE Access, 5, pp. 13951-13977.

Khan, B., & Singh, P. (2019). Economic operation of smart micro-grid: a meta-heuristic approach. In Handbook of research on smart power system operation and control (pp. 330-346). Hershey, PA: IGI Global. doi:10.4018/978-1-5225-8030-0.ch014

Khan, B., & Singh, P. (2019). Economic operation of smart micro-grid: a meta-heuristic approach. In Handbook of research on smart power system operation and control (pp. 330-346). Hershey, PA: IGI Global. doi:10.4018/978-1-5225-8030-0.ch014

Khan, B., & Tanwar, S. (2019). Issues associated with microgrid integration. In Handbook of research on smart power system operation and control (pp. 252-264). Hershey, PA: IGI Global. doi:10.4018/978-1-5225-8030-0.ch010

Kifle, Y., Khan, B., & Singh, J. (2018). Designing and modeling grid connected photovoltaic system: (Case study: EEU building at Hawassa city). *International Journal of Convergence Computing*, *3*(1), 20–34. doi:10.1504/IJCONVC.2018.091113

Kifle, Y., Khan, B., & Singh, P. (2018). Assessment and enhancement of distribution system reliability by renewable energy sources and energy storage, *Journal of Green Engineering, 8*(3), 2, pp. 219-262.

Kumar, M. L., Prasanna, H. M., & Ananthapadmanabha, T. (2015). Retracted: A literature review on distribution system state estimation.

Li, K. (1996). State estimation for power distribution system and measurement impacts. *IEEE Transactions on Power Systems*, *11*(2), 911–916. doi:10.1109/TPWRS.2009.2016599

Lin, W. M., & Teng, J. H. (1996). Distribution fast decoupled state estimation by measurement pairing. *IEE Proceedings. Generation, Transmission, and Distribution*, *143*(1), 43–48. doi:10.1049/ip-gtd:19960049

Lin, W. M., Teng, J. H., & Chen, S. J. (2001). A highly efficient algorithm in treating current measurements for the branch-current-based distribution state estimation. *IEEE Transactions on Power Delivery*, *16*(3), 433–439. doi:10.1109/61.924823

Lu, C. N., Teng, J. H., & Liu, W. H. (1995). Distribution system state estimation. *IEEE Transactions on Power Systems*, *10*(1), 229–240. doi:10.1109/59.373946

Majumdar, A., & Pal, B. C. (2016, January). A three-phase state estimation in unbalanced distribution networks with switch modelling. In *2016 IEEE First International Conference on Control, Measurement and Instrumentation (CMI)* (pp. 474-478). IEEE. 10.1109/CMI.2016.7413793

Makdisie, C., Haidar, B., & Alhelou, H. H. (2018). An optimal photovoltaic conversion system for future smart grids. In Handbook of research on power and energy system optimization (pp. 601–657). Hershey, PA: IGI Global. doi:10.4018/978-1-5225-3935-3.ch018

Manitsas, E., Singh, R., Pal, B. C., & Strbac, G. (2012). Distribution system state estimation using an artificial neural network approach for pseudo measurement modeling. *IEEE Transactions on Power Systems*, *27*(4), 1888–1896. doi:10.1109/TPWRS.2012.2187804

Meliopoulos, A. S., & Zhang, F. (1996). Multiphase power flow and state estimation for power distribution systems. *IEEE Transactions on Power Systems*, *11*(2), 939–946. doi:10.1109/59.496178

Mili, L., Cheniae, M. G., & Rousseeuw, P. J. (1994). Robust state estimation of electric power systems. *IEEE Transactions on Circuits and Systems. I, Fundamental Theory and Applications*, *41*(5), 349–358. doi:10.1109/81.296336

Molla, T., Khan, B., Moges, B., Alhelou, H. H., Zamani, R., & Siano, P. (2019). Integrated optimization of smart home appliances with cost-effective energy management system. *CSEE Journal of Power and Energy Systems*, *5*(2), 249–258.

Molla, T., Khan, B., & Singh, P. (2018). A comprehensive analysis of smart home energy management system optimization techniques, *Journal of Autonomous Intelligence, 1*(1), 15–21.

Naka, S., Genji, T., Yura, T., & Fukuyama, Y. (2003). A hybrid particle swarm optimization for distribution state estimation. *IEEE Transactions on Power Systems*, *18*(1), 60–68. doi:10.1109/TPWRS.2002.807051

Nanchian, S. (2015). State estimation for active distribution network.

Negash, K., Khan, B., Tesfaye, D., & Gayathri, M. (2017). Optimal placement of phasor measurement unit for system observability. In *Proceedings of IEEE International Conference on Computing Methodologies and Communication, (ICCMC 2017),* pp. 73-77, Erode, India.

Negash, K., Khan, B., & Yohannes, E. (2016). Artificial intelligence versus conventional mathematical techniques: A review for optimal placement of phasor measurement units [Springer]. *Journal of Technology and Economics of Smart Grids and Sustainable Energy*, *1*(1), 10. doi:10.100740866-016-0009-y

Nguyen, P. H., Venayagamoorthy, G. K., Kling, W. L., & Ribeiro, P. F. (2013). Dynamic state estimation for distribution networks with renewable energy integration. *Int. J. Smart Grid and Clean Energy*, *2*(3), 307–315. doi:10.12720gce.2.3.307-315

Niknam, T. (2008). Application of honey-bee mating optimization on state estimation of a power distribution system including distributed generators. *Journal of Zhejiang University. Science A*, *9*(12), 1753–1764. doi:10.1631/jzus.A0820047

Niknam, T., & Firouzi, B. B. (2009). A practical algorithm for distribution state estimation including renewable energy sources. *Renewable Energy*, *34*(11), 2309–2316. doi:10.1016/j.renene.2009.03.005

Niknam, T., Ranjbar, A. M., & Shirani, A. R. (2005). A new approach for distribution state estimation based on ant colony algorithm with regard to distributed generation. *Journal of Intelligent & Fuzzy Systems*, *16*(2), 119–131.

Njenda, T. C., Golshan, M. E. H., & Alhelou, H. H. (2018, November). WAMS based intelligent under frequency load shedding considering online disturbance estimation. In *2018 Smart Grid Conference (SGC)* (pp. 1-5). 10.1109/SGC.2018.8777779

Njenda, T. C., Golshan, M. E. H., & Alhelou, H. H. (2018, November). WAMS based under frequency load shedding considering minimum frequency prediction and extrapolated disturbance magnitude. In *2018 Smart Grid Conference (SGC)* (pp. 1-5).

Pau, M., Pegoraro, P. A., & Sulis, S. (2013). Efficient branch-current-based distribution system state estimation including synchronized measurements. *IEEE Transactions on Instrumentation and Measurement*, *62*(9), 2419–2429. doi:10.1109/TIM.2013.2272397

Picallo, M., Anta, A., De Schutter, B., & Panosyan, A. (2018, June). A two-step distribution system state estimator with grid constraints and mixed measurements. In 2018 Power Systems Computation Conference (PSCC) (pp. 1-7). IEEE. doi:10.23919/PSCC.2018.8442586

Primadianto, A., Lin, W. T., & Lu, C. N. (2016, November). Performance comparison of distribution system state estimation methods. In 2016 IEEE Innovative Smart Grid Technologies-Asia (ISGT-Asia) (pp. 1121-1126). IEEE doi:10.1109/ISGT-Asia.2016.7796543

Primadianto, A., & Lu, C. N. (2016). b). A review on distribution system state estimation. *IEEE Transactions on Power Systems*, *32*(5), 3875–3883. doi:10.1109/TPWRS.2016.2632156

Qi, J., Sun, K., Wang, J., & Liu, H. (2016). Dynamic state estimation for multi-machine power system by unscented Kalman filter with enhanced numerical stability. *IEEE Transactions on Smart Grid*, *9*(2), 1184–1196. doi:10.1109/TSG.2016.2580584

Ramesh, L., Chakraborthy, N., Chowdhury, S. P., & Chowdhury, S. (2012). Intelligent DE algorithm for measurement location and PSO for bus voltage estimation in power distribution system. *International Journal of Electrical Power & Energy Systems*, *39*(1), 1–8. doi:10.1016/j.ijepes.2011.10.009

Ray, P. K., & Subudhi, B. (2012). Ensemble-Kalman-filter-based power system harmonic estimation. *IEEE Transactions on Instrumentation and Measurement*, *61*(12), 3216–3224. doi:10.1109/TIM.2012.2205515

Roytelman, I., & Shahidehpour, S. M. (1993). State estimation for electric power distribution systems in quasi real-time conditions. *IEEE Transactions on Power Delivery*, *8*(4), 2009–2015. doi:10.1109/61.248315

Shabani, F., Prasad, N. R., & Smolleck, H. A. (1996, September). State estimation with aid of fuzzy logic. In *Proceedings of IEEE 5th International Fuzzy Systems* (Vol. 2, pp. 947–953). IEEE. doi:10.1109/FUZZY.1996.552306

Shih, K. R., & Huang, S. J. (2002). Application of a robust algorithm for dynamic state estimation of a power system. *IEEE Transactions on Power Systems*, *17*(1), 141–147. doi:10.1109/59.982205

Simon, D. (2006). *Optimal state estimation: Kalman, H infinity, and nonlinear approaches*. John Wiley & Sons. doi:10.1002/0470045345

Singh, P., & Khan, B. (2017). Smart microgrid energy management using a novel artificial shark optimization, Complexity, vol. 2017, Article ID 2158926. doi:10.1155/2017/2158926

Singh, P., Khan, B., Vidyarthi, A., Haes Alhelou, H., & Siano, P. (2019). Energy-aware online non-clairvoyant scheduling using speed scaling with arbitrary power function. *Applied Sciences*, *9*(7), 1467. doi:10.3390/app9071467

Singh, R., Pal, B. C., & Jabr, R. A. (2010). Distribution system state estimation through Gaussian mixture model of the load as pseudo-measurement. *IET Generation, Transmission, & Distribution*, *4*(1), 50–59. doi:10.1049/iet-gtd.2009.0167

Sinha, A. K., & Mondal, J. K. (1999). Dynamic state estimator using ANN based bus load prediction. *IEEE Transactions on Power Systems*, *14*(4), 1219–1225. doi:10.1109/59.801876

Taha, A. F., Qi, J., Wang, J., & Panchal, J. H. (2015). Dynamic state estimation under cyber attacks: A comparative study of Kalman filters and observers. *arXiv preprint arXiv:1508.07252*.

Thukaram, D., Jerome, J., & Surapong, C. (2000). A robust three-phase state estimation algorithm for distribution networks. *Electric Power Systems Research*, *55*(3), 191–200. doi:10.1016/S0378-7796(00)00088-2

Valverde, G., & Terzija, V. (2011). Unscented Kalman filter for power system dynamic state estimation. *IET Generation, Transmission, & Distribution*, *5*(1), 29–37. doi:10.1049/iet-gtd.2010.0210

Wang, H., & Schulz, N. N. (2004). A revised branch current-based distribution system state estimation algorithm and meter placement impact. *IEEE Transactions on Power Systems*, *19*(1), 207–213. doi:10.1109/TPWRS.2003.821426

Yeshalem, M. T., & Khan, B. (2017). Design of an off-grid hybrid PV/wind power system for remote mobile base station: A case study. *AIMS Energy*, *5*(1), 96–112. doi:10.3934/energy.2017.1.96

Zamani, R., Golshan, M. E. H., Alhelou, H. H., & Hatziargyriou, N. (2019). A novel hybrid islanding detection method using dynamic characteristics of synchronous generator and signal processing technique. *Electric Power Systems Research*, *175*, 105911. doi:10.1016/j.epsr.2019.105911

Zamani, R., Hamedani-Golshan, M. E., Haes Alhelou, H., Siano, P., & Pota, H. R. (2018). Islanding detection of synchronous distributed generator based on the active and reactive power control loops. *Energies*, *11*(10), 2819. doi:10.3390/en11102819

Zhang, H. (2018). Distribution grid state estimation based on smart meter data.

Zhang, J., Welch, G., Bishop, G., & Huang, Z. (2011, June). Reduced measurement-space dynamic state estimation (remedyse) for power systems. In 2011 IEEE Trondheim PowerTech (pp. 1-7). IEEE.

Zhou, N., Meng, D., Huang, Z., & Welch, G. (2014). Dynamic state estimation of a synchronous machine using PMU data: A comparative study. *IEEE Transactions on Smart Grid*, *6*(1), 450–460. doi:10.1109/TSG.2014.2345698

APPENDIX

Abbreviations

Table 2.

ADS	Active Distribution Systems	**FASE**	Forecasting-Aided Se
AMI	Advanced Metering Infrastructure	**KF**	Kalman Filter
DER	Distributed Energy Resources	**PMU**	Phasor Measurement Units
DSE	Dynamic State Estimator	**RTUs**	Remote Terminal Units
DSSE	Distribution System State Estimation	**RESs**	Renewable Energy Sources
DG	Distributed Generation	**SE**	State Estimation
DA	Distribution Automation	**SSE**	Static State Estimator
EMS	Energy Management System	**WLS**	Weighted Least Square

Chapter 16
Energy Storage System and Its Power Electronic Interface

Baseem Khan
https://orcid.org/0000-0002-0562-0933
Hawassa University, Hawassa, Ethiopia

Samuel Degarege
Hawassa University, Hawassa, Ethiopia

Fsaha Mebrahtu
Hawassa University, Hawassa, Ethiopia

Hassan Alhelou
Tisheen University, Syria

ABSTRACT

This chapter examines the modeling and simulation of energy storage (battery, flywheel, etc.) systems interfaced to the power grid by using power electronic device, like chopper module, Rectifier module, and filter circuits, which are essential to the load balance between supply and demand, and to eliminate harmonics and to ensure efficient, cost effective, and reliable operations. Energy storage system in power grid is the same as memory in computer system. Energy efficiency is a key performance indicator for energy storage system. The energy storage system is the most promising component to enhance the system reliability and flexibility.

INTRODUCTION

The grid scale energy storage system consists of three main components, such as energy storage medium itself (Electrical energy storage system (EESS)), power converters interface that links the storage medium to the grid (Rectifier module, chopper and filter) and controller that are planned to get a grid control and its objective is do not elaborated operation of the power converter grid interface or the performance of electrical energy storage system (Fares et al. 2015, Schewe, 2005). The major challenge of power convert-

DOI: 10.4018/978-1-7998-1230-2.ch016

ers during the designing are reliability, cost, efficiency and power density (AC to AC converter, DC to DC converter and filter) system and interacting with energy storage system. There are various practical applications of energy storage, but the requirements vary from one event to another event (Alhelou et al., 2019; Makdisie et al., 2018; Alhelou et al., 2018; Alhelou et al., 2016; Haes Alhelou et al., 2019; Njenda et al., 2018). Nevertheless, it is not each storage technology operates equally to be useful in any condition because of environmental factors, shortage of input source and also some other draw back to the load. In fact, the practicalities of energy storages rely on their technical characteristics, i.e. response times, efficiency, power rating and capacity for a certain application (Dötsch, 2007; Khan et al., 2019; Khan et al., 2018; Khan et al., 2017; Banteywalu et al., 2019; Anteneh et al., 2019; Molla et al., 2019, Molla et al., 2018, Jariso et al. 2018). A power electronic circuit manages the flow of electrical energy between the hydro-power grid and wind power grid (Hoffmann, 2010). There are two types of power electronic system discussed in this chapter; these are power module (converter) and filter. Therefore this chapter deals with the modeling of energy storage system with its power electronic interface with grid.

ENERGY STORAGE SYSTEM INTERFACING TO GRID

The converter of power electronics or module is an assembly including many power elements, mostly semiconductor device, properly connected to do the power conversion function. The functions of power converter positioned at the middle is to control the energy flow between a given power grid to the electrical energy storage system (De et al. 2013; Alhelou et al., 2019; Makdisie et al., 2018; Alhelou et al., 2018; Alhelou et al., 2016; Haes Alhelou et al., 2019; Njenda et al., 2018). A realistic converter design must approach 100% efficient. The energy storage system (ESS) is planned to balance with power supply as well as to store energy at the time of high peak power generation, which can be consequently exploited to supply electricity at the time of high demand (Bouzid et al. 2015; Khan et al. 2014; Khan et al. 2013; Khan et al. 2012; Negash et al., 2017; Negash et al., 2016; Jariso et al. 2017; Kifle et al. 2018; Yeshalem et al. 2017; Singh et al. 2017; Gupta et al., 2015). So, the main target of the power bidirectional inverter-rectifier to concurrently reserve the battery SOC within safe boundaries and to smooth the power profile exchanged with the grid, i.e., minimizing the grid power fluctuations and grid power peaks (Fini et al., 2016; Alhelou et al., 2018; Zamani et al., 2018; Alhelou et al., 2015; Njenda et al., 2018; Haes Alhelou et al., 2018; Haes Alhelou et al., 2019). Figure 1 presents the block diagram of energy storage integrated to the grid.

COMPLETE DESIGN FOR THE INTEGRATION OF GRID

The procedure of the interconnection is performed by using the flow chart given in figure 2.

Power Grid

The electrical grid becomes more flexible, collaborative and can deliver real-time feedback. Grid has the electricity production and delivery system along with consumption integrated with communication and information technology. The definition of electric power can be explained as the entire apparatus of wire and the machine that linked to the supply of electricity that's power plants with the end users

Figure 1. Block diagram of energy storage integrated to the grid

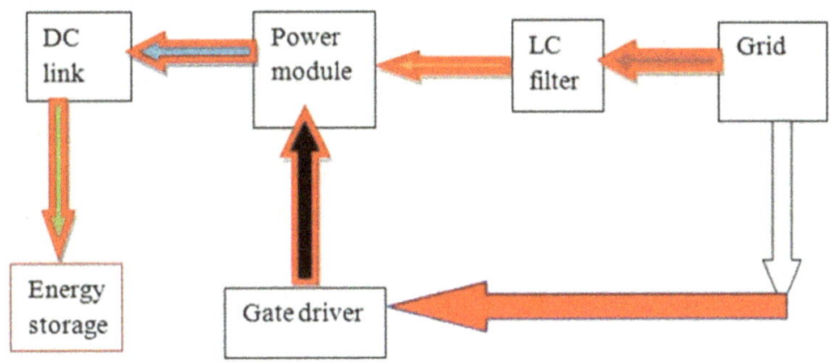

Figure 2. Interfacing of ESS to the grid

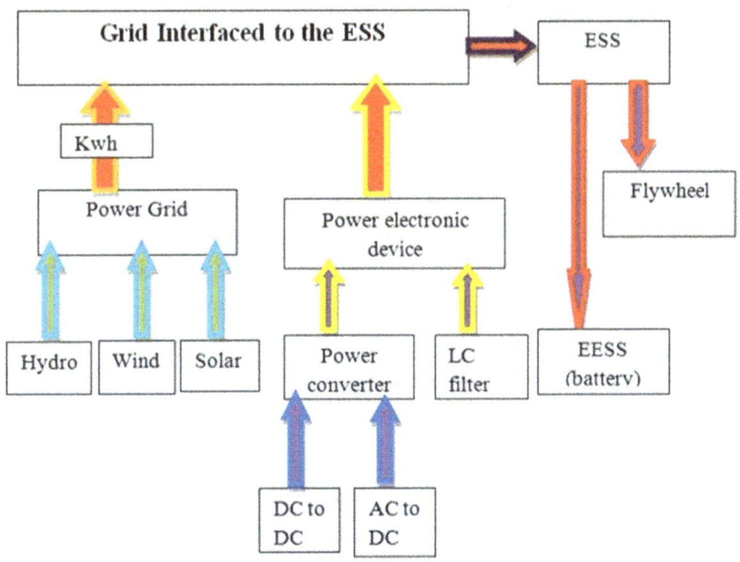

and their needs (Amin et al. 2003). The primary source of power plant converter is the chemical energy stored in cell, the radial energy in the sun light and wind pressure (Fini et al., 2016; Alhelou et al., 2018; Zamani et al., 2018; Alhelou et al., 2015; Njenda et al., 2018; Haes Alhelou et al., 2018; Haes Alhelou et al., 2019). When one talk about the grid, it denotes to the electrical transmission system, which transmits the power from the power plants to the substation located near the power plants (Amin et al. 2003). Nevertheless, the grid also involves the distribution facilities that distribute the electricity from the substations to the individual users.

Power Electronic Device

Power electronics is the application of solid-state electronics to the control and conversion of electric power. An AC/DC converter (rectifier) is the most typical power electronics device found in many consumer electronic devices, e.g. television sets, personal computers, battery chargers. The power converters are also used in energy storage system, which is used to solve the scarcity of electrical power in long distance transmission, in large load area and distributed micro grid of the new technology. Power converter also plays a significant role in the energy storage system. The new technology and power converters technology are used to speed up the development of large-scale energy storage system. The assumption of these new generation power electronic devices can greatly decrease the switching loss of the converter, enhance efficiency, decrease the volume and weight of the converter, and become adaptable to work at higher temperatures.

Power Converter

Power converter is an electrical or electro-mechanical device for converting electrical energy. It may be converting AC to DC or from DC to AC, or the voltage or frequency, or some combination of these, as per the requirement. It is the main and core technology of energy storage system and all the electrical materials generation plant, transmission and distribution are utilizing power converters. The operating mode of the power electronic converter is called the switching operation (Siewniak et al., 2014). Zhang (2015) proposed the power conversion system defines the operational condition of the entire energy storage system. The energy stored in the battery will pass through the power grid by using a single stage AC/DC converter. At the power generation system, the connection of grid is the only power source usually adopts this topology, which has the benefits of a simple structure and high efficiency but lacks flexibility in its capacity selection. Another structure is the AC-DC and DC-DC two-stage converter. Under this topology, the battery bank configuration of the energy storage system is more flexible, where the charging and discharging management is more perfect and reliable. Thus, it is suitable for coordinating with the new energies in a large-scale connection. The efficient power conversion of the DC-DC converter can remarkably reduce energy consumption. The application of the DC to DC converter in the BES has practical uses on achieving an efficient current limitation (Yarlagadda et al. 2013). Si based isolated bidirectional DC-DC converter much efficient than Si based converter (Zhao et al., 2013), which are lower power dissipation, high operating temperature and faster switching.

Energy Storage System

Energy storage system is the device that stores energy and generally called an accumulator or battery. Energy storage involves converting energy from different types that are difficult to store to more conveniently or economically storable forms. Some technologies provide short-term energy storage, while others can endure for much longer (Kawashima, 2011, BCG, 2011). Also, it plays a vital role in today's world, for instance next-generation of grid without energy storage is the same as a computer without a hard drive (Rehman et al. 2015). There are various types of energy storage mechanisms in electrical system such as pumped hydro storage (PHS) (Rehman et al. 2015). Compressed-Air Energy Storage (CAES) (Satkin et al., 2014), Battery Energy Storage (BES) (Mousavi, et al. 2014), Capacitor Storage (CS) (D. Lifshitz and G. Weiss, 2014), Super capacitors Energy Storage (SCES) (Xiao, 2015), Thermal

Table 1. Energy storage system

No.	Energy storage system	Example
1	Electrical energy storage system	BESS (lead acid, Nicd, NiMH, Li NaS etc.), capacitor storage, SCES.
2	Mechanical energy storage system	PHESS,FESS, etc.
3	Thermal energy storage system	CAESS

Energy Storage (TES) (Geth et al., 2015), and Flywheel Energy Storage System (FESS) technologies. ESS systems are classified into mechanical, electrochemical, chemical, electrical, and thermal energy storage systems. Table 1 presents the various types of energy storage system.

1. **Electrical Energy Storage System:** Electrical Energy Storage system, EES, is one of the main methods in the areas covered by the IEC. The next generation EES will become crucial in emerging IEC- relevant markets in the use of more renewable energy to get CO_2 reduction and for smart grids. Basically, EES has played three main tasks (IEA, 2009, Shinichi, 2011). The first one is EES reduce electricity cost by storing electricity achieved during the peak time, when its cost is lower, for use at peak times instead of electricity bought at higher prices. The second one is in order to improve the reliability of the power supply, EES systems support users, when power network failures happened due to natural disasters. The third role is to maintain and improve power quality, frequency and voltage (Perrin et al. 2014). Regarding emerging market needs, in on-grid areas, EES is expected to solve various problems such as excessive power fluctuation and undependable power supply, which are associated with the use of large amounts of renewable energy. In the off-grid domain, electric vehicles with batteries are the most promising technology to replace the fossil fuels by electricity from mostly renewable sources (VDE – ETG, 2008). The output of solar and wind power generation varies greatly depending on the weather and wind speeds, which can make them difficult to connect the grid. EES used for time shift can absorb this fluctuation more cost-effectively (Dufo-López et al., 2015). Facilitating EES has vital elements in many countries of the world (Balta et al., 2015), ranging from domestic community-scale devices to industrial utility-scale systems with power and energy capacities from some kW and kWh, respectively up to a few MW and MWh. There remain many technical and economic questions regarding the feasibility of large scale EES systems.
2. **Battery Energy Storage System:** Energy storage systems for batteries are the storage technology for electrical energy, which currently offers high investment advantages and is characterized by modularization, fast response time and a high marketing potential. With the technology innovation and successful development of the new battery, the efficiency, power density, energy density and battery life have improved significantly. The battery system is connected to a flexible installation and short construction cycles and has been successfully used in energy storage systems (Geth et al., 2015). Battery Energy Storage System (BESS) to take over and supplement specific applications such as power quality and voltage support.
3. **Multi-Agent Based Model:** Multi-agent is well fitted to simulate and analyze dynamic systems with complex interactions among the operation and design of batteries energy storage system and its components (UKPN, 2014).

4. **Flywheel Energy Storage System:** In flywheel energy storage (FESS), rotational energy is stored in an accelerated rotor or a massive rotating cylinder. The energy is held in the flywheel by keeping the rotating body at a constant speed. An increase in speed leads to a higher amount of stored energy. To accelerate the flywheel, electricity is supplied by a transmission device. Flywheel is known to be characterized by high self-discharge losses (Purvins et al., 2015) over long storage periods (several days). Therefore, in its present state, FES is best suited for direct uptake and release, rather than being stored for extended periods of time (Purvins et al., 2015). Accurate modeling and simulation work have been highly appreciated to understand and predict storage's behavior under changing demands.

Multi-Energy Demand Agent (MEDA): Forecasts the energy demands and records the real-time demand. The forecasting energy demands are based on weather, date and energy price from the real-time demand is stored as the database for more accurately prediction.

Hub Element Agents (HEA): Represent the performance of the units in energy hub. HEAs include micro-generators (GT, PV, and FC) energy storage devices battery, water tank, converters heat pump, absorption chiller, boiler and grid. The operation characteristic in detailed defined in each HEA (Li, 2015). It sends message to base on the environment data and get the optimal operation strategy from For example, the design, off-design condition performance and non-steady state condition performance as are defined in gas turbine agent GTA.

Integrated Dispatch Agent (IDA): Optimizes the energy carrier dispatch of the hub, according to the operation parameters provided by HEAs, cost information and the total energy demands of MEDA. The detailed optimization process and related objectives are further discussed in the optimization architecture.

ADVANTAGES

The following are some of the advantages, which authors analyze and review. The energy storage system and power electronic device interfaced to the power grid specifically done on electrical energy storage system because the storage of electrical energy has been identified as one of the ways to support a low carbon transition (Wade et al., 2010, Eyer et al., 2010) and highly applicable in the energy industry. The electrical energy storage could even improve production in renewable power plants, such as wind farms and solar power plants (Carbone, 2009).The main advantage of this system at EESS is the improvement in the reliability of the total system as at the time of failure also; electricity can be supplied with the help of battery energy storage system.

CONCLUSION

This chapter focused on integrating energy storage system to the grid. A network-scale energy storage system consists of three main components: the energy storage medium itself (Electrical energy storage system (EESS)), a power electronic interface that connects the storage medium to the grid and controller that are designed to achieve a grid control are discussed in this chapter. It does not target to take into account the detailed operation of the power electronic network interface or the performance of the electrical energy storage system. Power density, efficiency, cost and reliability are the biggest challenges in the development of power electronics (AC to AC converter, DC to DC converter and filter) system and interacting with energy storage system. The accuracy of models generally depends on the number and parameterization of the circuit components. A higher number of circuit elements can describe more complex dynamics, but at the cost of more computational effort.

REFERENCES

Rehman, S., Al-Hadhrami, L. M., & Alam, M. M. (2015). Pumped hydro energy storage system: A technological review. *Renewable & Sustainable Energy Reviews*, *44*, 586–598. doi:10.1016/j.rser.2014.12.040

Alhelou, H., Hamedani-Golshan, M. E., Zamani, R., Heydarian-Forushani, E., & Siano, P. (2018). Challenges and opportunities of load frequency control in conventional, modern and future smart power systems: A comprehensive review. *Energies*, *11*(10), 2497. doi:10.3390/en11102497

Alhelou, H. H. (2018). *Fault detection and isolation in power systems using unknown input observer. Advanced condition monitoring and fault diagnosis of electric machines* (p. 38). Hershey, PA: IGI Global.

Alhelou, H. H., Golshan, M., & Fini, M. (2018). Wind driven optimization algorithm application to load frequency control in interconnected power systems considering GRC and GDB nonlinearities. *Electric Power Components and Systems, 46*(11-12), 1223-1238.

Alhelou, H. H., & Golshan, M. E. H. (2016, May). Hierarchical plug-in EV control based on primary frequency response in interconnected smart grid. In *2016 24th Iranian Conference on Electrical Engineering (ICEE),* (pp. 561-566). IEEE. 10.1109/IranianCEE.2016.7585585

Alhelou, H. H., & Golshan, M. E. H. (2020). Decision-making-based optimal generation-side secondary-reserve scheduling and optimal LFC in deregulated interconnected power system. In *Decision making applications in modern power systems* (pp. 269–299). Academic Press. doi:10.1016/B978-0-12-816445-7.00011-6

Alhelou, H. H., Golshan, M. E. H., & Hatziargyriou, N. D. (2019). Deterministic dynamic state estimation-based optimal LFC for interconnected power systems using unknown input observer. *IEEE Transactions on Smart Grid*, 1. doi:10.1109/TSG.2019.2940199

Alhelou, H. H., Golshan, M. E. H., & Hatziargyriou, N. D. (2019). A decentralized functional observer based optimal LFC considering unknown inputs, uncertainties and cyber-attacks. *IEEE Transactions on Power Systems*.

Alhelou, H. H., Golshan, M. E. H., Zamani, R., Moghaddam, M. P., Njenda, T. C., Siano, P., & Marzband, M. (2019, June). An improved UFLS scheme based on estimated minimum frequency and power deficit. In 2019 IEEE Milan PowerTech (pp. 1-6). IEEE. doi:10.1109/PTC.2019.8810497

Alhelou, H. H., Golshan, M. H., & Askari-Marnani, J. (2018). Robust sensor fault detection and isolation scheme for interconnected smart power systems in presence of RER and EVs using unknown input observer. *International Journal of Electrical Power & Energy Systems*, *99*, 682–694. doi:10.1016/j.ijepes.2018.02.013

Alhelou, H. H., Hamedani-Golshan, M. E., Heydarian-Forushani, E., Al-Sumaiti, A. S., & Siano, P. (2018, September). Decentralized fractional order control scheme for LFC of deregulated nonlinear power systems in presence of EVs and RER. In *2018 International Conference on Smart Energy Systems and Technologies (SEST)* (pp. 1-6). IEEE. 10.1109/SEST.2018.8495858

Alhelou, H. S. H., Golshan, M. E. H., & Fini, M. H. (2015, December). Multi agent electric vehicle control based primary frequency support for future smart micro-grid. In *2015 Smart Grid Conference (SGC)* (pp. 22-27).

Anteneh, D., & Khan, B. (2019). Reliability enhancement of distribution substation by using network reconfiguration a case study at Debre Berhan Distribution Substation. *International Journal of Economy, Energy, & Environment*, *4*(2), 33.

Balta-Ozkan, N., Watson, T., Connor, P., Axon, C., Whitmarsh, L., Davidson, R., Spence, A., ... Taylor, G. (2015). Scenarios for the development of smart grids in the UK -Synthesis Report (UKERC: London, UK).

Banteywalu, S., Khan, B., De Smedt, V., & Leroux, P. (2019). A novel modular radiation hardening approach applied to a synchronous buck converter. *Electronics (Basel)*, *8*(5), 513. doi:10.3390/electronics8050513

Bouzid, A. M., Guerrero, J. M., Cheriti, A., Bouhamida, M., Sicard, P., & Benghanem, M. (2015) A survey on control of electric power distributed generation systems for microgrid applications. *Renew Sustain Energy Rev 2015,* 44, 7, 51 –66.

Carbone, R. (2009). Grid-connected photovoltaic systems with energy storage, International Conference on Clean Electrical Power 760–767. IEEE.

De, D., Klumpner, C., Patel, C., Ponggorn, K., Rashed, M., & Asher, G. (2013). Modelling and control of a multi-stage interleaved DC–DC converter with coupled inductors for super capacitor energy storage system. *IET Power Electronics*, *6*(7), 1360–1375. doi:10.1049/iet-pel.2012.0529

Doetsch, C., Berthold, S., Wolf, D., Smolinka, T., Tulbke, J., Bretschneider, P., & Radgen, P. (2007). Electrical energy storage from 100 kW–state of the art technologies, realisations, fields of use. In *Second International Renewable Energy Storage Conference (IRES II)*, Bonn, Germany.

Dufo-López, R., & Bernal-Agustín, J. L. (2015). Techno-economic analysis of grid connected battery storage. *Energy Conversion and Management*, *91*, 394–404. doi:10.1016/j.enconman.2014.12.038

Eyer, J., & Corey, G. (2010). Energy storage for the electricity grid: Benefits and market potential assessment guide, Report, Sandia National Laboratories.

Fares, R. L., & Webber, M. E. (2015). Combining a dynamic battery model with higher solution smart grid data to assess micro grid islanding lifetime. *Applied Energy*, *137*, 482–489. doi:10.1016/j.apenergy.2014.04.049

Fini, M. H., Yousefi, G. R., & Alhelou, H. H. (2016). Comparative study on the performance of many-objective and single-objective optimisation algorithms in tuning load frequency controllers of multi-area power systems. *IET Generation, Transmission, & Distribution*, *10*(12), 2915–2923. doi:10.1049/iet-gtd.2015.1334

Geth, F., Brijs, T., Kathan, J., Driesen, J., & Belmans, R. (2015). An overview of large-scale stationary electricity storage plants in Europe: Current status and new developments. *Renewable & Sustainable Energy Reviews*, *52*, 1212–1227. doi:10.1016/j.rser.2015.07.145

Gupta, K., Khan, B., & Mubeen, S. E. (2015). Available transfer capability enhancement by unified power flow controller. In *Proceedings IEEE International Conference on Signal Processing, Informatics, Communication, and Energy Systems (SPICES),* pp. 1, 4, 19-21.

Haes Alhelou, H., Hamedani-Golshan, M. E., Njenda, T. C., & Siano, P. (2019). A survey on power system blackout and cascading events: Research motivations and challenges. *Energies*, *12*(4), 682. doi:10.3390/en12040682

Haes Alhelou, H., Hamedani-Golshan, M. E., Njenda, T. C., & Siano, P. (2019). Wide-area measurement system-based optimal multi-stage under-frequency load-shedding in interconnected smart power systems using evolutionary computing techniques. *Applied Sciences*, *9*(3), 508. doi:10.3390/app9030508

Haes Alhelou, H., Hamedani Golshan, M. E., Njenda, T. C., & Siano, P. (2019). WAMS-based online disturbance estimation in interconnected power systems using disturbance observer. *Applied Sciences*, *9*(5), 990. doi:10.3390/app9050990

Hoffmann, C. (2010). *Design of transport- and storage capacities in energy supply systems with high shares of renewable energies*. Berlin, Germany: IRES.

Inage, S. I. (2011): Prospective on the decarbonised power grid, *IEC/MSB/EES Workshop*. Germany.

International Energy Agency (IEA): Prospects for large scale energy storage in decarbonized grids, Report, 2009.

Jariso, M., Khan, B., Tanwar, S., Tyagi, S., & Rishiwal, V. (2018). Hybrid energy system for upgrading the rural environment, IEEE GlobeCom 2018.

Jariso, M., Khan, B., Tesfaye, D., & Singh, J. (2017). Modelling and designing of stand-alone photovoltaic system (Case study: Addis Boder Health Center southwest Ethiopia). In *Proceedings IEEE International Conference on Electronics, Communication, and Aerospace Technology, ICECA 2017,* pp. 168-173, Coimbatore, India.

Kawashima, M. (2011). *Overview of electric power storage,* Internal paper of Tepco, 2011.

Khan, B., & Agnihotri, G. (2012). A novel transmission loss allocation method based on transmission usage. In *Proceedings IEEE Fifth Power India Conference*, Haryana, India. 10.1109/PowerI.2012.6479479

Khan, B., Agnihotri, G., & Gupta, G. (2013). A multipurpose matrices methodology for transmission usage, loss and reliability margin allocation in restructured environment, *Electrical & Computer Engineering: An International Journal, 2*(3).

Khan, B., Agnihotri, G., Rathore, P., Mishra, A., & Naidu, G. (2014). A cooperative game theory approach for usage and reliability margin cost allocation under contingent restructured market. *International Review of Electrical Engineering, 9*(4), 854–862.

Khan, B., Alhelou, H. H., & Mebrahtu, F. (2019). A holistic analysis of distribution system reliability assessment methods with conventional and renewable energy sources. *AIMS Energy, 7*(4), 413–429. doi:10.3934/energy.2019.4.413

Khan, B., & Singh, P. (2017). The current and future states of Ethiopia's energy sector and potential for green energy: A comprehensive study. *International Journal of Engineering Research in Africa, 33*, 115–139. doi:10.4028/www.scientific.net/JERA.33.115

Khan, B., & Singh, P. (2017). Optimal power flow techniques under characterization of conventional and renewable energy sources: A comprehensive analysis, *Journal of Engineering, vol. 2017,* Article ID 9539506.

Khan, B., & Singh, P. (2017). Selecting a meta-heuristic technique for smart micro-grid optimization problem: A comprehensive analysis, in IEEE Access, 5, pp. 13951-13977.

Khan, B., & Singh, P. (2019). Economic operation of smart micro-grid: a meta-heuristic approach. In Handbook of research on smart power system operation and control (pp. 330-346). Hershey, PA: IGI Global. doi:10.4018/978-1-5225-8030-0.ch014

Khan, B., & Singh, P. (2019). Economic operation of smart micro-grid: a meta-heuristic approach. In Handbook of research on smart power system operation and control (pp. 330-346). Hershey, PA: IGI Global. doi:10.4018/978-1-5225-8030-0.ch014

Khan, B., & Tanwar, S. (2019). Issues associated with microgrid integration. In Handbook of research on smart power system operation and control (pp. 252-264). Hershey, PA: IGI Global. doi:10.4018/978-1-5225-8030-0.ch010

Kifle, Y., Khan, B., & Singh, J. (2018). Designing and modeling grid connected photovoltaic system: (Case study: EEU building at Hawassa city). *International Journal of Convergence Computing, 3*(1), 20–34. doi:10.1504/IJCONVC.2018.091113

Kifle, Y., Khan, B., & Singh, P. (2018). Assessment and enhancement of distribution system reliability by renewable energy sources and energy storage, *Journal of Green Engineering, 8*(3), 2, pp. 219-262.

Li, G. (2015). Energy and exergy performance assessments for latent heat thermal energy storage systems. *Renewable & Sustainable Energy Reviews, 51*, 926–954. doi:10.1016/j.rser.2015.06.052

Lifshitz, D., & Weiss, G. (2015). Optimal control of a capacitor-type energy storage system. *IEEETransAutoControl, 60*(1), 216–220.

Makdisie, C., Haidar, B., & Alhelou, H. H. (2018). An optimal photovoltaic conversion system for future smart grids. In Handbook of research on power and energy system optimization (pp. 601–657). Hershey, PA: IGI Global. doi:10.4018/978-1-5225-3935-3.ch018

Molla, T., Khan, B., Moges, B., Alhelou, H. H., Zamani, R., & Siano, P. (2019). Integrated optimization of smart home appliances with cost-effective energy management system. *CSEE Journal of Power and Energy Systems*, *5*(2), 249–258.

Molla, T., Khan, B., & Singh, P. (2018). A comprehensive analysis of smart home energy management system optimization techniques, *Journal of Autonomous Intelligence, 1*(1), 15–21.

Mousavi, G. S. M., & Nikdel, M. (2014). Various battery models for various simulation studies and applications. *Renewable & Sustainable Energy Reviews*, *2014*, 477–485. doi:10.1016/j.rser.2014.01.048

Negash, K., Khan, B., Tesfaye, D., & Gayathri, M. (2017). Optimal placement of phasor measurement unit for system observability. *Proceedings of IEEE International Conference on Computing Methodologies and Communication, (ICCMC 2017),* pp. 73-77, Erode, India.

Negash, K., Khan, B., & Yohannes, E. (2016). Artificial intelligence versus conventional mathematical techniques: A review for optimal placement of phasor measurement units [Springer]. *Journal of Technology and Economics of Smart Grids and Sustainable Energy*, *1*(1), 10. doi:10.100740866-016-0009-y

Njenda, T. C., Golshan, M. E. H., & Alhelou, H. H. (2018, November). WAMS based intelligent under frequency load shedding considering online disturbance estimation. In *2018 Smart Grid Conference (SGC)* (pp. 1-5). 10.1109/SGC.2018.8777779

Njenda, T. C., Golshan, M. E. H., & Alhelou, H. H. (2018, November). WAMS based under frequency load shedding considering minimum frequency prediction and extrapolated disturbance magnitude. In *2018 Smart Grid Conference (SGC)* (pp. 1-5).

Perrin, M. et al. (2014). Investigation on storage technologies for intermittent renewable energies: Evaluation and recommended R&D strategy, INVESTIRE, Final Technical Report of European Community project ENK5-CT-2000-20336, 2004.

Purvins, A., & Sumner, M. (2013). Optimal management of stationary lithium-ion battery system in electricity distribution grids. *Journal of Power Sources*, *242*, 742–755. doi:10.1016/j.jpowsour.2013.05.097

Satkin, M., Noorollahi, Y., Abbaspour, M., & Yousefi, H. (2014). Multi-criteria site selection model for wind-compressed air energy storage power plants in Iran. *Renewable & Sustainable Energy Reviews*, *32*, 579–590. doi:10.1016/j.rser.2014.01.054

Schewe, P. F. (2005). The grid: A journey through the heart of our electrified world (Joseph Henry Press).

Schwunk, S. (2011). Battery systems for storing renewable energy. *Germany: Fraunhofer--Institut für Solare Energie.*

Siewniak, P., & Grzesik, B. (2014). A brief review of models of DC –DC power electronic converters for analysis of their stability. *International Journal of Electronics*, *101*(10), 1405–1426. doi:10.1080/00207217.2013.836696

Singh, P., & Khan, B. (2017). Smart microgrid energy management using a novel artificial shark optimization, Complexity, vol. 2017, Article ID 2158926. doi:10.1155/2017/2158926

Singh, P., Khan, B., Vidyarthi, A., Haes Alhelou, H., & Siano, P. (2019). Energy-aware online non-clairvoyant scheduling using speed scaling with arbitrary power function. *Applied Sciences*, *9*(7), 1467. doi:10.3390/app9071467

The Boston Consulting Group (BCG). (2011). Revisiting energy storage, Report.

UK Power Networks, UKPN. (2014). Smarter network storage, Six monthly report.

VDE - ETG Energy Storage Task Force: Energy storage in power supply systems with a high share of renewable energy sources Significance - state of the art - need for action, Report, Dec. 2008.

Wade, N. S., Taylor, P. C., Lang, P. D., & Jones, P. R. (2010). Evaluating the benefits of an electrical energy storage system in a future smart grid. *Energy Policy*, *38*(11), 7180–7188. doi:10.1016/j.enpol.2010.07.045

Xiao, X. (2015). Basic problems of the new complex AC-DC power grid with multiple energy resources and multiple conversions. *Trans China Elect Rotech Soc*, *30*, 1–14.

Yarlagadda, S., Hartley, T. T., & Husain, I. (2013). A battery management system using an active charge equalization technique based on a DC/DC converter topology. *IEEE Transactions on Industry Applications*, *49*(6), 2720–2729. doi:10.1109/TIA.2013.2264794

Yeshalem, M. T., & Khan, B. (2017). Design of an off-grid hybrid PV/wind power system for remote mobile base station: A case study. *AIMS Energy*, *5*(1), 96–112. doi:10.3934/energy.2017.1.96

Zamani, R., Golshan, M. E. H., Alhelou, H. H., & Hatziargyriou, N. (2019). A novel hybrid islanding detection method using dynamic characteristics of synchronous generator and signal processing technique. *Electric Power Systems Research*, *175*, 105911. doi:10.1016/j.epsr.2019.105911

Zamani, R., Hamedani-Golshan, M. E., Haes Alhelou, H., Siano, P., & Pota, H. R. (2018). Islanding detection of synchronous distributed generator based on the active and reactive power control loops. *Energies*, *11*(10), 2819. doi:10.3390/en11102819

Zhao, B., Song, Q., & Liu, W. (2014). Experimental comparison of isolated bidirectional DC – DC converters based on all-Si and all-SiC power devices for next-generation power conversion application. *IEEE Trans Ind Electron, 61,* 13, 89 –93.

KEY TERMS AND DEFINITION

Battery Energy Storage: A battery energy storage system is a system that stores energy via the use of a battery technology for it to be used at a later time.

Flywheel Storage: The flywheel system operates on the principle of kinetic energy for storing and releasing energy depending on load demand.

State of charge (SOC): It is the level of charge of an electric battery relative to its capacity.

APPENDIX

List of Abbreviations

BES Battery Energy Storage
CAES Compressed Air Energy Storage
CS Capacitor Storage
EESS Electrical Energy Storage System
ESS Energy storage system
FESS Flywheel Energy Storage System
LC Inductive capacitive filter
PHS Pumped Hydro Storage
SCES Supper Capacitor Energy Storage
SOC State of Charge
TES Thermal Energy Storage

Chapter 17
Applied Power Electronics:
Inverters, UPSs

Carlo Joseph Makdisie
Tishreen University, Syria

Marah Fadl Mariam
Tishreen University, Syria

ABSTRACT

Power Electronics is the study of switching electronic devices and circuits to convert and control the flow of electrical energy. Power Electronics is the basic technology of switching power supplies, power converters, power inverters, motor drives, and motor soft starters. Inverters are devices that transform DC input into fixed AC output through changing the trigger angle, in the ideal statues the output wave form is perfect sine wave, however in the practical case it has some higher harmonics that need to be reduced. Inverters are an important part of UPS systems, fully explained in this chapter, where UPSs are uninterruptible Power Supply in which the inverters are included as an essential part, that can provide emergency power to a load when the input power source or main power source fails.

INTRODUCTION

Solar Electric Generating Stations (SEGS). (1989) helped understanding the Solar electric systems that we discussed in hybrid UPSs, however Makdisie, C. (2016) explained more about the conventional UPSs and its applications, we concluded of Chlodnicki, Z., & Koczara, W. (n.d.) the techniques of using supercapacitors storage.

Makdisie, C. (2008) introduced us the fundamental construction of the inverters and the UPSs. However, in Panda, B., Panda, B., & Hota, P. K explains the efficiency of the interfaces used in the UPSs. Hybrid UPS systems are defined in Eep. (n.d.). back to the detailed demonstration of the inverters in Chlodnicki, Z., Koczara, W., & Al-Khayat, N, to the recent classification of power electronics the interior construction in Makdisie, C. (2008).

DOI: 10.4018/978-1-7998-1230-2.ch017

Vodorozov, V. (2010) & Rashid, M. H. (ed.). (2011) elucidated the fully explained applications of hybrid UPS systems, and its usage in modern power grids.

BACKGROUND

In our chapter, we studied the fully functional inverters, and its various modes of operation. As well as the recent PWM techniques with the methods of reducing the high harmonic affects. In the end we explained the fundamental application of the UPS systems in addition to the construction of its interfaces and elements (Alhelou et al., 2019; Makdisie et al., 2018; Alhelou et al., 2018; Alhelou et al., 2016; Haes Alhelou et al., 2019; Njenda et al., 2018).

Inverters:

An inverter is a static device that converts DC power into AC power at desired output voltage and frequency.

The inverter applications involves:

- Adjustable – speed AC drives,
- Induction Heating,
- Aircraft power supplies,
- UPS.

Classification of Inverters:

- According to the method of Commutation
  - Line Commutated Inverter
  - Force Commutated Inverter
- According to the method of Connections
  - Series Inverter
  - Parallel Inverter
  - Bridge Type Inverter
- According to the nature of DC source feeding the Inverter
  - Voltage source Inverter
    - VSI has a constant voltage at its input terminals.
    - Its output voltage does not depend on load.
    - Its output current depends on the type of load.
  - Current Source Inverter
    - CSI has a constant current at its input terminals.
    - ITS output current does not depend on load.
    - Its output voltage depends on the type of load.

Many semiconductor elements are used in the internal construction of inverters like BJT, MOSFET, IGBT, MCT, SIC, GTO, SCRs.

Figure 1. Single phase half bridge inverter.

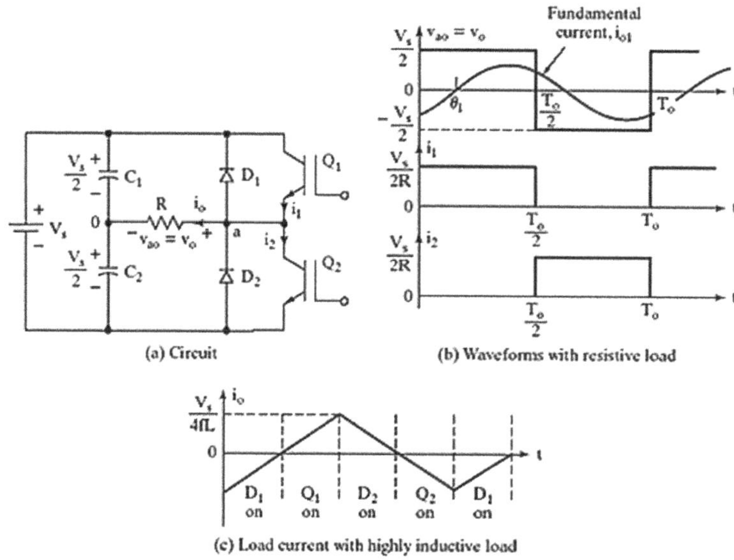

Operation principle of single-phase half bridge inverters:

Figure (1) configure a single-phase half bridge inverter where the voltage source is a DC source V_S, two transistors and two diodes.

And the operation systems vary depending on the type of the load:

Pure resistive load:

This case can be studied on two phases:

The first phase starts with a trigger pulse on the base of the transistor Q1 and switches off with a trigger pulse on the base of Q2 which is at the moment $t = \frac{T_o}{2}$, and the voltage in this phase is equal to $\left(\frac{V_S}{2}\right)$, and the current is equal to $\left(\frac{V_S}{2R}\right)$.

The second phase starts at the end of the first one, and switches off at the moment $t = T_o = \frac{1}{f}$.

The voltage applied on the load in this phase is equal to $\left(-\frac{V_S}{2}\right)$, and the current $\left(-\frac{V_S}{2R}\right)$, figure (1 – b) shows the voltage and current wave forms of the inverters elements.

The effective value of the output voltage can be calculated from the equation:

$$V_o = \left(\frac{2}{T_o}.\int_0^{\frac{T_o}{2}} \frac{V_S^2}{4}.dt\right)^{\frac{1}{2}} = \frac{V_S}{2} ....(1)$$

Fourier series can be used to get the instantaneous value of the output voltage:

$$V_{o(t)} = \sum_{n=1}^{\infty} \frac{2.V_S}{n.\pi}.\sin(n\omega t).........(2)$$

Where:
n: is a real odd number ($n = 1,3,5,..$)
$\text{É} = 2\text{À}f_o$: the frequency of the inverter's output voltage and its measuring unit is: $\left(\frac{\text{rad}}{\text{sec}}\right)$.
By substituting $n = 1$ in the equation (2) the basic value of the output voltage can be deduced:

$$V_{1(t)} = \frac{2.V_S}{\text{À}}.\sin(\text{É}t).........(3)$$

And the effective basic value is:

$$V_1 = \frac{2.V_S}{\sqrt{2}.\text{À}} = 0.45V_S.........(4)$$

An observation could be made is that the conduction period of the transistors is equal to $\left(\frac{T_o}{2}\right)$ sec.

1) Pure inductive load:

In this case the current can't change instantaneously along with the voltage, so two diodes are added to the inverter $D1, D2$, to secure a path for a current reversed to the voltage source (regenerative).

When the transistor Q1 is turned off at the moment $t = \frac{T_o}{2}$, the load current will continue to pass with the same direction through D2, although the voltage polarity is negative.

The diode D1 is operated in the same way as mentioned above when the transistor Q2 is turned off.

The figure (40 – c) shows the form of the inductive current, it is obvious that the phase different between the basic value of the voltage and the load current is $90°$, and the conduction period of the transistors and the diodes is $\frac{T_o}{4}$ sec. (Chlodnicki, Z., & Koczara, W. (n.d.). Supercapacitors (2005)).

2) R- L Load:

The conduction period in this case is in the range $\frac{T_o}{4} < t < \frac{T_o}{2}$, considering the time needed to reversed recovery of the semiconductor elements properties t_q, at the end of the turned off period, the maximum turn on time of the transistors working on a pure resistive load is equal to $\left(-\frac{T_o}{2} - t_q\right)$.

The instantaneous current in this case is:

$$i_o = \sum_{n=1}^{\infty} \frac{2.V_S}{n.\pi.\sqrt{R^2 + (n\omega L)^2}}.\sin(n\omega t - \theta_n)\text{.........}(5)$$

Where:
n: is a real odd number ($n = 1,3,5,..$)

$$Q_n = arc\tan\left(\frac{n.\omega.L}{R}\right)$$

If we assumed that I_{o1} is the first order component (basic value) of the load current, then the output power is:

$$P_{o1} = V_1.I_{o1}.\cos\theta_1 = I_{o1}^2.R$$

$$= \left[\frac{2.V_S}{\sqrt{2}.\pi.\sqrt{R^2 + (\omega.L)^2}}\right]^2.R\text{.........}(6)$$

The output power caused by the higher harmonics are dissipated mostly in the form thermal power, which is not useful, so we can consider the only useful one is equal to the basic output power P_{o1}. (Vodorozov, V. (2010)).

Determination factors of the distortion ratio of the output wave forms of the inverters:

The output wave forms of the practical inverters always contain higher harmonics as well as the basic harmonic, so the determination of the work efficiency of the inverter depends on the following factors:

1- Harmonic Factor of the nth harmonic HF_n:

$$HF_n = \frac{V_n}{V_1}\text{.........}(7)$$

2- Total harmonic distortion (T H D):

$$THD = \frac{\sqrt{\sum_{n=2,3,..}^{\infty} V_n^2}}{V_1}\text{.........}(8)$$

3- Distortion factor (DF):

Figure 2. Single phase full bridge inverters

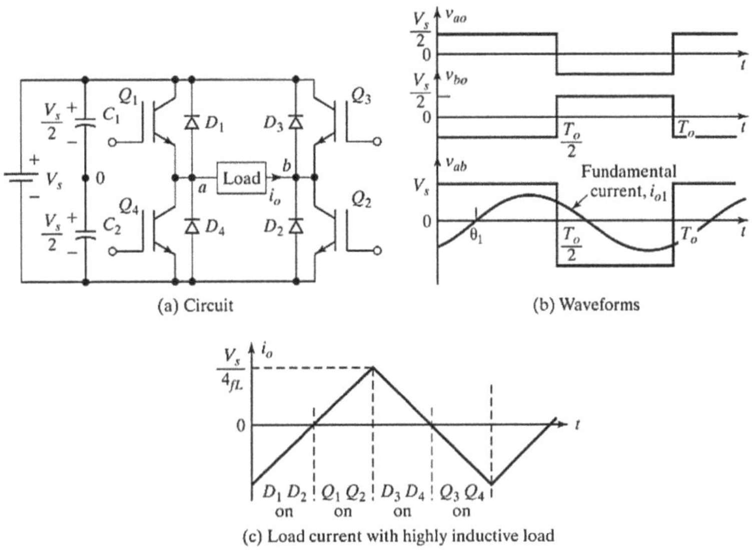

The signal contains high harmonics that are to be attenuated. The efficiency of the attenuation has second order equation (divided over n^2), so we can express the DF value without considering the value of load filters, as following:

$$DF = \frac{\sqrt{\sum_{n=2,3,..}^{\infty}\left(\frac{V_n}{n^2}\right)^2}}{V_1} \dots\dots\dots(9)$$

While the distortion factor of nth can be expressed as following:

$$DF_n = \frac{V_n}{V_1 . n^2} \dots\dots\dots\dots(10)$$

Single phase full bridge inverters:

This type of inverters is configured of four transistors as it is shown in figure (2).

When a trigger pulse is given to each pair of the transistors Q1,Q2, the voltage applied on the load is equal to V_S, however when a trigger pulse is given to Q3,Q4 then the voltage is equal to $-V_S$.

The output voltage of the waveform in figure (2 – b) is calculated with the equation:

$$V_o = \left(\frac{2}{T_o} . \int_0^{\frac{T_o}{2}} V_S^2 . dT\right)^{\frac{1}{2}} = V_S \dots\dots\dots(11)$$

Fourier series can be used to get the instantaneous value of the output voltage:

$$V_{C(t)} = \sum_{n=1;3;5;\ldots}^{\infty} \frac{4.V_S}{n.\pi}.\sin\left(n.\omega.t\right)\ldots\ldots\ldots(12)$$

By substituting $\mathrm{n} = 1$ in the equation (12) the basic value of the output voltage can be deduced:

$$\mathrm{V}_{1(\mathrm{t})} = \frac{4.\mathrm{V}_\mathrm{S}}{\text{À}}.\sin\left(\text{É}\mathrm{t}\right)\ldots\ldots\ldots(13)$$

And the effective basic value is:

$$\mathrm{V}_1 = \frac{4.\mathrm{V}_\mathrm{S}}{\sqrt{2}.\text{À}} = 0.45\mathrm{V}_\mathrm{S}\ldots\ldots\ldots(14)$$

The instantaneous current in this case (R – L) load is:

$$i_o = \sum_{n=1}^{\infty} \frac{4.V_S}{n.\pi.\sqrt{R^2 + \left(n\omega L\right)^2}}.\sin\left(n\omega t - \theta_n\right)\ldots\ldots\ldots(15)$$

Where:
n: is a real odd number ($n = 1,3,5,..$)

$$Q_n = arc\tan\left(\frac{n.\omega.L}{R}\right)$$

D1 and D2 – freewheel reversed diodes, due to the diodes D1 and D2 are in conduction mode, then the power direction will be from the load to the source (regenerative).

Figure (2 – c) shows the signal form of inductive (R –L) load current.

Three phase inverters:

The three phase inverters are usually used in feeding the high power loads. These types of inverters are configured of three single phased half bridge or full bridge inverters in parallel with the DC power supply.

Figure (3 – a) shows the connection of the single-phase inverters that build the three phase inverter. The inverter shown in figure (3) concludes three single phase transformers, 12 transistors, and 12 diodes. So a balanced system of the three phases can be achieved through controlling the turn on and off times of the transistors. (Rashid, M. H. (ed.). (2011)).

Two working modes can be distinguished in the three-phase inverter:

Figure 3. Single phase inverters that builds the three-phase inverter

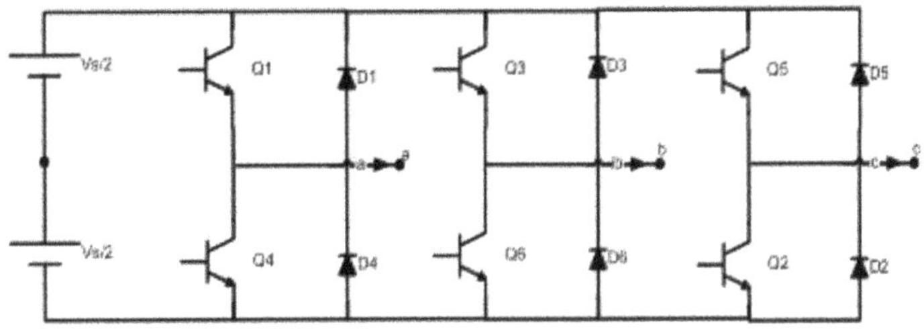

(a)

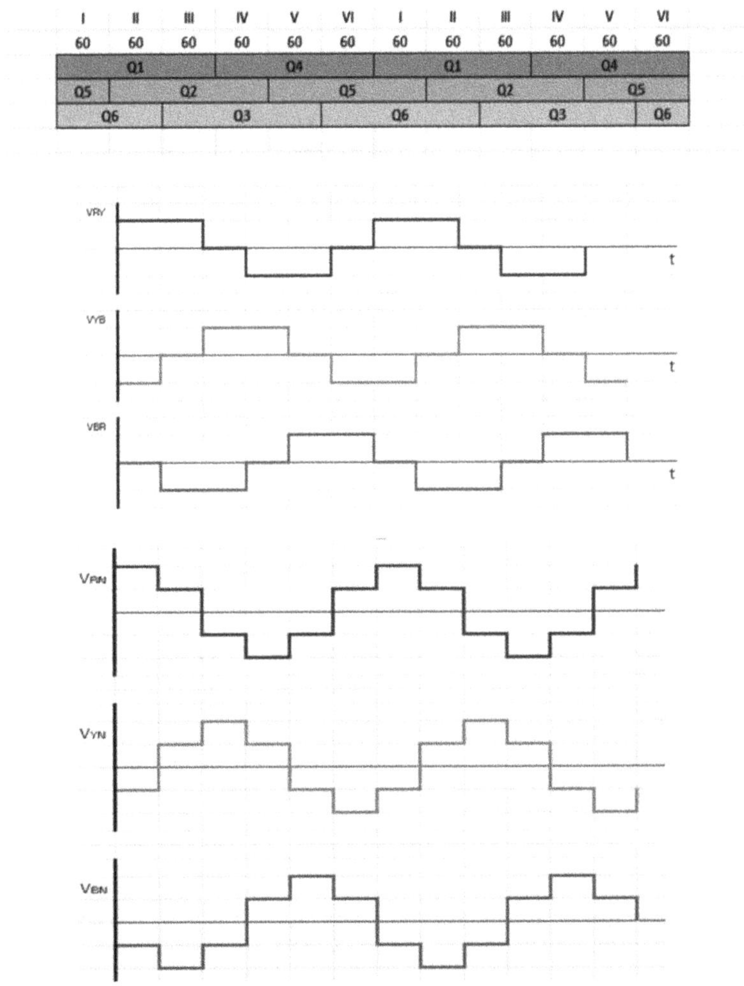

(b)

Figure 4. 180° operation mode inverter

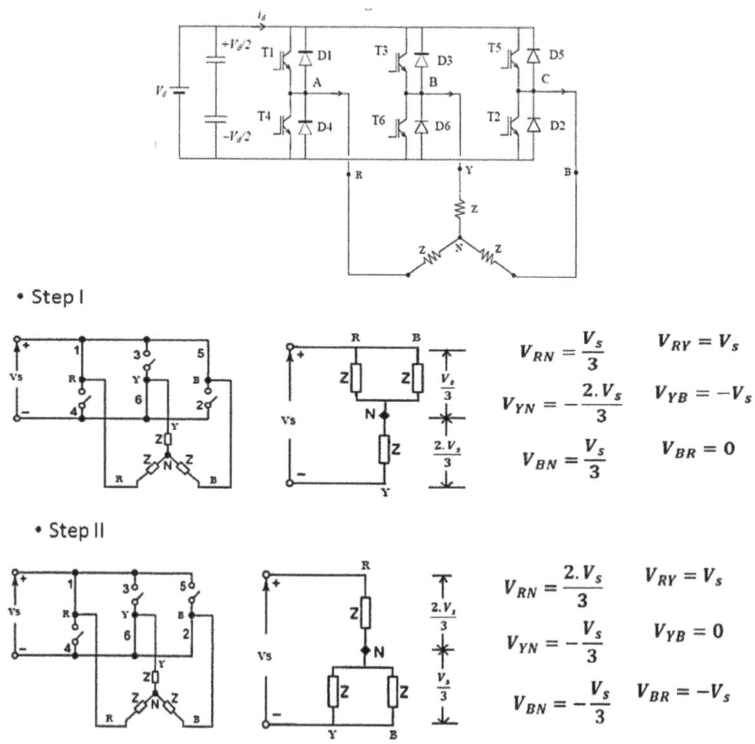

180° Operation Mode

Figure 3 shows the pulses that control the transistors and the output line to line voltage.

However, figure 4 shows the configuration of the inverter.

Table 1 configure the main transistors and its working time, with the detailed explanation to each step.

It is obvious from figure 5 the circuits that are equal to each phase in the first half wave of the inverter's work, and with a pure resistive load with a star connection in the outlet.

With the output voltage shown for each phase in figure (3 – b). (Panda, B., Panda, B., & Hota, P. K)

In the first step when transistors1,5,6 are working:

$$\text{Å} \le \text{wt} < \frac{\pi}{3},\ R_{eq} = R + \frac{R}{2} = \frac{3R}{2}$$

$$i_1 = \frac{V_s}{R_{eq}} = \frac{2V_s}{3R},\ U_{an} = U_{cn} = \frac{i_1 R}{2} = \frac{V_s}{3}$$

$$U_{bn} = -i_1.R = \frac{-2V_s}{3}$$

Table 1. Transistors operation time.

Step	Devices Conducting
I	T1, T5 & T6
II	T1, T2 & T6
III	T1, T2 & T3
IV	T4, T2 & T3
V	T4, T5 & T3
VI	T4, T5 & T6

Figure 5. Circuits that are equal to each phase.

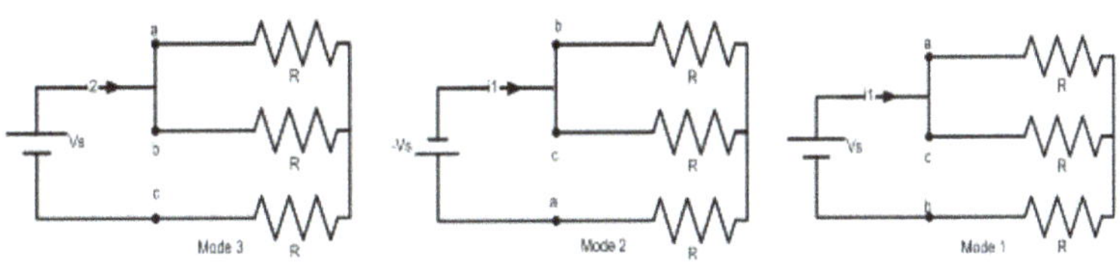

However when transistors 1,2,6 are working:

$$\frac{\pi}{3} \le wt < \frac{2\pi}{3},\ R_{eq} = \frac{3R}{2}$$

$$i_2 = \frac{V_s}{R_{eq}} = \frac{2V_s}{3R},\ U_{an} = i_2.R = \frac{2V_s}{3}$$

$$U_{bn} = U_{cn} = \frac{-i_2.R}{2} = \frac{-V_s}{3}$$

And when transistors 1,2,3 are working:

Table 2. 180^0 mode line and the phase voltage for each step

Step	Phase Voltage			Line Voltage		
	V_{AN}	V_{BN}	V_{CN}	V_{AB}	V_{BC}	V_{CA}
I	Vs/3	-2Vs/3	Vs/3	Vs	- Vs	0
II	2Vs/3	-Vs/3	-Vs/3	Vs	0	- Vs
III	Vs/3	Vs/3	-2Vs/3	0	Vs	- Vs
IV	-Vs/3	2Vs/3	-Vs/3	- Vs	Vs	0
V	-2Vs/3	Vs/3	Vs/3	- Vs	0	Vs
VI	-Vs/3	-Vs/3	2Vs/3	0	- Vs	Vs

$$\frac{2\pi}{3} \leq wt < \pi$$

$$R_{eq} = \frac{3R}{2}$$

$$i_3 = \frac{V_s}{R_{eq}} == \frac{2V_s}{3R}$$

$$U_{an} = U_{bn} = \frac{i_3.R}{2} = \frac{V_s}{3}$$

$$U_{cn} = -i_3.R = \frac{-2V_s}{3}$$

Fourier series of the output linear voltage of the inverters can be established:

$$U_{ab} = \sum_{1,3,5....}^{\infty} \left(\frac{4V_s}{n\pi} \cos\frac{n\pi}{6} \sin n(wt + \frac{\pi}{6}) \right) ....(16)$$

$$U_{bc} = \sum_{1,3,5....}^{\infty} \left(\frac{4V_s}{n\pi} \cos\frac{n\pi}{6} \sin n(wt - \frac{\pi}{2}) \right) ....(17)$$

$$U_{bc}=\sum_{1,3,5\ldots}^{\infty}\left(\frac{4V_s}{n\pi}\cos\frac{n\pi}{6}\sin n(wt-\frac{7\pi}{6})\right)\ldots\ldots\ldots(18)$$

It is obvious from equations 15, 16, 18 that the third harmonics and its multiplications of the linear voltages is equal to zero. (Makdisie, C. (2016)).

$$V_L=\left[\frac{2}{2\pi}\int_0^{\frac{2\pi}{3}}V_s^2d(wt)\right]^{\frac{1}{2}}=\sqrt{\frac{2}{3}}V_s=0.8165V_s\ldots\ldots\ldots(19)$$

And the effective value for the n harmonic:

$$V_{Ln}=\frac{4V_s}{\sqrt{2}.n.\pi}\cos\frac{n\pi}{6}\ldots\ldots\ldots(20)$$

So for the basic harmonic the effective value is:

$$V_{Ln}=\frac{4V_s}{\sqrt{2}.\pi}\cos\frac{\pi}{6}=0.7797V_s\ldots\ldots\ldots(21)$$

And the line to earth (phasic) voltage is:

$$V_p=\frac{V_L}{\sqrt{3}}=\frac{\sqrt{2}V_s}{3}=0.4714V_s\ldots\ldots\ldots(22)$$

However, when the inverter is working on an R- L Load then the diodes secure a pass for the reversed currents which are passing through the load during the switching period. (Chlodnicki, Z., Koczara, W., & Al-Khayat, N.)

The current for this case is defined by the equation (16), by the assumption of $V_{phase}=\frac{V_{Line}}{\sqrt{3}}$, with an angle delay of $30°$.

$$i_a=\sum_{n=1,3,5\ldots}^{\infty}\left[\frac{4V_s}{\sqrt{3}.n.\pi\sqrt{R^2+(n.w.L)^2}}\cos\frac{n\pi}{6}\right]\sin(nwt-\theta_a)\ldots\ldots(23)$$

Where: $\theta_n=arc\tan\left(\frac{n.w.L}{R}\right)$

Figure 6. 120° operation mode inverter.

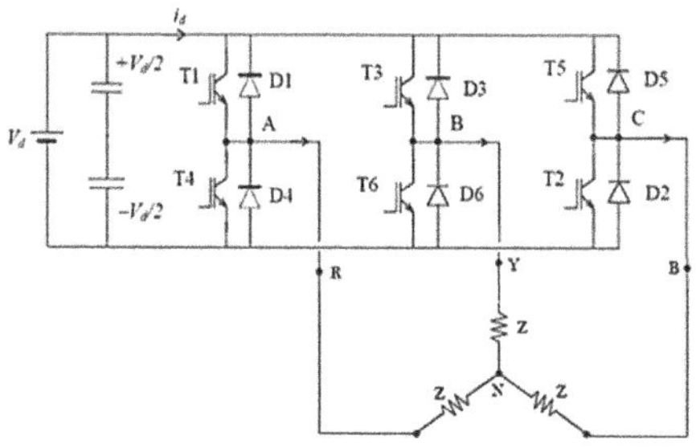

(a)

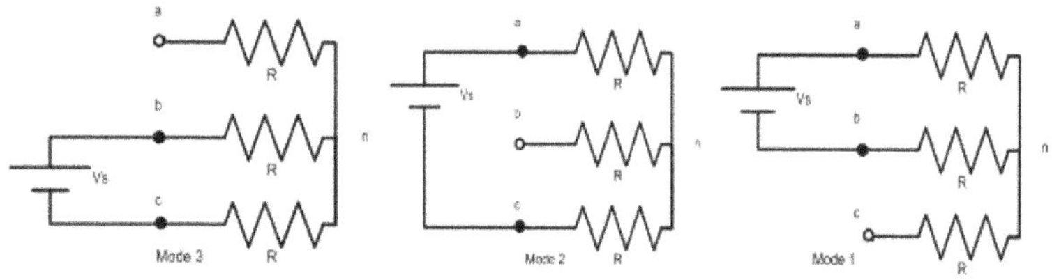

(b)

Table 3. 120° mode line and the phase voltage for each step.

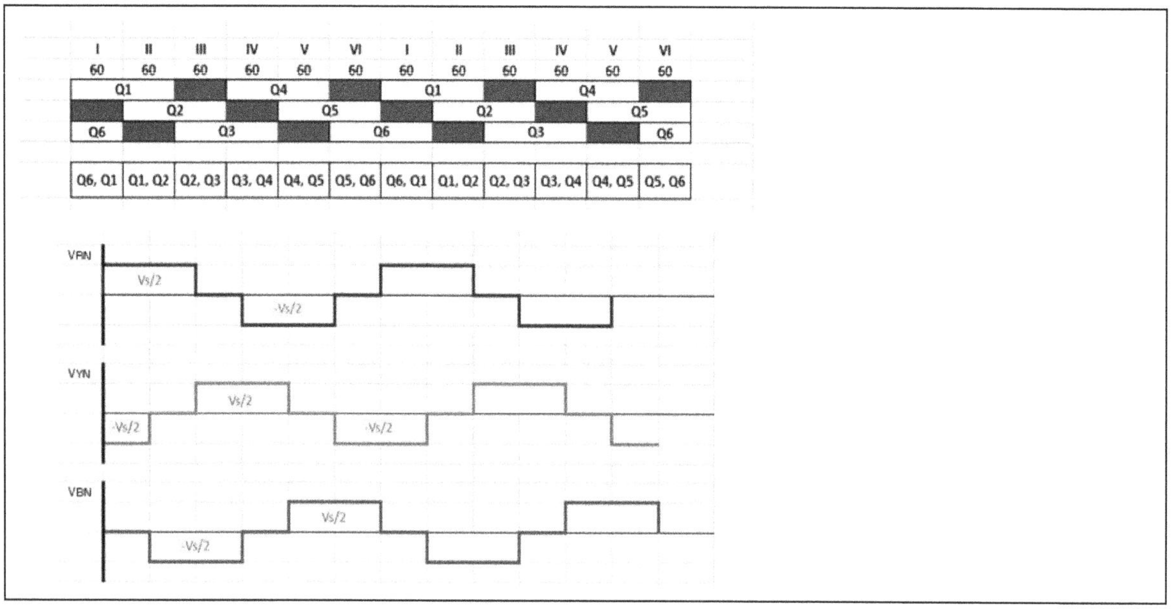

Figure 7. 120° mode voltage wave forms

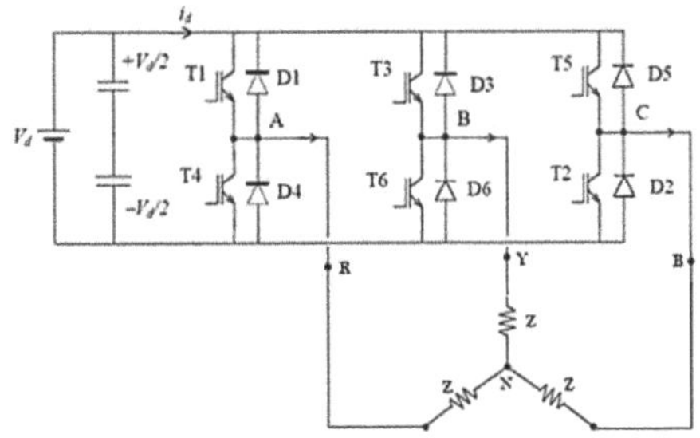

(a)

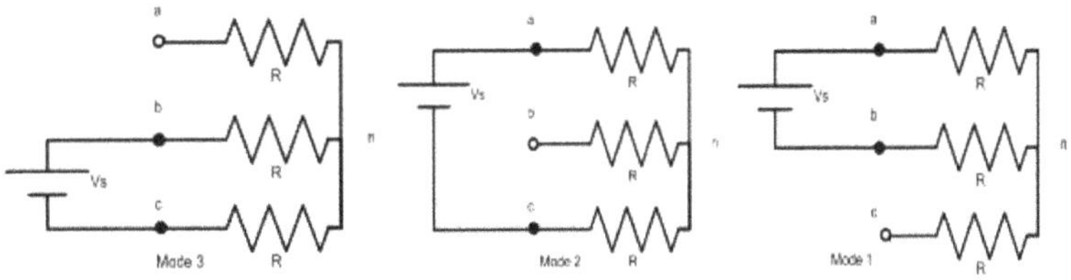

(b)

• Step II

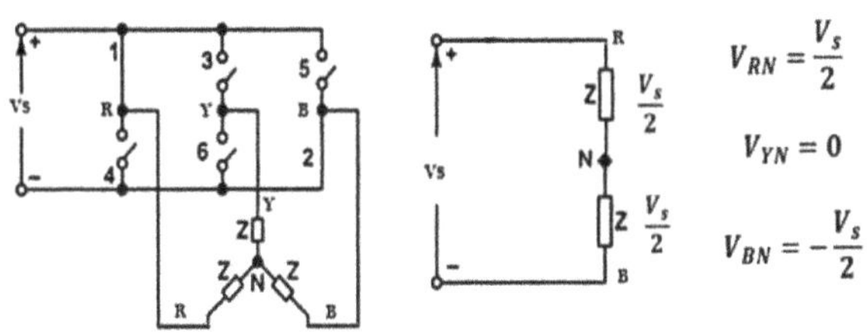

$$V_{RN} = \frac{V_s}{2} \qquad V_{RY} = \frac{V_s}{2}$$

$$V_{YN} = 0 \qquad V_{YB} = \frac{V_s}{2}$$

$$V_{BN} = -\frac{V_s}{2} \qquad V_{BR} = -V_s$$

• Step I

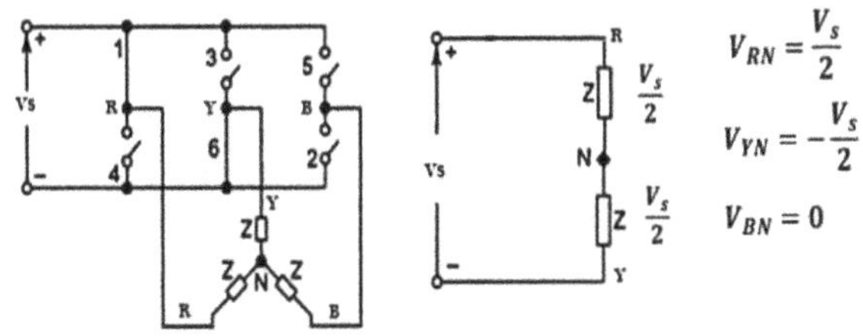

$$V_{RN} = \frac{V_s}{2} \qquad V_{RY} = V_s$$

$$V_{YN} = -\frac{V_s}{2} \qquad V_{YB} = -\frac{V_s}{2}$$

$$V_{BN} = 0 \qquad V_{BR} = -\frac{V_s}{2}$$

120° operation mode:
Figure (6 –a) shows the inverter working on the mode 120 degree.
However, figure (6 – b) shows the different modes to connect the load with the outlet of the inverter
Table 3 configure the main transistors and its working time, with the detailed explanation to each step.
And the voltage wave forms and values are in figure 7 and table 4
The enhancement of output voltage of the inverters:

Table 4. 120° mode voltage wave values.

Step	Phase Voltage			Line Voltage		
	V_{RN}	V_{YN}	V_{BN}	V_{RY}	V_{YB}	V_{BR}
I	Vs/2	-Vs/2	0	Vs	-Vs/2	-Vs/2
II	Vs/2	0	-Vs/2	Vs/2	Vs/2	- Vs
III	0	Vs/2	-Vs/2	-Vs/2	Vs	-Vs/2
IV	-Vs/2	Vs/2	0	- Vs	Vs/2	Vs/2
V	-Vs/2	0	Vs/2	-Vs/2	-Vs/2	Vs
VI	0	-Vs/2	Vs/2	Vs/2	- Vs	Vs/2

The inverters are designed in most of the industrial applications with controlled or regulated output voltage (fixed frequency) due to the following reasons:

1. Control the operation systems of electrical loads.
2. When the supply provided from uncontrolled voltage sources.
3. To achieve the ratio v/f = constant on the inverter output.

External Control of DC Input Voltage

By controlling the gain of the inverter we can achieve the regulation of the output wave forms of inverters.

The implementation of pulse width modulation (PWM) is considered the best and favorable method to control the inverter gain (Fini et al., 2016; Alhelou et al., 2018; Zamani et al., 2018; Alhelou et al., 2015; Njenda et al., 2018; Haes Alhelou et al., 2018; Haes Alhelou et al., 2019).

Pulse Width Modulated Inverters:

- These inverters can produce ac voltages of variable magnitude as well as variable frequency.
- The quality of output voltage can also be greatly enhanced, when compared with those of square wave inverters.
- Pulse Width Modulation is the process of changing the width of the pulses to obtain variation in the o/p voltage with reduced harmonic content.
- The main aim of using different PWM techniques is to generate a sinusoidal output voltage of desired fundamental frequency and magnitude.

Figure 8. Voltage of the inverter with single pulse modulation

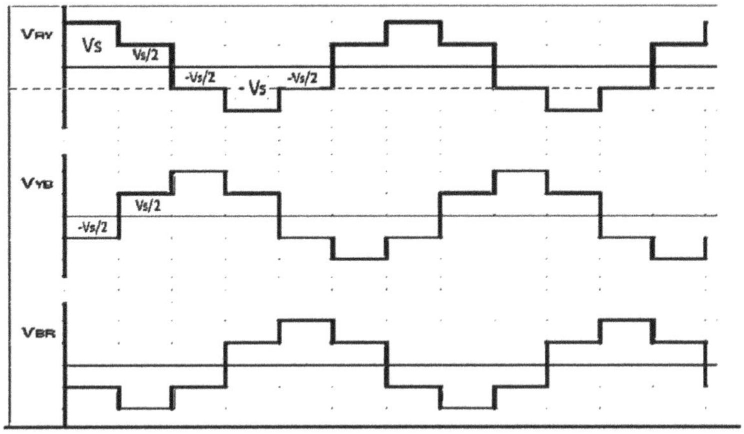

Different types of PWM inverters:

- Single pulse modulation (PWM).
- Multiple pulse modulation (MPWM).
- Sinusoidal pulse modulation (SMPWM).
- Modified sinusoidal pulse modulation (MSPWM)
- Space vector pulse width modulation (SVM)

Single pulse modulation:

- Frequency of the reference signal determines the frequency of output voltage.
- The ratio of A_r to $A_{C,}$ called modulation index, controls the output voltage.
- The output voltage of the inverter with single pulse modulation is given by,

$$V_o = \sum_{n=1,3,5}^{\infty} \frac{4V_s}{n\pi} \sin\frac{n\pi}{2} \sin nd \sin n\omega t$$

$$V_o = \frac{4V_s}{\pi}\left(\sin d \sin \omega t - \frac{1}{3}\sin 3d \sin 3\omega t + \frac{1}{5}\sin 5d \sin 5\omega t \ldots.. \right)$$

$$V_{o1} = \frac{4V_s}{\pi}\sin\frac{\pi}{2}\sin d \sin \omega t = \frac{4V_s}{\pi}\sin d \sin \omega t$$

$$V_{o1m} = \frac{4V_s}{\pi}\sin d$$

Figure 9. Controlling the frequency of the reference signal

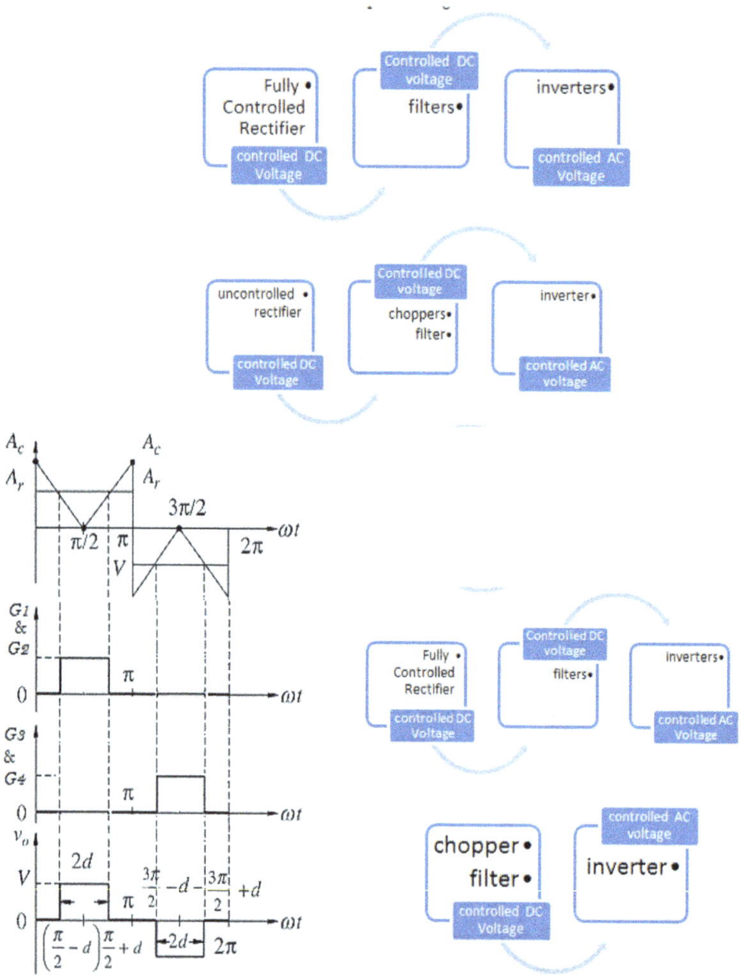

Figure 10. Voltage of the reference controlled signal

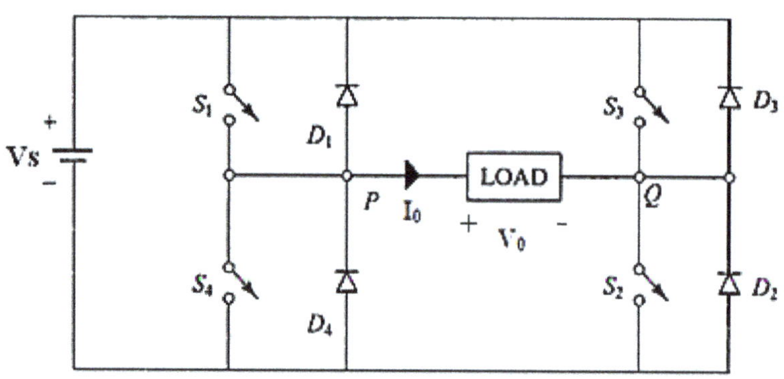

Figure 11. Pulse width.

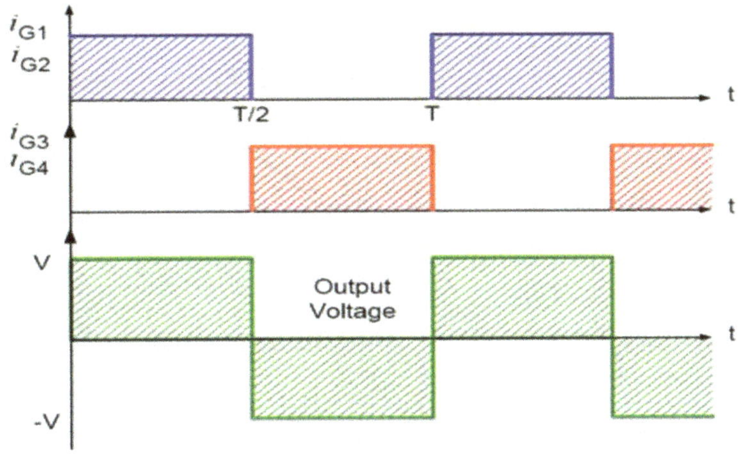

- If nd = π or d=π/n, then n^{th} harmonic will be eliminated from the inverter output voltage.
- For example, for eliminating third harmonic, 3d = π. i.e pulse width, $2d = 2\pi / 3 = 120^0$.

Conclusions from Single PWM (Khan et al. 2014; Khan et al. 2013; Khan et al. 2012; Negash et al., 2017; Negash et al., 2016; Jariso et al. 2017; Kifle et al. 2018; Yeshalem et al. 2017; Singh et al. 2019; Gupta et al., 2015):

- 3rd, 5th & 7th harmonics dominate when the voltage is reduced.
- A large amount of harmonics is introduced at lower output voltages.
- Harmonic content can be reduced by having many pulses in each half cycle of output voltage.

Multiple Pulse Modulation:

- In this method, many pulses having equal widths are produced per every half cycle.
- The gating signals are produced by comparing reference signal with triangular carrier wave.

$$_3 = \frac{\grave{A} - 2d}{3} + \frac{d}{2}$$

- Frequency of the reference signal determines the frequency of output voltage.
- The ratio of A_r to $A_{C,}$ called modulation index, controls the output voltage.

The output voltage waveform can be expressed in Fourier series as,

Figure 12. Multiple pulse modulation

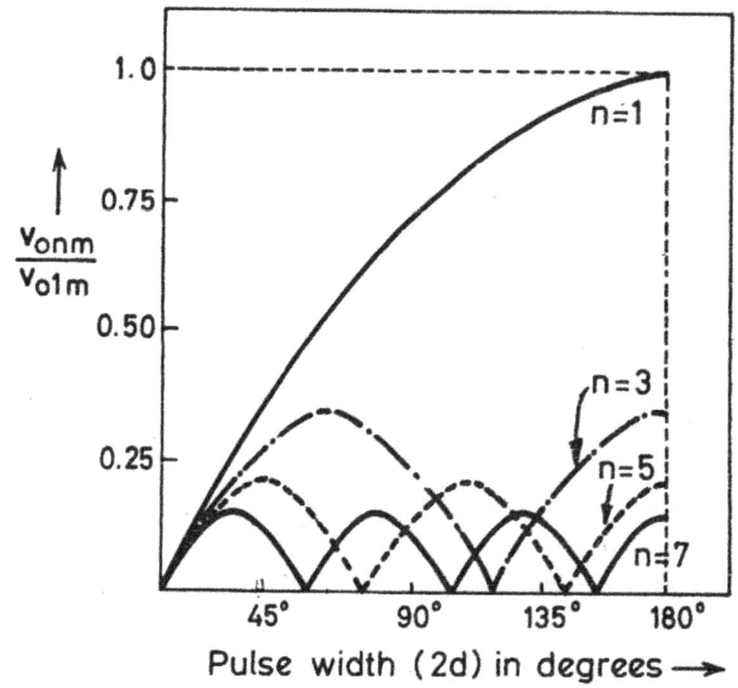

Figure 13. Multiple pulse modulation reference signal

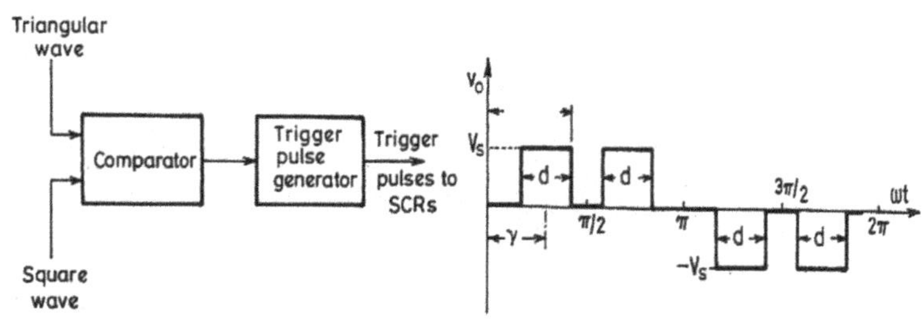

$$V_o = \sum_{n=1,3,5}^{\infty} \frac{8V_s}{n\pi} \sin n^3 \sin\frac{nd}{2} \sin n\omega t \ldots\ldots\ldots(24)$$

$$V_o = \frac{8V_s}{\pi}\begin{pmatrix} \sin\gamma \sin\frac{d}{2}\sin\omega t - \frac{1}{3}\sin 3\gamma \sin\frac{3d}{2}\sin 3\omega t \\ +\frac{1}{5}\sin 5\gamma \sin\frac{5d}{2}\sin 5\omega t \ldots\ldots \end{pmatrix}\ldots(25)$$

Figure 14. Sinusoidal pulse modulation

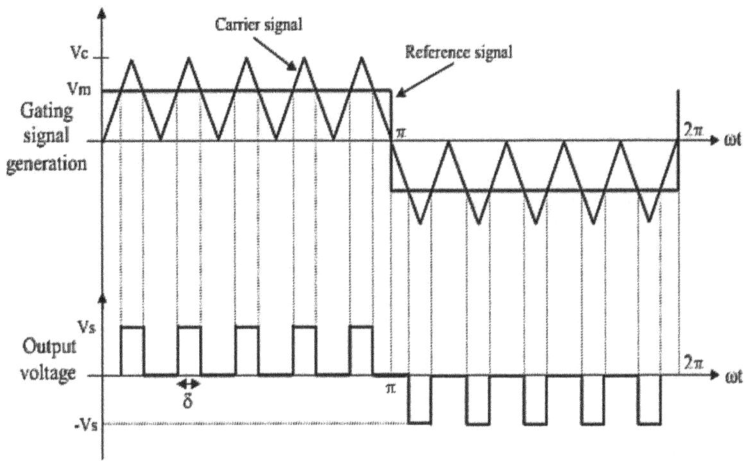

$$V_{o1} = \frac{8V_s}{\pi} \sin^3 \sin\frac{d}{2} \sin \omega t$$

$$V_{olm} = \frac{8V_s}{\pi} \sin^3 \sin\frac{d}{2} \ldots\ldots\ldots(26)$$

It is seen from the above that the fundamental component of output voltage is low for two pulse modulation than it is for single pulse modulation. But lower order harmonics are eliminated and higher order harmonics are increased. But higher order harmonics can be filtered easily. This scheme is advantageous than single pulse modulation. But large number of pulses per half cycle requires frequent turn on and turn off thyristors. This will increase switching losses. (Eep. (n.d.))

Sinusoidal Pulse Modulation:

- In this method, several pulses per half cycle are used as in the case of multiple pulse modulation.
- But width of each pulse is modulated proportional to the amplitude of sine wave.
- Gate pulses are generated by comparing sinusoidal reference signal with triangular carrier signal.
- Frequency of reference signal (f_r) decides the frequency of output voltage.
- The ratio of V_r/V_c is called the modulation index which controls the output voltage.
- Number of pulses per half cycle depends on the carrier frequency (f_c).

Figure (14) shows the sinusoidal pulse modulation wave forms:

Modulation index is the ratio of peak magnitudes of the modulating waveform and the carrier waveform.

$$M_i = \frac{V_m}{V_c}$$

M_i controls the harmonic content in the output voltage

Figure 15. Pulses per half cycle

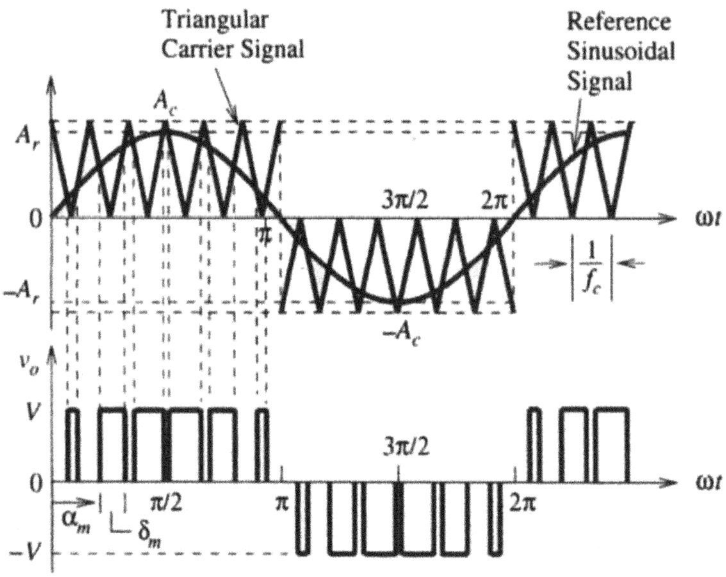

Figure 16. Transformer Connections

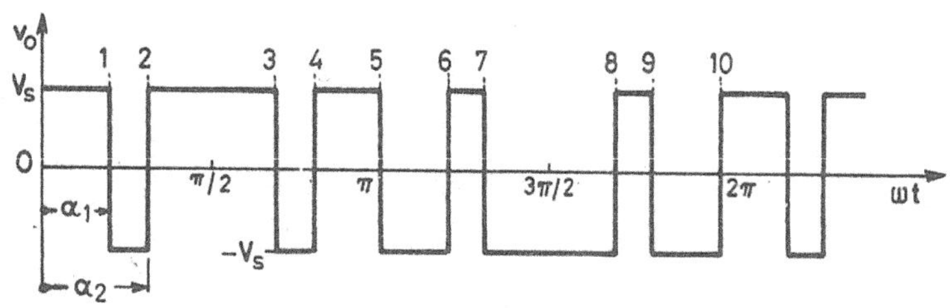

By increasing the number of pulses (N) per half cycle, the lower order harmonics get cancelled. But higher order harmonics will get increased. Higher order harmonics can be filtered out easily.

Higher value of N results in more switching losses and leads to reduction of efficiency of inverter.

Harmonic Reduction by PWM:

Several pulses per half cycle reduces the lower order harmonics as shown in figure (15).

As the waveform is symmetrical during every quarter cycle, a_n=0.

$$b_n = \frac{4}{\pi} V_S \left[\int_0^{\alpha 1} \sin n\omega t.d\omega t - \int_{\alpha 1}^{\alpha 2} \sin n\omega t.d\omega t + \int_{\alpha 2}^{\frac{\pi}{2}} \sin n\omega t.d\omega t \right] \ldots\ldots(27)$$

Figure 17. The combined output of two or more inverters

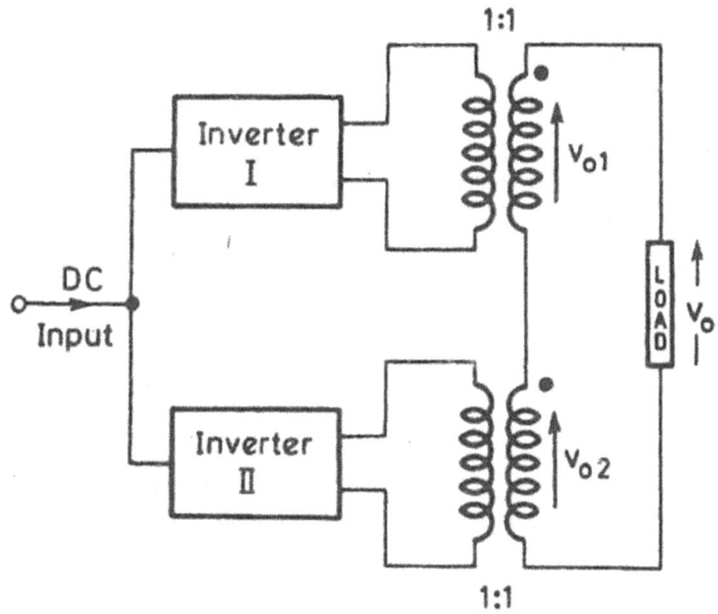

Figure 18. Harmonic reduction by stepped wave inverters

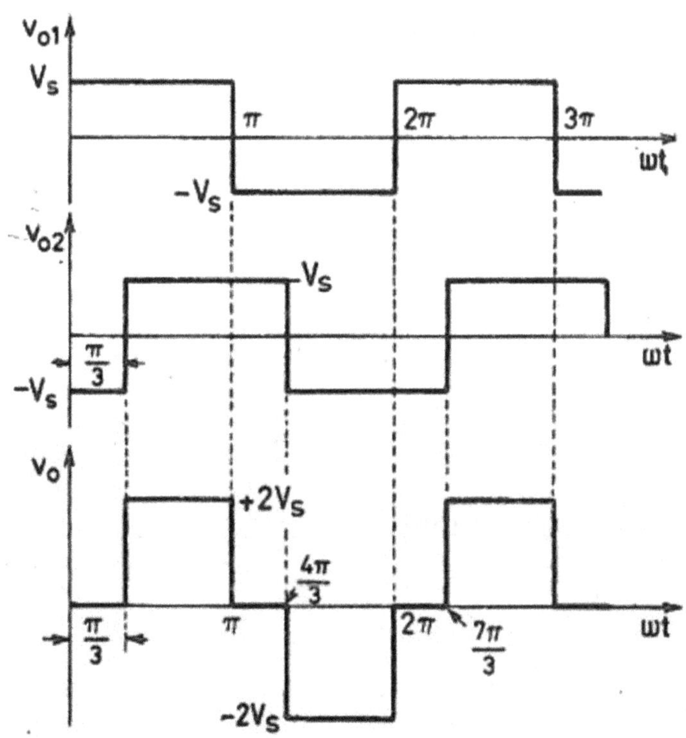

$$b_n = \frac{4V_S}{\pi}\left[\frac{1-2\cos(n\alpha_1)+2\cos(n\alpha_2)}{n}\right]\text{.........}(28)$$

From equations 59 & 60 and If 3rd and 5th harmonics are to be eliminated we can conclude that $\alpha_1 = 23.62°, \alpha_2 = 33.304°$

Harmonic Reduction by Transformer Connections:

The output of two or more inverters are combined using transformers to get a net voltage with reduced harmonic content. The voltage waveform should be similar but phase shifted from each other as in figure 16

Harmonic Reduction by Stepped Wave Inverters:

• In this method, pulses of different widths and heights are super imposed to get a resultant stepped wave with reduced harmonic content.

UPS systems:

An uninterruptible power supply or uninterruptible power source (UPS) is an electrical apparatus that provides emergency power to a load when the input power source or main power source fails. A UPS differs from an auxiliary, emergency power system or standby generator in that, it will provide near-instantaneous protection from input power interruptions and other disturbances by supplying energy stored in batteries, supercapacitors, or flywheels (Khan et al., 2019; Kifle et al. 2018; Khan et al., 2018; Khan et al., 2017; Banteywalu et al., 2019; Anteneh et al., 2019; Molla et al., 2019, Molla et al., 2018, Jariso et al., 2018).

It is known for us, that power converters and semiconductor devices produce distortion and disturbances or Electro Magnetic Interferences (EMI), this leads to the injection of the current high harmonics into the supply system, a UPS device can reduce the effects of these harmonics, when the UPS is connected as an interface between the source and the electronic converter (loads).

Disturbances types:

In practical applications the voltage behavior is not similar to the ideal behavior due to the disturbances occurring in the power line, the followings are types of these disturbances:

- over voltage.
- Voltage drop.
- Black outs.
- Spike pulses.
- Chopped waves.
- High harmonics.
- Electro Magnetic Interferers (EMI).

The influences of the disturbances on the delicate equipment:

That depends on:

1. The type of the disturbance and its value in the power line.
2. The equipment's quality and design.
3. The power regulators used.

Figure 19. The tolerance in the computer system for a power line disturbance

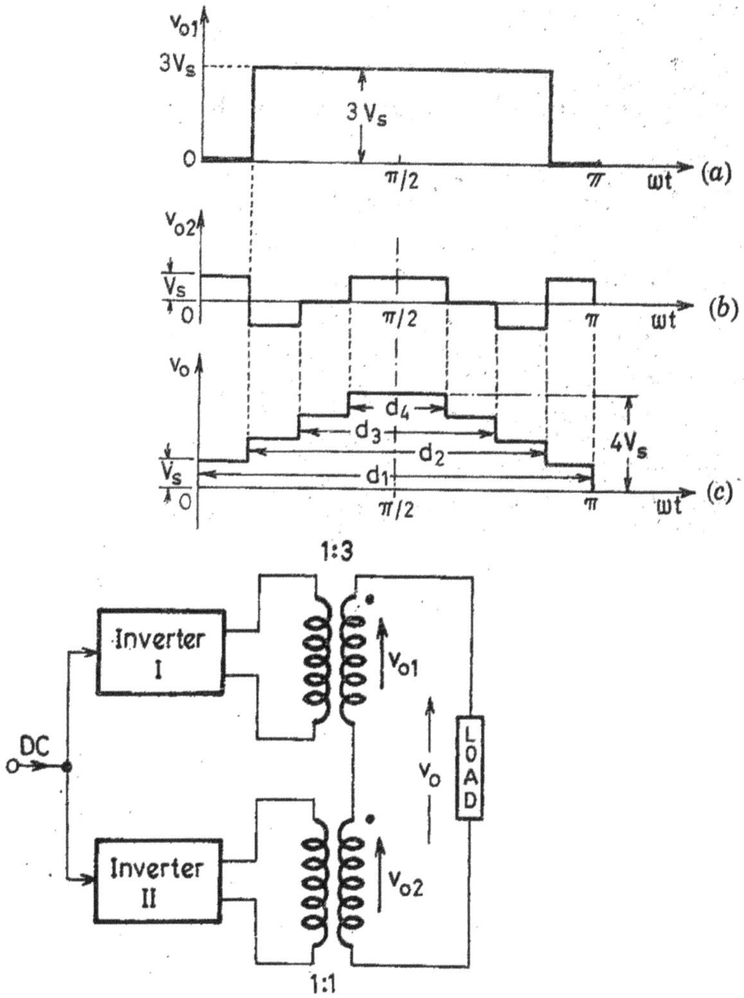

Table 5. Typical power quality identified by major computer manufacturers

Measured value		Maximum limit
Voltage controlling		+5, -10 to +10%, -15%
Voltage disturbances	Minimum voltage limit	-25 to -30% for less than 0.5 sec, with -100% acceptable for 4-20 ms
	High transient voltage	+150 to 200% for less than 0.2 ms
Voltage harmonics distortions		3-5 % with linear load
noise		Unsteady
Frequency variety		60 Hz ±0.5 Hz to ±1 Hz
Frequency variety rate		1 Hz/s (slow rate)

Figure 20. Continuous power supply system

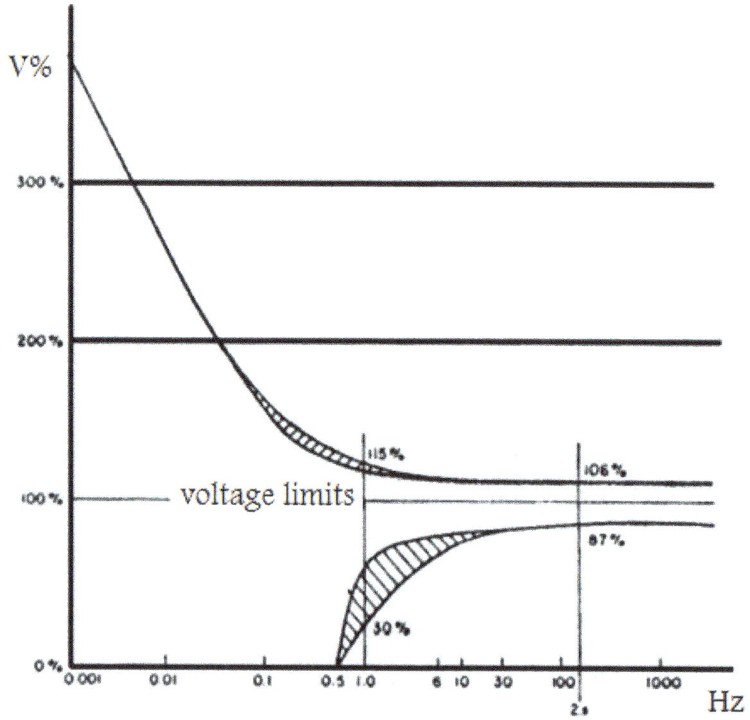

The effect of interruptions depends on the power system, duration of interruption, and the design of the device. For example, the power supply module can be designed in a PC so that when a break is less than 100ms, the output of the source continues to provide power to the digital circuitry group with the same nominal value, and no significant impact on the computer's operation is felt. For interruptions longer than 100ms, a logical signal inside the computer allows the CPU to have an additional 50ms period to back up the existing information. After this period, all power supply output voltage decreases rapidly. Figure(19) shows the tolerance in the computer system for a power line disturbance. Backup is initiated and a shutdown occurs for some time. Table (5) shows typical power quality identified by major computer manufacturers. In the case of critical applications, operation is not acceptable; therefore, the supply reserve source is provided by UPSs.

It is necessary to use UPS to feed critical loads, such as computers used to control important production processes, medical devices, etc.

This system provides protection against power failure as well as regulating voltage in high voltage situations (above the allowed value) and low voltage (under the allowed value). This system is also effective in reducing harmonic and transient disorders in the power line.

Figure (20) shows a diagram of the continuous power supply system (uninterrupted-UPS). The rectifier is used to convert the single (or three phase AC current) into DC, which provides the capacitor with power, and also provides battery rows with the capacity to keep it charged. In the normal operation mode, the capacitor is supplied with power by the rectifier. In case of power failure, the power comes from the battery compartment. The inverter produces a single-phase or three-phase sinusoidal waveform depending on the UPS, and the output voltage is filtered before it is applied to the load.

Figure 21. Box diagram of UPSs.

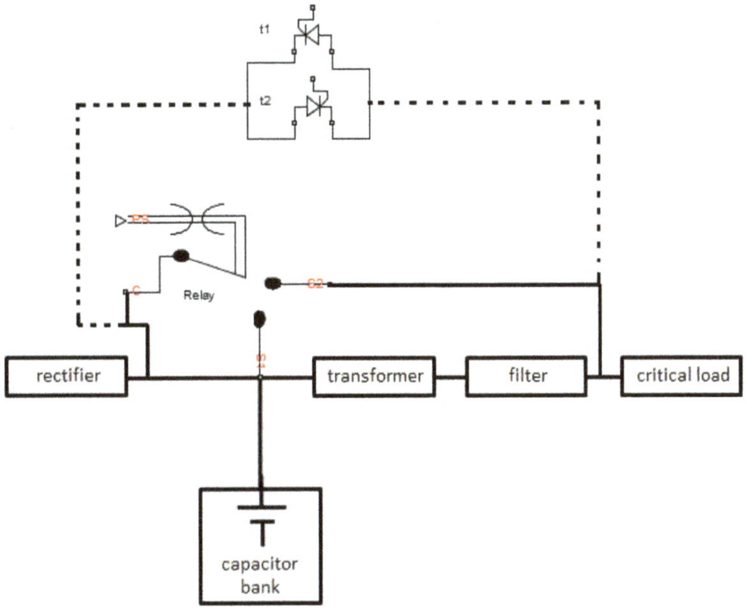

Figure 22. Parallel connection of loads with the UPSs.

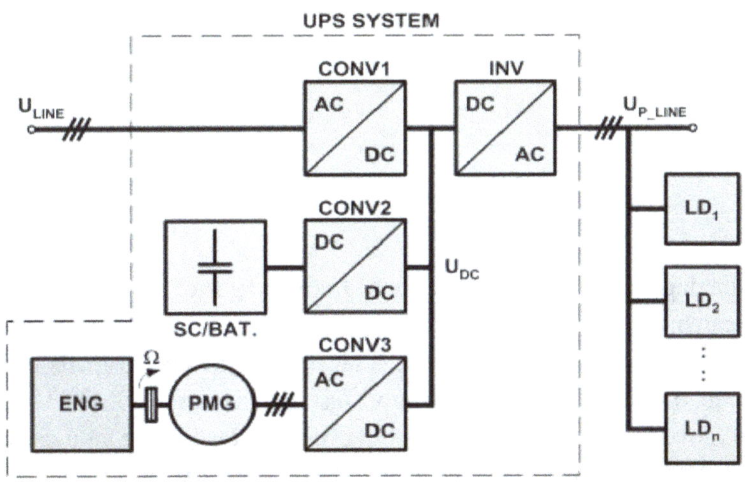

The rectifiers: are electromagnetic or electric circuit that contains semiconductor electronic components (switches) that convert AC voltage into DC voltage

Speed control systems are a modern technology that can be applied to UPS based on electronic power adapters. The highly efficient, high-speed power generation systems ensure long life for the diesel engine. Another modern technique is to store energy in superconductors. An important advantage of super capacitors is the ability to provide high current in charging and discharging conditions. Moreover, deep

Figure 23. Rectifiers used in UPS Systems

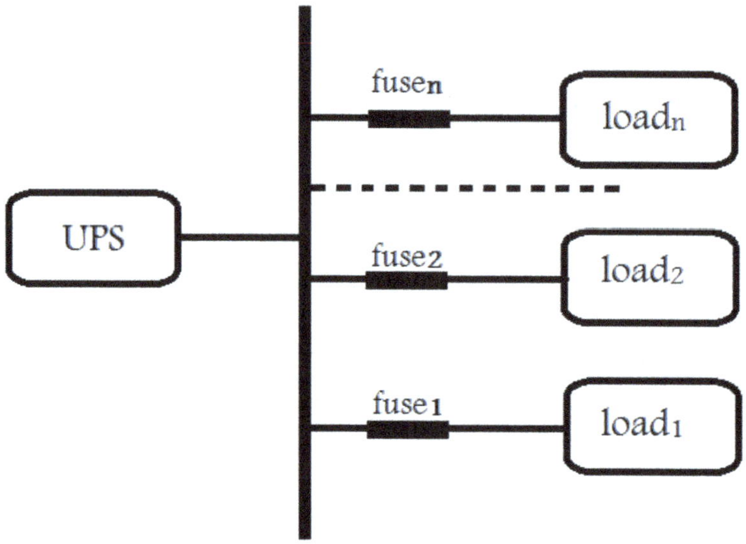

Figure 24. Rectifier with the Battery charging circuit

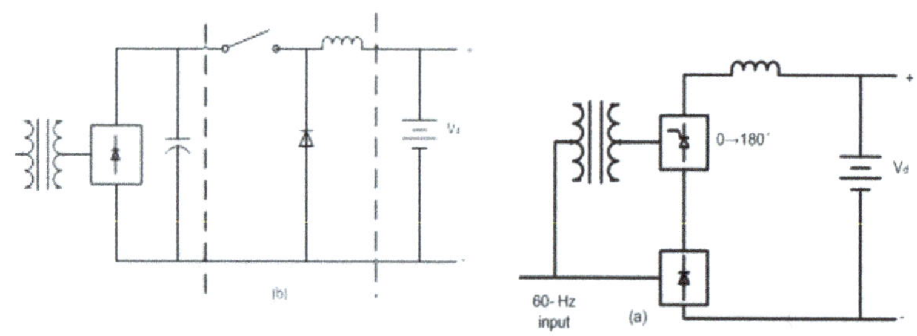

discharge does not damage capacitors. The number of advantages of applying super capacitors is still under study and research.

Most UPS systems above a few kilowatts feed a number of connected loads in parallel. As shown in Figure (22), each load is fed through a protection fuse. In the case of a short circuit in a load, it is necessary for the UPS to disconnect the corresponding fuse (located in the fault circuit) and to continue feeding the rest of the loads; The nominal current of UPS in the short circuit state must be sufficient to separate the fuse corresponding to the failing load. In this context, the UPS system with a rotary type which has great potential to withstand short circuit current far outweighs the type of UPS with electronic power circuits.

rectifiers used in UPS Systems:

There are two types that can be used in this case like shown in figure (23):

However the rectifier with the Battery charging circuit is shown in figure (24):

There are many types of battery systems, including conventional lead-acid batteries used in UPS applications.

Figure 25. Battery charging.

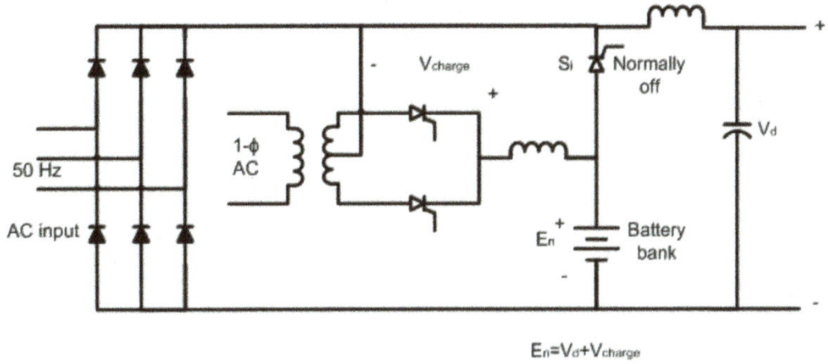

Figure 26. Battery charger used in UPS

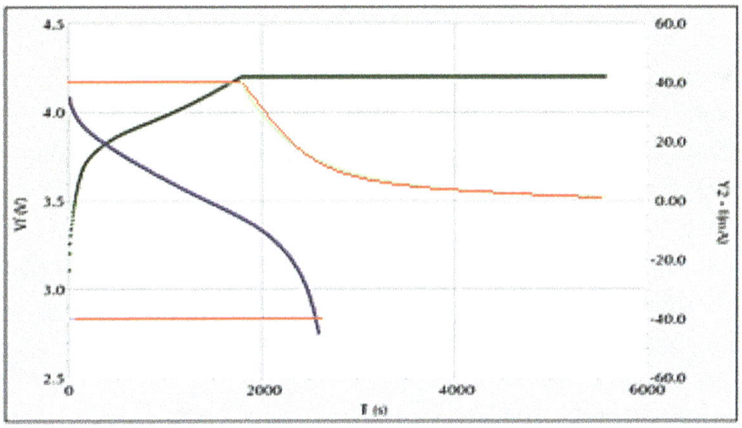

Figure 27. Voltage regulators

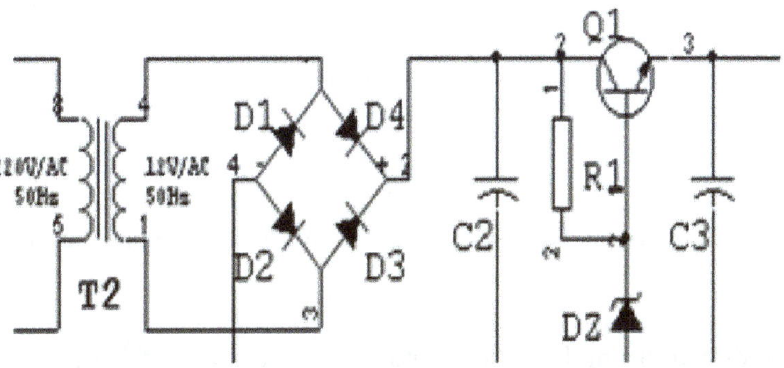

Figure 28.The oscillator

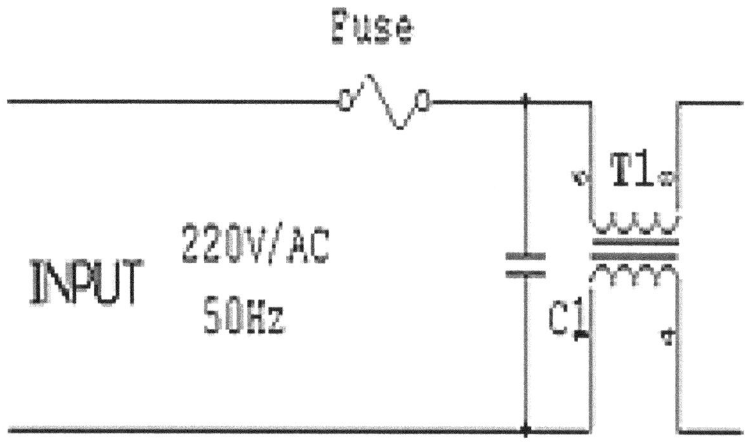

In the normal operating system, when the line tension is present, the battery is charged slightly to compensate for the battery self-discharge. This requires the application of a weak and constant charging voltage through the battery. The battery is constantly discharged by a small amount of current and thus maintains its full charging position.

If the line is disconnected, the battery feeds the load, and the battery capacity is expressed in ampere-hours(Ah). It is the output of the static discharge current in the time period after which the battery voltage drops below a certain voltage level called the final discharge voltage.

Hybrid UPS based on supercapacitors storage:

The battery charger used in UPS:

The circuit shown in figure (26) works on charging the 12 V battery, a 220 V(AC) voltage source is applied on the rectifier that consists of four diodes, so a 12 V output is got.

Overall study case:

As shown in figure (32) the whole circuit of the UPS System is configured of:

Voltage regulators:

AC voltage regulators are a type of converters that transform the fixed AC voltage into variable or controlled.

Usually these regulators are used to control the speed of induction motors and it is equal to an auto-transformer, as shown in figure (27) .

The battery charging circuit:

As shown in figure (26), the battery charging circuit is configured of four diodes to form the bridge rectifiers, and the zener diode is used to voltage regulation and isolating the supply when the battery is fully charged.

The oscillator:

This circuit is fed directly from the battery, and it produces 50 Hz frequency, suitable for the grid frequency (required), and it is shown in figure (28)

The current step up circuit:

The circuit shown in figure (29) is configured from connecting the transistors with the same base, the same emitter, and the same collector.

The main advantage of this circuit is stepping up the current to high levels.

Figure 29. The current step up circuit

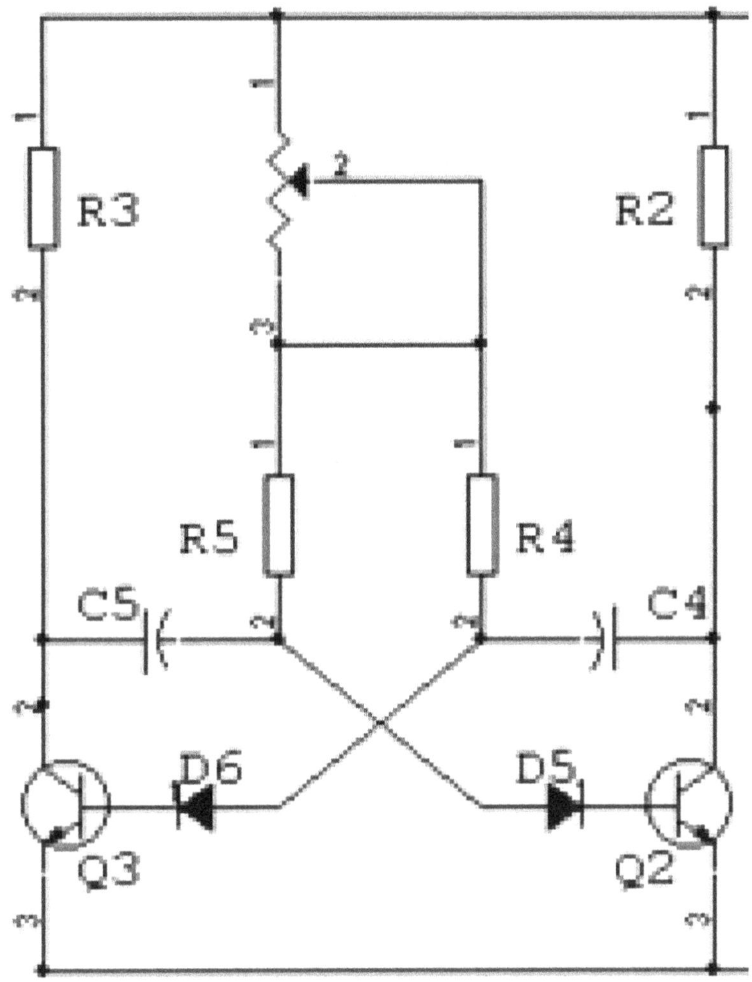

The transformer:

The circuit shown in figure (30) transforms the 12 VAC into 220 VAC, then passes it to the filter to get a similar voltage to the grid to supply the loads.

Relay:

Figure (31) shows the Relay circuit which is responsible for switching the connections in case of an electrical turn off between two positions (power supply or UPS). Recently there are modern switches used instead of relay such as Triac. (Makdisie, C. (2008)).

And the overall circuit diagram of the UPS systems:

Algorithm of the control strategy of the laboratory hybrid UPS system:

The main feature of the Hybrid UPS systems is that they are capable of generating electricity from Solar, Batteries, Grid or Emergency Generator in a controlled manner. (Solar Electric Generating Stations (SEGS). (1989)).

Figure 30. The transformer

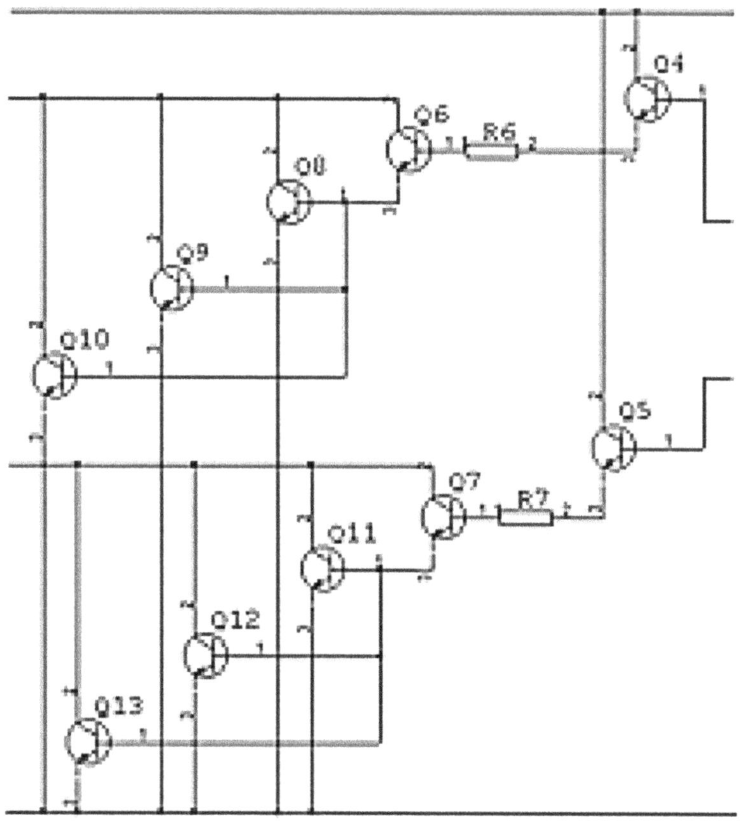

Figure 31. Relay

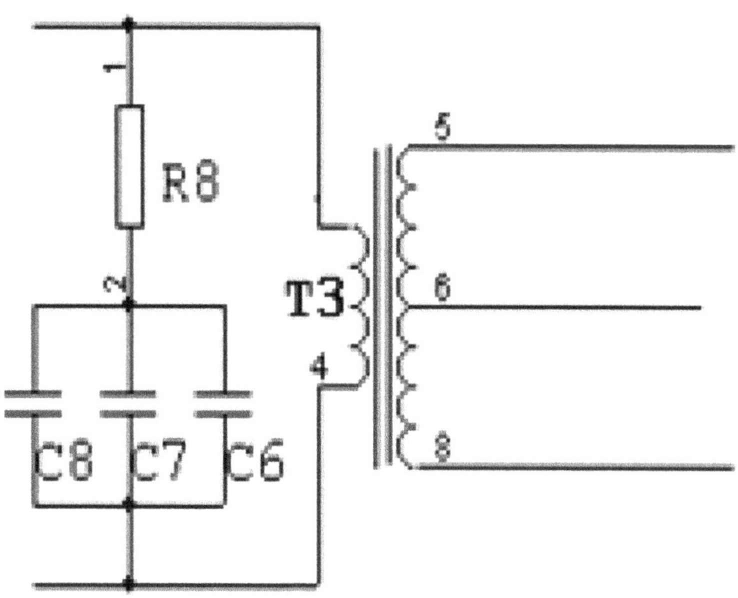

Figure 32. Circuit diagram of the UPS systems

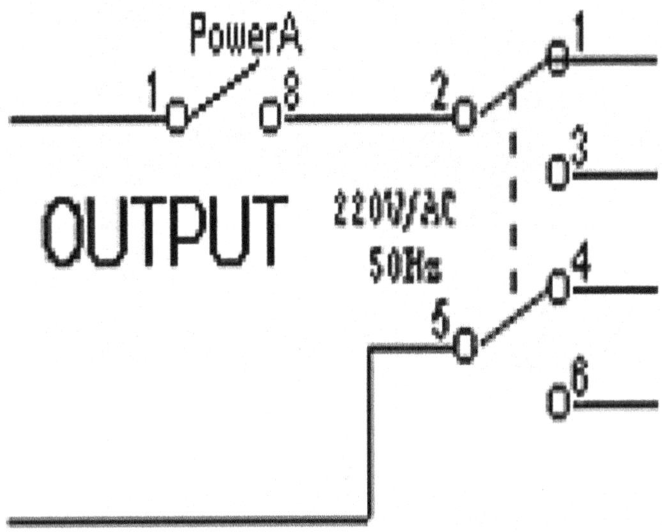

Figure 33. Algorithm of the control strategy of the laboratory hybrid UPS system

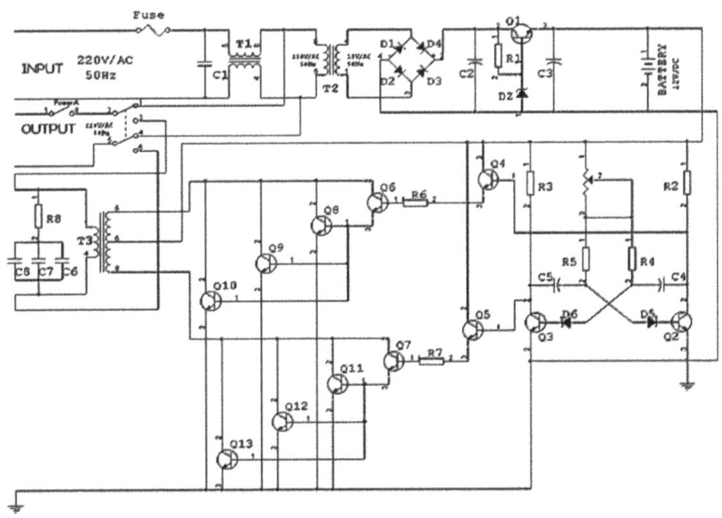

Figure 34. The reality of Hybrid UPSs

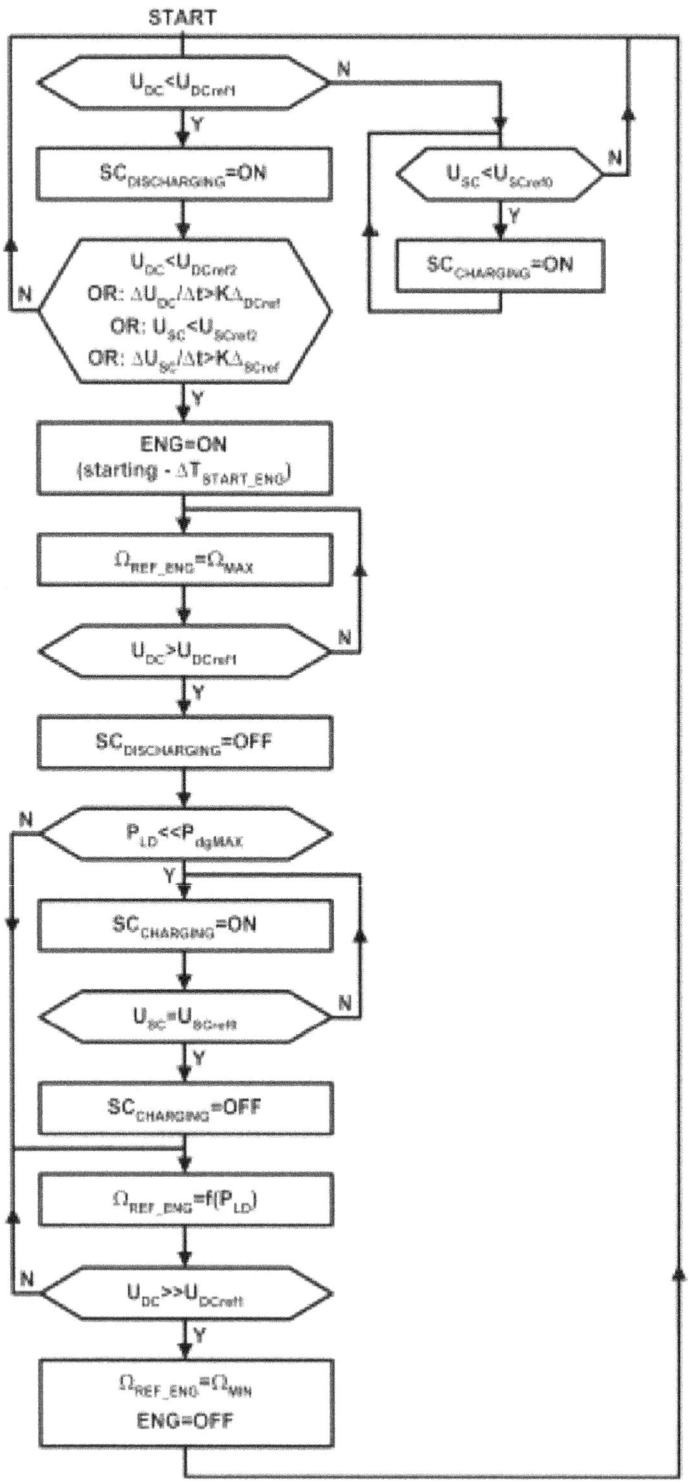

Figure 35. Grid unavailability

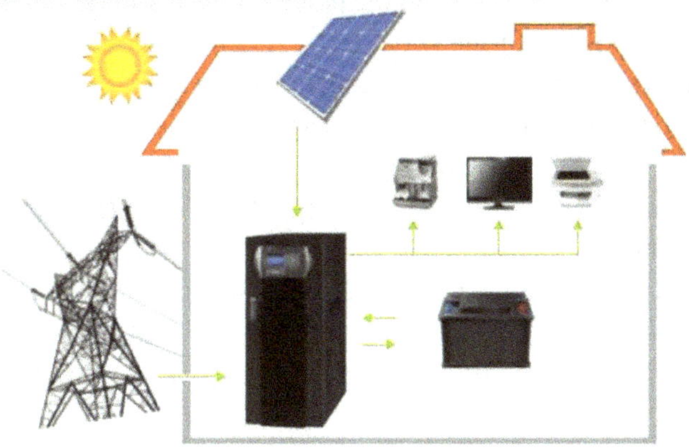

Figure 36. Unavailability of the solar or battery group.

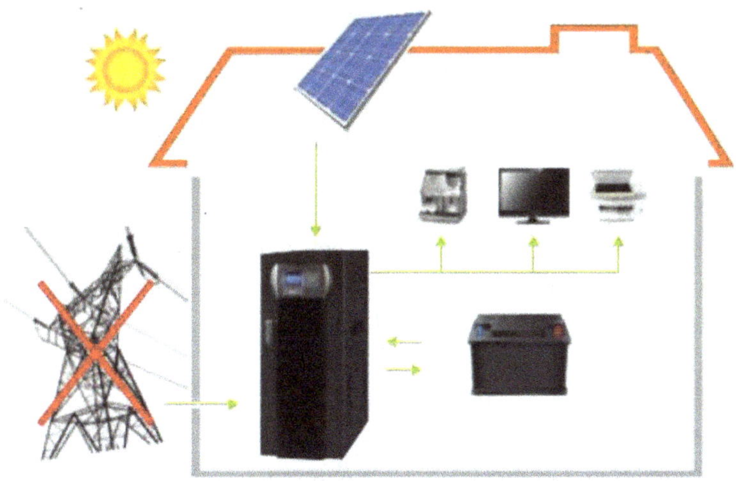

1. Uninterruptible power by solar energy, grid and battery
2. Return of investment
3. MPPT Algorithm
4. Solar energy storage
5. Intelligent controller
6. 100% stabilized output power
7. Emergency generator

The reality of Hybrid UPSs is shown in figure (34).

However, there might be some problems such as grid unavailability:

In case of power failure, the requested energy is coming from the solar panels and/or batteries. The backup time vary with the connected load and the power of panels/batteries. The backup time vary with the connected load and the power of panels/ batteries. As in figure (35):

Or there might be unavailability of the solar or battery group:

Hybrid UPS system automatically starts the emergency generator when the solar energy, batteries and grid are unavailable, as shown in figure (36).

Study case of calculation method of Solar panels, batteries and transformers to convert a house to fully operated with solar energy

Solar Energy Calculation – by applying the following steps:

Calculate the amount of energy consumed per day, by applying the following equations:

Number × Machine capacity × Working hours = The energy consumed by the device per day.

For example:

4 lumps × 10 Watt × 6 hours = 240Watt per hour

1 TV × 80 Watt × 6 hours = 480 Watt per hour

1 PC × 80 Watt × 4 hours = 320 Watt per hour

1 Receiver × 20 Watt × 6 hours = 120 Watt per hour

240+480+320+120=1160 watt per day, that means 1.16 kWatt per hour

Calculation of losses during the installation: 30%

Of course, there is a loss during the installation of any electrical system, and losses may reach up to 30% due to the conductivity and quality of wires and the resistance of batteries used as well as the efficiency of solar panels, and therefore should be added to this loss of total energy consumed in the day by applying the following equation:

Total power required = total energy consumed per day x 1.3

To apply it on the example:

The energy to be generated = 1.3 × 1160 = 1508 Wh

Calculation of solar panels number:

For solar energy, the energy to be generated should be divided by the solar radiation rate per day for the area where panels will be installed, between 4 and 6.3, one of the highest in the world.

Number of panels = energy needed of the panels / plate energy that we want to buy,

For example, if we want to buy panels with a capacity of 100 watts each:

The number of solar panels = 377/100 = 3.77 → equals approximately 4 panels 100 watts. Or two 200-watt panels.

Calculation of the batteries number:

Battery capacity (Ah) = (Power to be generated × Number of cloudy days (where panel charges will be lost) 1.3 x (30% of battery capacity required to maintain) /Voltage

For the example the system voltage is 12 V = 1508×2 (assuming the cloudy days are 2) ×1.3 / 12 = 3920.8/12 = 326.76 Ah.

However if the system voltage is 24 = 1508×2 (assuming the cloudy days are 2) ×1.3 / 24 = 3920.8/24 = 163.36 Ah.

For the number of batteries we apply the following relationship:

Number of batteries = Battery capacity / Battery size to be purchased.

So the number of batteries in our example is 80/163.36=2.042 and is roughly equal to 80 amps or two batteries of 100 amps per hour.

Solar regulator size: Calculated as follows:

Number of Planned Panels Installed in the System × Top charging amp for the panel.

In our example

The size of the solar regulator = 2 x 8.5 = about 17 A. The voltage is 24 volts.

We prefer to double the size to take reserve in the future if we want to expand the system to work longer or to add other devices.

Calculate the size of the inverter (from battery to 220 volt)

The inverter size depends on the total power of devices peak time

In our example: 4 Lumps × 10W + 1 TV x 80 W + 1PC x 80 W + 1 Receiver x 20 W = 220 watts.

30% should be taken as an efficiency factor for the performance of the switches and it varies according to the manufacturing company and its efficiency:

Inverter size = 220 x 1.3 = 286 watts.

Note: You must consider the power of the inverter to be able to give a high start if used to run a refrigerator, an old TV, water compressor..

So we need

Two 200 Watt Panels

Two 100 amp Battery

20 Amp Solar Regulator-

connecting wires

300 W or higher inverter.

To summarize:

There are many types of disturbances associated with power lines, and regulators provide an effective way to protect sensitive electronic loads from these disturbances except power cuts or frequency deviations.

UPS systems are used for the very critical loads, so that the power to load feed is uninterrupted even in the case of power failure. The storage capacity of the battery is measured depending on the occurrence of a specific period of interruption.

CONCLUSIONS AND FUTURE WORKS

In this chapter, the elements of power electronics, inverters and UPS were studied, some electrical power systems using these elements were modeled, and the design of a solar power system using UPS systems was studied. In the end there are many points that may require deeper study, based on the modeling results in the chapter.

REFERENCES

Alhelou, H., Hamedani-Golshan, M. E., Zamani, R., Heydarian-Forushani, E., & Siano, P. (2018). Challenges and opportunities of load frequency control in conventional, modern and future smart power systems: A comprehensive review. *Energies*, *11*(10), 2497. doi:10.3390/en11102497

Alhelou, H. H. (2018). *Fault detection and isolation in power systems using unknown input observer. In Advanced condition monitoring and fault diagnosis of electric machines* (p. 38). Hershey, PA: IGI Global.

Alhelou, H. H., Golshan, M., & Fini, M. (2018). Wind driven optimization algorithm application to load frequency control in interconnected power systems considering GRC and GDB nonlinearities. *Electric Power Components and Systems, 46*(11-12), 1223-1238.

Alhelou, H. H., & Golshan, M. E. H. (2016, May). Hierarchical plug-in EV control based on primary frequency response in interconnected smart grid. In *2016 24th Iranian Conference on Electrical Engineering (ICEE),* (pp. 561-566). IEEE. 10.1109/IranianCEE.2016.7585585

Alhelou, H. H., & Golshan, M. E. H. (2020). Decision-making-based optimal generation-side secondary-reserve scheduling and optimal LFC in deregulated interconnected power system. In *Decision making applications in modern power systems* (pp. 269–299). Academic Press. doi:10.1016/B978-0-12-816445-7.00011-6

Alhelou, H. H., Golshan, M. E. H., & Hatziargyriou, N. D. (2019). Deterministic dynamic state estimation-based optimal LFC for interconnected power systems using unknown input observer. *IEEE Transactions on Smart Grid*, 1. doi:10.1109/TSG.2019.2940199

Alhelou, H. H., Golshan, M. E. H., & Hatziargyriou, N. D. (2019). A decentralized functional observer based optimal lfc considering unknown inputs, uncertainties and cyber-attacks. *IEEE Transactions on Power Systems.*

Alhelou, H. H., Golshan, M. E. H., Zamani, R., Moghaddam, M. P., Njenda, T. C., Siano, P., & Marzband, M. (2019, June). An improved UFLS scheme based on estimated minimum frequency and power deficit. In 2019 IEEE Milan PowerTech (pp. 1-6). IEEE. doi:10.1109/PTC.2019.8810497

Alhelou, H. H., Golshan, M. H., & Askari-Marnani, J. (2018). Robust sensor fault detection and isolation scheme for interconnected smart power systems in presence of RER and EVs using unknown input observer. *International Journal of Electrical Power & Energy Systems*, *99*, 682–694. doi:10.1016/j.ijepes.2018.02.013

Alhelou, H. H., Hamedani-Golshan, M. E., Heydarian-Forushani, E., Al-Sumaiti, A. S., & Siano, P. (2018, September). Decentralized fractional order control scheme for LFC of deregulated nonlinear power systems in presence of EVs and RER. In *2018 International Conference on Smart Energy Systems and Technologies (SEST)* (pp. 1-6). IEEE. 10.1109/SEST.2018.8495858

Alhelou, H. S. H., Golshan, M. E. H., & Fini, M. H. (2015, December). Multi agent electric vehicle control based primary frequency support for future smart micro-grid. In *2015 Smart Grid Conference (SGC)* (pp. 22-27).

Anteneh, D., & Khan, B. (2019). Reliability enhancement of distribution substation by using network reconfiguration a case study at Debre Berhan Distribution Substation. *International Journal of Economy, Energy, & Environment*, *4*(2), 33.

Chlodnicki, Z., & Koczara, W. (n.d.). Supercapacitor *(2005).* Storage application for reduction drive negative impact on supply grid. *IEEE Compatibility in Power Electronics.* doi:10.1109/cpe.2005.1547549

Chlodnicki, Z., Koczara, W., & Al-Khayat, N. (2007). Hybrid UPS based on supercapacitor energy storage and adjustable speed generator. In *Proceedings 2007.* Gdansk, Poland: Compatibility in Power Electronics; doi:10.1109/cpe.2007.4296512

Eep. (n.d.). *Uninterruptible static power supplies and the protection of persons.* Retrieved from https://electrical-engineering-portal.com/download-center/books-and-guides/schneider-electric/uninterruptible-static-power

Fini, M. H., Yousefi, G. R., & Alhelou, H. H. (2016). Comparative study on the performance of many-objective and single-objective optimisation algorithms in tuning load frequency controllers of multi-area power systems. *IET Generation, Transmission, & Distribution*, *10*(12), 2915–2923. doi:10.1049/iet-gtd.2015.1334

Haes Alhelou, H., Hamedani-Golshan, M. E., Njenda, T. C., & Siano, P. (2019). A survey on power system blackout and cascading events: Research motivations and challenges. *Energies*, *12*(4), 682. doi:10.3390/en12040682

Haes Alhelou, H., Hamedani-Golshan, M. E., Njenda, T. C., & Siano, P. (2019). Wide-area measurement system-based optimal multi-stage under-frequency load-shedding in interconnected smart power systems using evolutionary computing techniques. *Applied Sciences*, *9*(3), 508. doi:10.3390/app9030508

Haes Alhelou, H., Hamedani Golshan, M. E., Njenda, T. C., & Siano, P. (2019). WAMS-based online disturbance estimation in interconnected power systems using disturbance observer. *Applied Sciences*, *9*(5), 990. doi:10.3390/app9050990

Jariso, M., Khan, B., Tanwar, S., Tyagi, S., & Rishiwal, V. (2018). Hybrid energy system for upgrading the rural environment, IEEE GlobeCom 2018.

Jariso, M., Khan, B., Tesfaye, D., & Singh, J. (2017). Modelling and designing of stand-alone photovoltaic system (Case study: Addis Boder Health Center southwest Ethiopia). In *Proceedings IEEE International Conference on Electronics, Communication, and Aerospace Technology, ICECA 2017,* pp. 168-173, Coimbatore, India.

Khan, B., & Agnihotri, G. (2012). A novel transmission loss allocation method based on transmission usage. In *Proceedings IEEE Fifth Power India Conference*, Haryana, India. 10.1109/PowerI.2012.6479479

Khan, B., Agnihotri, G., & Gupta, G. (2013). A multipurpose matrices methodology for transmission usage, loss, and reliability margin allocation in restructured environment, *Electrical & Computer Engineering: An International Journal, 2*(3).

Khan, B., Agnihotri, G., Rathore, P., Mishra, A., & Naidu, G. (2014). A cooperative game theory approach for usage and reliability margin cost allocation under contingent restructured market. *International Review of Electrical Engineering*, *9*(4), 854–862.

Khan, B., Alhelou, H. H., & Mebrahtu, F. (2019). A holistic analysis of distribution system reliability assessment methods with conventional and renewable energy sources. *AIMS Energy*, *7*(4), 413–429. doi:10.3934/energy.2019.4.413

Khan, B., & Singh, P. (2017). The current and future states of Ethiopia's energy sector and potential for green energy: A comprehensive study. *International Journal of Engineering Research in Africa*, *33*, 115–139. doi:10.4028/www.scientific.net/JERA.33.115

Khan, B., & Singh, P. (2017) Optimal power flow techniques under characterization of conventional and renewable energy sources: a comprehensive analysis, *Journal of Engineering, vol. 2017,* Article ID 9539506.

Khan, B., & Singh, P. (2017). Selecting a meta-heuristic technique for smart micro-grid optimization problem: a comprehensive analysis. In IEEE Access, 5, pp. 13951-13977.

Khan, B., & Singh, P. (2019). Economic operation of smart micro-grid: a meta-heuristic approach. In Handbook of research on smart power system operation and control (pp. 330-346). Hershey, PA: IGI Global. doi:10.4018/978-1-5225-8030-0.ch014

Khan, B., & Singh, P. (2019). Economic operation of smart micro-grid: A meta-heuristic approach. In Handbook of research on smart power system operation and control (pp. 330-346). Hershey, PA: IGI Global. doi:10.4018/978-1-5225-8030-0.ch014

Khan, B., & Tanwar, S. (2019). Issues associated with microgrid integration. In Handbook of research on smart power system operation and control (pp. 252-264). Hershey, PA: IGI Global. doi:10.4018/978-1-5225-8030-0.ch010

Kifle, Y., Khan, B., & Singh, J. (2018). Designing and modeling grid connected photovoltaic system: (Case study: EEU building at Hawassa city). *International Journal of Convergence Computing*, *3*(1), 20–34. doi:10.1504/IJCONVC.2018.091113

Kifle, Y., Khan, B., & Singh, P. (2018). Assessment and enhancement of distribution system reliability by renewable energy sources and energy storage, *Journal of Green Engineering, 8*(3), 2, pp. 219-262.

Makdisie, C. (2008). *Power electronics / 2*. Tishreen University Publications.

Makdisie, C. (2016). *Power electronics / 1*. Tishreen University Publications.

Makdisie, C., Haidar, B., & Alhelou, H. H. (2018). An optimal photovoltaic conversion system for future smart grids. In Handbook of research on power and energy system optimization (pp. 601–657). Hershey, PA: IGI Global. doi:10.4018/978-1-5225-3935-3.ch018

Molla, T., Khan, B., Moges, B., Alhelou, H. H., Zamani, R., & Siano, P. (2019). Integrated optimization of smart home appliances with cost-effective energy management system. *CSEE Journal of Power and Energy Systems*, *5*(2), 249–258.

Molla, T., Khan, B., & Singh, P. (2018). A comprehensive analysis of smart home energy management system optimization techniques, *Journal of Autonomous Intelligence, 1*(1), 15–21.

Negash, K., Khan, B., Tesfaye, D., & Gayathri, M. (2017). Optimal placement of phasor measurement unit for system observability. In *Proceedings of IEEE International Conference on Computing Methodologies and Communication, (ICCMC 2017),* pp. 73-77, Erode, India.

Negash, K., Khan, B., & Yohannes, E. (2016). Artificial intelligence versus conventional mathematical techniques: A review for optimal placement of phasor measurement units [Springer]. *Journal of Technology and Economics of Smart Grids and Sustainable Energy, 1*(1), 10. doi:10.100740866-016-0009-y

Njenda, T. C., Golshan, M. E. H., & Alhelou, H. H. (2018, November). WAMS based intelligent under frequency load shedding considering online disturbance estimation. In *2018 Smart Grid Conference (SGC)* (pp. 1-5). 10.1109/SGC.2018.8777779

Njenda, T. C., Golshan, M. E. H., & Alhelou, H. H. (2018, November). WAMS based under frequency load shedding considering minimum frequency prediction and extrapolated disturbance magnitude. In *2018 Smart Grid Conference (SGC)* (pp. 1-5).

Rashid, M. H. (Ed.). (2011). *Power electronics handbook: Devices, circuits, and applications.* Burlington, MA: Elsevier.

Singh, P., & Khan, B. (2017). Smart microgrid energy management using a novel artificial shark optimization. *Complexity, 2017,* 2158926.

Singh, P., Khan, B., Vidyarthi, A., Haes Alhelou, H., & Siano, P. (2019). Energy-aware online non-clairvoyant scheduling using speed scaling with arbitrary power function. *Applied Sciences, 9*(7), 1467. doi:10.3390/app9071467

Solar Electric Generating Stations (SEGS). (1989). Solar electric generating stations (SEGS). *IEEE Power Engineering Review, 9*(8), 4–8. doi:10.1109/MPER.1989.4310850

Vodorozov, V. (2010). Introduction to power electronics (1st ed.). London, UK: London Business School.

Yeshalem, M. T., & Khan, B. (2017). Design of an off-grid hybrid PV/wind power system for remote mobile base station: A case study. *AIMS Energy, 5*(1), 96–112. doi:10.3934/energy.2017.1.96

Zamani, R., Golshan, M. E. H., Alhelou, H. H., & Hatziargyriou, N. (2019). A novel hybrid islanding detection method using dynamic characteristics of synchronous generator and signal processing technique. *Electric Power Systems Research, 175,* 105911. doi:10.1016/j.epsr.2019.105911

Zamani, R., Hamedani-Golshan, M. E., Haes Alhelou, H., Siano, P., & Pota, H. R. (2018). Islanding detection of synchronous distributed generator based on the active and reactive power control loops. *Energies, 11*(10), 2819. doi:10.3390/en11102819

Chapter 18
Applied Power Electronics:
Rectifiers, Choppers, Regulators

Carlo Joseph Makdisie
Tishreen University, Syria

Marah Fadl Mariam
Tishreen University, Syria

ABSTRACT

Most of the electric machines had a conventional design for speed –control. Previously, the speed regulation of these motors was done via traditional or mechanical contacts, for example: inserting resistors to the armature circuit or controlling the excited circuit of DC motor, and other methods of control. These classical methods, however, lead to non-linearity in mechanical or electromechanical characteristics [$\omega = f(M)$ or $\omega = f(I)$], which in turn lead to increased power losses as the result of the non-soft regulation of speed, as well as the great inertia of classical control methods that rely on mechanical and electromagnetic devices.

INTRODUCTION

(Thorborg, K. 1988) explained the main idea for power electronics, and the importance of its applications. However (Makdisie, C. 2016) classified some of the power electronics types and the variety of its internal construction.

(Fredsted, J. 1998) specified the converters and its ideal and practical functions.

The main idea in (Histand, Michael B. 1999) was to determine the measurement methods of power electronic, and from (Mohan, N., Undeland, T. M., & Robbins, W. P. 2007 And Holtz, J., & Boelkens, U. 1989) the fundamental construction of the basic power electronic elements was understood.

Some important information about the drives suitable for power electronics. To go back and study more about converters from (Vodorozov, V. 2010). and more about the converters operation and applications in (Makdisie, C. 2008). To summarize the final classifications and most recent applications of power electronics in general in (Salam Z. 2003).

DOI: 10.4018/978-1-7998-1230-2.ch018

BACKGROUND

In this chapter, the recent classification of rectifiers, converters, and regulators are explained. Empowering this chapter with MATLAB / Simulink simulation, and thorough representation of each circuit's eave forms, and the equations representing each one.

The electronics revolution began when the silicon transistor was invented, and then the invention of the Thyristor in 1956, which was developed and produced commercially. Since then, the power electronics have been deployed in many industrial applications and the importance of their development has increased with the continuous progress in computer science, microelectronics and advanced control systems of production lines in order to deal with high capacities of transferred power, control it and increase the efficiency . Nowadays, power electronic converters are used in every factory, facility, or industrial device, as well as domestic apparatus, such as vacuum cleaners, blenders, mixers, blending devices.

The speed equation for a DC motor is given by the following basic relationship:

$$w = \frac{V}{K.\phi} = \frac{I.\sum R}{K.\phi} \dots\dots\dots(1)$$

It is clear from the previous equation, that the speed is varied or regulated by either changing the magnetic flux (¦ - the I_f current), the applied voltage V_a or changing the resistance of the armature circuit R_a.

The important question which is highlighted: Why the speed control is required? What is the goal?

If the speed exceeds the required limit, this will lead to distortion of the productive parts (ie, poor productivity), so they do not meet the required specifications. If the speed is less than the required value, this will lead to decreasing the efficiency of the actuator machine, which effects directly on productivity, cost and other factors. Therefore the use of power electronics converters is necessary to control the speed of the driven machine and set it at a suitable value for the accuracy of the production process.

There is also another purpose to use the power electronics converter; as an interface between the power supply and the load, and provide small size- weight efficiency, supply quality, control requirements.

Depending on the function of power electronics converters, they are divided into several main types:

1.Rectifiers: and it is divided into two types:

a. Uncontrolled Rectifiers.
b. Controlled Rectifiers.

These circuits of both types convert the AC voltage into DC value, either constant /fixed (Uncontrolled) or variable (controlled).

The rectifiers: It is an electromagnetic or electric circuit that contains semiconductor electronic components (switches) that convert AC voltage into DC voltage.

If without a regulation of the output voltage by the semiconductor switches (Diodes), then it is called the uncontrolled rectifier circuit .

Figure 1. Rectifiers

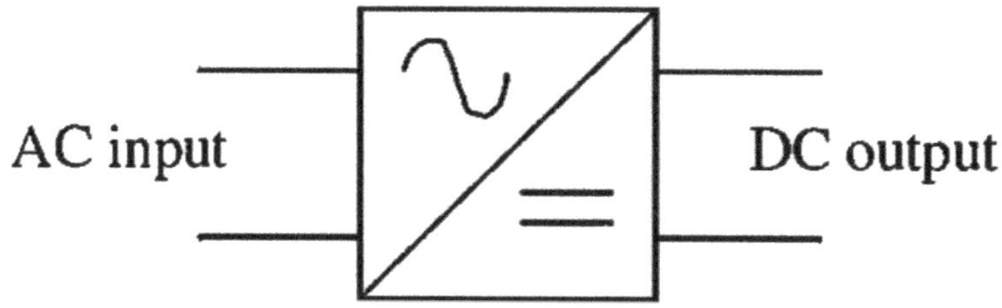

When the voltage is switched and regulated (controlled) by the semi-conductor electronic components, then it is called the controlled rectifier circuit. We can generally represent the circuit according to the following Figure:

Classification of rectifier circuits:

First, by the number of phases, mono-phase or multiphase (3-phase).

Second, according to the output waveform, half wave rectifier or full wave rectifier.

Note: We will study the rectifier circuits assuming that the ideal switch has resistance equal to zero in the forward direction (forward biased- the voltage drop across the switch is almost zero or too small) when the switch is closed (on state), while its resistance in the reversed direction is infinite when it is opened (off state) (Vodorozov, V. (2010)).

The Controlled Rectifiers are characterized by simplicity, high efficiency and low cost, and are therefore widely used in controlling industrial applications requiring variable speeds. The controlled rectifiers as mentioned above into four types:

1. Half Wave Controlled Rectifier.
2. Full Wave Semi Controlled Rectifier.
3. Full Wave Controlled Rectifier.
4. Dual Converter.

In this section we will focus only on the controlled rectifiers, considering uncontrolled rectifiers are operated at firing angle $(\alpha = 0)$.

Rectifier with pure resistive load (R-load)

This circuit is not practical, but it helps to understand the principle of the rectifier operation. In positive half wave of the input voltage the switch is forward biased, then when apply the firing pulse to the gate of Thyristor (switch) it will conduct allowing the current to pass through the load so we can get the voltage across the terminals of the load $\left(0 \leq \omega t \leq \pi\right)$.

At moment $\left(\omega t = \pi\right)$ the Thyristor will be switched off and the voltage equal to zero.

Now in the period of negative half wave the Thyristor will not conduct (off state), because this switch is reversed biased (switch is opened).

The reversed biased will continue up to $\left(\omega t = \pi\right)$. And after this moment the switch reconducts because it is forward biased again.

The currents and voltage output wave forms of 1-ph half wave rectifier is shown below in figure(3)

We can find the average value of the DC output voltage V_o by the following known equation:

Figure 2. Rectifier with pure resistive load

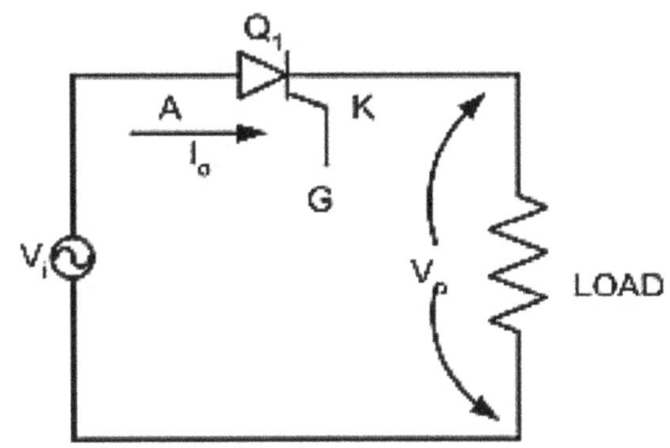

Figure 3. Rectifier with pure resistive load waveforms

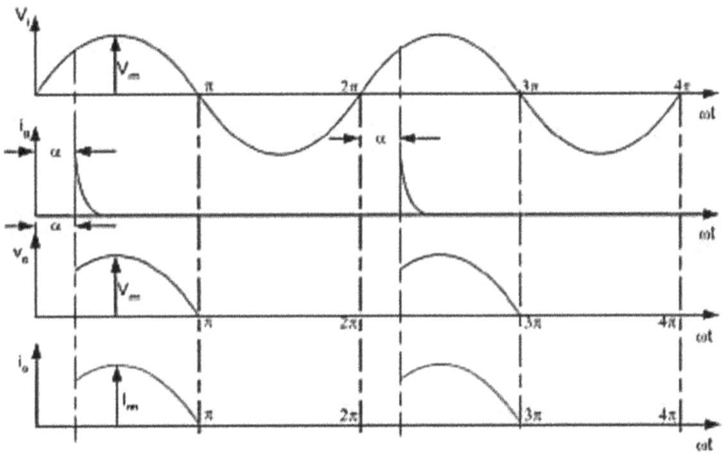

$$V_o = \frac{1}{2\pi}\int_{\alpha}^{\pi} V_m.\sin(\omega t)\,d(\omega t) = \frac{V_m}{2\pi}.(1+\cos\alpha)\ldots\ldots\ldots(2)$$

It is clear from the previous equation (1) that the voltage is varied by varying angle α .

Considering that the load current is not continues, so this circuit is not used in industrial application and its use is limited only for small power circuits.

Rectifier with (R - L) Load:

Most of the industrial loads contains an induction part of the load, so we must also study the work of the rectifiers with R, L –Load.

The circuit above shows 1- ph half wave rectifier with a resistance and inductance (R+L).

When the Thyristor is triggered (turned on), the current of the load doesn't change instantly like it did with the pure resistance load, and that is the result of the induction which fights this change.

Figure 4. Rectifier with (R - L) Load

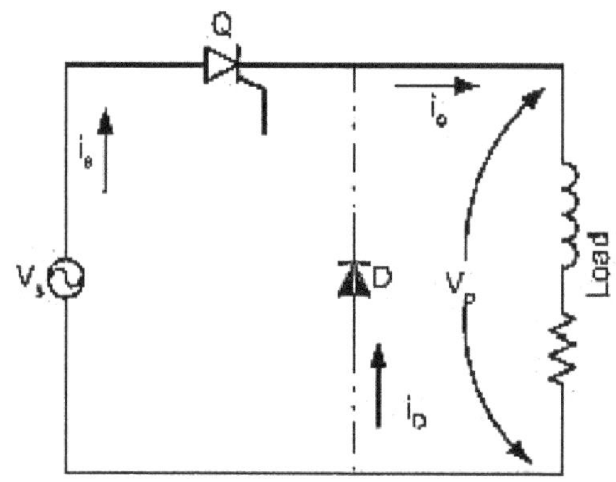

Figure 5. Rectifier with (R - L) Load waveforms

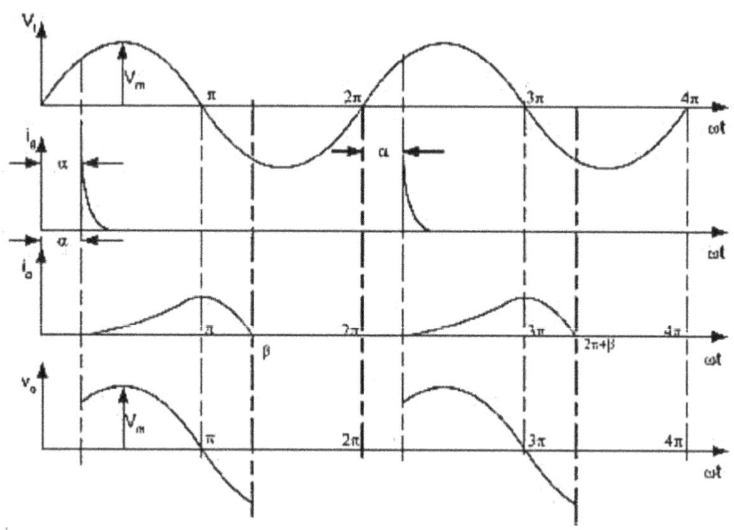

The moment the source voltage drops to zero, and the negative wave starts; the Thyristor current is higher than zero, and so the Thyristor does not switch off and as a result there will be a part of the negative wave shown in the Figure. To get rid of this occasion we can use a freewheel diode, which helps with reducing the disconnection in the load current wave.

We can find the average value of the DC output voltage V_o without the free wheel diode by the following known equation:

$$V_o = \frac{1}{2\pi} \cdot \int_{\alpha}^{\beta} V_m \sin(\omega t) \, d(\omega t) = \frac{V_m}{2\pi} \cdot (\cos\alpha - \cos\beta) \ldots\ldots\ldots (3)$$

Figure 6. Single Phase Full wave- Semi controlled Rectifier

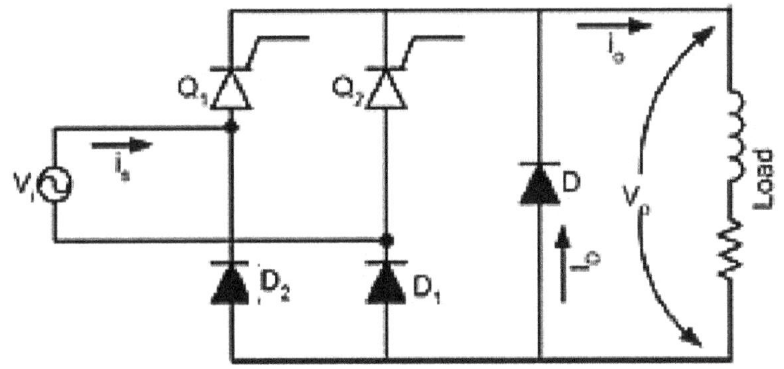

The β is the switching off angle of the Thyristor, however when we want to calculate it with the existing of the free wheel diode we can use equation (1).

Single Phase Full wave- Semi controlled Rectifier:

It contains four electronic elements, two Thyristors and two diodes, in addition to the free wheel diode.

In the first positive half wave the Thyristor Q1 is switched on at α angle and the current passes it to the load through the diode d1 and continuo until the voltage becomes zero and that happens when $\omega t = \pi$, during the period $(\pi + \alpha > \omega t > \pi)$ the voltage is negative, and the free wheel diode is forwarded so it becomes a shot circuit with the load and the load current passes through it.

However during the period $(2\pi > \omega t > \pi + \alpha)$, the Thyristor Q2 is switched on at an angle $(\pi + \alpha)$, and the current passes to the load through the diode D2, until the point the voltage reaches zero at $\omega t = 2\pi$, then the load current goes again through the free wheel diode and the Thyristor is switched off, and the cycle is repeated Figure (6)

We can find the average value of the DC output voltage V_o by the following known equation:

$$V_o = \frac{1}{\pi}.\int_{\alpha}^{\pi} V_m.\sin(\omega t)\, d(\omega t) = \frac{V_m}{\pi}.(1+\cos\alpha).........(4)$$

As shown in equation (3), the voltage can be controlled by the angle α, and it is obvious that the output voltage must always be positive as well as the load current, so this kind of converters cannot be used in applications that needs to reverse the voltage nor the current.

Single Phase Full Wave Controlled Rectifier

The Single Phase Full Wave Controlled Rectifier is formed by four Thyristors as in figure (8), and here we will focus on the high induction load. (Mohan, N., Undeland, T. M., & Robbins, W. P. (2007)).

Each of the Thyristors Q1 and Q2 is switched on in the first positive half wave, and they are triggered (switched on) at αangle, due to the induction load (because the induction load stores the power in the form of magnetic field in the first quarter and gives it back in the next quarter). As a result the current keeps passing even after ωt is higher than π, even though the source voltage is negative, and it continues until the Thyristors Q3 and Q4 are triggered at an angle of $2\pi + \alpha$. And again the cycle is repeated

Figure 7. Single phase full wave- semi controlled rectifier waveforms

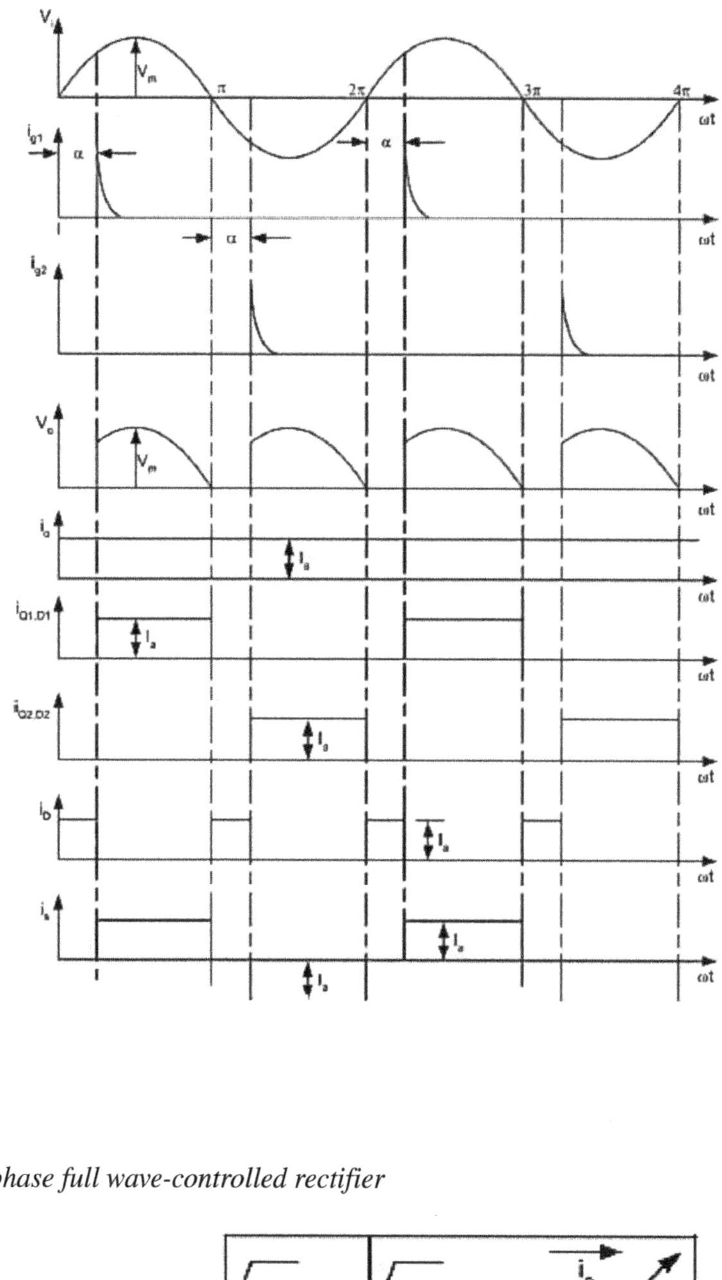

Figure 8. Single phase full wave-controlled rectifier

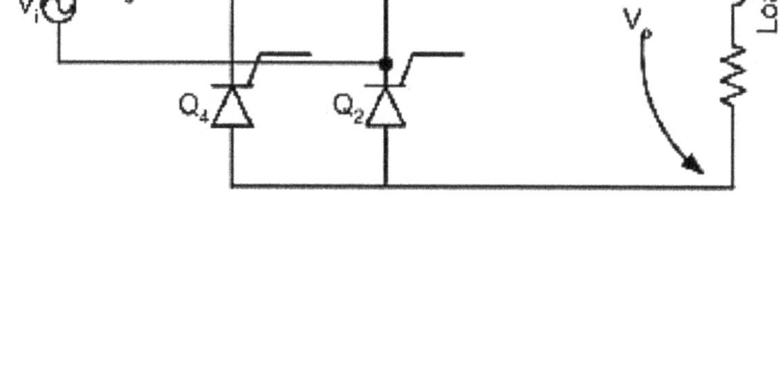

as in figure (9) (Khan et al., 2019; Kifle et al. 2018; Khan et al., 2018; Khan et al., 2017; Anteneh et al., 2019; Molla et al., 2019, Molla et al., 2018, Jariso et al. 2018)

We can find the average value of the DC output voltage V_o by the following known equation:

$$V_o = \frac{1}{\pi} . \int_{\alpha}^{\pi+\alpha} V_m . \sin(\omega t) d(\omega t) = \frac{2V_m}{\pi} . \cos\alpha \dots\dots\dots (5)$$

As we can observe, the output voltage can be controlled through the angle α, and it can be negative or positive depending on the trigger angle, if the angle is less than 90⁰ the voltage is positive, otherwise it is negative.

However, the load current is always positive, so these kinds of converters are used in applications that might need to reverse the voltage only.

3-Phase Half Wave Controlled Rectifier

We call these type of converters (Central Point 3-Ph Converters), it can be used in industrial applications due to the high output power comparing to the single phase converters, as the frequency of the output voltage wave is usually high, so the filtering of the load current is simpler.

Now this converter is configured from three Thyristors placed between the source and the load as in figure (10), in this situation the Thyristor Q1 is triggered when the angle $\omega t = \frac{\pi}{6} + \alpha$ and so there will be a voltage V_{an} applied on the load. And it continues until Thyristor Q2 is triggered when $\omega t = \frac{5\pi}{6} + \alpha$, with a difference of $\frac{2\pi}{3}$ from the last switch, and so on.

And so we can observe that the number of frequencies in the output wave is less than what it is in the single wave converters, and we can calculate the average output voltage by integration of the waveform in the period from $\frac{\pi}{6} + \alpha$ to $\frac{5\pi}{6} + \alpha$ through the following equation:

$$V_o = \frac{3}{2\pi} . \int_{\frac{\pi}{6}+\alpha}^{\frac{5\pi}{6}+\alpha} V_m . \sin(\omega t) d(\omega t) = \frac{3\sqrt{3}V_m}{2\pi} \cos\alpha \dots\dots\dots (6)$$

3- Phase Full Wave - Semi Controlled Rectifier

This converter is configured of three Thyristors and three diodes placed as a bridge circuit in addition of a free wheel diode as we can see in figure (12), and these kind of converters are used in applications which need positive voltage and current, and it can reach $120\,kW$, and we can observe that the Power Factor decreases with increasing the trigger angle, and the average voltage is calculated by the equation:

Figure 9. Single phase full wave-controlled rectifier waveforms

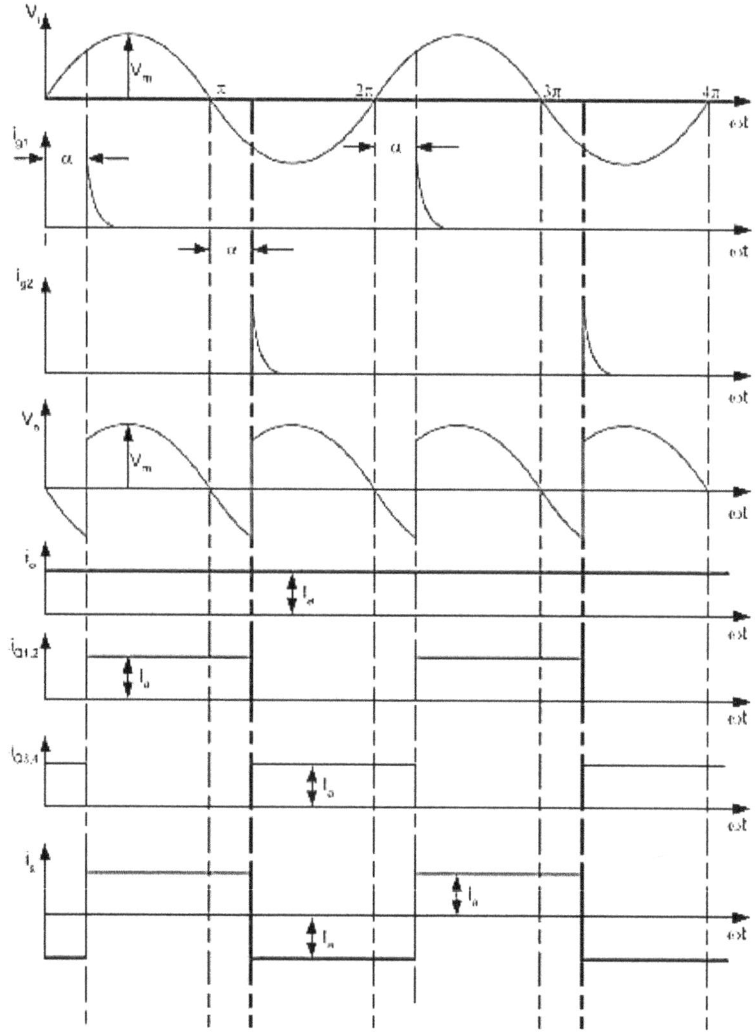

$$V_o = \frac{3\sqrt{3}V_m}{2\pi}(1+\cos\alpha).........(7)$$

3- Phase Full Wave – Full Controlled Rectifier

Now from the phrase (full wave) we can realize that this converter is configured from six Thyristors forming a bridge circuit as in figure (13), it can be used in applications that needs positive or negative voltage according to the trigger angle, and the average voltage is calculated by the following equation:

$$V_o = \frac{3\sqrt{3}V_m}{\pi}(\cos\alpha).........(8)$$

Figure 10. 3-Phase half wave-controlled rectifier

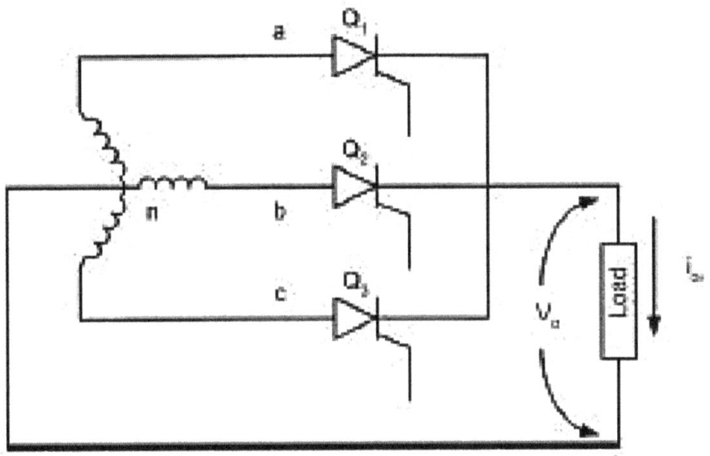

Figure 11. 3-Phase half wave-controlled rectifier waveforms

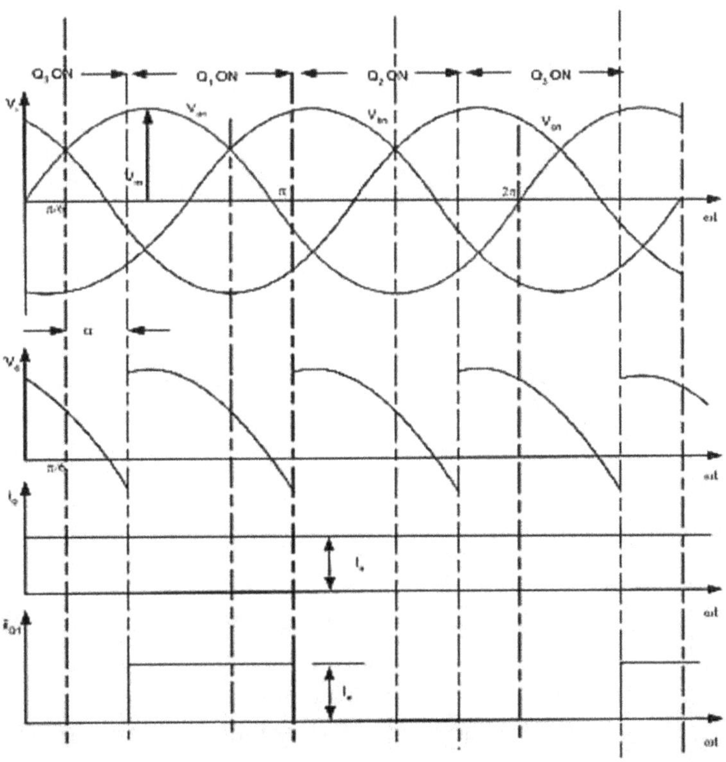

Figure 12. 3- Phase full wave - semi controlled rectifier

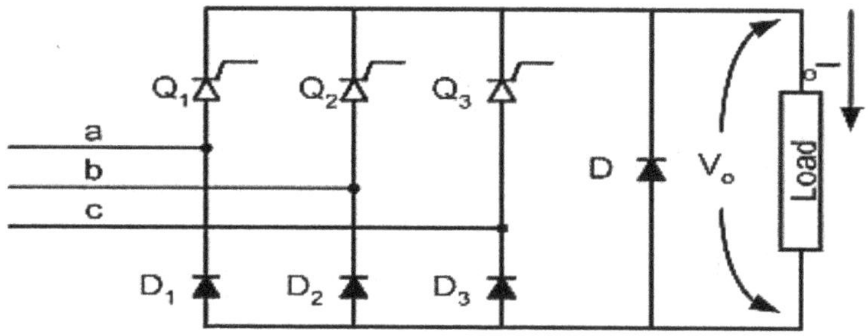

Figure 13. 3- Phase Full wave – full controlled rectifier

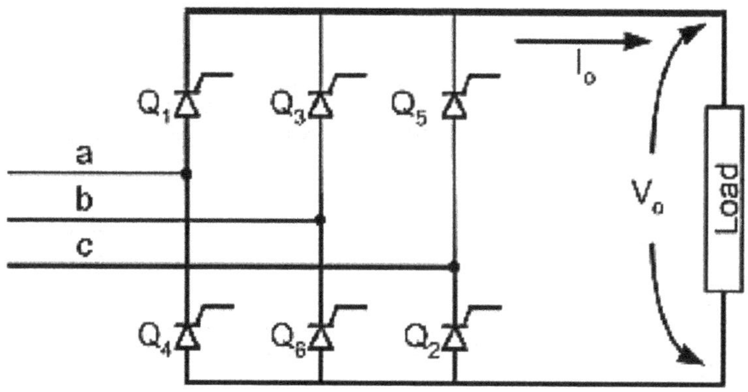

3-phase Full Bridge Controlled Rectifier

In figure (14) shown two sets of Thyristors, as if it is two centered point rectifiers, the first one is configured of three Thyristors (V_{c1}, V_{c3}, V_{c5}) as we can call it the cathode group, and the second one of (V_{c2}, V_{c4}, V_{c6}) (the anode group), the main frequency of the output DC voltage u_d is six times higher than the input voltage that is due to the existing of six Thyristors in the circuit(six pulses per cycle in the output waveform), in this case two Thyristors at least are operated at the same time, one of the anode group the second of the cathode group, which leads to connect the load with two phases of the transformer at any time (Fini et al., 2016; Alhelou et al., 2018; Zamani et al., 2018; Alhelou et al., 2015; Njenda et al., 2018; Haes Alhelou et al., 2018; Haes Alhelou et al., 2019).

The operation principle of this type of rectifiers is that the current of the main electrical network comes from the cathode group then passes through the transformer then goes back through the anode group.

Figure 14. 3-phase full bridge-controlled rectifier

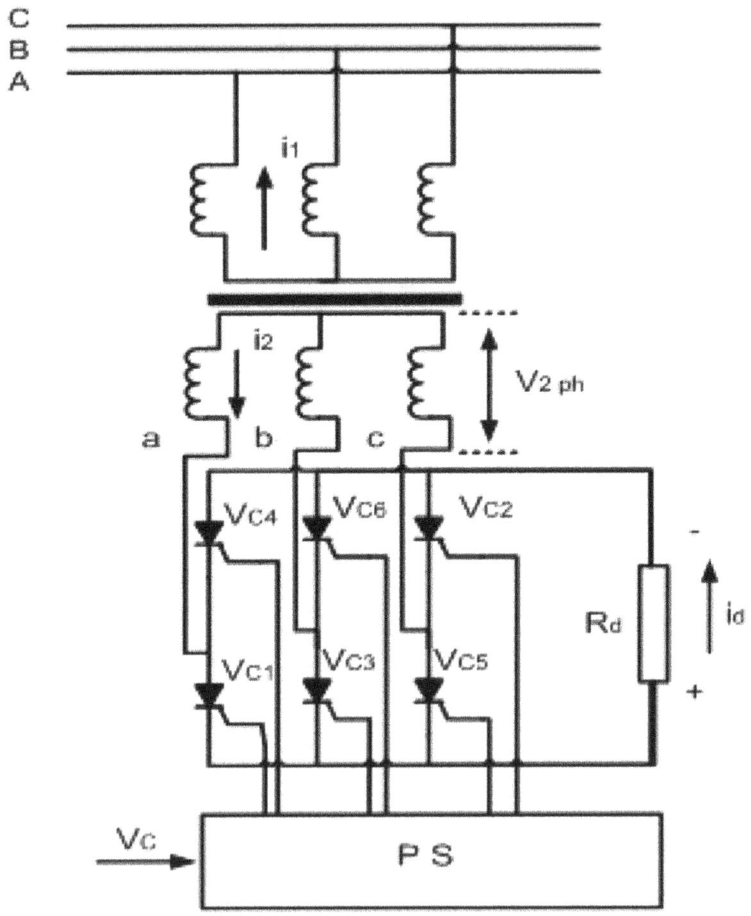

3-phase Full Bridge Controlled Rectifier with R-L Load

The analyzing of this bridge rectifier is built on the assumption that the current is filtered ideally, and the load current is continued through the controlling period, and the main average DC voltage value is calculated with the following equation:

$$U_{d\alpha} = U_{do} \cos\alpha \dots\dots\dots (9)$$

The reversed maximum voltage is:

$$U_{Rmax} = \sqrt{2}.\sqrt{3}.U_2 \dots\dots\dots (10)$$

The forward maximum voltage on the Thyristor is:

Figure 15. 3-phase full bridge-controlled rectifier with R-l load

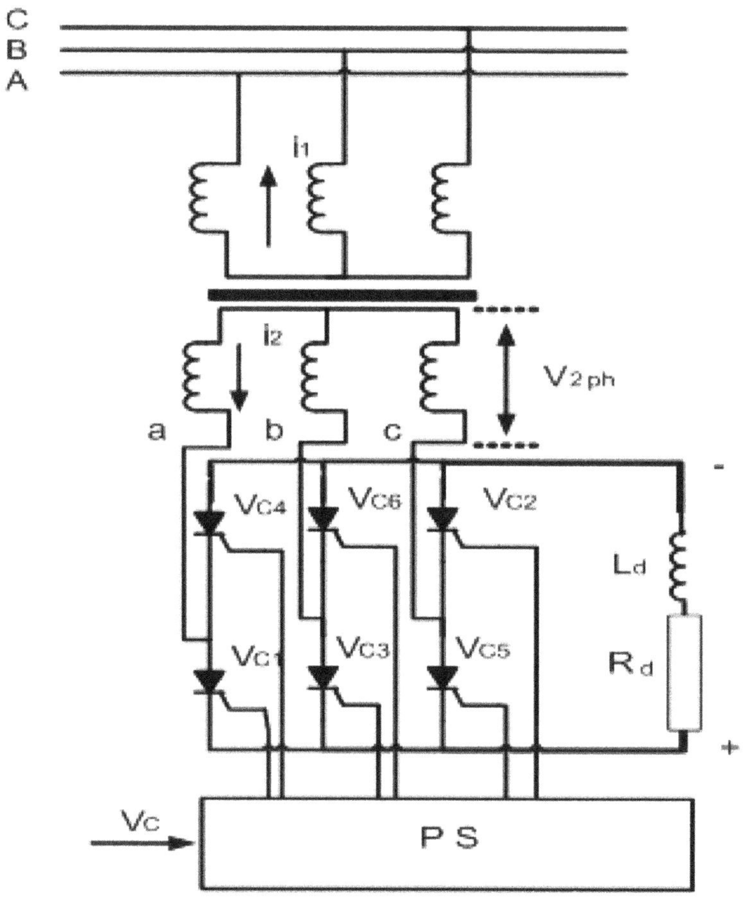

$$U_{F\,max} = \sqrt{2}.\sqrt{3}.U_2.\sin\alpha \,.........(11)$$

and here is a study of the various examples of the waveforms:
Resistive load:

$\alpha = 0°$

Resistive load:

$\alpha = 30°$

Resistive load: $\alpha = 60°$
Resistive load:

$\alpha = 90°$

Figure 16. Waveforms when $\alpha = 0°$

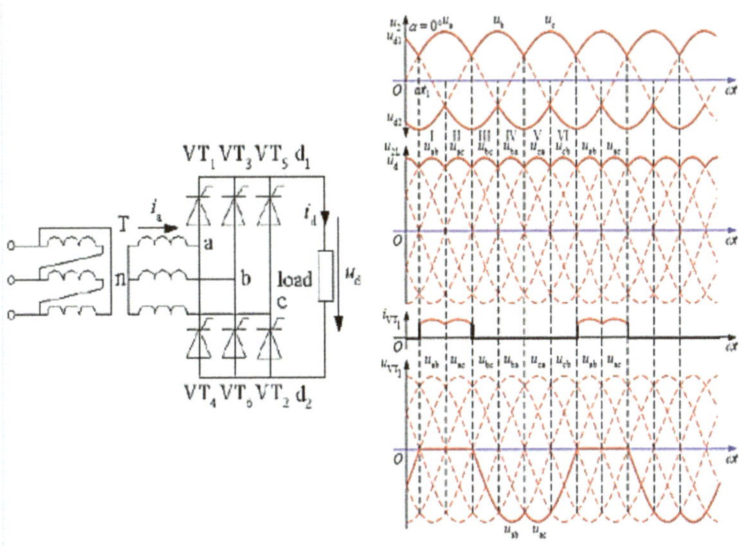

Figure 17. Waveforms when $\alpha = 30°$

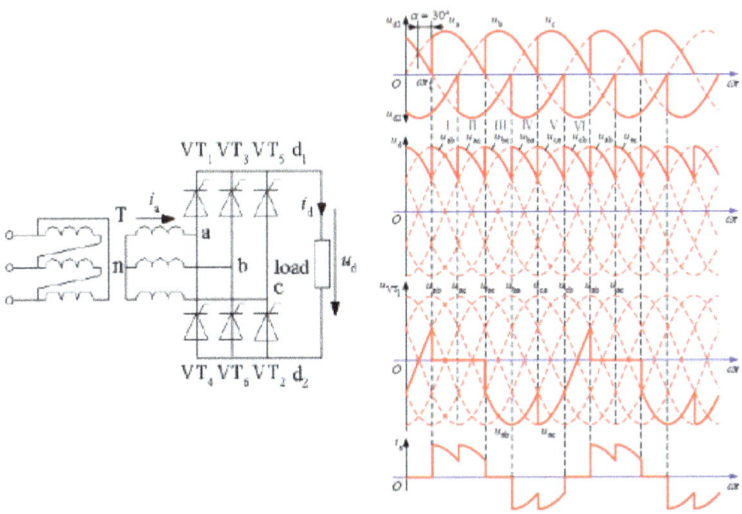

Inductive load:

$\alpha = 0°$

Inductive load:

Figure 18. Waveforms when $\alpha = 60°$

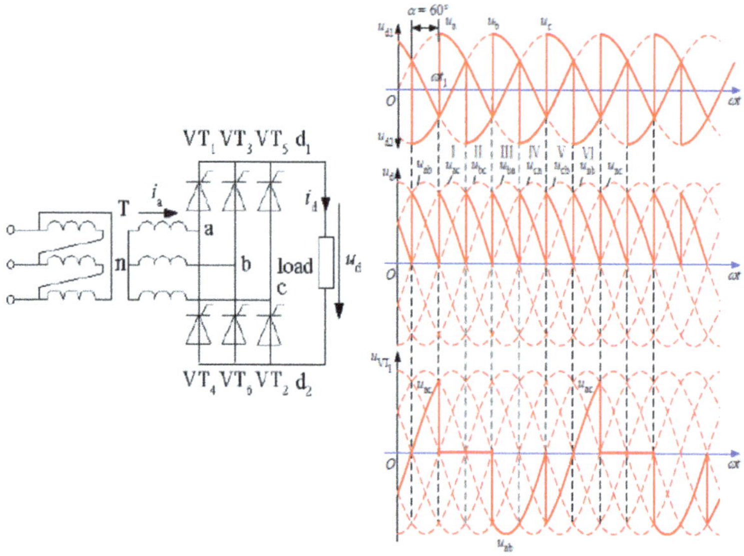

Figure 19. Waveforms when $\alpha = 90°$

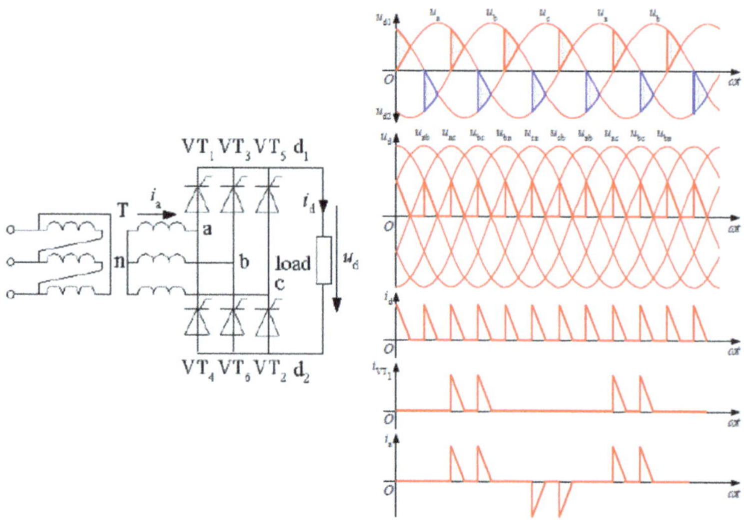

$\alpha = 30°$

Inductive load:

$\alpha = 90°$

Figure 20. Waveforms for inductive load when $\alpha = 0°$

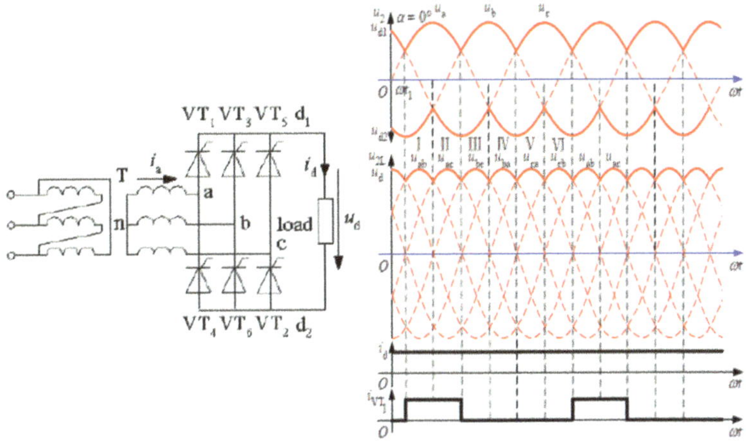

Figure 21. Waveforms for inductive load when $\alpha = 30°$

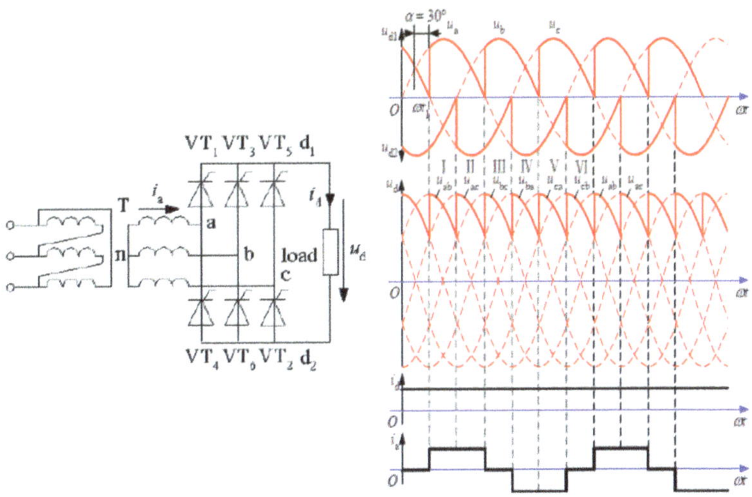

Applications of rectifiers:

Controlled rectifiers are used in different industrial applications, for observing the performance of DC motors and controlling its speed, through transforming AC current into DC with controlled voltage, using Thyristors. Where we can change the voltage by changing the trigger angle α of the Thyristors.

Controlling the voltage applied to the motor armature:

The speed of the motor varies proportionally with the voltage of the armature through changing the trigger angle of the thyristors in the rectifier circuit that connects the source with the load, so when the load increases then the load voltage must be increased through decreasing the trigger angle as shown in figure (23).(Makdisie, C. (2008))

Controlling the flux current:

Figure 22. Waveforms for inductive load when $\alpha = 90°$

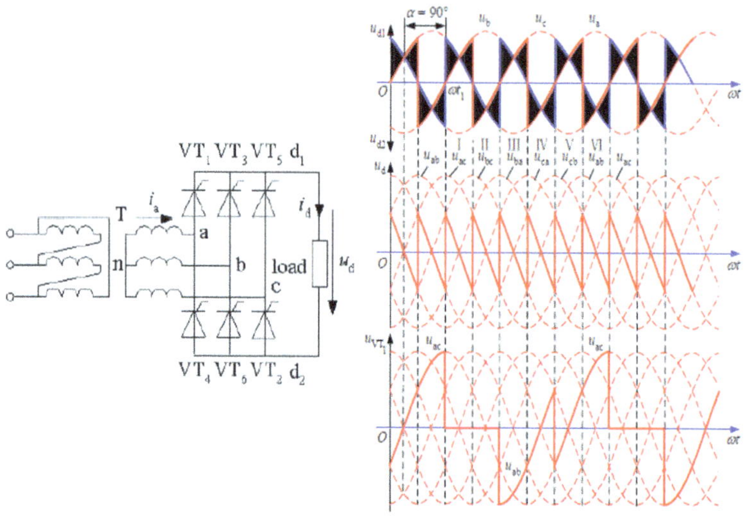

Figure 23. Controlling the voltage applied to the motor armature

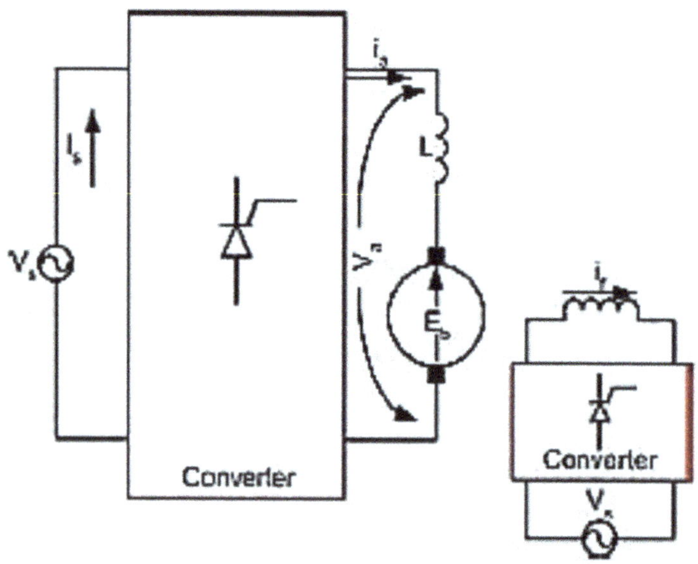

In this case the speed varies reversely with the load current, so when the speed should be decreased, the current must increase through decreasing the trigger angle of the thyristor, and the circuit is shown in figure(24).

DC – Choppers:

Figure 24. Controlling the flux current

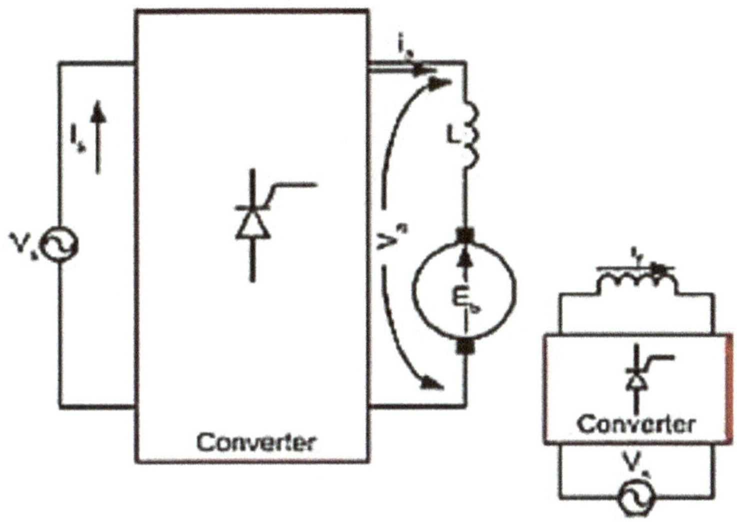

Figure 25. DC choppers

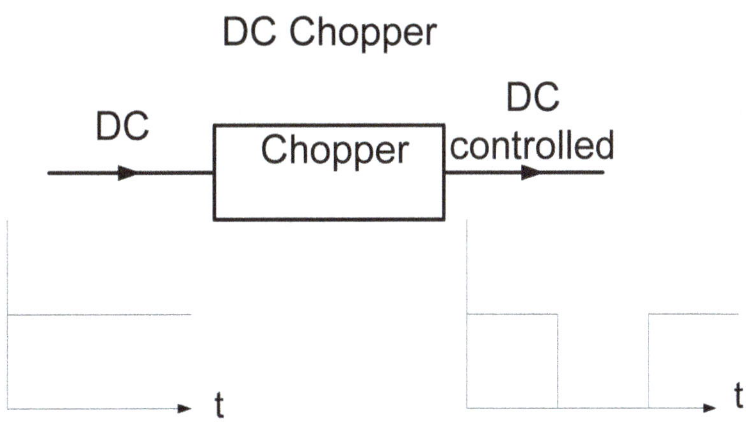

DC choppers are semiconductor switching devices which convert the fixed DC voltage into controlled DC voltage, it is used to control the power of traction and propulsion like in electrical vehicles and subway metro and transportations in mines.

The essential applications of chopper converters:

- •DC speed motors control.
- •Switched filters.
- •Variable frequency drives.
- •Battery chargers.

Figure 26. Step down chopper

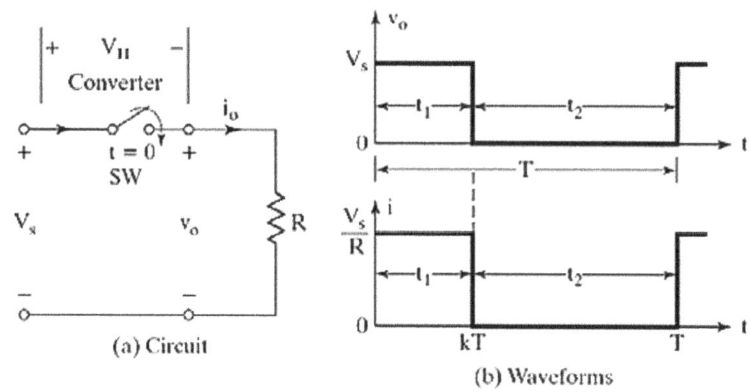

- •Railway traction.
- •Switched mode power supply.
- •Electronic amplifiers class D.

We can classify DC choppers into two main types:

Step down choppers (buck), and step up choppers (boost).

step down chopper (Buck-Chopper):

Figure (26) shows a buck chopper, where the switch [SW] represents the chopper, and this switch can be any of the electronic devices (SCR, BJT, MOSFET, IGBT, GTO…).

The voltage drop on this switch is between $0.5 \rightarrow 2$ [V], but we will presume it's not there to simplify our study(Alhelou et al., 2019; Makdisie et al., 2018; Alhelou et al., 2018; Alhelou et al., 2016; Haes Alhelou et al., 2019; Njenda et al., 2018).

The switch [SW] will work for a period of time we shall call it switch on [$t_{on}(t_1)$], so the input voltage passes to the load, and the switch off time is $t_{off}(t_2)$, so the period T is the sum of t1 & t2 ($T = t_{on} + t_{off}$).

The average output voltage of the above figure is given through the following equation:

$$V_a = V_0 = \frac{1}{T}\int_0^{t_1} v_s.dt = k.V_s = \frac{t_1}{T}.V_s \,.........(12)$$

Where: $k = \frac{t1}{T}$ is the duty cycle constant and it is always $K \leq 1$, and as a result $vo < vs$, so we call this chopper a step down (buck) chopper.

And the average load current is:

$$I_o = \frac{V_o}{R} = \frac{K.V_s}{R} \,.........(13)$$

Figure 27. Buck chopper with (R – L) load

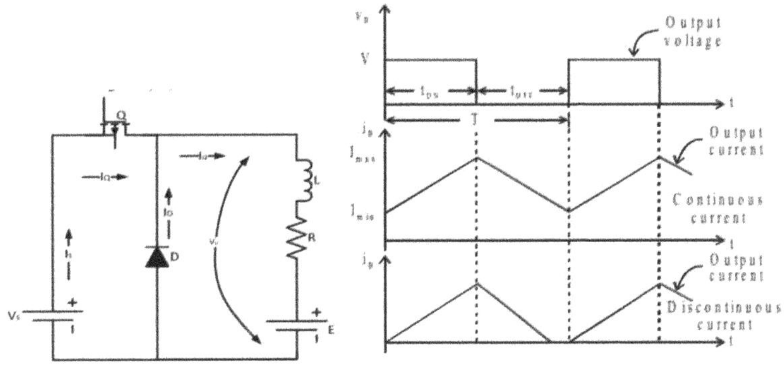

Figure 28. The turned on status

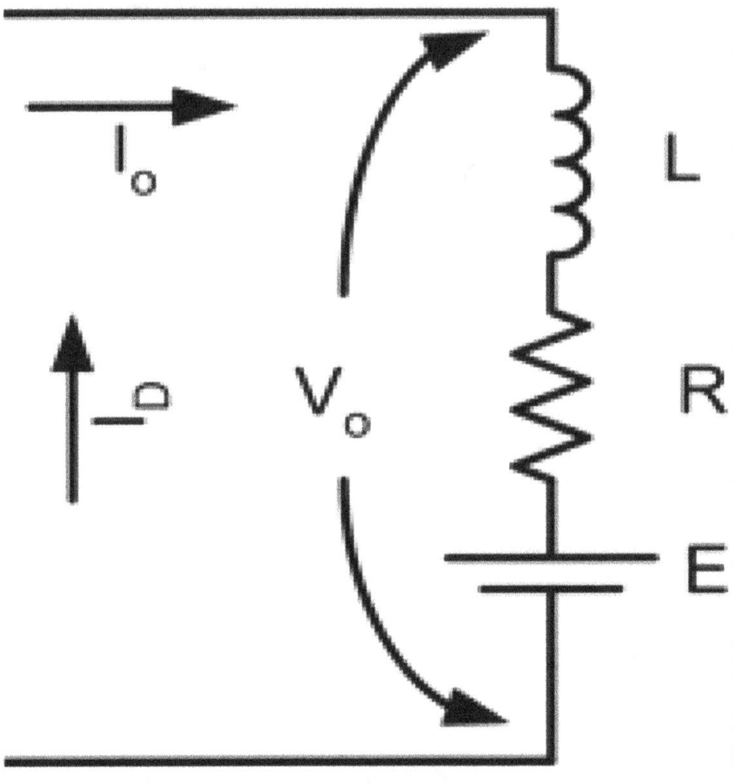

1-2- buck chopper with (R – L) load:

The operating condition of this type of choppers will be framed in this study in each of turned on and off status, when the chopper is turned on the load will be fed, however when the chopper is turned off the load current will continue flowing with the same direction through the free wheel diode D due

Figure 29. The turned off status

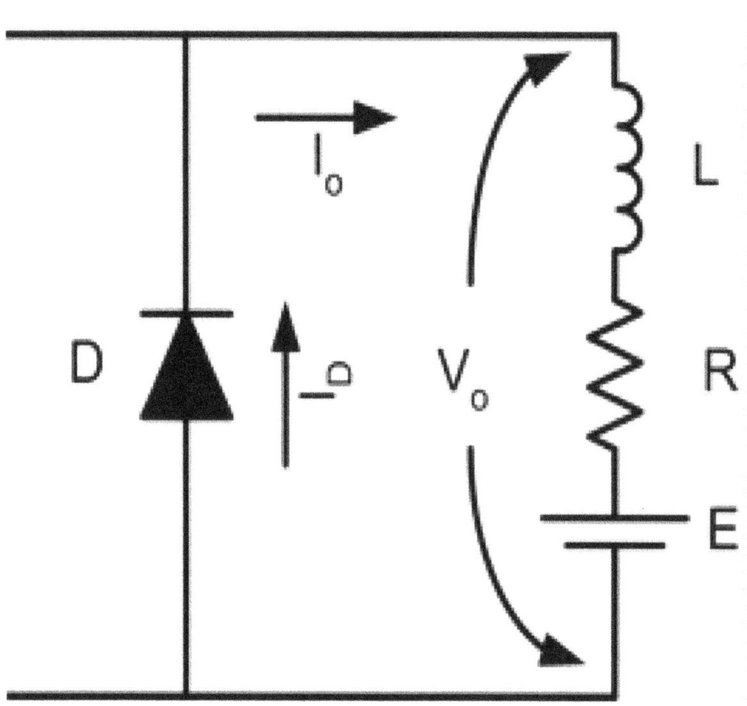

to the stored energy in the induction L, the current can be continues or chopped depending on the value of L and K.

If the current continues then the value of this current is changed between $I_{min} \div I_{max}$, the output waves is shown in figure (27).

To Analyze this circuit chopper, we have two states:

the turned on status of the chopper:

figure (28) shows the chopper in the turned on status where $0 < t < t_1$

$$Vs = R.i_1 + \frac{L.di_1}{dt} + E.........(14)$$

We can solve this differential equation with the following condition:

$$i_{1(t=0)} = I_1.........(15)$$

And the conclusion is:

Figure 30. Step up chopper

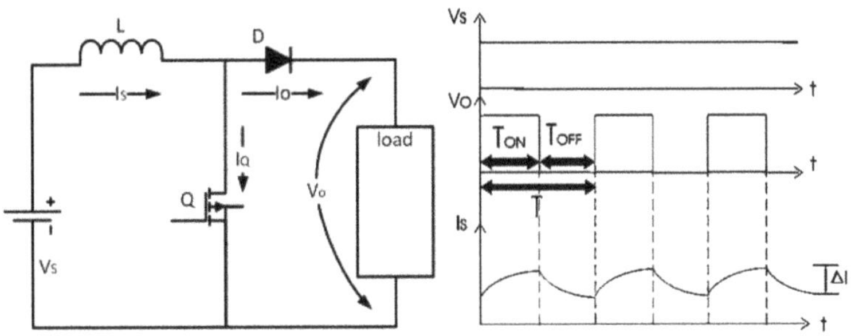

$$e^{-\frac{tR}{L}} + \frac{Vs - E}{R\left(1 - e^{-\frac{tR}{L}}\right)}.I_1 = i_{1(t)} \dots\dots\dots(16)$$

The current value is calculated by compensation the value $t = t1$:

$$I_2 = i_{1(t=t1=k.T)} \dots\dots\dots(17)$$

the turned off status of the chopper:

In this situation the current will circulate through the freewheel diode D, and the differential equation which is expressed the load current is:

$$0 = R.i2 + L\frac{di2}{dt} + E \dots\dots\dots(18)$$

And we can solve this equation using the following condition:

$$i_{2(t=0)} = I_2 \dots\dots\dots(19)$$

As the result is: $e^{-\frac{tR}{L}} - \frac{E}{R\left(1 - e^{-\frac{tR}{L}}\right)}.I_2 = i_{2(t)} \dots\dots\dots(20)$

The current value at the end of the turned off cycle when $t = t2$ is:

$$I_3 = i_{2(t=t2=(1-K)T)} = I_1 \dots\dots\dots(21)$$

The range of the load current change is: $\Delta I = I_2 - I_1$

Figure 31. Simulation of boost chopper

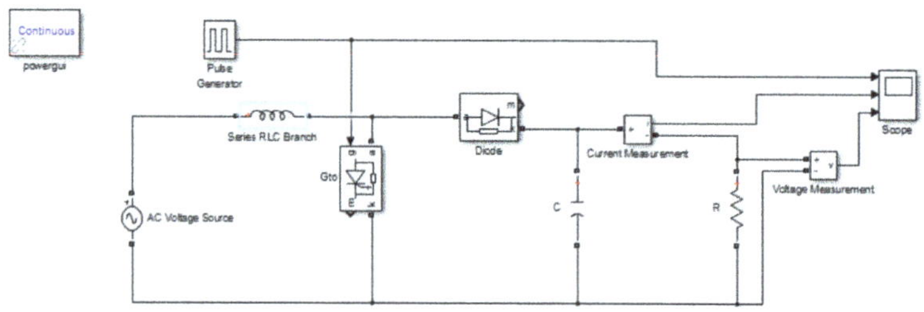

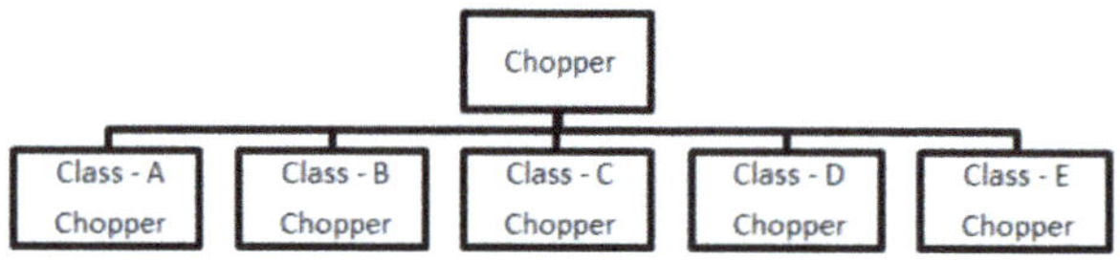

$$\Delta I = \frac{V_s}{R} \cdot \frac{1 - e^{-\frac{KTR}{L}} + e^{-\frac{TR}{L}} - e^{-(1-K)\frac{TR}{L}}}{1 - e^{-\frac{TR}{L}}} \ldots\ldots\ldots(22)$$

Where:

$$k = \frac{t_1}{t1_{+}t_2} = \frac{t_1}{T} \Rightarrow t_1 = k.T\ ,\ t_2 = T - t_1 = T - k.T = T(1-k)$$

step up chopper (Boost- Chopper):

As it is shown in figure (30), these types of choppers are allowed to obtain an output voltage higher than the input voltage value. $V_o > V_s$

We always have in this kind of choppers induction coil (L) at the input of the chopper circuit, as shown in figure (30).

Operation principle:

within the switch on period $0 < t < t1$, the current of the induction coil increases to the maximum value at the end of this period, and during this period the coil will store electromagnetic power.

At the moment $t = t1$, the switch SW is turned off for a period of time $t2$, the stored energy in the coil (L) will be discharged, and the current starts to decrease passing through the diode D1 and the Load.

The output waveform is shown in figure (30).

During the switched on period the voltage value on the coil is:

$$U_L = \mathrm{L}\frac{\mathrm{di}}{\mathrm{dt}} \ldots\ldots\ldots(23)$$

And the current varies between two values $\mathrm{I2}\ \&\ I1$:

Figure 32. Class A choppers

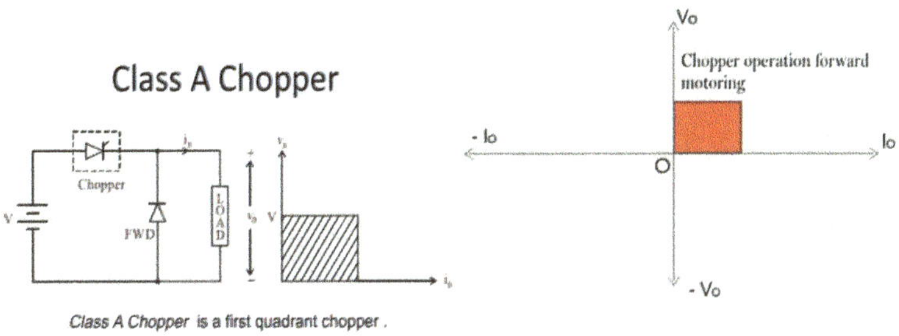

Figure 33. Class A waveforms

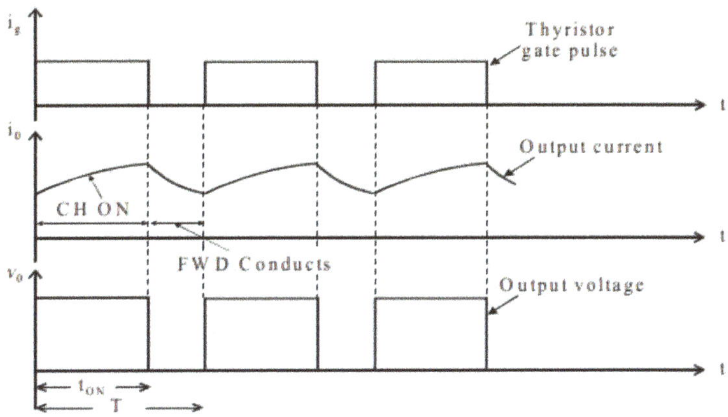

$$◆I = I2 - I1 = \frac{Vs}{L} t1 \ldots\ldots\ldots (24)$$

Through the switch on period of SW, the load circuit is shorted, and $V_s = U_L$.

During the period t2 the instant load voltage is:

When a capacitive filter is connected to the load, we can presume that the output voltage is constant and $Va = Vo$.

From the equation (24) we can observe that the load voltage is configured by the duty cycle (k), where the voltage increases with the increase of (k), however when $k = 0$ the voltage will be minimum and $Vo = Vs$.

And here is a study case of the boost choppers:

By simulating the boost chopper using MATLAB /Simulink, it is obvious from figure (31) that the main electronic element is the GTO thyristor (Khan et al. 2014; Khan et al. 2013; Khan et al. 2012; Negash et al., 2017; Negash et al., 2016; Jariso et al. 2017; Kifle et al. 2018; Yeshalem et al. 2017; Singh et al. 2019; Gupta et al., 2015)

Figure 34. Class B chopper

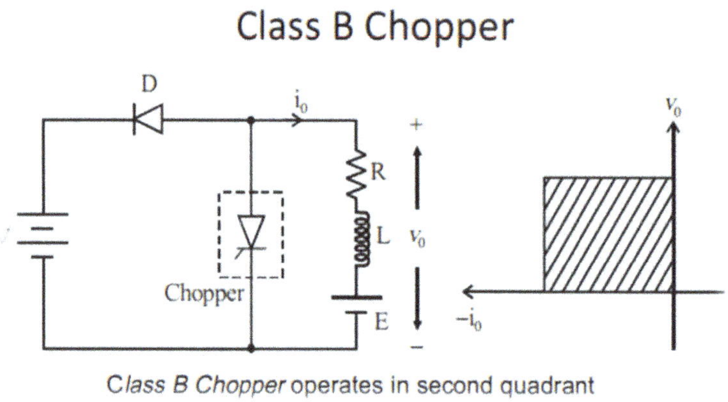

Figure 35. Class B chopper operation quarter

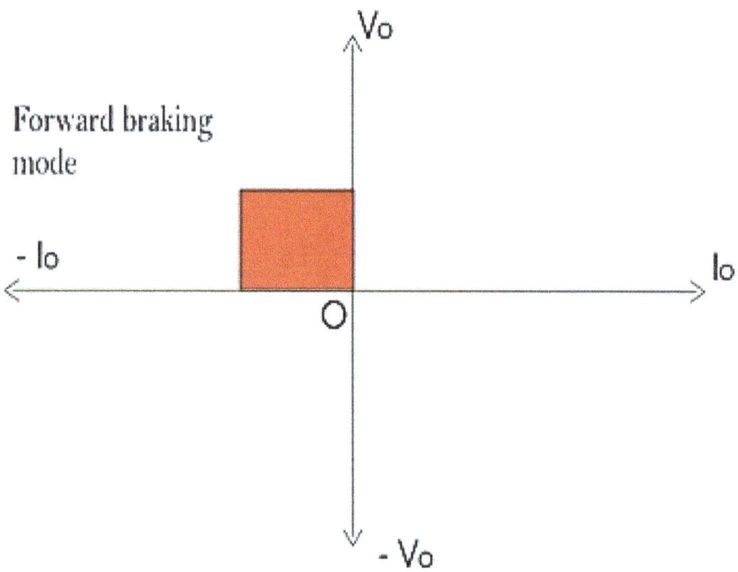

This study case is similar to the circuit shown in figure (30), however instead of the transistor shown in figure (30), a GTO is used.

The capacitor used in figure (31) is used for filtering and softening the waveform of the output measurements of the load.

Classification of the choppers by their work in the four quadrants:

Figure 36. Class C chopper

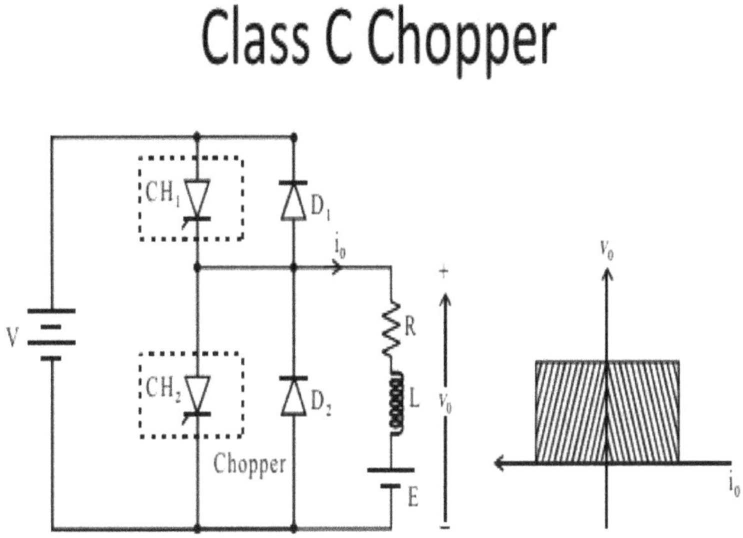

Figure 37. Class D chopper

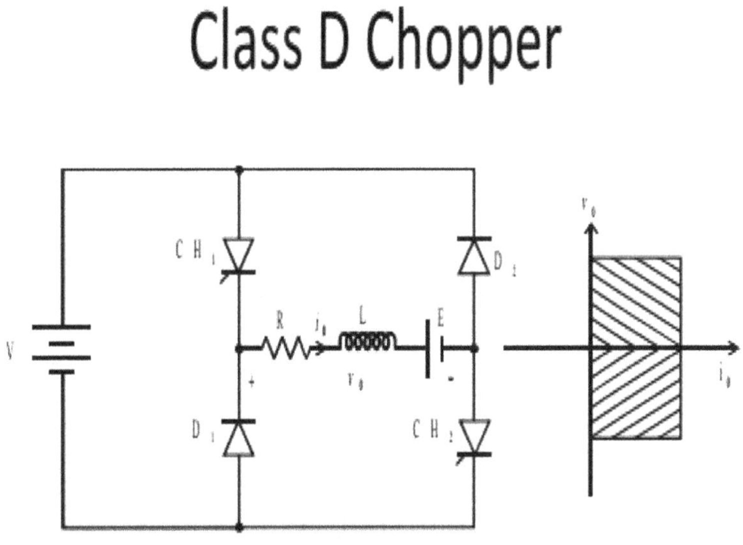

Class A- Chopper.
Class B- Chopper.
Class C- Chopper.
Class D- Chopper.
Class E- Chopper.
Class A choppers:

Figure 38. Class E chopper

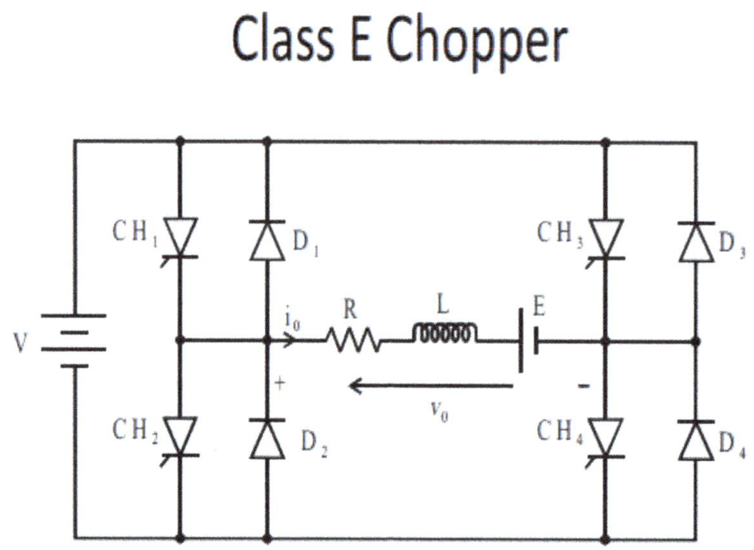

Figure 39. Class E chopper operation quarters

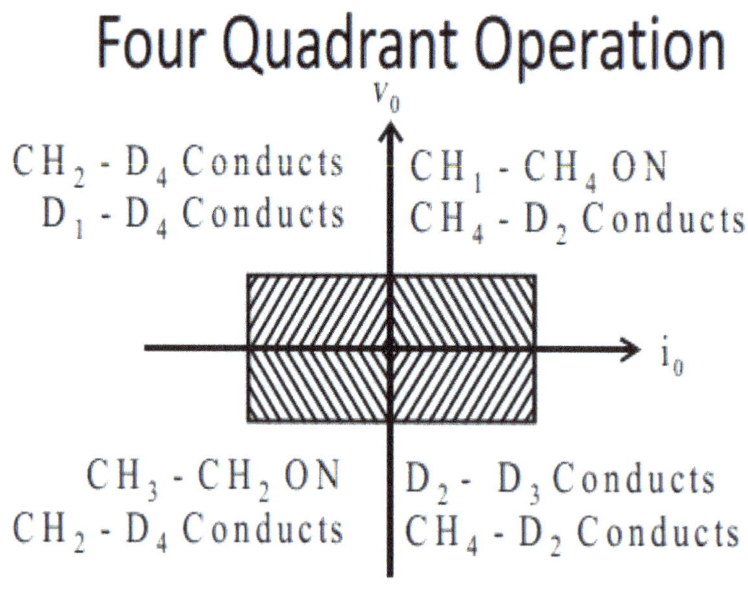

This type of choppers is also called single quadrant chopper because the direction of both the voltage and current of the load is always positive.

Figure 40. Single phase AC voltage regulators

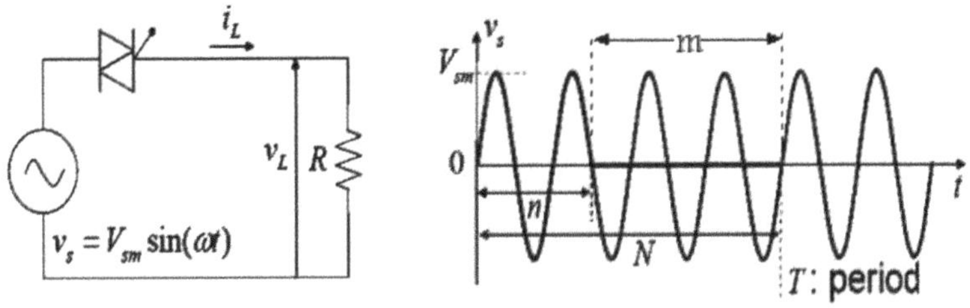

When the switch is turned on, the source voltage is feeding the load, however when the switch is turned off then the output voltage is zero and the load current keeps flowing with the same direction through the free wheel diode, class A chopper is considered a buck chopper and used to control the speed of DC motors. (Salam Z. (2003)).

Class B chopper:

The load current in this type of choppers is negative, however the voltage is still positive. It is also called a single quadrant chopper, but it works in the second quadrant.

As understood the class B chopper works as an inverter, because it allows energy pass from the load to the source (like breaks in DC motors).

Figure(34) the circuit configuring the class B chopper, where E is reversed electrical power that is considered part of the load while the chopper is working in the breaking period. So the class B chopper is a boost chopper.

Class C chopper:

This chopper is called a two-quadrants chopper, because the current can be positive or negative however the voltage can only be positive.

This chopper has all the qualities of both A&B choppers as it is shown in figure (36).

If the switches S1&D2 are turned on then the chopper works as a converter.

If the switches S2&D1 are turned on then the chopper works as an inverter.

Class D chopper:

Class D chopper is also a two-quadrant chopper, the difference is that in class D chopper the current is always positive, however the voltage can be negative or positive.

When the switches S1&S2 are turned on then the chopper works as a converter.

When these switches are turned off the current will preserve its direction because the induction load and the two diodes D1&D2 creates a pass for a positive current and a negative voltage.

Class E chopper:

This kind of chopper is called a four-quadrant chopper, both the voltage and the current can be negative or positive, depending on the working system of the chopper, so it is a merge of class C&E choppers.

Figure (39) shows the current and the voltage of the chopper in the four quadrants.

AC voltage regulators:

AC voltage regulators are a type of converters that transform the fixed AC voltage into variable or controlled.

Usually these regulators are used to control the speed of induction motors and it is equal to an autotransformer.

Figure 41.

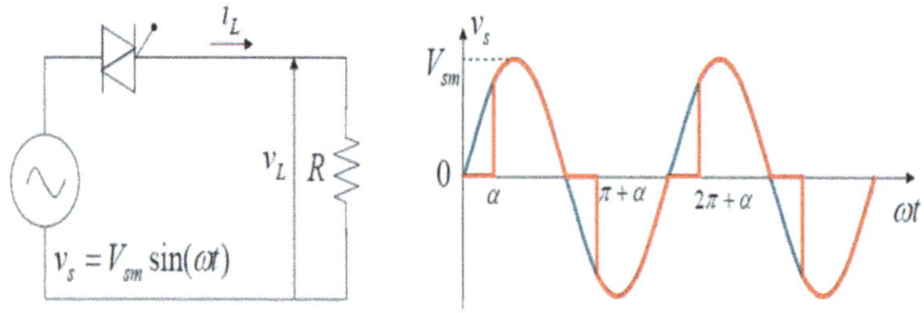

Figure 42. Waveforms

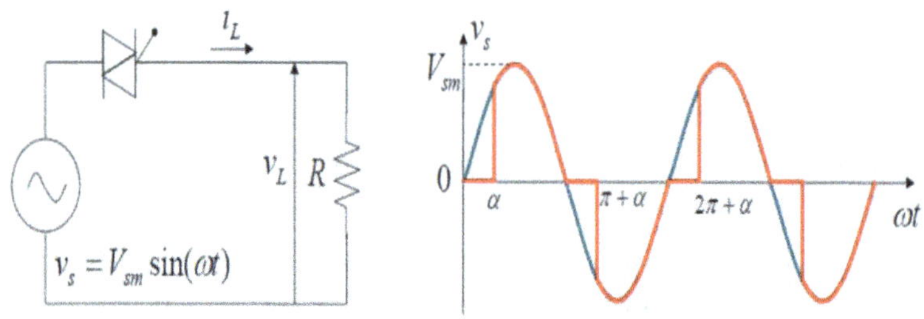

These regulators are natural switches due to the voltage source that switches at every zero passing point. And because these regulators are configured of controlled semiconductors; both the current and the voltage waves are not a sine wave (unconnected), that leads to the need of correcting the power factor and decreasing the harmonics. (Holtz, J., & Boelkens, U. (1989)).

Single phase AC voltage regulators:

Whereas the output voltage is controlled through changing the trigger moment, or delaying the α angle of the triac or the angle of the SCR as it is obvious from the figure (40).

Controlling the On/Off situations with an R – load:

where:

N: number of half cycles within the Time (T).

n: number of half cycles within the whole turn on period.

$$V_{orms} = \left[\frac{n}{2\pi(n+m)} \int_0^{2\pi} \sqrt{2}.V_{Sm}.\sin\theta d\theta \right]^{\frac{1}{2}}$$

$$= V_{Sm}.\sqrt{\frac{n}{n+m}}$$

Figure 43. Simulation results using MATLAB

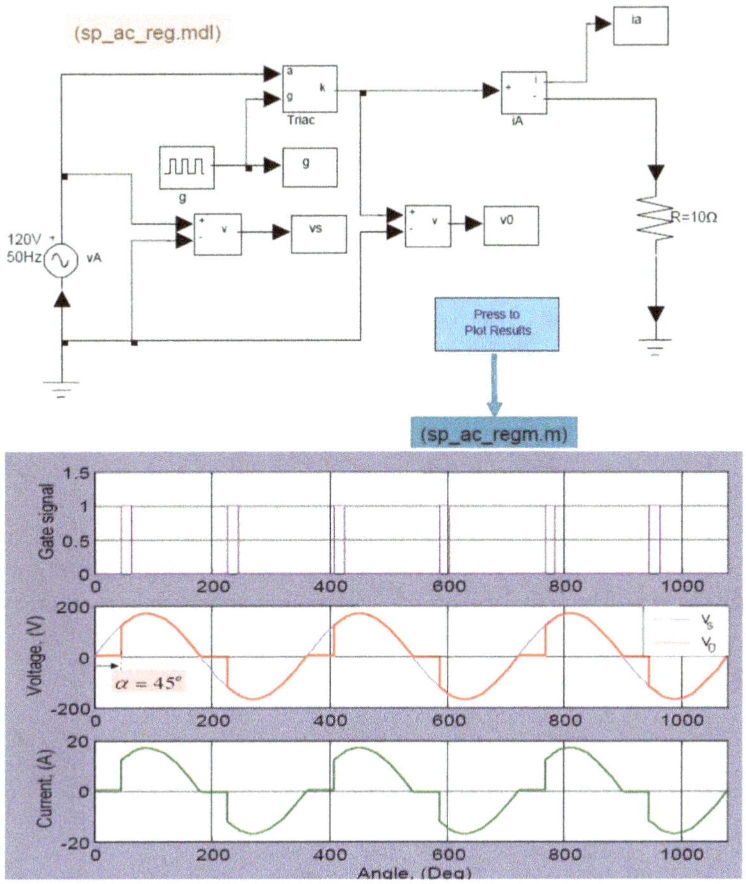

Figure 44. Inductive load

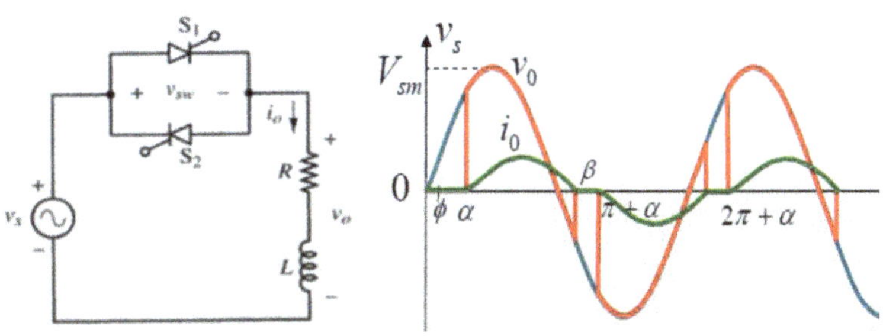

$$V_{orms} = V_{srms}\sqrt{\frac{n}{N}} = \frac{V_{sm}}{\sqrt{2}}\sqrt{\frac{n}{N}} = V_S\sqrt{k}\,.........(26)$$

Figure 45. SIMULINK Simulation for inductive load

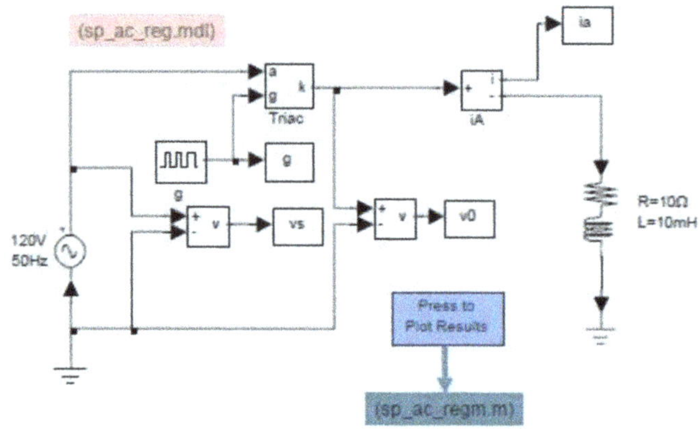

k: duty cycle.

The triac is triggered on at α and $\alpha+\pi$, when the triac starts passing at the moment α the voltage is feeding the resistive load R which makes V=Vo, and the load current starts increasing from zero then again starts dropping, so the positive half wave will pass through the load.

Single phase AC voltage regulators

$$V_{orms}=\left[\frac{1}{2\pi}\int_{\alpha}^{\pi}\left(\sqrt{2}.V_{Sm}.\sin\theta\right)^{2}d\theta\right]^{\frac{1}{2}}$$

$$\sin^{2}\theta=\frac{1-\cos 2\theta}{2}$$

$$V_{orms}=\sqrt{2}.V_{sm}.\left[\frac{1}{2\pi}\int_{\alpha}^{2\pi}1-\cos 2\theta d\theta\right]^{\frac{1}{2}}$$

$$V_{orms}=V_{s}\sqrt{\frac{1}{\pi}\left(\pi-\alpha+\frac{\sin(2\alpha)}{2}\right)}\ldots\ldots\ldots(27)$$

Simulation results using MATLAB:

Inductive load:

Figure(44) shows a single-phase regulator with R – L Load and the operation mode is:

The Thyristor S1 is triggered at a moment α however the Thyristor $\alpha+\pi$.

Figure 46. Simulation when $\alpha > \theta$

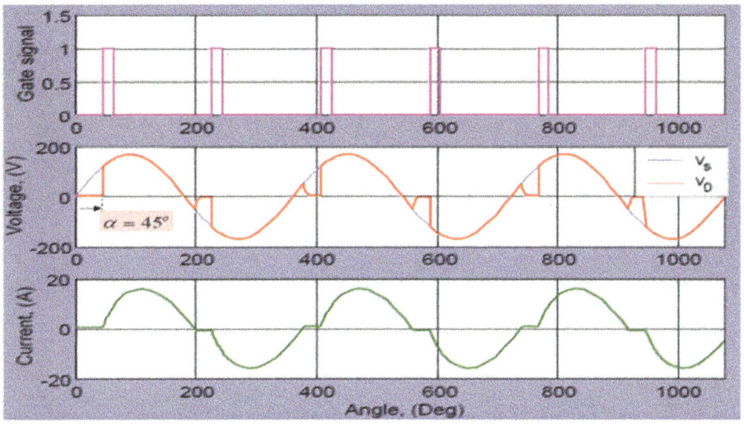

When S1 is turned on at the moment α, the source voltage is applied on the R – L Load which makes V=Vo and the load current starts increasing from zero, then again at a moment β the current starts decreasing back to zero.

When the Thyristor S2 starts passing at a moment Ét = À+±, the negative current half wave will pass through the load, as it is obvious from figure (44).

And the main equations configuring this status:

$$V = V_o = R.i_o + L\frac{di}{dt}\text{.........}(28)$$

$$\sqrt{2}V_p.\sin wt = R.i_o + L\frac{di}{dt}\text{.........}(29)$$

Which is a differential equation can be solved with the condition:

$$i_o = i_{steady} + i_{transient}\text{.........}(30)$$

$$i_o = \frac{\sqrt{2}V_p}{Z}\sin(wt - \varphi) + A.e^{-\frac{t}{\tau}}\text{.........}(31)$$

Where: $\varphi = \arctan\frac{wL}{R}, Z = \sqrt{R^2 + (wL)^2}, \tau = \frac{L}{R}$

A is defined from the beginning conditions, so:

When $\text{wt} = \pm \rightarrow \text{i}_o = 0 \Rightarrow$

Figure 47. Simulation when $\alpha < \theta$

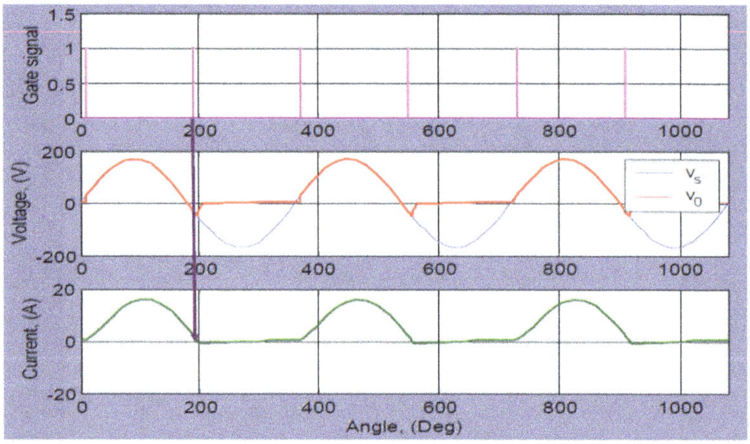

Figure 48. Simulation when $\pm = 10 < \theta = 17.44$

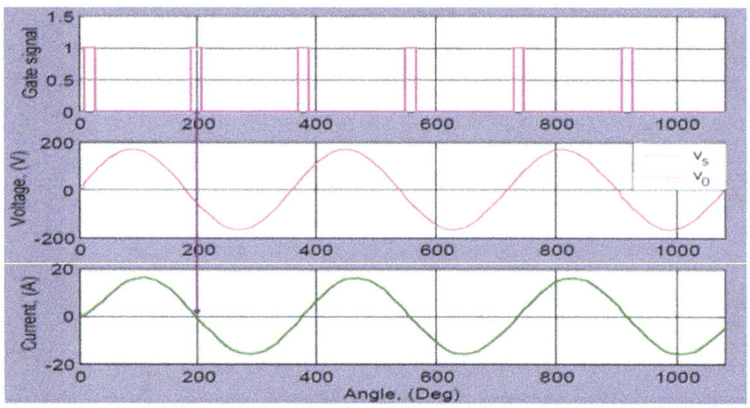

From the equation (31):

$$A = \frac{-\sqrt{2}.V_p}{Z} \sin(\alpha - \varphi).e^{\left(\frac{R}{WL}\right)\alpha} \quad ....(32)$$

By solving equation (31) through (32):

$$i_o = \frac{\sqrt{2}.V_p}{Z} \left[\sin(wt - \varphi) - \sin(\alpha - \varphi).e^{\left(\frac{R}{WL}\right)(\alpha - wt)} \right] ....(33)$$

SIMULINK Simulation:
Three cases can be distinguished:

Figure 49. Simulation when $\alpha = \theta = 17.44$;

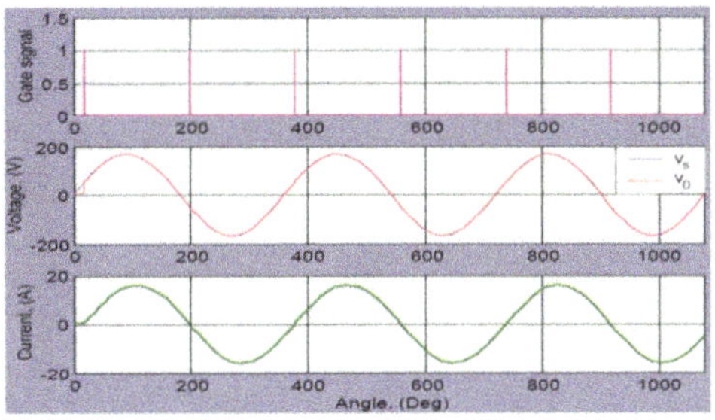

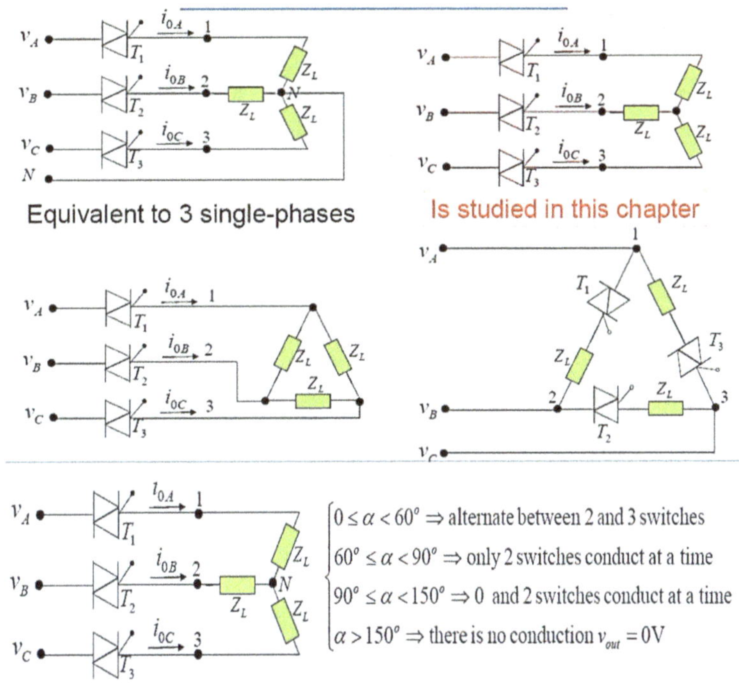

$\alpha > \theta$

$$V_{orms} = V_s \cdot \sqrt{\frac{1}{\pi} \cdot \left(\beta - \alpha + \frac{\sin 2\alpha}{2} - \frac{\sin 2\beta}{2} \right)} \ldots\ldots\ldots(34)$$

$\alpha = 45 > \theta = 17.44$

$\alpha < \theta$

This case must be avoided because there cannot be any passing to the load where ² >±+À

Figure 50. General simulation of 3-ph regulators

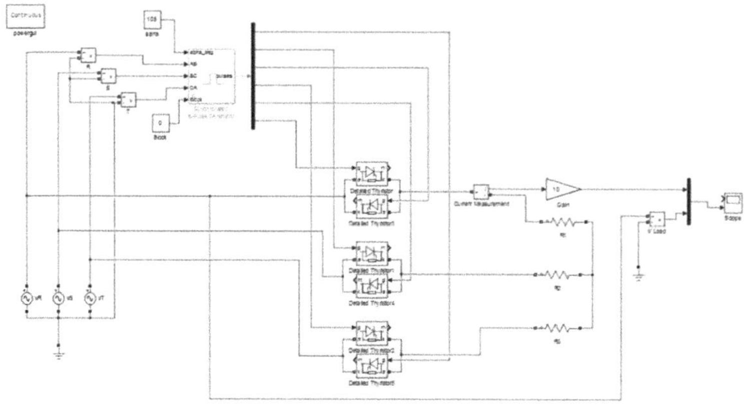

$\alpha = 10 < \theta = 17.44$

However if the trigger pulses are big enough, a continues trigger can be achieved where $\pm = 10 < \theta = 17.44$

1) $\alpha = \theta = 17.44$; Where ² = ± + À

$$V_{orms} = \frac{V_{sm}}{\sqrt{2}} = V_{srms} \,.........(35)$$

$$\mathrm{i_o} = \frac{\sqrt{2}.\mathrm{V_p}}{\mathrm{Z}} \sin(\mathrm{wt} - \text{Æ}).........(36)$$

Three phases regulators:

The three phase regulators, it can be used for high power loads such as the relatively big induction motors that drives electric fans and pumps.

$$V_{AN} = \sqrt{2} V_S \sin(\theta).........(37)$$

$$V_{AN} = \sqrt{2} V_S \sin\left(\theta - \frac{2\pi}{3}\right).........(38)$$

$$V_{AN} = \sqrt{2} V_S \sin\left(\theta + \frac{2\pi}{3}\right).........(39)$$

Figure 51. AC sources in the simulation

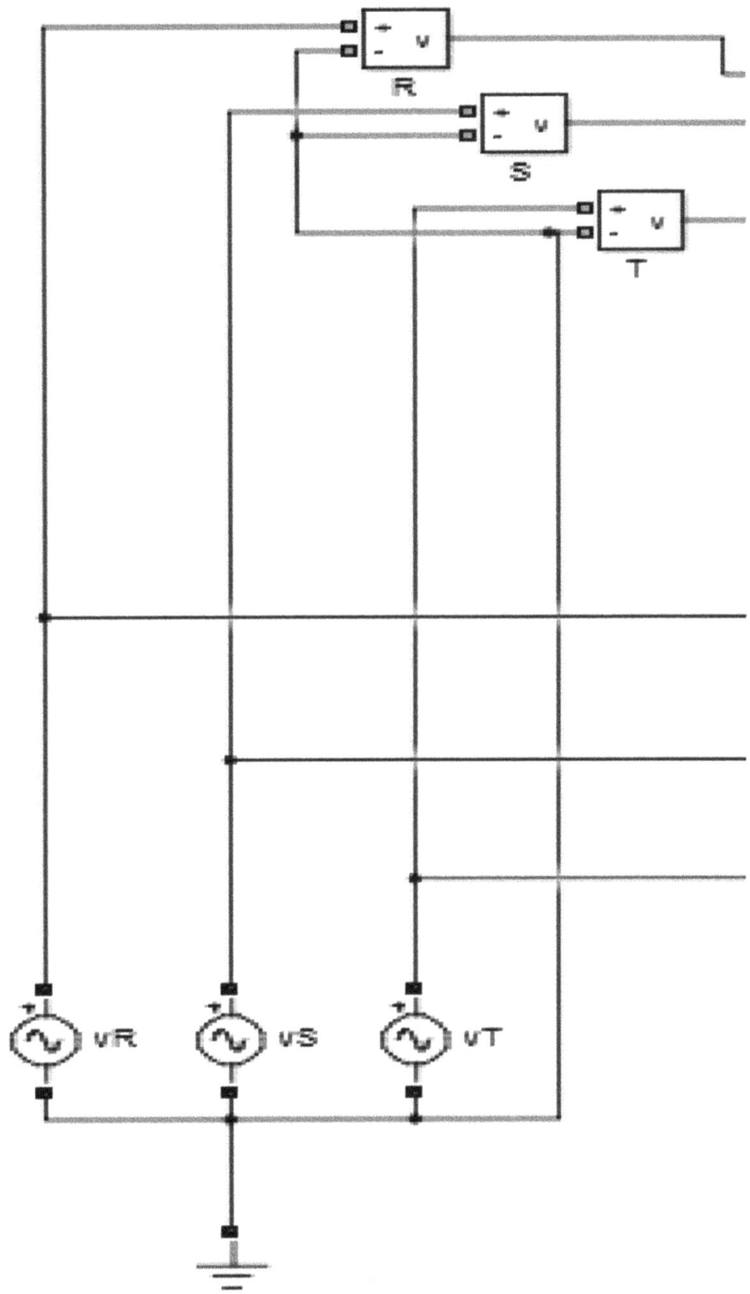

Now when it comes to three phase regulators simulation/using MATLAB – Simulink, it can be a little tricky, so the chapter will explain each step. (Histand, Michael B. (1999)).

The general simulation is shown in figure (43):

Three AC voltage sources are used with a voltage of 310 line to ground, a voltage measurement is used between each two lines to we can get a voltage of 220 as an effective value as shown in figure (43).

Figure 52. 6-pulse generator

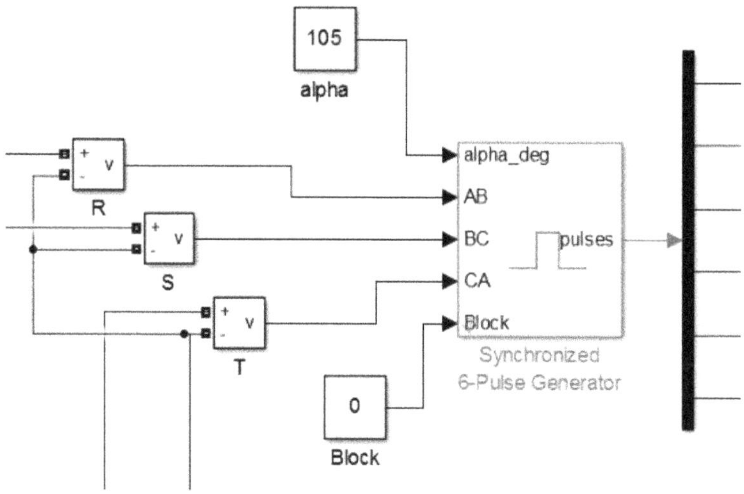

The output of the voltage measurement is applied on the synchronized 6-pulse generator, which parameters are 3 voltage inputs and the α angle which will be changed to monitor the different modes, and a block element which is logical element to latch the pulses if needed, and it would be set to zero so the pulses won't be latched as figure (45) shows.

The output of the synchronized 6-pulse generator is a single line, so we connect it to a demux bar element (accumulator) to obtain 6 lines, one line to each gate of the 3-ph regulator.

If we assume that the lines are (R – S – T) then the synchronization of the pulses or the Thyristors would be $\left(R, T', S, R', T, S'\right)$, as shown in figure (53).

Three pure resistances are connected (star connection), each one is 10, as we can see in figure (50) the scope only monitoring the first phase voltage, using a current measurement multiplied by a gain of the resistor's value, so it can be a voltage measurement, with the main source voltage to compare.

Each of the thyristors has the parameters as in figure (54):

As we can see in figure (54), the inductance L_{on} is zero, so the wave forms would describe a pure resistive load.

We can observe that the snubber resistance is 1000Ω, it is raised to reduce the disturbances in the wave forms.

Three modes are created in this situation:

Mode 2/3 $\left(0 \le \alpha < 60°\right)$

For this situation an alpha (α) angle of 15° is chosen, and the wave forms of the simulation:

$$V_o = \sqrt{\frac{1}{2\pi} \cdot \int_0^{2\pi} v_{an}^2 . d\left(\omega t\right)} \ldots\ldots\ldots (40)$$

Figure 53. Synchronization of the pulses

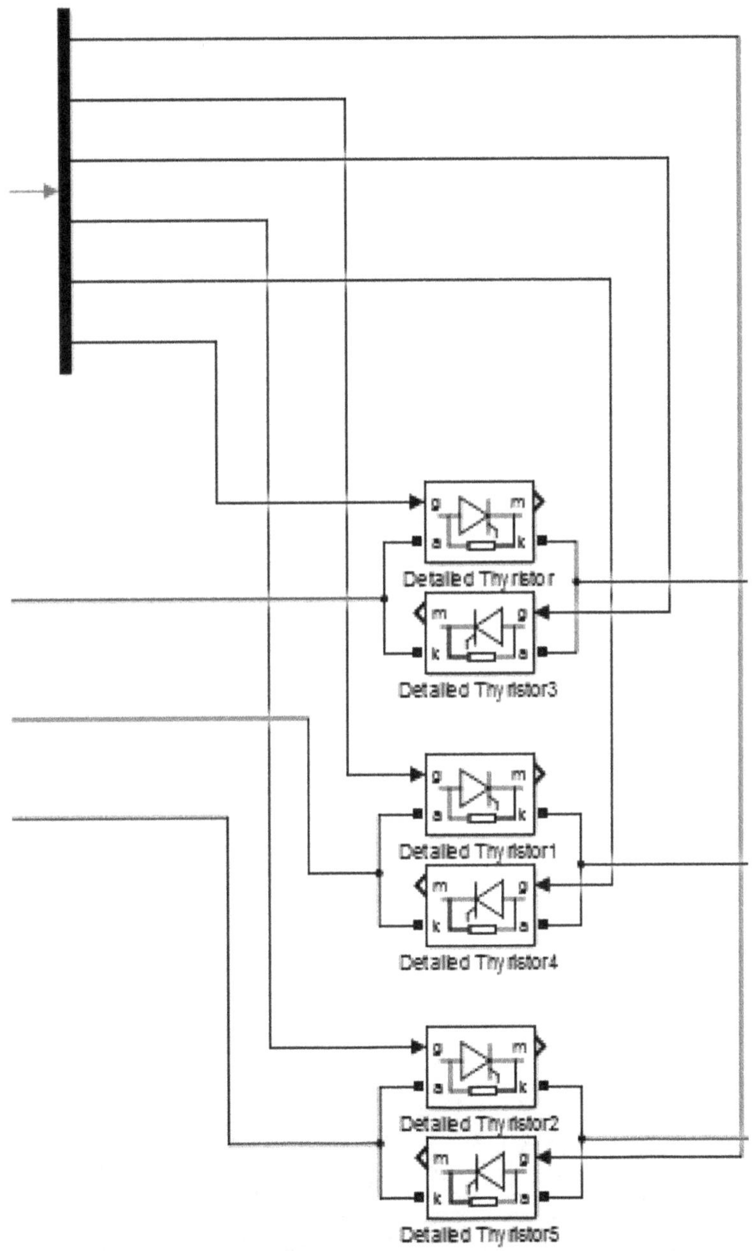

$$= \sqrt{\frac{1}{\pi}} \cdot \left[\int_{\alpha}^{\frac{\pi}{3}} v_a^2 .d(\omega t) + \int_{\frac{\pi}{3}}^{\frac{\pi}{3}+\alpha} \frac{v_{ab}^2}{4} .d(\omega t) + \int_{\frac{\pi}{3}+\alpha}^{\frac{2\pi}{3}} v_a^2 .d(\omega t) + \int_{\frac{2\pi}{3}}^{\frac{2\pi}{3}+\alpha} \frac{v_{ac}^2}{4} .d(\omega t) + \int_{\frac{2\pi}{3}+\alpha}^{\pi} v_a^2 .d(\omega t) \right]$$

Figure 54. Thyristors parameters

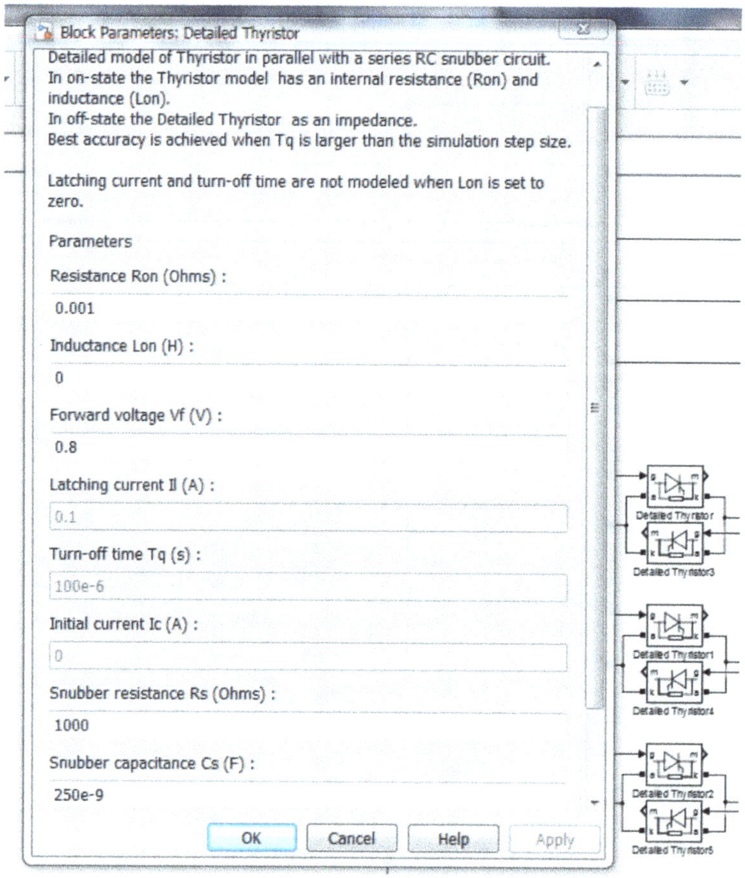

Figure 55. Simulation results when $(\alpha = 15°)$

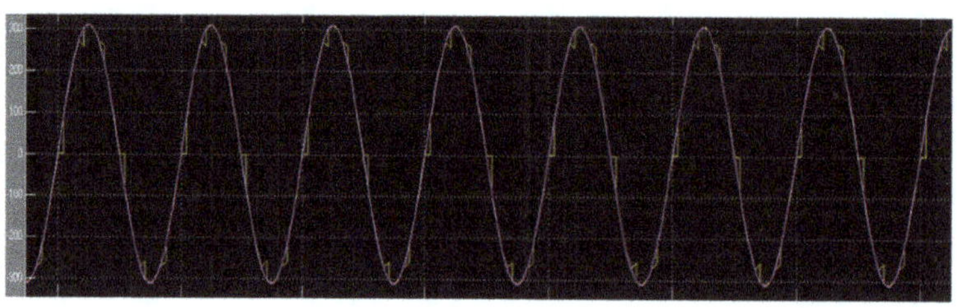

Figure 56. Simulation results when $(\alpha = 60°)$

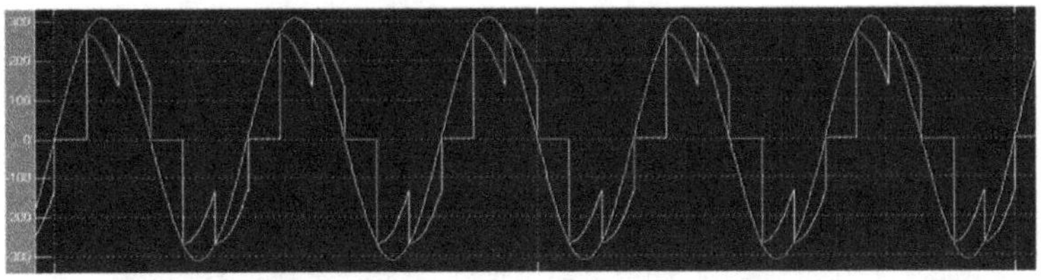

Figure 57. Simulation results when $(\alpha = 75°)$

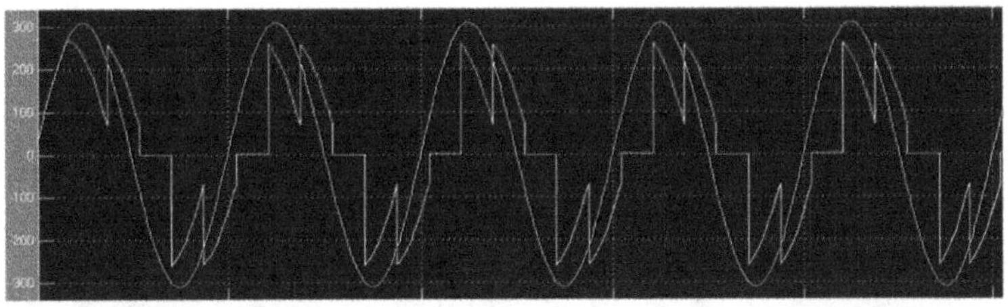

$$V_o = \sqrt{6}V_s \sqrt{\frac{1}{\pi}.\left[\begin{array}{c}\displaystyle\int_{\alpha}^{\frac{\pi}{3}} \frac{\sin^2(\omega t)}{3}.d(\omega t) + \int_{\frac{\pi}{3}}^{\frac{\pi}{3}+\alpha} \frac{\sin^2\left(\omega t+\frac{\pi}{6}\right)}{4}.d(\omega t) \\ \displaystyle + \int_{\frac{\pi}{3}+\alpha}^{\frac{2\pi}{3}} \frac{\sin^2(\omega t)}{3}.d(\omega t) + \int_{\frac{2\pi}{3}}^{\frac{2\pi}{3}+\alpha} \frac{\sin^2\left(\omega t-\frac{\pi}{6}\right)}{4}.d(\omega t) + \int_{\frac{2\pi}{3}+\alpha}^{\pi} \frac{\sin^2(\omega t)}{3}.d(\omega t)\end{array}\right]}$$

Figure 58. Simulation results when $(\alpha = 90°)$

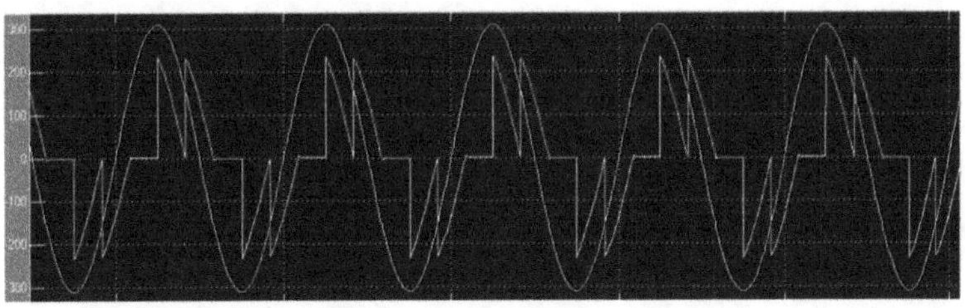

Figure 59. Simulation results when $(\alpha = 105°)$

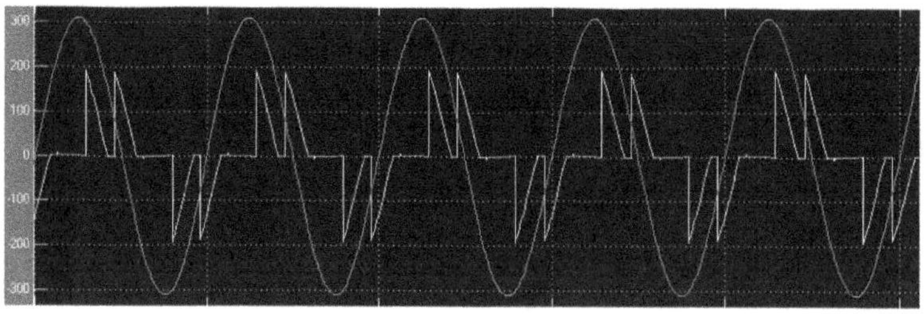

Figure 60. Simulation results when $(\alpha = 135°)$

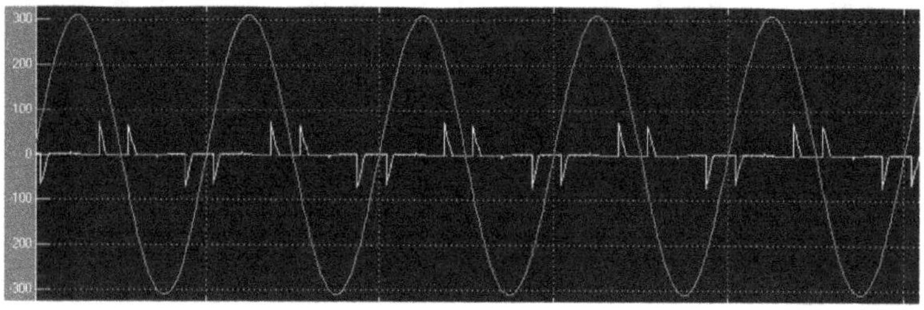

(41)

$$V_o = \sqrt{6}V_s.\sqrt{\frac{1}{\pi}.\left(\frac{\pi}{6} - \frac{\alpha}{4} + \frac{\sin 2\alpha}{8}\right)}.........(42)$$

$$\Rightarrow I_o = \frac{V_o}{R}, P_o = 3.I_o.V_o.........(43)$$

Mode 2/2 $(60 \le \alpha < 90°)$:

Before showing the wave forms of mode 2/2 we shall see the wave forms of the critical angle $(\alpha = 60°)$ in figure (56).

Now an angle of $(\alpha = 75°)$ determine the mode 2/2 in figure (57)

$$V_o = \sqrt{\frac{1}{2\pi}.\int_0^{2\pi} v_{an}^2\, d(\omega t)}.........(44)$$

$$= \sqrt{\frac{1}{\pi}\left[\int_{\alpha}^{\frac{\pi}{3}+\alpha} \frac{v_{ab}^2}{4}.d(\omega t) + \int_{\frac{\pi}{3}+\alpha}^{\frac{2\pi}{3}+\alpha} \frac{v_{ac}^2}{4}.d(\omega t)\right]} \ldots\ldots\ldots(45)$$

In this case there is another critical angle which is $(\alpha = 90°)$, we can observe in figure (58) the zero passing point in the middle of the wave.

Mode 0/2 $(90 \leq \alpha < 150°)$

We can observe a disconnection in this wave form obvious in figure (59) that shows the angle $(\alpha = 105°)$, however it is more obvious as the angle increases like in figure (60)

CONCLUSIONS AND FUTURE WORKS

This chapter shows us the importance of using the elements of applied power electronics, which has shown us its importance in many areas, including speed regulation and electric drive systems.

Several electrical systems using power electronics have also been simulated.

Finally, some values and results are illustrated from this simulation, in which the elements of power electronics should be used.

REFERENCES

Alhelou, H., Hamedani-Golshan, M. E., Zamani, R., Heydarian-Forushani, E., & Siano, P. (2018). Challenges and opportunities of load frequency control in conventional, modern and future smart power systems: A comprehensive review. *Energies, 11*(10), 2497. doi:10.3390/en11102497

Alhelou, H. H. (2018). *Fault detection and isolation in power systems using unknown input observer. In Advanced condition monitoring and fault diagnosis of electric machines* (p. 38). Hershey, PA: IGI Global.

Alhelou, H. H., Golshan, M., & Fini, M. (2018). Wind driven optimization algorithm application to load frequency control in interconnected power systems considering GRC and GDB nonlinearities. *Electric Power Components and Systems, 46*(11-12), 1223-1238.

Alhelou, H. H., & Golshan, M. E. H. (2016, May). Hierarchical plug-in EV control based on primary frequency response in interconnected smart grid. In *2016 24th Iranian Conference on Electrical Engineering (ICEE),* (pp. 561-566). IEEE. 10.1109/IranianCEE.2016.7585585

Alhelou, H. H., & Golshan, M. E. H. (2020). Decision-making-based optimal generation-side secondary-reserve scheduling and optimal LFC in deregulated interconnected power system. In *Decision making applications in modern power systems* (pp. 269–299). Academic Press. doi:10.1016/B978-0-12-816445-7.00011-6

Alhelou, H. H., Golshan, M. E. H., & Hatziargyriou, N. D. (2019). Deterministic dynamic state estimation-based optimal LFC for interconnected power systems using unknown input observer. *IEEE Transactions on Smart Grid*, 1. doi:10.1109/TSG.2019.2940199

Alhelou, H. H., Golshan, M. E. H., & Hatziargyriou, N. D. (2019). A decentralized functional observer based optimal lfc considering unknown inputs, uncertainties and cyber-attacks. *IEEE Transactions on Power Systems*.

Alhelou, H. H., Golshan, M. E. H., Zamani, R., Moghaddam, M. P., Njenda, T. C., Siano, P., & Marzband, M. (2019, June). An improved UFLS scheme based on estimated minimum frequency and power deficit. In 2019 IEEE Milan PowerTech (pp. 1-6). IEEE. doi:10.1109/PTC.2019.8810497

Alhelou, H. H., Golshan, M. H., & Askari-Marnani, J. (2018). Robust sensor fault detection and isolation scheme for interconnected smart power systems in presence of RER and EVs using unknown input observer. *International Journal of Electrical Power & Energy Systems*, *99*, 682–694. doi:10.1016/j.ijepes.2018.02.013

Alhelou, H. H., Hamedani-Golshan, M. E., Heydarian-Forushani, E., Al-Sumaiti, A. S., & Siano, P. (2018, September). Decentralized fractional order control scheme for LFC of deregulated nonlinear power systems in presence of EVs and RER. In *2018 International Conference on Smart Energy Systems and Technologies (SEST)* (pp. 1-6). IEEE. 10.1109/SEST.2018.8495858

Alhelou, H. S. H., Golshan, M. E. H., & Fini, M. H. (2015, December). Multi agent electric vehicle control based primary frequency support for future smart micro-grid . In *2015 Smart Grid Conference (SGC)* (pp. 22-27).

Anteneh, D., & Khan, B. (2019). Reliability enhancement of distribution substation by using network reconfiguration a case study at Debre Berhan Distribution Substation. *International Journal of Economy, Energy, & Environment*, *4*(2), 33.

Danfoss. (2011). *Danfoss power electronics global managers meeting (Pe Gmm): Steigenberger Hotel Hamburg, Germany*. Hamburg.

Fini, M. H., Yousefi, G. R., & Alhelou, H. H. (2016). Comparative study on the performance of many-objective and single-objective optimisation algorithms in tuning load frequency controllers of multi-area power systems. *IET Generation, Transmission, & Distribution*, *10*(12), 2915–2923. doi:10.1049/iet-gtd.2015.1334

Fredsted, J. (1998). *Handbook frequency converter – Facts worth knowing about frequency converters*. Danfoss Publications.

Haes Alhelou, H., Hamedani-Golshan, M. E., Njenda, T. C., & Siano, P. (2019). A survey on power system blackout and cascading events: Research motivations and challenges. *Energies*, *12*(4), 682. doi:10.3390/en12040682

Haes Alhelou, H., Hamedani-Golshan, M. E., Njenda, T. C., & Siano, P. (2019). Wide-area measurement system-based optimal multi-stage under-frequency load-shedding in interconnected smart power systems using evolutionary computing techniques. *Applied Sciences*, *9*(3), 508. doi:10.3390/app9030508

Haes Alhelou, H., Hamedani Golshan, M. E., Njenda, T. C., & Siano, P. (2019). WAMS-based online disturbance estimation in interconnected power systems using disturbance observer. *Applied Sciences*, *9*(5), 990. doi:10.3390/app9050990

Histand, M. B. (1999). *Introduction to mechatronics and Measurement Systems. McGraw-Hill Education*. ISE Editions.

Holtz, J., & Boelkens, U. (1989). Direct frequency convertor with sinusoidal line currents for speed-variable AC motors. *IEEE Transactions on Industrial Electronics*, *36*(4), 475–479. doi:10.1109/41.43005

Jariso, M., Khan, B., Tanwar, S., Tyagi, S., & Rishiwal, V. (2018). Hybrid energy system for upgrading the rural environment, IEEE GlobeCom 2018.

Jariso, M., Khan, B., Tesfaye, D., & Singh, J. (2017). Modelling and designing of stand-alone photovoltaic system (Case study: Addis Boder Health Center southwest Ethiopia). In *Proceedings IEEE International Conference on Electronics, Communication, and Aerospace Technology, ICECA 2017,* pp. 168-173, Coimbatore, India.

Khan, B., & Agnihotri, G. (2012). A novel transmission loss allocation method based on transmission usage. In *Proceedings IEEE Fifth Power India Conference*, Haryana, India. 10.1109/PowerI.2012.6479479

Khan, B., Agnihotri, G, & Gupta, G. (2013). A multipurpose matrices methodology for transmission usage, loss and reliability margin allocation in restructured environment, *Electrical & Computer Engineering: An International Journal, 2*(3).

Khan, B., Agnihotri, G., Rathore, P., Mishra, A., & Naidu, G. (2014). A cooperative game theory approach for usage and reliability margin cost allocation under contingent restructured market. *International Review of Electrical Engineering*, *9*(4), 854–862.

Khan, B., Alhelou, H. H., & Mebrahtu, F. (2019). A holistic analysis of distribution system reliability assessment methods with conventional and renewable energy sources. *AIMS Energy*, *7*(4), 413–429. doi:10.3934/energy.2019.4.413

Khan, B., & Singh, P. (2017). The current and future states of Ethiopia's energy sector and potential for green energy: A comprehensive study. *International Journal of Engineering Research in Africa*, *33*, 115–139. doi:10.4028/www.scientific.net/JERA.33.115

Khan, B., & Singh, P. (2017). Optimal power flow techniques under characterization of conventional and renewable energy sources: A comprehensive analysis, *Journal of Engineering, vol. 2017,* Article ID 9539506.

Khan, B., & Singh, P. (2017). Selecting a meta-heuristic technique for smart micro-grid optimization problem: a comprehensive analysis, in IEEE Access, 5, pp. 13951-13977, 2017.

Khan, B., & Singh, P. (2019). Economic operation of smart micro-grid: a meta-heuristic approach. In Handbook of research on smart power system operation and control (pp. 330-346). Hershey, PA: IGI Global. doi:10.4018/978-1-5225-8030-0.ch014

Khan, B., & Singh, P. (2019). Economic operation of smart micro-grid: A meta-heuristic approach. In Handbook of research on smart power system operation and control (pp. 330-346). Hershey, PA: IGI Global. doi:10.4018/978-1-5225-8030-0.ch014

Khan, B., & Tanwar, S. (2019). Issues associated with microgrid integration. In Handbook of research on smart power system operation and control (pp. 252-264). Hershey, PA: IGI Global. doi:10.4018/978-1-5225-8030-0.ch010

Kifle, Y., Khan, B., & Singh, J. (2018). Designing and modeling grid connected photovoltaic system: (Case study: EEU building at Hawassa city). *International Journal of Convergence Computing*, *3*(1), 20–34. doi:10.1504/IJCONVC.2018.091113

Kifle, Y., Khan, B., & Singh, P. (2018). Assessment and enhancement of distribution system reliability by renewable energy sources and energy storage, *Journal of Green Engineering, 8*(3), 2, pp. 219-262.

Makdisie, C. (2008). *Power electronics / 2.* Tishreen University Publications.

Makdisie, C. (2016). *Power electronics / 1.* Tishreen University Publications.

Makdisie, C., Haidar, B., & Alhelou, H. H. (2018). An optimal photovoltaic conversion system for future smart grids. In Handbook of research on power and energy system optimization (pp. 601–657). Hershey, PA: IGI Global. doi:10.4018/978-1-5225-3935-3.ch018

Mohan, N., Undeland, T. M., & Robbins, W. P. (2007). *Power electronics: converters, applications, and design.* New Delhi, India: Wiley India.

Molla, T., Khan, B., Moges, B., Alhelou, H. H., Zamani, R., & Siano, P. (2019). Integrated optimization of smart home appliances with cost-effective energy management system. *CSEE Journal of Power and Energy Systems*, *5*(2), 249–258.

Molla, T., Khan, B., & Singh, P. (2018). A comprehensive analysis of smart home energy management system optimization techniques, *Journal of Autonomous Intelligence, 1*(1), 15–21.

Negash, K., Khan, B., Tesfaye, D., & Gayathri, M. (2017). Optimal placement of phasor measurement unit for system observability. In *Proceedings of IEEE International Conference on Computing Methodologies and Communication, (ICCMC 2017),* pp. 73-77, Erode, India.

Negash, K., Khan, B., & Yohannes, E. (2016). Artificial intelligence versus conventional mathematical techniques: A review for optimal placement of phasor measurement units [Springer]. *Journal of Technology and Economics of Smart Grids and Sustainable Energy*, *1*(1), 10. doi:10.100740866-016-0009-y

Njenda, T. C., Golshan, M. E. H., & Alhelou, H. H. (2018, November). WAMS based intelligent under frequency load shedding considering online disturbance estimation. In *2018 Smart Grid Conference (SGC)* (pp. 1-5). 10.1109/SGC.2018.8777779

Njenda, T. C., Golshan, M. E. H., & Alhelou, H. H. (2018, November). WAMS based under frequency load shedding considering minimum frequency prediction and extrapolated disturbance magnitude. In *2018 Smart Grid Conference (SGC)* (pp. 1-5).

Salam, Z. (2003). *Power electronics and drives.* UTM-JB.

Singh, P., & Khan, B. (2017). Smart microgrid energy management using a novel artificial shark optimization, Complexity, vol. 2017, Article ID 2158926. doi:10.1155/2017/2158926

Singh, P., Khan, B., Vidyarthi, A., Haes Alhelou, H., & Siano, P. (2019). Energy-aware online non-clairvoyant scheduling using speed scaling with arbitrary power function. *Applied Sciences*, *9*(7), 1467. doi:10.3390/app9071467

Thorborg, K. (1988). *Power electronics*. New York: Prentice Hall.

Vodorozov, V. (2010). *Introduction to power electronics* (1st ed.).

Yeshalem, M. T., & Khan, B. (2017). Design of an off-grid hybrid PV/wind power system for remote mobile base station: A case study. *AIMS Energy*, *5*(1), 96–112. doi:10.3934/energy.2017.1.96

Zamani, R., Golshan, M. E. H., Alhelou, H. H., & Hatziargyriou, N. (2019). A novel hybrid islanding detection method using dynamic characteristics of synchronous generator and signal processing technique. *Electric Power Systems Research*, *175*, 105911. doi:10.1016/j.epsr.2019.105911

Zamani, R., Hamedani-Golshan, M. E., Haes Alhelou, H., Siano, P., & Pota, H. R. (2018). Islanding detection of synchronous distributed generator based on the active and reactive power control loops. *Energies*, *11*(10), 2819. doi:10.3390/en11102819

KEY TERMS AND DEFINITIONS

Chopper: is a basically static power electronics device which converts fixed DC voltage/power to variable DC voltage or power. It is nothing but a high-speed switch which connects and disconnects the load from source at a high rate to get variable or chopped voltage at the output.

Power electronics: is the technology associated with the efficient conversion, control and conditioning of electric power by static means from its available input form into the desired electrical output form.

Rectifier: is an electrical device that converts alternating current, which periodically reverses direction, to direct current, which flows in only one direction.

Voltage regulator: is a system designed to automatically maintain a constant voltage level. A voltage regulator may use a simple feed-forward design or may include negative feedback. It may use an electromechanical mechanism, or electronic components.

Compilation of References

ABB. (2018). Power factor correction and harmonic filtering in electrical plants, Technical application papers, July, 2018.

ABB. (2018, July). Power factor correction and harmonic filtering in electrical plants, Technical application papers.

Abbasi, A. R., & Seifi, A. R. (2015). Considering cost and reliability in electrical and thermal distribution networks reinforcement planning. *Energy*, *84*, 25–35. doi:10.1016/j.energy.2015.01.113

Abdel Aty, A. (2012). Framework for monitoring and maintenance subways stations using building information modeling. Construction Research Congress, 2320-2328.

Abdel Aty, A. (2013). *Maintaining subways infrastructures using BIM.* (Master Thesis, Cairo University).

Abdelaziz, A. Y., Ali, E. S., & Abd Elazim, S. M. (2016). Optimal sizing and locations of capacitors in radial distribution systems via flower pollination optimization algorithm and power loss index, *Engineering Science and Technology, an International Journal, Vol. 19*(1), pp. 610-618.

Abdin, I. F., & Zio, E. (2018). An integrated framework for operational flexibility assessment in multi-period power system planning with renewable energy production. *Applied Energy*, *222*, 898–914. doi:10.1016/j.apenergy.2018.04.009

Achraf, A., Abdelkarim, M., & Lotfi, K. (2018). Assisted power management of a stand-alone renewable multi-source system, *Energy, 145*(15), 195-205.

Adefarati, T., & Bansal, R. C. (2016). Integration of renewable distributed generators into the distribution system: A review. *IET Renewable Power Generation*, *10*(7), 873–884. doi:10.1049/iet-rpg.2015.0378

Adefarati, T., & Bansal, R. C. (2017). Reliability assessment of distribution system with the integration of renewable distributed generation. *Applied Energy*, *185*(Part 1), 158–171. doi:10.1016/j.apenergy.2016.10.087

Adefarati, T., & Bansal, R. C. (2017). The impacts of PV-wind-diesel-electric storage hybrid system on the reliability of a power system. *Energy Procedia*, *105*, 616–621. doi:10.1016/j.egypro.2017.03.364

Adegboye, B. A., & Ekundayo, K. R. (2011). Reliability assessment of 4.2 MW single shaft typhoon gas fired turbine power generation station (2003–2008). *Advanced Materials Research*, *367*, 143–150. doi:10.4028/www.scientific.net/AMR.367.143

Adly Girgis, S. B. (n.d.). Effect of distributed generation on protective device coordination in distribution system. IEEE, (pp. 115-119).

Aghaebrahimi, M. R., & Mehdizadeh, M. (2011). A new procedure in reliability assessment of wind–diesel islanded grids. *Electric Power Components and Systems, 39*(14), 1563–76.

Ahadi, A., Kang, S. K., & Lee, J. H. (2016). A novel approach for optimal combination of wind, PV and energy storage system in diesel - free isolated communities. *Applied Energy*, *170*, 101–115. doi:10.1016/j.apenergy.2016.02.110

Ahmad, F., Rasool, A., Ozsoy, E., Sekar, R., Sabanovic, A., & Elitaş, M. (2018). Distribution system state estimation-A step towards smart grid. *Renewable & Sustainable Energy Reviews*, *81*, 2659–2671. doi:10.1016/j.rser.2017.06.071

Akagi, H. (2016). Modern active filters and traditional passive filters. *Bulletin of the Polish Academy of Sciences. Technical Sciences*, *54*(3), 255–269.

Akasah. B. (2007) Generic process model for maintenance management of school buildings, Malaysia University.

Akshata, A. S., Harshada, V. H., Madhuri, S., & Mandasmita, R. K. (2012). Seminar report on microgrid.

Alemu, A. (2012). Reliability assessment of Bahir Dar town distribution system, (MSc. Thesis), Bahir Dar University, 2012.

Alhassan, B., & Jrad, F. (2015). *Enhancing maintenance management for governmental buildings using building information modelling (BIM),* (Master Thesis, Teshreen University, Latakia).

Alhelou, H. H., & Golshan, M. E. H. (2016, May). Hierarchical plug-in EV control based on primary frequency response in interconnected smart grid. In *2016 24th Iranian Conference on Electrical Engineering (ICEE),* (pp. 561-566). IEEE. 10.1109/IranianCEE.2016.7585585

Alhelou, H. H., Golshan, M. E. H., Zamani, R., Moghaddam, M. P., Njenda, T. C., Siano, P., & Marzband, M. (2019, June). An improved UFLS scheme based on estimated minimum frequency and power deficit. In 2019 IEEE Milan PowerTech (pp. 1-6). IEEE. doi:10.1109/PTC.2019.8810497

Alhelou, H. H., Golshan, M., & Fini, M. (2018). Wind driven optimization algorithm application to load frequency control in interconnected power systems considering GRC and GDB nonlinearities. *Electric Power Components and Syst.*

Alhelou, H. H., Golshan, M., & Fini, M. (2018). Wind driven optimization algorithm application to load frequency control in interconnected power systems considering GRC and GDB nonlinearities. *Electric Power Components and Systems, 46*(11-12), 1223-1238.

Alhelou, H. S. H., Golshan, M. E. H., & Fini, M. H. (2015, December). Multi agent electric vehicle control based primary frequency support for future smart micro-grid. In *2015 Smart Grid Conference (SGC)* (pp. 22-27). IEEE.

Alhelou, H. S. H., Golshan, M. E. H., & Fini, M. H. (2015, December). Multi agent electric vehicle control based primary frequency support for future smart micro-grid. In *2015 Smart Grid Conference (SGC),* (pp. 22-27). IEEE.

Alhelou, H. H. (2018). *Fault detection and isolation in power systems using unknown input observer. Advanced condition monitoring and fault diagnosis of electric machines* (p. 38). Hershey, PA: IGI Global.

Alhelou, H. H. (2018). *Fault detection and isolation in power systems using unknown input observer. In Advanced condition monitoring and fault diagnosis of electric machines* (p. 38). Hershey, PA: IGI Global.

Alhelou, H. H., & Golshan, M. E. H. (2020). Decision-making-based optimal generation-side secondary-reserve scheduling and optimal LFC in deregulated interconnected power system. In *Decision making applications in modern power systems* (pp. 269–299). Academic Press. doi:10.1016/B978-0-12-816445-7.00011-6

Alhelou, H. H., Golshan, M. E. H., & Hatziargyriou, N. D. (2019). A decentralized functional observer based optimal lfc considering unknown inputs, uncertainties and cyber-attacks. *IEEE Transactions on Power Systems.*

Alhelou, H. H., Golshan, M. E. H., & Hatziargyriou, N. D. (2019). A decentralized functional observer based optimal LFC considering unknown inputs, uncertainties and cyber-attacks. *IEEE Transactions on Power Systems.*

Alhelou, H. H., Golshan, M. E. H., & Hatziargyriou, N. D. (2019). A decentralized functional observer based optimal LFC considering unknown inputs, uncertainties, and cyber-attacks. *IEEE Transactions on Power Systems*.

Alhelou, H. H., Golshan, M. E. H., & Hatziargyriou, N. D. (2019). A decentralized functional observer-based optimal LFC considering unknown inputs, uncertainties, and cyber-attacks. *IEEE Transactions on Power Systems*.

Alhelou, H. H., Golshan, M. E. H., & Hatziargyriou, N. D. (2019). Deterministic dynamic state estimation-based optimal LFC for interconnected power systems using unknown input observer. *IEEE Transactions on Smart Grid*, 1. doi:10.1109/TSG.2019.2940199

Alhelou, H. H., Golshan, M. H., & Askari-Marnani, J. (2018). Robust sensor fault detection and isolation scheme for interconnected smart power systems in presence of RER and EVs using unknown input observer. *International Journal of Electrical Power & Energy Systems*, *99*, 682–694. doi:10.1016/j.ijepes.2018.02.013

Alhelou, H. H., Hamedani-Golshan, M. E., Heydarian-Forushani, E., Al-Sumaiti, A. S., & Siano, P. (2018, September). Decentralized fractional order control scheme for LFC of deregulated nonlinear power systems in presence of EVs and RER. In *Proceedings 2018 International Conference on Smart Energy Systems and Technologies (SEST)* (pp. 1-6). IEEE. 10.1109/SEST.2018.8495858

Alhelou, H. S. H., Golshan, M. E. H., & Fini, M. H. (2015, December). Multi agent electric vehicle control based primary frequency support for future smart micro-grid . In *2015 Smart Grid Conference (SGC)* (pp. 22-27).

Alhelou, H. S. H., Golshan, M. E. H., & Fini, M. H. (2015, December). Multi agent electric vehicle control based primary frequency support for future smart micro-grid. In *Proceedings Smart Grid Conference (SGC)* (pp. 22-27).

Alhelou, H., Hamedani-Golshan, M. E., Zamani, R., Heydarian-Forushani, E., & Siano, P. (2018). Challenges and opportunities of load frequency control in conventional, modern and future smart power systems: A comprehensive review. *Energies*, *11*(10), 2497. doi:10.3390/en11102497

Ali, A., Nor, N. M., Ibrahim, T., & Romlie, M. F. (2017). Sizing and placement of battery-coupled distributed photovoltaic generations. *Journal of Renewable and Sustainable Energy*, *9*(5). doi:10.1063/1.4995531

Alkuhayli, A. A., Raghavan, S., & Chowdhury, B. H. (2012). Reliability evaluation of distribution systems containing renewable distributed generations, in *North American Power Symposium (NAPS)*, 2012, pp. 1-6. 10.1109/NAPS.2012.6336324

Alkuhayli, A. A. Reliability evaluation of distribution system contained renewable distributed generations. In *Proceedings IEEE North American Power Symposium (NAPS)*, Champaign, IL. IEEE. pp. 1–6.

Allirani, S., & Thurai, R. (2015). *V. B. development of space vector pulse width modulation algorithm for voltage source inverter using dsPIC controller 30f4011(IJARCCE), 2*(3), pp. 01-05.

Al-Muhaini, M., Heydt, G. T., & Huynh, A. (2010). The reliability of power distribution systems as calculated using system theoretic concepts. *IEEE PES General Meeting*, Providence, RI, pp. 1-8. 10.1109/PES.2010.5588116

Al-Muhaini, M., & Heydt, G. T. (2013). Evaluating future power distribution system reliability including distributed generation. *IEEE Transactions on Power Delivery*, *28*(4), 2264–2272. doi:10.1109/TPWRD.2013.2253808

Amanifar, O., & Golshan, M. E. H. (2012, May). Optimal DG allocation and sizing for mitigating voltage sag in distribution systems with respect to economic consideration using Particle Swarm Optimization. In 2012 Proceedings of 17th Conference on Electrical Power Distribution (pp. 1-8). IEEE.

Ansal, V., Ravikumar, K., & Parthiban, P. (2016). Transformer less dynamic voltage restorer for voltage sag mitigation, *2016 Biennial International Conference on Power and Energy Systems: Towards Sustainable Energy (PESTSE)*, Bangalore, India, 2016, pp. 1-4.

Anteneh, D., & Khan, B. (2019). Reliability enhancement of distribution substation by using network reconfiguration a case study at Debre Berhan Distribution Substation, *International Journal of Economy, Energy, & Environment*, *4*(2), 33.

Anteneh, D., & Khan, B. (2019). Reliability enhancement of distribution substation by using network reconfiguration A case study at Debre Berhan distribution substation, *International Journal of Economy, Energy, & Environment*, *4*(2), 33.

Anteneh, D., & Khan, B. (2019). Reliability enhancement of distribution substation by using network reconfiguration a case study at Debre Berhan Distribution Substation. *International Journal of Economy, Energy & Environment*, *4*(2), 33.

Anteneh, D., & Khan, B. (2019). Reliability enhancement of distribution substation by using network reconfiguration a case study at Debre Berhan distribution substation. *International Journal of Economy, Energy, & Environment*, *4*(2), 33.

Anteneh, D., & Khan, B. (2019). Reliability enhancement of distribution substation by using network reconfiguration a case study at Debre Berhan Distribution Substation. *International Journal of Economy, Energy, & Environment*, *4*(2), 33.

Anteneh, D., & Khan, B. (2019). Reliability enhancement of distribution substation by using network reconfiguration, A case study at Debre Berhan Distribution Substation. *International Journal of Economy, Energy, & Environment*, *4*(2), 33.

Anteneh, D., & Khan, B. (2019). Reliability enhancement of distribution substation by using network reconfiguration: A case study at Debre Berhan Distribution Substation. *International Journal of Economy, Energy, & Environment*, *4*(2), 33.

Arditi, N. (1999). Designing buildings for maintenance: Designers' perspective, Technical Papers, 117–132.

Arefi, A., Haghifam, M. R., Fathi, S. H., Niknam, T., & Olamaei, J. (2009, June). A novel algorithm based on honey bee mating optimization for distribution harmonic state estimation including distributed generators. In 2009 IEEE Bucharest PowerTech (pp. 1-7). IEEE.] doi:10.1109/PTC.2009.5282101

Arifujjaman, M. D. (2015). A comprehensive power loss, efficiency, reliability and cost calculation of a 1 MW/500 kWh battery-based energy storage system for frequency regulation application. *Renewable Energy*, *74*, 158–169. doi:10.1016/j.renene.2014.07.046

Awadalla, M., Omer, M., & Mohamed, A. (2015). Single-tuned filter design for harmonic mitigation and optimizationwith capacitor banks. In *Proceedings of the 1st International Conference on Computing, Control, Networking, Electronics, and Embedded Systems Engineering, (ICCNEEE '15)*, pp. 242–247, IEEE, Khartoum, Sudan.

Babaei, E., & Kangarlu, M. F. (2015). Comparison four topologies for three-phase dynamic voltage restorer. In *Proceedings 2015 International Conference on Renewable Energy Research and Applications (ICRERA)*, Palermo, Sicily. pp. 1527-1532. 10.1109/ICRERA.2015.7418663

Badran, O., Mekhilef, S., Mokhlis, H., & Dahalan, W. (2017). Optimal reconfiguration of distribution system connected with distributed generations: A review of different methodologies. *Renewable & Sustainable Energy Reviews*, *73*, 854–867. doi:10.1016/j.rser.2017.02.010

Bahmani-Firouzi, B., & Azizipanah-Abarghooee, R. (2014). Optimal sizing of battery energy storage for micro-grid operation management using a new improved bat algorithm. In *Electric Power Energy Systems, 56,* 42-54.

Bajpai, R. S., & Gupta, R. (2011, June). Series compensation to mitigate harmonics and voltage sags/swells in distributed generation based on symmetrical components estimation, *IEEE International Symposium* Industrial Electronics (ISIE*)*, 1639–1644.

Balta-Ozkan, N., Watson, T., Connor, P., Axon, C., Whitmarsh, L., Davidson, R., Spence, A., ... Taylor, G. (2015). Scenarios for the development of smart grids in the UK -Synthesis Report (UKERC: London, UK).

Banaei, M. R., Hosseini, S. H., & Khajee, M. D. (2006, August). Mitigation of voltage sag using adaptive neural network with dynamic voltage restorer. In Proceedings Power Electronics and Motion Control Conference, IPEMC, pp. 1–5.

Banteywalu, S., Khan, B., De Smedt, V., & Leroux, P. (2019). A novel modular radiation hardening approach applied to a synchronous buck converter. *Electronics (Basel)*, *8*(5), 513. doi:10.3390/electronics8050513

Baradaran, R. H., & Seyed-Sheneva, J. (2012, March-April). Electrical distribution system reliability improvement by optimal placement of fault indicators using immune algorithm. *International Journal of Engineering Research and Applications*, *2*(2), 1383–1390.

Baran, M. E., Jung, J., & McDermott, T. E. (2009, July). Including voltage measurements in branch current state estimation for distribution systems. In 2009 IEEE Power & Energy Society General Meeting (pp. 1-5). IEEE. doi:10.1109/PES.2009.5275479

Baran, M. E., & Kelley, A. W. (1994). State estimation for real-time monitoring of distribution systems. *IEEE Transactions on Power Systems*, *9*(3), 1601–1609. doi:10.1109/59.336098

Baran, M. E., & Kelley, A. W. (1995). A branch-current-based state estimation method for distribution systems. *IEEE Transactions on Power Systems*, *10*(1), 483–491. doi:10.1109/59.373974

Barchi, G. (2015). *Algorithms and performance analysis for synchrophasor and grid state estimation* (Doctoral dissertation, University of Trento).

Barker, P. P., & De Mello, R. W. (2000, July). Determining the impact of distributed generation on power systems. I. Radial distribution systems. In *2000 Power Engineering Society Summer Meeting* (Cat. No. 00CH37134) (Vol. 3, pp. 1645-1656). IEEE.

Beena, V., Jayaraju, M., & Sebin, D. K. (2018). Active and reactive power control of single phase transformerless grid connected inverter for distributed generation system. *International Journal of Applied Engineering Research*, *13*(1), 150–157.

Bhandari, B., Lee, K. T., Lee, C. S., Song, C., Maskey, R. K., & Ahn, S. (2014). A novel off-grid hybrid power system comprised of solar photovoltaic, wind, and hydro energy sources. *Applied Energy*, *133*, 236–242. doi:10.1016/j.apenergy.2014.07.033

Bhullar, S., & Ghosh, S. (2018). Optimal integration of multi distributed generation sources in radial distribution networks using a hybrid algorithm. *Energies*, *12*(4-5), 2–15.

Billinton, R., & Allan, R. N. (1996). *Reliability evaluation of power systems* (2nd ed.). New York: Plenum Press. doi:10.1007/978-1-4899-1860-4

Billinton, R., & Gao, Y. (2006). Multistate wind energy conversion system models for adequacy assessment of generating systems incorporating wind energy. *IEEE Transactions on Energy Conversion*, *23*(1), 163–170. doi:10.1109/TEC.2006.882415

Bin Akasah. (2007). *Generic process model for maintenance management of school buildings*. Malaysia University.

Blaauwbroek, N. (2018). Flexibility in active distribution network management: A new horizon in monitoring and control for grid supportive demand side management.

Boktor, C. G., Youssef, A. R., Ali, A. H., & Kamel, S. (2017). Optimal distribution power flow including shunt capacitor allocation based on voltage deviation and power loss minimization. In *Proceedings 2017 Nineteenth International Middle East Power Systems Conference (MEPCON)*, Menoufia University, 2017. 10.1109/MEPCON.2017.8301288

Bollen, M., & Hassan, F. (2011). *Integration of distributed generation in the power system.* IEEE Press. doi:10.1002/9781118029039

Borges, C. L. T. (2012). An overview of reliability models and methods for distribution systems with renewable energy distributed generation. *Renewable & Sustainable Energy Reviews, 16*(6), 4008–4015. doi:10.1016/j.rser.2012.03.055

Bornand, E. (1991) Le guide de maintenance, Berne.

Bouzid, A. M., Guerrero, J. M., Cheriti, A., Bouhamida, M., Sicard, P., & Benghanem, M. (2015) A survey on control of electric power distributed generation systems for microgrid applications. *Renew Sustain Energy Rev 2015,* 44, 7, 51 –66.

Bove, R., Bucher, M., & Ferretti, F. (2012). Integrating large shares of wind energy in macro economical cost-effective way. *Energy, 43,* 438-47.

Brown, R., Gupta, S., Christie, R., Venkata, S., & Fletcher, R. (1997). Automated primary distribution system design: Reliability and cost optimization. *IEEE Transactions on Power Delivery, 12*(2), 1017–1022. doi:10.1109/61.584431

Brumsickle, W. E., Schneider, R. S., Luckjiff, G. A., Divan, D. M., & McGranaghan, M. F. (2001). Dynamic sag correctors: Cost-effective industrial power line conditioning. *IEEE Transactions on Industry Applications, 37*(1), 212–217. doi:10.1109/28.903150

Cabrane, Z, Ouassaid, M, & Maaroufi, M. (2016). Analysis and evaluation of battery-supercapacitor hybrid energy storage system for photovoltaic Installation. *International Journal of Hydrogen Energy - Elsevier,* 1-11.

Carbone, R. (2009). Grid-connected photovoltaic systems with energy storage, International Conference on Clean Electrical Power 760–767. IEEE.

Carmen, L. T., & Borges, D. F. (2006). Optimal distributed generation allocation for reliability, losses, and voltage improvement. *Electrical Power and Energy Systems,* 413-420.

Celik, M. K., & Liu, W. H. (1999, January). A practical distribution state calculation algorithm. In *IEEE Power Engineering Society. 1999 Winter Meeting (Cat. No. 99CH36233)* (Vol. 1, pp. 442-447). IEEE] 10.1109/PESW.1999.747496

Čepin, M. (2011). Assessment of power system reliability, Springer-Verlag London, 2011.

Chakravorty, J. (2015). Network reconfiguration of distribution system using fuzzy controlled evolutionary programming, *International Journal Of Engineering Science & Advanced Technology, 2*(2), 176–182, 2015.

Chan, T. F. (2001). Synchronous machines. Encyclopedia of Life Support Systems, Polythecnic University, Hong Kong. Energy or Technologies, Inc. (n.d.). Retrieved from http://www.energyor.com/energyor/fuel_cell.cfm

Changfu, Z., Lei, Z., Xiaosong, H., Zhenpo, W., Torsten, W., & Michael, P. (2018). A review of fractional-order techniques applied to lithium-ion batteries, lead-acid batteries, and supercapacitors. *Journal of Power Sources, 390,* 286–296. doi:10.1016/j.jpowsour.2018.04.033

Chattopadhyay, S., Sengupta, S., & Mitra, M. (2015). *Electric power quality.* Springer.

Chaudhary, M. K., & Prakash, S. (2019). Investigation of blade geometry and airfoil for small wind turbine blade, *Advanced Science, Engineering in Medicine, 11*(5), 448–452.

Che, L., Khodayar, M., & Shahidehpour, M. (2014). Only connect: Microgrids for distribution system restoration. *IEEE Power & Energy Magazine, 12*(1), 70–81. doi:10.1109/MPE.2013.2286317

Chen, F., Huang, G., & Fan, Y. (2015). A linearization and parameterization approach to tri-objective linear programming problems for power generation expansion planning. *Energy, 87,* 240–250. doi:10.1016/j.energy.2015.04.104

Chlodnicki, Z., & Koczara, W. (n.d.). Supercapacitor *(2005).* Storage application for reduction drive negative impact on supply grid. *IEEE Compatibility in Power Electronics.* doi:10.1109/cpe.2005.1547549

Chlodnicki, Z., Koczara, W., & Al-Khayat, N. (2007). Hybrid UPS based on supercapacitor energy storage and adjustable speed generator. In *Proceedings 2007.* Gdansk, Poland: Compatibility in Power Electronics; doi:10.1109/cpe.2007.4296512

Chowdhury, A. A., Agarwal, S. K., & Koval, D. O. (2013). Reliability modelling of distributed generation in conventional distribution systems planning and analysis. *IEEE Transactions on Industry Applications*, *39*(5), 1493–1501. doi:10.1109/TIA.2003.816554

Christeen, G. B., Abdel-Raheem, Y., Ali, A. H., & Kamel, S. (2017). Optimal distribution power flow including shunt capacitor allocation based on voltage deviation and power loss minimization. In *2017 Nineteenth International Middle East Power Systems Conference (MEPCON*), Menoufia University, 2017.

Chusovitin, P., Polyakov, I., & Pazderin, A. (2016, November). Three-phase state estimation model for distribution grids. In *2016 IEEE International Conference on the Science of Electrical Engineering (ICSEE)* (pp. 1-5). IEEE. 10.1109/ICSEE.2016.7806063

Danfoss. (2011). *Danfoss power electronics global managers meeting (Pe Gmm): Steigenberger Hotel Hamburg, Germany*. Hamburg.

Das, J. C. (2015). Power system harmonics and passive filter designs, IEEE Press.

de la O Serna, J. A., & Rodriguez-Maldonado, J. (2011). Instantaneous oscillating phasor estimates with Taylor $^ K $-Kalman filters. *IEEE Transactions on Power Systems*, *26*(4), 2336–2344. doi:10.1109/TPWRS.2011.2157539

de Queiroz, L. M. O., & Forrer, J. (2012). Assessing the overall performance of Brazilian electric distribution companies. George Washington University, Washington, DC.

Debnath, K. B., & Mourshed, M. (2018). Forecasting methods in energy planning models. *Renewable & Sustainable Energy Reviews*, *88*, 297–325. doi:10.1016/j.rser.2018.02.002

De, D., Klumpner, C., Patel, C., Ponggorn, K., Rashed, M., & Asher, G. (2013). Modelling and control of a multi-stage interleaved DC–DC converter with coupled inductors for super capacitor energy storage system. *IET Power Electronics*, *6*(7), 1360–1375. doi:10.1049/iet-pel.2012.0529

Della Giustina, D., Pau, M., Pegoraro, P. A., Ponci, F., & Sulis, S. (2014). Electrical distribution system state estimation: Measurement issues and challenges. *IEEE Instrumentation & Measurement Magazine*, *17*(6), 36–42. doi:10.1109/MIM.2014.6968929

Deng, Y., He, Y., & Zhang, B. (2002). A branch-estimation-based state estimation method for radial distribution systems. *IEEE Transactions on Power Delivery*, *17*(4), 1057–1062. doi:10.1109/TPWRD.2002.803800

Dezaki, H. H., Abyaneh, H. A., Garmrudi, M., Mahdinia, H., & Mazlumi, K. (2014). A novel approach based on reliability sensitivity analysis to allocate protective devices. *Turkish Journal of Electrical Engineering & Computer Sciences*, *22*, 315–326. doi:10.3906/elk-1202-86

Ding, H., Swangyan, S., Xianzhong, D., & Jun, G. (2002). A novel dynamic voltage restorer and its unbalanced control strategy based on space vector PWM. *Electrical Power and Energy Systems*, *24*(9), 693–699. doi:10.1016/S0142-0615(02)00004-2

Ding, Y., Wang, P., Goel, L., Loh, P. C., & Wu, Q. (2011). Long-term reserve expansion of power systems with high wind power penetration using universal generating function methods. *IEEE Transactions on Power Systems*, *26*(2), 766–774. doi:10.1109/TPWRS.2010.2054841

Doetsch, C., Berthold, S., Wolf, D., Smolinka, T., Tulbke, J., Bretschneider, P., & Radgen, P. (2007). Electrical energy storage from 100 kW–state of the art technologies, realisations, fields of use. In *Second International Renewable Energy Storage Conference (IRES II)*, Bonn, Germany .

Dogaheh, Z., & Dogaheh, M. (2017). Optimizing the performance of smart grids in relation with residential energy centers equipping with solar power units (PV), *Sindhological Studies, 2017 (1),* 14-22.

Dohn, R. L. (2011). The business case for micro grids. White paper: The new face of energy modernization.

Dorji, T. (2009). Reliability assessment of distribution systems a case study on Wang due distribution system in Bhutan, (MSc. Thesis), NTNU, Norway.

Dragicevi, T., Lu, X., Vasquez, J. C., & Guerrero, J. M. (2016a). DC micro grids part I: A review of control strategies and stabilization techniques. *IEEE Transactions on Power Electronics*, *31*(7), 4876–4891.

Dragicevi, T., Lu, X., Vasquez, J. C., & Guerrero, J. M. (2016b). DC micro grids part II: A review of power architectures, applications, and standardization issues. *IEEE Transactions on Power Electronics*, *31*(5), 3528–3549. doi:10.1109/TPEL.2015.2464277

Duan, D. L., Ling, X. D., Wu, X. Y., & Zhong, B. (2015). Reconfiguration of distribution network for loss reduction and reliability improvement based on an enhanced genetic algorithm. *International Journal of Electrical Power & Energy Systems*, *64*, 88–95. doi:10.1016/j.ijepes.2014.07.036

Dufo-López, R., & Bernal-Agustín, J. L. (2015). Techno-economic analysis of grid connected battery storage. *Energy Conversion and Management*, *91*, 394–404. doi:10.1016/j.enconman.2014.12.038

Dugan, R. C., McGranaghan, M. F., & Beaty, H. W. (1996). Electrical power systems quality. New York: McGraw-Hill.

Durgaprasad, G., & Thakur, S. S. (1998). Robust dynamic state estimation of power systems based on M-estimation and realistic modeling of system dynamics. *IEEE Transactions on Power Systems*, *13*(4), 1331–1336. doi:10.1109/59.736273

Eep. (n.d.). *Uninterruptible static power supplies and the protection of persons.* Retrieved from https://electrical-engineering-portal.com/download-center/books-and-guides/schneider-electric/uninterruptible-static-power

Elsheikh, A., Helmy, Y., Abouelseoud, Y., & Elsherif, A. (2014, November). Optimal capacitor placement and sizing in radial electric power systems. *Alexandria Engineering Journal*, *10*(4), 809–816. doi:10.1016/j.aej.2014.09.012

Energy efficiency and demand side management, Course Module, University of Warwick, REEEP, April, 2005.

Ervural, B. C., Evren, R., & Delen, D. (2018). A multi-objective decision-making approach for sustainable energy investment planning. *Renewable Energy*, *126*, 387–402. doi:10.1016/j.renene.2018.03.051

Esau, Z., & Jayaweera, D. (2014). Reliability assessment in active distribution networks with detailed effects of PV systems. *J Mod Power Syst Clean Energy*, *2*(1), 59–68. doi:10.100740565-014-0046-2

Eyer, J., & Corey, G. (2010). Energy storage for the electricity grid: Benefits and market potential assessment guide, Report, Sandia National Laboratories.

Fares, R. L., & Webber, M. E. (2015). Combining a dynamic battery model with higher solution smart grid data to assess micro grid islanding lifetime. *Applied Energy*, *137*, 482–489. doi:10.1016/j.apenergy.2014.04.049

Farhoodnea, M., Mohamed, A., & Shareef, H. (2014, May). A comparative study on the performance of custom power devices for power quality improvement. In *Proceedings 2014 IEEE Innovative Smart Grid Technologies-Asia (ISGT ASIA),* (pp. 153-157). IEEE.

Ferreira, G. D., Bretas, A. S., & Cardoso, G. (2010, September). Optimal distribution protection design considering momentary and sustained reliability indices. In *2010 Modern Electric Power Systems* (pp. 1-8). IEEE.

Fini, M. H., Yousefi, G. R., & Alhelou, H. H. (2016). Comparative study on the performance of many-objective and single-objective optimisation algorithms in tuning load frequency controllers of multi-area power systems. *IET Generation, Transmission, & Distribution, 10*(12), 2915–2923. doi:10.1049/iet-gtd.2015.1334

Flores-Quiroz, A., Palma-Behnke, R., Zakeri, G., & Moreno, R. (2016). A column generation approach for solving generation expansion planning problems with high renewable energy penetration. *Electric Power Systems Research, 136*, 232–241. doi:10.1016/j.epsr.2016.02.011

Francy, R. (2013). *Enhanced power system state estimation techniques for the incorporation of variable energy* (Doctoral dissertation, Masdar Institute of Science and Technology).

Frangopol, D. (2003). Reliability deterioration and lifetime maintenance cost optimization. University of Colorado, Boulder.

Fredsted, J. (1998). *Handbook frequency converter – Facts worth knowing about frequency converters*. Danfoss Publications.

Gaikwad, D., & Mehraeen, S. (2012, April). Reactive power considerations in reliability analysis of photovoltaic systems. In Proceedings 2012 IEEE Green Technologies Conference, pp. 1–6. doi:10.1109/GREEN.2012.6200982

Gao, Y., & Yu, N. (2017, April). State estimation for unbalanced electric power distribution systems using AMI data. In 2017 IEEE Power & Energy Society Innovative Smart Grid Technologies Conference (ISGT) (pp. 1-5). IEEE. doi:10.1109/ISGT.2017.8085999

Gao, Y., & Billinton, R. (2009). Adequacy assessment of generating systems containing wind power considering wind speed correlation. *IET Renewable Power Generation, 3*(2), 217–226. doi:10.1049/iet-rpg:20080036

Gaur, G., Mehta, N., Khanna, R., & Kaur, S. (2017). Demand side management in a smart grid environment. In *Proceedings IEEE International Conference on Smart Grid and Smart Cities*, PEC University of Technology, Chandigarh, India, 2017. 10.1109/ICSGSC.2017.8038581

Gelagaev, R., Vermeyen, P., & Driesen, J. (2008, September). State estimation in distribution grids. In *2008 13th International Conference on Harmonics and Quality of Power* (pp. 1-6). IEEE. 10.1109/ICHQP.2008.4668803

Geng, L., & Lu, Z. (2016). A novel model for home energy management system based on internet of things. *IEEE International Conference on Power and Renewable Energy*, 2016.

Gers, J. M. (2013). *Distribution system analysis and automation*. IET. doi:10.1049/PBPO068E

Geth, F., Brijs, T., Kathan, J., Driesen, J., & Belmans, R. (2015). An overview of large-scale stationary electricity storage plants in Europe: Current status and new developments. *Renewable & Sustainable Energy Reviews, 52*, 1212–1227. doi:10.1016/j.rser.2015.07.145

Ghahremani, E., & Kamwa, I. (2011). Dynamic state estimation in power system by applying the extended Kalman filter with unknown inputs to phasor measurements. *IEEE Transactions on Power Systems, 26*(4), 2556–2566. doi:10.1109/TPWRS.2011.2145396

Ghajar, R. F., & Billinton, R. (2006). Economic costs of power interruptions: A consistent model and methodology. *International Journal of Electrical Power & Energy Systems, 28*(1), 29–35. doi:10.1016/j.ijepes.2005.09.003

Ghosh, A. K., Lubkeman, D. L., Downey, M. J., & Jones, R. H. (1997). b). Distribution circuit state estimation using a probabilistic approach. *IEEE Transactions on Power Systems, 12*(1), 45–51. doi:10.1109/59.574922

Ghosh, A. K., Lubkeman, D. L., & Jones, R. H. (1997). Load modeling for distribution circuit state estimation. *IEEE Transactions on Power Delivery, 12*(2), 999–1005. doi:10.1109/61.584427

GIZ. (2017). *Analysis of Indian electricity distribution systems for the integration of high shares of rooftop PV, final report*. GIZ.

Gokhale, S. S., & Kale, V. S. (2016). An application of a tent map initiated chaotic firefly algorithm for optimal overcurrent relay coordination. *International Journal of Electrical Power & Energy Systems, 78*, 336–342. doi:10.1016/j.ijepes.2015.11.087

Gomez-Exposito, A., & Abur, A. (2004). Power system state estimation: theory and implementation. Boca Raton, FL: CRC Press.

Gonen, T. (2013). Electric power distribution engineering (3rd ed.). Boca Raton, FL: CRC Press.

Gonen, T. (2014). Electric power distribution engineering, Third Ed. Boca Raton, FL: CRC Press.

Gonzalez-Longatt, F. M. (2008, Junio). Impacto de la Generación Distribuida en el Comportamiento de los Sistemas de Potencia. Universidad Central de Venezuela.

González, M., Cárdenas, V., & Espinosa, G. (2014). Advantages of the passivity-based control in dynamic voltage restorers for power quality improvement. *Simulation Modelling Practice and Theory, 47*, 221–235. doi:10.1016/j.simpat.2014.06.009

Granado, P. C. D., Pang, Z., & Wallace, S. W. (2016). Synergy of smart grids and hybrid distributed generation on the value of energy storage. *Applied Energy, 170*, 476–488. doi:10.1016/j.apenergy.2016.01.095

Guerrero, J. M., Vasquez, J. C., Matas, J., de Vicuña, L. G., & Castilla, M. (2011). Hierarchical control of droop-controlled AC and DC microgrids: A general approach toward standardization. *IEEE Transactions on Industrial Electronics, 58*(1), 158–172. doi:10.1109/TIE.2010.2066534

Gupta, K., Khan, B., & Mubeen, S. E. (2015). Available transfer capability enhancement by unified power flow controller. In *Proceedings IEEE International Conference on Signal Processing, Informatics, Communication and Energy Systems (SPICES)*, 2015, pp. 1, 4, 19-21.

Gupta, K., Khan, B., & Mubeen, S. E. (2015). Available transfer capability enhancement by unified power flow controller. In *Proceedings IEEE International Conference on Signal Processing, Informatics, Communication, and Energy Systems (SPICES)*, 2015, pp. 1, 4, 19-21.

Gupta, K., Khan, B., & Mubeen, S. E. (2015). Available transfer capability enhancement by unified power flow controller. In *Proceedings IEEE International Conference on Signal Processing, Informatics, Communication, and Energy Systems (SPICES)*, pp. 1, 4, 19-21.

Gupta, K., Khan, B., & Mubeen, S. E. (2015). Available transfer capability enhancement by unified power flowcontroller, *IEEE International Conference on Signal Processing, Informatics, Communication and Energy Systems (SPICES)*, pp. 1, 4, 19-21.

Gupta, K., Khan, B., & Mubeen, S. E. (2015). Available transfer capability enhancement by unified power flowcontroller, *IEEE International Conference on Signal Processing, Informatics, Communication, and Energy Systems (SPICES)*, pp. 1, 4, 19-21.

Gupta, K., Khan, B., & Mubeen, S. E. (2015). Available transfer capability enhancement by unified power flowcontroller. *IEEE International Conference on Signal Processing, Informatics, Communication, and Energy Systems (SPICES)*, 2015, pp. 1, 4, 19-21.

Gupta, K., Khan, B., & Mubeen, S. E. (2015). Available transfer capability enhancement by unified power flowcontroller. In *IEEE International Conference on Signal Processing, Informatics, Communication, and Energy Systems (SPICES)*, pp. 1, 4, 19-21.

Gupta, K., Khan, B., & Mubeen, S. E. (2015). Available transfer capability enhancement by unified power flowcontroller. In *Proceedings IEEE International Conference on Signal Processing, Informatics, Communication, and Energy Systems (SPICES)*, pp. 1, 4, 19-21.

Gupta, K., Khan, B., & Mubeen, S. E. (2015, February). Available transfer capability enhancement by unified power flowcontroller. In *Proceedings IEEE International Conference on Signal Processing, Informatics, Communication, and Energy Systems (SPICES)*, 2015, pp. 1, 4, 19-21.

Haddad, *R., & Hassan, B.* (2001). *Develop a methodology to analyze the factors affecting maintenance costs in government hospitals,* Tishreen University Magazine.

Haes Alhelou, H., Hamedani Golshan, M. E., Njenda, T. C., & Siano, P. (2019). WAMS-based online disturbance estimation in interconnected power systems using disturbance observer. *Applied Sciences*, *9*(5), 990. doi:10.3390/app9050990

Haes Alhelou, H., Hamedani-Golshan, M. E., Njenda, T. C., & Siano, P. (2019). A survey on power system blackout and cascading events: Research motivations and challenges. *Energies*, *12*(4), 682. doi:10.3390/en12040682

Haes Alhelou, H., Hamedani-Golshan, M. E., Njenda, T. C., & Siano, P. (2019). Wide-area measurement system-based optimal multi-stage under-frequency load-shedding in interconnected smart power systems using evolutionary computing techniques. *Applied Sciences*, *9*(3), 508. doi:10.3390/app9030508

Häger, M. F. S. (2006). The impact of distributed energy resources on distribution-system protection. Ludvika, Sweden.

Hajipour, E., Bozorg, M., & Fotuhi-Firuzabad, M. (2015). Stochastic capacity expansion planning of remote micro grids with wind farms and energy storage. *IEEE Trans. Sust. Energy*, *6*(2), 491–498. doi:10.1109/TSTE.2014.2376356

Halpin, S. M., Ribeiro, P. F., & Dai, J. J. (1998). Frequency-domain harmonic analysis methods, IEEE Power Engineering Society: Tutorial in Harmonics Modeling and Simulation, pp. 49–54.

Hashim, T. J., & Mohamed, A. (2013). Fuzzy logic based coordinated voltage control for distribution network with distributed generations. *World Academy of Science, Engineering, and Technology*, 79, 24-29.

Hatziargyriou, N. (Ed.). (2014). *Microgrids: Architectures and control.* John Wiley & Sons.

Haughton, D. A., & Heydt, G. T. (2012). A linear state estimation formulation for smart distribution systems. *IEEE Transactions on Power Systems*, *28*(2), 1187–1195. doi:10.1109/TPWRS.2012.2212921

Hemmati, R., Saboori, H., & Jirdehi, M. A. (2016). Multistage generation expansion planning incorporating large scale energy storage systems and environmental pollution. *Renewable Energy*, *97*, 636–645. doi:10.1016/j.renene.2016.06.020

Hesam, H. (2008). Distribution system protection es586b: Power system protection, University of Western Ontario.

Hina Fathima, A., & Palanisamy, K. (2015). Optimization in microgrids with hybrid energy systems – a review. *Renewable & Sustainable Energy Reviews*, *45*, 431–446. doi:10.1016/j.rser.2015.01.059

Hippert, H. S., Pedreira, C. E., & Souza, R. C. (2001). Neural networks for short-term load forecasting: A review and evaluation. *IEEE Transactions on Power Systems*, *16*(1), 44–55. doi:10.1109/59.910780

Histand, M. B. (1999). *Introduction to mechatronics and Measurement Systems. McGraw-Hill Education.* ISE Editions.

Hoffmann, C. (2010). *Design of transport- and storage capacities in energy supply systems with high shares of renewable energies.* Berlin, Germany: IRES.

Holtz, J., & Boelkens, U. (1989). Direct frequency convertor with sinusoidal line currents for speed-variable AC motors. *IEEE Transactions on Industrial Electronics*, *36*(4), 475–479. doi:10.1109/41.43005

Hossam-Eldin, A. A., Abdallah, E. N., & El-Nozahy, M. S. (2009, July). A modified genetic based technique for solving the power system state estimation problem. In *International Conference on World Academy of Science, Engineering and Technology* (pp. 311-320).

Howlader, A. M., & Senjyu, T. (2016). A comprehensive review of low voltage ride through capability strategies for the wind energy conversion systems. *Renewable & Sustainable Energy Reviews*, *56*, 643–658. doi:10.1016/j.rser.2015.11.073

Hua, B., Baldick, R., & Wang, J. (2018). Representing operational flexibility in generation expansion planning through convex relaxation of unit commitment. *IEEE Transactions on Power Systems*, *33*(2), 2272–2281. doi:10.1109/TPWRS.2017.2735026

Huang, J., Gupta, V., & Huang, Y. F. (2012, March). Electric grid state estimators for distribution systems with microgrids. In *2012 46th Annual Conference on Information Sciences and Systems (CISS)* (pp. 1-6). IEEE. 10.1109/CISS.2012.6310739

Huang, Y. F., Werner, S., Huang, J., Kashyap, N., & Gupta, V. (2012b). State estimation in electric power grids: Meeting new challenges presented by the requirements of the future grid. *IEEE Signal Processing Magazine*, *29*(5), 33–43. doi:10.1109/MSP.2012.2187037

Huang, Z., Schneider, K., & Nieplocha, J. (2007, December). Feasibility studies of applying Kalman filter techniques to power system dynamic state estimation. In *2007 International Power Engineering Conference (IPEC 2007)* (pp. 376-382). IEEE.

Huber, P. J. (1992). Robust estimation of a location parameter. In *Breakthroughs in statistics* (pp. 492–518). New York, NY: Springer. doi:10.1007/978-1-4612-4380-9_35

IEC 61000-3-3. (2017). Limitation of voltage changes, voltage fluctuations and flicker in public low-voltage supply systems, for equipment with rated current ≤16A per phase and not subject to conditional connection, 2017.

IEEE guide for voltage sag indices, IEEE 1564, 2014.

IEEE recommended practice for electric power distribution for industrial plants. (1994, April). In IEEE Std. 141-1993, pp. 1-768.

IEEE recommended practice for electric power systems in commercial buildings. (1991, September). In IEEE Std. 241-1990, pp. 1-768.

IEEE. (1998). Recommended practice for industrial and commercial power systems analysis (Brown book). In IEEE Std 399-1997, pp. 1-488.

IEEE. (2009) Recommended practice for monitoring electric power quality. In IEEE Std 1159-2009 (Revision of IEEE Std 1159-1995), pp. 1-94.

IEEE. (2009). IEEE application guide for IEEE Std 1547(TM). IEEE Standard for Interconnecting Distributed Resources with Electric Power Systems. IEEE Std 1547.

IEEE. (2014) Recommended practice and requirements for harmonic control in electric power systems. In IEEE Std 519-2014 (Revision of IEEE Std 519-1992), pp. 1-29.

Inage, S. I. (2011): Prospective on the decarbonised power grid, *IEC/MSB/EES Workshop*. Germany.

Information Technology Industry Council (ITIC), Note, 2000.

IngenierosA. M. (2011). grupoglobalgreen. Retrieved from www.grupoglobalgreen.es

International Energy Agency (IEA): Prospects for large scale energy storage in decarbonized grids, Report, 2009.

Ipinnimo, O., Chowdhury, S., Chowdhury, S. P., & Mitra, J. (2012). Intelligent voltage dip detection in power networks with distributed generation (DG). In Proceedings 2012 North American Power Symposium (NAPS), pp. 1-6. IEEE.

Ipinnimoa, O., Chowdhurya, S., Chowdhurya, S. P., & Mitrab, J. (2012). A review of voltage dip mitigation techniques with distributed generation in electricity networks.

Jamila, S., Seifeddine, B. E., Mohamed, B., & Mohamed, F. M. (2018). Optimal sizing of energy storage systems using frequency-separation-based energy management for fuel cell hybrid electric vehicles. *IEEE Transactions on Vehicular Technology*, *10*, 9337–9346.

Jariso M., Khan B, Tanwar S., Tyagi S., & Rishiwal, V. (2018). Hybrid energy system for upgrading the rural environment, *IEEE GlobeCom 2018*.

Jariso, M., Khan, B, Tanwar, S., Tyagi, S., & Rishiwal, V. (2018). Hybrid energy system for upgrading the rural environment, *IEEE GlobeCom 2018*.

Jariso, M., Khan, B., Tanwar, S., Tyagi, S., & Rishiwal, V. (2018). Hybrid energy system for upgrading the rural environment, IEEE GlobeCom 2018.

Jariso, M., Khan, B., Tanwar, S., Tyagi, S., & Rishiwal, V. (2018, December). Hybrid energy system for upgrading the rural environment. In *2018 IEEE Globecom Workshops* (pp. 1-6). IEEE.

Jariso, M., Khan, B., Tesfaye, D., & Singh, J. (2017). Modelling and designing of stand-alone photovoltaic system (case study: Addis Boder Health Center south west Ethiopia). In *Proceedings IEEE International Conference on Electronics, Communication, and Aerospace Technology, ICECA 2017,* pp. 168-173, Coimbatore, India.

Jariso, M., Khan, B., Tesfaye, D., & Singh, J. (2017). Modelling and designing of stand-alone photovoltaic system (Case study: Addis Boder Health Center southwest Ethiopia), *IEEE International conference on Electronics, Communication, and Aerospace Technology, ICECA 2017,* pp. 168-173, Coimbatore, India.

Jariso, M., Khan, B., Tesfaye, D., & Singh, J. (2017). Modelling and designing of stand-alone photovoltaic system (Case study: Addis Boder Health Center southwest Ethiopia), *IEEE International Conference on Electronics, Communication, and Aerospace Technology, ICECA 2017,* pp. 168-173, Coimbatore, India.

Jariso, M., Khan, B., Tesfaye, D., & Singh, J. (2017). Modelling and designing of stand-alone photovoltaic system (Case study: Addis Boder Health Center southwest Ethiopia). In *Proceedings IEEE International Conference on Electronics, Communication, and Aerospace Technology, ICECA 2017,* pp. 168-173, 20-21. Coimbatore, India.

Jariso, M., Khan, B., Tesfaye, D., & Singh, J. (2017). Modelling and designing of stand-alone photovoltaic system (Case study: Addis Boder Health Center southwest Ethiopia). In *Proceedings IEEE International conference on Electronics, communication, and aerospace Technology, ICECA 2017,* pp. 168-173, Coimbatore, India.

Jariso, M., Khan, B., Tesfaye, D., & Singh, J. (2017). Modelling and designing of stand-alone photovoltaic system (Case study: Addis Boder Health Center southwest Ethiopia). In *Proceedings IEEE International Conference on Electronics, Communication, and Aerospace Technology, ICECA 2017,* pp. 168-173, Coimbatore, India.

Jariso, M., Khan, B., Tesfaye, D., & Singh, J. (2017). Modelling and designing of stand-alone photovoltaic system (Case study: Addis Boder Health Center southwest Ethiopia. In *Proceedings IEEE International Conference on Electronics, Communication, and Aerospace Technology, ICECA 2017,* pp. 168-173, Coimbatore, India.

Jegadeesan, M., & Venkatesh, N. (2014, March). Estimation and mitigation of current and voltage harmonics in distribution system using distributed generation, *International Journal of Innovative Research in Science, Engineering, and Technology, 3*(Special Issue), 3.

Jenkins, N., Ekanayake, J. B., & Strbac, G. (2010). *"Distributed generation", Renewable Energy Series 1.* IET.

Jihane, K., & Mohamed, C. (2019). Study of the different structures of hybrid systems in renewable energies: A review. *Energy Procedia, 157,* 323–330. doi:10.1016/j.egypro.2018.11.197

Jundi, J., Guangya, Y., Arne, H. N., & Peter, R.-H. (2018). Study of control strategies of power electronics during faults in microgrids, *Hybrid-Renewable Energy Systems in Microgrids,* 109-146.

Kalia, C., Kumar, D., & Rana, P. (2014). Analysis of power quality improvement by using BFO-fuzzy controlled DVR, *International Journal of Application or Innovation in Engineering & Management, 3*(8).

Kalman, R. E. (1960). A new approach to linear filtering and prediction problems. *Journal of Basic Engineering, 82*(1), 35–45. doi:10.1115/1.3662552

Kannan, S., Baskar, S., McCalley, J. D., & Murugan, P. (2009). Application of NSGA-II algorithm to generation expansion planning. *IEEE Transactions on Power Systems, 24*(1), 454–461. doi:10.1109/TPWRS.2008.2004737

Karunanithi, K., Saravanan, S., Prabakar, B. R., Kannan, S., & Thangaraj, C. (2017). Integration of demand and supply side management strategies in generation expansion planning. *Renewable & Sustainable Energy Reviews, 73,* 966–982. doi:10.1016/j.rser.2017.01.017

Kawashima, M. (2011). *Overview of electric power storage,* Internal paper of Tepco, 2011.

Kersting, W. H. (2001). Distribution system modelling and analysis, Fourth Ed. Boca Raton, FL: CRC Press. doi:10.1201/9781420041736

Khan B., Agnihotri G, & Gupta G. (2013). A multipurpose matrices methodology for transmission usage, loss and reliability margin allocation in restructured environment, *Electrical & Computer Engineering: An International Journal, 2*(3).

Khan, B., & Singh, P. (2017) Optimal power flow techniques under characterization of conventional and renewable energy sources: a comprehensive analysis, *Journal of Engineering, vol. 2017,* Article ID 9539506.

Khan, B., & Singh, P. (2017). Optimal power flow techniques under characterization of conventional and renewable energy sources: A comprehensive analysis, *Journal of Engineering, 2017,* Article ID 9539506.

Khan, B., & Singh, P. (2017). Optimal power flow techniques under characterization of conventional and renewable energy sources: a comprehensive analysis, *Journal of Engineering, vol. 2017,* Article ID 9539506.

Khan, B., & Singh, P. (2017). Optimal power flow techniques under characterization of conventional and renewable energy sources: A comprehensive analysis, *Journal of Engineering, vol. 2017,* Article ID 9539506.

Khan, B., & Singh, P. (2017). Selecting a meta-heuristic technique for smart micro-grid optimization problem: a comprehensive analysis, in IEEE Access, 5, pp. 13951-13977, 2017.

Khan, B., & Singh, P. (2017). Selecting a meta-heuristic technique for smart micro-grid optimization problem: a comprehensive analysis, in IEEE Access, 5, pp. 13951-13977.

Khan, B., & Singh, P. (2017). Selecting a meta-heuristic technique for smart micro-grid optimization problem: A comprehensive analysis, in IEEE Access, 5, pp. 13951-13977.

Khan, B., & Singh, P. (2017). Selecting a meta-heuristic technique for smart micro-grid optimization problem: A comprehensive analysis, In IEEE Access, 5, pp. 13951-13977.

Khan, B., & Singh, P. (2017). Selecting a meta-heuristic technique for smart micro-grid optimization problem: a comprehensive analysis. In IEEE Access, 5, pp. 13951-13977, 2017.

Khan, B., & Singh, P. (2017). Selecting a meta-heuristic technique for smart micro-grid optimization problem: a comprehensive analysis. In IEEE Access, 5, pp. 13951-13977.

Khan, B., & Singh, P. (2017). Selecting a meta-heuristic technique for smart micro-grid optimization problem: A comprehensive analysis. In IEEE Access, 5, pp. 13951-13977.

Khan, B., & Singh, P. (2017). Selecting a meta-heuristic technique for smart micro-grid optimization problem: A comprehensive analysis. In IEEE Access, vol. 5, pp. 13951-13977.

Khan, B., & Singh, P. (2019). Economic operation of smart micro-grid: A meta-heuristic approach. In Handbook of research on smart power system operation and control (pp. 330-346). Hershey, PA: IGI Global. doi:10.4018/978-1-5225-8030-0.ch014

Khan, B., & Tanwar, S. (2019). Issues associated with microgrid integration. In Handbook of research on smart power system operation and control (pp. 252-264). Hershey, PA: IGI Global. doi:10.4018/978-1-5225-8030-0.ch010

Khan, B., Agnihotri, G, & Gupta, G. (2013). A multipurpose matrices methodology for transmission usage, loss and reliability margin allocation in restructured environment, *Electrical & Computer Engineering: An International Journal, 2*(3).

Khan, B., Agnihotri, G, & Gupta, G. (2013). A multipurpose matrices methodology for transmission usage, loss, and reliability margin allocation in restructured environment. *Electrical & Computer Engineering: An International Journal, 2*(3).

Khan, B., Agnihotri, G, Gupta G. (2013). A multipurpose matrices methodology for transmission usage, loss, and reliability margin allocation in restructured environment, *Electrical & Computer Engineering: An International Journal, 2*(3), September 2013.

Khan, B., Agnihotri, G, Gupta, G. (2013). A multipurpose matrices methodology for transmission usage, loss and reliability margin allocation in restructured environment, *Electrical & Computer Engineering: An International Journal, 2*(3).

Khan, B., Agnihotri, G., & Gupta, G. (2013) A multipurpose matrices methodology for transmission usage, loss, and reliability margin allocation in restructured environment, *Electrical & Computer Engineering: An International Journal, 2*(3).

Khan, B., Agnihotri, G., & Gupta, G. (2013). A multipurpose matrices methodology for transmission usage, loss and reliability margin allocation in restructured environment, *Electrical & Computer Engineering: An International Journal, 2*(3), September 2013.

Khan, B., Agnihotri, G., & Gupta, G. (2013). A multipurpose matrices methodology for transmission usage, loss and reliability margin allocation in restructured environment, *Electrical & Computer Engineering: An International Journal, 2*(3).

Khan, B., Agnihotri, G., & Gupta, G. (2013). A multipurpose matrices methodology for transmission usage, loss, and reliability margin allocation in restructured environment, *Electrical & Computer Engineering: An International Journal, 2*(3).

Khan, S. (2007). Industrial power systems. Boca Raton, FL: CRC Press.

Khan, B., & Agnihotri, G. (2012). A novel transmission loss allocation method based on transmission usage. In *Proceedings IEEE Fifth Power India Conference*, Haryana, India. 10.1109/PowerI.2012.6479479

Khan, B., Agnihotri, G., Rathore, P., Mishra, A., & Naidu, G. (2014). A cooperative game theory approach for usage and reliability margin cost allocation under contingent restructured market. *International Review of Electrical Engineering*, *9*(4), 854–862.

Khan, B., Alhelou, H. H., & Mebrahtu, F. (2019). A holistic analysis of distribution system reliability assessment methods with conventional and renewable energy sources. *AIMS Energy*, *7*(4), 413–429. doi:10.3934/energy.2019.4.413

Khan, B., & Singh, P. (2017). The current and future states of Ethiopia's energy sector and potential for green energy: A comprehensive study. *International Journal of Engineering Research in Africa*, *33*, 115–139. doi:10.4028/www.scientific.net/JERA.33.115

Khomami, H., & Javidi, M. (2013). An efficient home energy management system for automated residential demand response. In *13th IEEE International Conference on Environment and Electrical Engineering (EEEIC)*.

Kifle, Y., Khan, B., & Singh, P. (2018). Assessment and enhancement of distribution system reliability by renewable energy sources and energy storage, *Journal of Green Engineering, 8*(3), 2, pp. 219-262.

Kifle, Y., Khan, B., & Singh, J. (2018). Designing and modeling grid connected photovoltaic system: (Case study: EEU building at Hawassa City). *International Journal of Convergence Computing*, *3*(1), 20–34. doi:10.1504/IJCONVC.2018.091113

Krim, Y., Abbes, D., Krim, S., & Mimouni, M. F. (2019). A flexible control strategy of a renewable active generator to participate in system services under grid faults. *Int Trans Electr. Energ. Syst.*, 1-24.

Kumar, M. L., Prasanna, H. M., & Ananthapadmanabha, T. (2015). Retracted: A literature review on distribution system state estimation.

Kumar, M., Nallagownden, P., & Elamvazuthi, I. (2017). Multi-objective PSO based optimal placement of solar power DG in radial distribution system. *J. Electrical Systems*, *13*(2), 322–331.

Kumar, M., Nallagownden, P., & Elamvazuthi, I. (2017). Optimal placement and sizing of renewable distributed generations and capacitor banks into radial distribution systems. *Energies*, 10, 1–25.

Kuo, Y. C., Liang, T. J., & Chen, J. F. (2001). Novel maximum-power-point-tracking controller for photovoltaic energy conversion system. *IEEE Transactions on Industrial Electronics*, *48*(3), 594–601. doi:10.1109/41.925586

Kuzlu, M., Pipattanasomporn, M., & Rahman, S. (2014). Communication network requirements for major smart grid applications in HAN, NAN, and WAN. *Computer Networks*, *67*, 74–88. doi:10.1016/j.comnet.2014.03.029

LaCommare, K., & Eto, J. H. (2004). *Understanding the cost of power interruptions to U. S. electricity consumers, technical report*. Lawrence Berkeley National Laboratory. doi:10.2172/834270

Lalitha, M. P., Suresh Babu, P., & Adivesh, B. (2016). Optimal distributed generation and capacitor placement for loss minimization and voltage profile improvement using symbiotic organisms search algorithm. *International Journal of Electrical Engineering*, *9*(3), 250–262.

Levitin, G., Mazal-Tov, S., & Elmakis, D. (1995). Genetic algorithm for optimal sectionalizing in radial distribution systems with alternative supply. *Electric Power Systems Research*, 35(3), 149-155.

Lidula, N. W. A., & Rajapakse, A. D. (2011). Microgrids research: A review of experimental microgrids and test systems. *Renewable & Sustainable Energy Reviews*, *15*(1), 186–202. doi:10.1016/j.rser.2010.09.041

Lifshitz, D., & Weiss, G. (2015). Optimal control of a capacitor-type energy storage system. *IEEETransAutoControl*, *60*(1), 216–220.

Li, G. (2015). Energy and exergy performance assessments for latent heat thermal energy storage systems. *Renewable & Sustainable Energy Reviews*, *51*, 926–954. doi:10.1016/j.rser.2015.06.052

Li, K. (1996). State estimation for power distribution system and measurement impacts. *IEEE Transactions on Power Systems*, *11*(2), 911–916. doi:10.1109/TPWRS.2009.2016599

Lin, W. M., & Teng, J. H. (1996). Distribution fast decoupled state estimation by measurement pairing. *IEE Proceedings. Generation, Transmission, and Distribution*, *143*(1), 43–48. doi:10.1049/ip-gtd:19960049

Lin, W. M., Teng, J. H., & Chen, S. J. (2001). A highly efficient algorithm in treating current measurements for the branch-current-based distribution state estimation. *IEEE Transactions on Power Delivery*, *16*(3), 433–439. doi:10.1109/61.924823

Li, S., Coit, D. W., & Felder, F. (2016). Stochastic optimization for electric power generation expansion planning with discrete climate change scenarios. *Electric Power Systems Research*, *140*, 401–412. doi:10.1016/j.epsr.2016.05.037

Liu, X., & Li, P. (2007). The effect of DVR on distribution system with distributed generation. In *Proceedings of International Conference on Electrical Machines and Systems*, pp. 277–281.

Li, Y., & Zio, E. (2012). A multi-state model for the reliability assessment of distributed generation system via universal generating function. *Reliability Engineering & System Safety*, *106*, 28–36. doi:10.1016/j.ress.2012.04.008

Locatelli, G., Palerma, E., & Mancini, M. (2015). Assessing the economics of large energy storage plants with an optimisation methodology. *Energy*, *83*, 15–28. doi:10.1016/j.energy.2015.01.050

Lu, C. N., Teng, J. H., & Liu, W. H. (1995). Distribution system state estimation. *IEEE Transactions on Power Systems*, *10*(1), 229–240. doi:10.1109/59.373946

Lucian, I. D. M. A. (2014). Effects of distributed generation on electric power system. *Science Direct,* 681-686.

Lu, X., Bahramirad, S., Wang, J., & Chen, C. (2015). Bronzeville community micro grids: A reliable, resilient, and sustainable solution for integrated energy management with distribution systems. *The Electricity Journal*, *28*(10), 29–42. doi:10.1016/j.tej.2015.11.009

Lyshevski, S. E., Ehawary, M. E., Bolea, L., Nasar, A., Kesrtin, W. H., & Crow, C. S. M. (2002). Power quality. Boca Raton, FL: CRC Press.

Mahat, P. (2010, September). *Control and operation of islanded distribution system.* (Doctoral dissertation, Aalborg University, Denmark).

Mahesh, K., Nallagownden, P., & Elamvazuthi, I. (2017). Advanced pareto front non-dominated sorting multi-objective particle swarm optimization for optimal placement and sizing of distributed generation. *Energies*, 9, 1–23.

Mahmoodi, M., Shamsi, P., & Fahimi, B. (2015). Economic dispatch of a hybrid microgrid with distributed energy storage. *IEEE Transactions on Smart Grid*, *6*(6), 2607–2614. doi:10.1109/TSG.2014.2384031

Majumdar, A., & Pal, B. C. (2016, January). A three-phase state estimation in unbalanced distribution networks with switch modelling. In *2016 IEEE First International Conference on Control, Measurement and Instrumentation (CMI)* (pp. 474-478). IEEE.] 10.1109/CMI.2016.7413793

Makdisie, C., Haidar, B., & Alhelou, H. H. (2018). An optimal photovoltaic conversion system for future smart grids. In Handbook of research on power and energy system optimization (pp. 601–657). Hershey, PA: IGI Global. doi:10.4018/978-1-5225-3935-3.ch018

Makdisie, C. (2008). *Power electronics / 2.* Tishreen University Publications.

Makdisie, C. (2016). *Power electronics / 1*. Tishreen University Publications.

Malheiro, A., Catrol, P. M., Lima, R. M., & Estanqueiro, A. (2015). Integrated sizing and scheduling of wind/PV/diesel/battery systems. *Renewable Energy*, *83*, 646–657. doi:10.1016/j.renene.2015.04.066

Manickavasagam, M., Anjos, M. F., & Rosehart, W. D. (2015). Sensitivity-based chance- constrained generation expansion planning. *Electric Power Systems Research*, *127*, 32–40. doi:10.1016/j.epsr.2015.05.011

Manitsas, E., Singh, R., Pal, B. C., & Strbac, G. (2012). Distribution system state estimation using an artificial neural network approach for pseudo measurement modeling. *IEEE Transactions on Power Systems*, *27*(4), 1888–1896. doi:10.1109/TPWRS.2012.2187804

Martin-Arnedo, J. A. (2009, October). Impacts of distributed generation on protection and power quality. In *IEEE Power & Energy Society General Meeting.*

McGranaghan, M. F. (1993). Voltage sags in industrial systems. *IEEE Trans., 29*(2), pp 397-403.

Mehrizi-Sani, A., & Iravani, R. (2010). Potential-function based control of a microgrid in islanded and grid-connected modes. *IEEE Transactions on Power Systems*, *25*(4), 1883–1891. doi:10.1109/TPWRS.2010.2045773

Meliopoulos, A. S., & Zhang, F. (1996). Multiphase power flow and state estimation for power distribution systems. *IEEE Transactions on Power Systems*, *11*(2), 939–946. doi:10.1109/59.496178

Merlin, A., & Back, H. (1975). Search for a minimal loss operating spanning tree configuration for an urban power distribution system. In *Proceedings of the 5th Power System Computation Conference (PSOC)*, Cambridge, pp. 1-18.

Metha, V. K., & Metha, R. (1982). *Principle of power system, S.* Chand Publications.

Mili, L., Cheniae, M. G., & Rousseeuw, P. J. (1994). Robust state estimation of electric power systems. *IEEE Transactions on Circuits and Systems. I, Fundamental Theory and Applications*, *41*(5), 349–358. doi:10.1109/81.296336

Mohammadi, S., Momtazpour, M., & Sanaei, E. (2013). Optimization-based home energy management in the presence of solar energy and storage. In *Proceedings 21st Iranian Conference on Electrical Engineering (ICEE)*, Mashhad, pp. 1-6. 10.1109/IranianCEE.2013.6599649

Mohammed Shazly, A., Cerrada Aurelio, G., Abdel-Moamen, M. A., & Hasanin, B. (2013, January). Dynamic voltage restorer (DVR) system for compensation of voltage sags, state-of-the-art review. *International Journal of Computational Engineering Research*, *3*(1), 177–183.

Mohan, N., Undeland, T. M., & Robbins, W. P. (2007). *Power electronics: converters, applications, and design*. New Delhi, India: Wiley India.

Mohsenian-Rad, H. (2012). *Introduction to smart grid*. Texas Tech University.

Molla, T., Khan, B., Moges, B., Alhelou, H. H., Zamani, R., & Siano, P. (2019). Integrated optimization of smart home appliances with cost-effective energy management system. *CSEE Journal of Power and Energy Systems*, *5*(2), 249–258.

Molla, T., Khan, B., & Singh, P. (2018). A comprehensive analysis of smart home energy management system optimization techniques, *Journal of Autonomous Intelligence, 1(1)*, 15–21.

Molla, T., Khan, B., & Singh, P. (2018). A Comprehensive Analysis of Smart Home Energy Management System Optimization Techniques, *Journal of Autonomous Intelligence, 1(1)*, 15–21.

Moshtagh, J., & Ghasemi, S. (2013). Power restoration in distribution network using MST algorithms, *Journal of Operation and Automation in Power Engineering, 1*(1).

Mousavi, G. S. M., & Nikdel, M. (2014). Various battery models for various simulation studies and applications. *Renewable & Sustainable Energy Reviews*, *2014*, 477–485. doi:10.1016/j.rser.2014.01.048

Nagy, S. A., Ibrahim, I. S., Ahmed, M. K., Adail, A. S., & Soliman, S. (2013). Network reconfiguration for loss reduction in electrical distribution system using genetic algorithm. *Arab Journal of Nuclear Science and Applications*, *46*(1), 78–87.

Naka, S., Genji, T., Yura, T., & Fukuyama, Y. (2003). A hybrid particle swarm optimization for distribution state estimation. *IEEE Transactions on Power Systems*, *18*(1), 60–68. doi:10.1109/TPWRS.2002.807051

Nallagownden, P., Kumar, M., & Elamvazuthi, I. (2017). Optimal placement and sizing of renewable distributed generations and capacitor banks into radial distribution systems.

Nanchian, S. (2015). State estimation for active distribution network.

Nasir, S. S., Jamian, J. J., & Mustafa, M. W. (2018). Minimizing harmonic distortion impact at distribution system with considering large-scale EV load behaviour using modified lightning search algorithm and pareto-fuzzy approach, Vol. 2018, Article ID 6587493.

Negash, K., Khan, B., Tesfaye, D., & Gayathri, M. (2017). Optimal placement of phasor measurement unit for system observability. In *Proceedings of IEEE International Conference on Computing Methodologies and Communication, (ICCMC 2017),* pp. 73-77, Erode, India.

Negash, K., Khan, B., Tesfaye, D., & Gayathri, M. (2017). Optimal placement of phasor measurement unit for system observability. In *Proceedings of IEEE International Conference on Computing Methodologies and Communication, (ICCMC 2017),* pp. 73-77, July 18, 19, 2017, Erode India.

Negash, K., Khan, B., Tesfaye, D., & Gayathri, M. (2017). Optimal placement of phasor measurement unit for system observability. *Proceedings of IEEE International Conference on Computing Methodologies and Communication, (ICCMC 2017),* pp. 73-77, Erode, India.

Negash, K., Khan, B., & Yohannes, E. (2016). Artificial intelligence versus conventional mathematical techniques: a review for optimal placement of phasor measurement units [Springer]. *Journal of Technology and Economics of Smart Grids and Sustainable Energy*, *1*(1), 10. doi:10.100740866-016-0009-y

Nematollah, D., & Rahman, D. (2011). Optimization of recloser placement to improve reliability by genetic algorithm. *Energy and Power Engineering*, *3*(04), 508–512. doi:10.4236/epe.2011.34061

Ngaopitakkul, A., Pothisarn, C., Bunjongjit, S., Suechoey, B., Thammart, C., & Nawikavatan, A. (2013). A reliability impact and assessment of distributed generation integration to distribution system. *Energy and Power Engineering*, *5*(04), 1043–1047. doi:10.4236/epe.2013.54B199

Nguyen, P. H., Venayagamoorthy, G. K., Kling, W. L., & Ribeiro, P. F. (2013). Dynamic state estimation for distribution networks with renewable energy integration. *Int. J. Smart Grid and Clean Energy*, *2*(3), 307–315. doi:10.12720gce.2.3.307-315

Nikmehr, N., & Najafi Ravadanegh, S. (2015). Optimal power dispatch of multi-microgrids at future smart distribution grids. *IEEE Transactions on Smart Grid*, *6*(4), 1648–1657. doi:10.1109/TSG.2015.2396992

Niknam, T. (2008). Application of honey-bee mating optimization on state estimation of a power distribution system including distributed generators. *Journal of Zhejiang University. Science A*, *9*(12), 1753–1764. doi:10.1631/jzus.A0820047

Niknam, T., & Firouzi, B. B. (2009). A practical algorithm for distribution state estimation including renewable energy sources. *Renewable Energy*, *34*(11), 2309–2316. doi:10.1016/j.renene.2009.03.005

Niknam, T., Ranjbar, A. M., & Shirani, A. R. (2005). A new approach for distribution state estimation based on ant colony algorithm with regard to distributed generation. *Journal of Intelligent & Fuzzy Systems*, *16*(2), 119–131.

Njenda, T. C., Golshan, M. E. H., & Alhelou, H. H. (2018, November). WAMS based Under Frequency Load Shedding Considering Minimum Frequency Prediction and Extrapolated Disturbance Magnitude. In *2018 Smart Grid Conference (SGC)* (pp. 1-5). IEEE.

Njenda, T. C., Golshan, M. E. H., & Alhelou, H. H. (2018, November). WAMS based under frequency load shedding considering minimum frequency prediction and extrapolated disturbance magnitude. In *2018 Smart Grid Conference (SGC),* (pp. 1-5). IEEE.

Njenda, T. C., Golshan, M. E. H., & Alhelou, H. H. (2018, November). WAMS based intelligent under frequency load shedding considering online disturbance estimation. In *Smart Grid Conference (SGC)* (pp. 1-5). 10.1109/SGC.2018.8777779

Njenda, T. C., Golshan, M. E. H., & Alhelou, H. H. (2018, November). WAMS based under frequency load shedding considering minimum frequency prediction and extrapolated disturbance magnitude. In *Smart Grid Conference (SGC)* (pp. 1-5).

Ochoa, L. F., Padilha-Feltrin, A., & Harrison, G. (2006). Evaluating distributed generation impacts with a multi objective index. *IEEE Transactions on Power Delivery*, *21*(3), 1452–1458. doi:10.1109/TPWRD.2005.860262

Ogunjuyigbe, A. S. O., Ayodele, T. R., & Akinola, O. A. (2016). Optimal allocation and sizing of PV/Wind/Split-diesel/Battery hybrid energy system for minimizing life cycle cost, carbon emission and dump energy of remote residential building. *Applied Energy*, *171*, 153–171. doi:10.1016/j.apenergy.2016.03.051

Olagunju, R. E. (2012). Predictive modelling for sustainable residential building maintenance in developing countries: A Nigeria case. *Interdisciplinary Journal of Contemporary Research in Business*, *4*(6), 1273–1283.

Olagunju, R. E. (2012). Sustainability of residential buildings in Nigeria: An appraisal of the factors that influence maintenance of residential buildings 'standards. *Civil and Environmental Research*, *2*(4), 20–29.

Olken, M. (2009). Beyond the gridlock. *IEEE Power & Energy Magazine*, *7*(2), 4–6. doi:10.1109/MPE.2009.931830

Pal, R., & Gupta, S. (2015, June). State of the art: Dynamic voltage restorer for power quality improvement. *Electrical & Computer Engineering: An International Journal*, *4*(2), 79–98.

Papic, I. (2000). Power quality improvement using distribution static compensator with energy storage system. In Proceedings *9th International Conference on Harmonics and Quality of Power*, Orlando, FL, pp. 916-920, vol. 3. 10.1109/ICHQP.2000.896851

Park, H., & Baldick, R. (2015). Stochastic generation capacity expansion planning reducing greenhouse gas emissions. *IEEE Transactions on Power Systems*, *30*(2), 1–9. doi:10.1109/TPWRS.2014.2386872

Pau, M., Pegoraro, P. A., & Sulis, S. (2013). Efficient branch-current-based distribution system state estimation including synchronized measurements. *IEEE Transactions on Instrumentation and Measurement*, *62*(9), 2419–2429. doi:10.1109/TIM.2013.2272397

Peña Asensio, A., Arnaltes Gómez, S., Rodriguez-Amenedo, J., García Plaza, M., Eloy-García Carrasco, J., & Alonso-Martínez de las Morenas, J. (2018). A voltage and frequency control strategy for stand-alone full converter wind energy conversion systems. *Energies, 11*(3), 474.

Perdana, Y. S., Muyeen, S. M., Al-Durra, A., Helmo, K. M-P, & Marcelo, G. S. (2018). Direct connection of supercapacitor-battery hybrid storage system to the grid-tied photovoltaic system, *IEEE Transactions on Sustainable Energy,* 1-10.

Perrin, M. et al. (2014). Investigation on storage technologies for intermittent renewable energies: Evaluation and recommended R&D strategy, INVESTIRE, Final Technical Report of European Community project ENK5-CT-2000-20336, 2004.

Picallo, M., Anta, A., De Schutter, B., & Panosyan, A. (2018, June). A two-step distribution system state estimator with grid constraints and mixed measurements. In 2018 Power Systems Computation Conference (PSCC) (pp. 1-7). IEEE. doi:10.23919/PSCC.2018.8442586

Pippia, T., Sijs, J., & De Schutter, B. (2018, December). A parametrized model predictive control approach for microgrids. In *2018 IEEE Conference on Decision and Control (CDC)* (pp. 3171-3176). IEEE.

Pogaku, N., & Green, T. C. (2006, May 11). Harmonic mitigation throughout a distribution system: A distributed-generator-based solution. *IEE Proceedings. Generation, Transmission, and Distribution*, *153*(3), 350–358. doi:10.1049/ip-gtd:20050086

Power, P. (2009). *The cost of using biomass for power generation and potential for decreased greenhouse gas emissions.* Retrieved from https://sequestration.mit.edu/pdf/AmandaCuellar_Thesis_June2012.pdf

Primadianto, A., Lin, W. T., & Lu, C. N. (2016, November). Performance comparison of distribution system state estimation methods. In 2016 IEEE Innovative Smart Grid Technologies-Asia (ISGT-Asia) (pp. 1121-1126). IEEE doi:10.1109/ISGT-Asia.2016.7796543

Primadianto, A., & Lu, C. N. (2016). b). A review on distribution system state estimation. *IEEE Transactions on Power Systems*, *32*(5), 3875–3883. doi:10.1109/TPWRS.2016.2632156

Priya, S. S., & Sagar, B. K. (2018). An advanced droop control method for proportional reactive power sharing. *International Research Journal of Engineering and Technology*, *5*(7), 1735–1743.

Purvins, A., & Sumner, M. (2013). Optimal management of stationary lithium-ion battery system in electricity distribution grids. *Journal of Power Sources*, *242*, 742–755. doi:10.1016/j.jpowsour.2013.05.097

Qayyum, F. A., Naeem, M., Khwaja, A. S., Anpalagan, A., Guan, L., & Venkatesh, B. (2015). Appliance scheduling optimization in smart home networks, Special section on smart grids, IEEE. doi:10.1109/ACCESS.2015.2496117

Qi, J., Sun, K., Wang, J., & Liu, H. (2016). Dynamic state estimation for multi-machine power system by unscented Kalman filter with enhanced numerical stability. *IEEE Transactions on Smart Grid*, *9*(2), 1184–1196. doi:10.1109/TSG.2016.2580584

Qi, L., Bo, S., Yuchen, P., Ying, H., Tianhong, W., Liangzhen, Y., & Weirong, C. (2019). A state machine control based on equivalent consumption minimization for fuel cell/ supercapacitor hybrid tramway. *IEEE Transactions on Transportation Electrification*, *5*(2), 552–564. doi:10.1109/TTE.2019.2915689

Rahman, M. S., Hossain, M. J., & Lu, J. (2016). Coordinated control of three-phase AC and DC type EV–ESSs for efficient hybrid microgrid operations. *Energy Conversion and Management*, *122*, 488–503. doi:10.1016/j.enconman.2016.05.070

Rajesh, K., Bhuvanesh, A., Kannan, S., & Thangaraj, C. (2016). Least cost generation expansion planning with solar power plant using differential evolution algorithm. *Renewable Energy*, *85*, 677–686. doi:10.1016/j.renene.2015.07.026

Ramesh, L., Chakraborthy, N., Chowdhury, S. P., & Chowdhury, S. (2012). Intelligent DE algorithm for measurement location and PSO for bus voltage estimation in power distribution system. *International Journal of Electrical Power & Energy Systems*, *39*(1), 1–8. doi:10.1016/j.ijepes.2011.10.009

Rani, P. A., & Sivakumar, R. (2014, January). Improvement of power quality using DVR in distribution systems. *International Journal of Innovative Research in Science, Engineering, and Technology*, *3*(Special Issue), 1.

Rashid, M. H. (Ed.). (2011). *Power electronics handbook: Devices, circuits, and applications*. Burlington, MA: Elsevier.

Rauf, A. M., & Khadkikar, V. (2015, May). An enhanced voltage SAG compensation scheme for dynamic voltage restorer. *IEEE Transactions on Industrial Electronics*, *62*(5), 2683–2692. doi:10.1109/TIE.2014.2362096

Ray, P. K., & Subudhi, B. (2012). Ensemble-Kalman-filter-based power system harmonic estimation. *IEEE Transactions on Instrumentation and Measurement*, *61*(12), 3216–3224. doi:10.1109/TIM.2012.2205515

Reddy, P. D. P., Reddy, V. C. V., & Manohar, T. G. (2017). *Whale optimization algorithm for optimal sizing of renewable resources for loss reduction in distribution systems* (pp. 1–13). Renewables Wind, Water, and Soloar.

Rehman, S., Al-Hadhrami, L. M., & Alam, M. M. (2015). Pumped hydro energy storage system: A technological review. *Renewable & Sustainable Energy Reviews*, *44*, 586–598. doi:10.1016/j.rser.2014.12.040

RETSCREEN International, Renewable Energy Project Analysis: Ret screen Engineering and Cases Textbook. Available at http://www.retscreen.net

Rovai, C. H., & Doorwar, A. (2014). An overview of various control techniques of DVR. In *Proceedings 2014 International Conference on Circuits, Power, and Computing Technologies (ICCPCT-2014)*, Nagercoil, India. pp. 53-57. 10.1109/ICCPCT.2014.7054844

Roytelman, I., & Shahidehpour, S. M. (1993). State estimation for electric power distribution systems in quasi real-time conditions. *IEEE Transactions on Power Delivery*, *8*(4), 2009–2015. doi:10.1109/61.248315

Ruan, B., Yang, Q., Fang, X., & Yan, W. (2014). Demand response under real-time pricing for domestic energy system with DGs. In *Proceedings International Conference on Power System Technology (POWERCON 2014)*, Chengdu, 10.1109/POWERCON.2014.6993601

Saboori, H., Hemmati, R., & Jirdehi, M. A. (2015). Reliability improvement in radial electrical distribution network by optimal planning of energy storage systems. *Energy, 93*, 2299-2312.

Saboori, H., Hemmati, R., & Jirdehi, M. A. (2015). Reliability improvement in radial electrical distribution network by optimal planning of energy storage systems. *Energy*, *93*, 2299–2312. doi:10.1016/j.energy.2015.10.125

Sadigh, A. K., & Smedley, K. M. (2016). Fast and precise voltage sag detection method for dynamic voltage restorer (DVR) application. *Electric Power Systems Research*, *130*, 192–207. doi:10.1016/j.epsr.2015.08.002

Saini, S. (2004). Conservation vs. generation-The significance of demand side management (DSM), its tools and techniques. *Energy Management,* 52-54.

Salam, Z. (2003). *Power electronics and drives*. UTM-JB.

Saremi, S., Mirjalili, S., & Lewis, A. (2017). Grasshopper optimization algorithm: Theory and applications. *Advances in Engineering Software*, *105*, 30–47. doi:10.1016/j.advengsoft.2017.01.004

Satkin, M., Noorollahi, Y., Abbaspour, M., & Yousefi, H. (2014). Multi-criteria site selection model for wind-compressed air energy storage power plants in Iran. *Renewable & Sustainable Energy Reviews*, *32*, 579–590. doi:10.1016/j.rser.2014.01.054

Saumil, A. P. (2007). *Fuzzy logic based power load balancing using fuzzy logic,* (M. E. thesis), Gujarat University, India.

Savaghebi, M., Vasquez, J. C., Jalilian, A., Guerrero, J. M., & Lee, T. L. (2012). Selective compensation of voltage harmonics in grid-connected microgrids. *Mathematics and Computers in Simulation*, *91*, 211–228. doi:10.1016/j.matcom.2012.05.015

Schewe, P. F. (2005). The grid: A journey through the heart of our electrified world (Joseph Henry Press).

Schneider electric (SE). (2015). Guide for the design and production of LV power factor correction cubicles, Panel builder guide.

Schwunk, S. (2011). Battery systems for storing renewable energy. *Germany: Fraunhofer--Institut für Solare Energie.*

Selve, V., Baskar, G., & Pradheeba, R. (2014, March). Distribution feeder reconfiguration for reliability evaluation by using binary particle swarm optimization. *International Journal of Innovative Research in Science, Engineering, and Technology*, *3*(Special Issue), 3.

Shabani, F., Prasad, N. R., & Smolleck, H. A. (1996, September). State estimation with aid of fuzzy logic. In *Proceedings of IEEE 5th International Fuzzy Systems* (Vol. 2, pp. 947–953). IEEE. doi:10.1109/FUZZY.1996.552306

Shah Arifur Rahman, B. D. (n.d.). *Impact of distributed generation on power system protection.* IEEE, (pp. 1-9). Pakistan.

Shahidehpour, M., & Khodayar, M. (2013). Cutting campus energy costs with hierarchical control: The economical and reliable operation of a micro grid. *IEEE Electr. Mag.*, *1*(1), 40–56. doi:10.1109/MELE.2013.2273994

Shah, R., Mithulananthan, N., Bansal, R. C., & Ramachandaramurthy, V. K. (2015). A review of key power system stability challenges for large-scale PV integration. *Renewable & Sustainable Energy Reviews*, *41*, 1423–1436. doi:10.1016/j.rser.2014.09.027

Shahryaria, E., Shayeghi, H., Mohammadi-ivatloo, B., & Moradzadeh, M. (2019). A copula-based method to consider uncertainties for multi-objective energy management of microgrid in presence of demand response. *Energy*, *175*, 879–890. doi:10.1016/j.energy.2019.03.129

Shahsavari, A., Sadeghi-Mobarakeh, A., Stewart, E. M., Cortez, E., Alvarez, L., Megala, F., & Mohsenian-Rad, H. (2017). Distribution grid reliability versus regulation market efficiency: An analysis based on micro-PMU data. IEEE Transactions on Smart Grid, 8(6), 2916-2925.

Shih, K. R., & Huang, S. J. (2002). Application of a robust algorithm for dynamic state estimation of a power system. *IEEE Transactions on Power Systems*, *17*(1), 141–147. doi:10.1109/59.982205

Shiyu Liu, Y. Y. (n.d.). *Impact of distributed generation on electric power distribution system reliability*. IEEE. China.

Shokrzadeh, S., Jozani, M. J., Bibeau, E., & Molinski, T. (2015). A statistical algorithm for predicting the energy storage capacity for baseload wind power generation in the future electric grids. *Energy, 89,* 793-802.

Short, T., & Swayne, T. (2012). Assessment of Transmission and Distribution Losses in New York PID071178 (NYSERDA 15464), The New York State Energy Research and Development Authority, New York.

Siewniak, P., & Grzesik, B. (2014). A brief review of models of DC –DC power electronic converters for analysis of their stability. *International Journal of Electronics*, *101*(10), 1405–1426. doi:10.1080/00207217.2013.836696

Simon, D. (2006). *Optimal state estimation: Kalman, H infinity, and nonlinear approaches*. John Wiley & Sons. doi:10.1002/0470045345

Singh, P., & Khan, B. (2017). Smart microgrid energy management using a novel artificial shark optimization, Complexity, vol. 2017, Article ID 2158926.

Singh, P., & Khan, B. (2017). Smart microgrid energy management using a novel artificial shark optimization, Complexity, Vol. 2017, Article ID 2158926. doi:10.1155/2017/2158926

Singh, S., Rai, V., Kumar, A., & Sahay, K. B. (2016). Simulation and comparison of DVR and D-STATCOM for voltage sag mitigation. In *Proceedings 2016 IEEE 6th International Conference on Power Systems (ICPS),* New Delhi, India. pp. 1-6. 10.1109/ICPES.2016.7584238

Singh, B., Al-Haddad, K., & Chandra, A. (2012, October). A review of active filters for power quality improvement. *IEEE Transactions on Industrial Electronics*, *46*(5), 960–971. doi:10.1109/41.793345

Singh, P., & Khan, B. (2017). Smart microgrid energy management using a novel artificial shark optimization. *Complexity*, *2017*, 2158926.

Singh, P., Khan, B., Vidyarthi, A., Haes Alhelou, H., & Siano, P. (2019). Energy-aware online non-clairvoyant scheduling using speed scaling with arbitrary power function. *Applied Sciences*, *9*(7), 1467. doi:10.3390/app9071467

Singh, R., Pal, B. C., & Jabr, R. A. (2010). Distribution system state estimation through Gaussian mixture model of the load as pseudo-measurement. *IET Generation, Transmission, & Distribution*, *4*(1), 50–59. doi:10.1049/iet-gtd.2009.0167

Singh, U. (2014). *Radial distribution system reconfiguration for loss minimization Using exhaustive search method.* India: Thapar University.

Sinha, A. K., & Mondal, J. K. (1999). Dynamic state estimator using ANN based bus load prediction. *IEEE Transactions on Power Systems*, *14*(4), 1219–1225. doi:10.1109/59.801876

Sivaraman, P., Sharmeela, C., & Kothari, D. P. (2017, September). Enhancing the voltage profile in distribution system with 40 GW of solar PV rooftop in Indian grid by 2022: a review. 1st International Conference on Large Scale Grid Integration Renewable Energy in India, New Delhi, India.

Sivaraman, P., & Sharmeela, C. (2020). Solar Micro-Inverter. In J. Zbitou, C. Pruncu, & A. Errkik (Eds.), *Handbook of Research on Recent Developments in Electrical and Mechanical Engineering* (pp. 283–303). Hershey, PA: IGI Global.

Sivaraman, P., Sharmeela, C., & Kothari, D. P. (2017). Enhancing the voltage profile in distribution system with 40 GW of solar PV rooftop in Indian grid by 2022: A review. *1st International Conference on Large Scale Grid Integration Renewable Energy in India*, New Delhi, India.

Slootweg, J. G., & Kling, W. L. (n.d.). *Impacts of distributed generation on power system transient stability.* IEEE, (pp. 862-867).

Smart Building (SB). Retrieved from https://www.rcrwireless.com/20160725/business/smart-building-tag31-tag99

Solar Electric Generating Stations (SEGS). (1989). Solar electric generating stations (SEGS). *IEEE Power Engineering Review*, *9*(8), 4–8. doi:10.1109/MPER.1989.4310850

Somnath, D., & Ashok, K. A. (2018). Power flow control of PV-wind-battery hybrid renewable energy systems for stand-alone application. *International Journal of Renewable Energy Research-IJRER*, *8*(1), 36–43.

Soomro, D. M., & Almelian, M. M. (2015). Optimal design of a single tuned passive filter to mitigate harmonics in power frequency. *Journal of Engineering and Applied Sciences (Asian Research Publishing Network)*, *10*(19), 9009–9014.

Souza, J., Rider, M. J., & Mantovani, J. R. S. (2015). Planning of distribution systems using mixed-integer linear programming models considering network reliability. *J. Control Autom. Electr. Syst.*, *26*(2), 170–179. doi:10.100740313-014-0165-z

Sreedharan, P., Farbes, J., Cutter, E., Woo, C. K., & Wang, J. (2016). Micro grid and renewable generation integration: University of California, San Diego. *Applied Energy*, *169*, 709–720. doi:10.1016/j.apenergy.2016.02.053

Srikanth, K., & Mahesh, K. M. (2019). Design and stability analysis of DC microgrid with hybrid energy storage system. *IEEE Transactions on Sustainable Energy*, *10*(3), 1603–1612. doi:10.1109/TSTE.2019.2891255

Sustainable energy regulation and policymaking training manual-demand side management module 14.

Sustainable energy regulation and policymaking training manual-supply side management module 13.

Taha, A. F., Qi, J., Wang, J., & Panchal, J. H. (2015). Dynamic state estimation under cyber attacks: A comparative study of Kalman filters and observers. *arXiv preprint arXiv:1508.07252*.

Teng, J. H., & Lu, C. N. (2008). Value-based distribution feeder automation planning. *International Journal of Electrical Power & Energy Systems*, *28*(3), 186–194. doi:10.1016/j.ijepes.2005.11.008

Thang, V. V., & Minh, N. D. (2017). Optimal allocation and sizing of capacitors for distribution systems reinforcement based on minimum life cycle cost and considering uncertainties. *The Open Electrical & Electronic Engineering Journal*, *11*(18), 165–176. doi:10.2174/1874129001711010165

The Boston Consulting Group (BCG). (2011). Revisiting energy storage, Report.

The smart grids European technology platform (SGETP). Retrieved from http://www.smartgrids.eu/ ETP SmartGrids.

Thomas Gallery. L. M. (n.d.). Impact of distributed generation on distribution network protection. ESBI Engineering & Facility Management, Ireland.

Thorborg, K. (1988). *Power electronics*. New York: Prentice Hall.

Thukaram, D., Jerome, J., & Surapong, C. (2000). A robust three-phase state estimation algorithm for distribution networks. *Electric Power Systems Research*, *55*(3), 191–200. doi:10.1016/S0378-7796(00)00088-2

Time based rate programs (TBRP). (n.d.). Retrieved from https://www.smartgrid.gov/recovery_act/time_based_rate_programs.html

tutorvista. (n.d.). Retrieved from http://www.tutorvista.com/physics/wind-electric-power

UK Power Networks, UKPN. (2014). Smarter network storage, Six monthly report.

ul Hassan, C. A., Khan, M. S., Ghafar, A., Aimal, S., Asif, S., & Javaid, N. (2018). Energy optimization in smart grid using grey wolf optimization algorithm and bacterial foraging algorithm, Advances in Intelligent Networking and Collaborative Systems, Springer International Publishing, AG 2018, Islamabad, Pakistan.

Valverde, G., & Terzija, V. (2011). Unscented Kalman filter for power system dynamic state estimation. *IET Generation, Transmission, & Distribution*, *5*(1), 29–37. doi:10.1049/iet-gtd.2010.0210

VDE - ETG Energy Storage Task Force: Energy storage in power supply systems with a high share of renewable energy sources Significance - state of the art - need for action, Report, Dec. 2008.

VenkataBramhanaidu, D., & Sreenivasulu, A. (2017). Implementation of space vector pulse width modulation technique on three phase two level inverter, *IOSR-JEEE, 12*(3) Ver. II May – June. 2017.

Vodorozov, V. (2010). Introduction to power electronics (1st ed.). London, UK: London Business School.

Vodorozov, V. (2010). *Introduction to power electronics* (1st ed.).

Vu Van Thong, J. D. (2004, January). Interconnection of distributed generators and their influences on power system. Katholieke Univeristeit Leuven, Belgium.

Wade, N. S., Taylor, P. C., Lang, P. D., & Jones, P. R. (2010). Evaluating the benefits of an electrical energy storage system in a future smart grid. *Energy Policy*, *38*(11), 7180–7188. doi:10.1016/j.enpol.2010.07.045

Wakileh, G. J. (2001). Power systems harmonics: Fundamentals, analysis, and filter design. Springer.

Wang, H., & Schulz, N. N. (2004). A revised branch current-based distribution system state estimation algorithm and meter placement impact. *IEEE Transactions on Power Systems, 19*(1), 207–213. doi:10.1109/TPWRS.2003.821426

Wang, P., & Billinton, R. (2002). Reliability cost/worth assessment of distribution systems incorporating time-varying weather conditions and restoration resources. *IEEE Transactions on Power Delivery, 17*(1), 260–265. doi:10.1109/61.974216

Wang, Z., Chen, B., Wang, J., Begovic, M., & Chen, C. (2015b). Coordinated energy management of networked microgrids in distribution systems. *IEEE Transactions on Smart Grid, 6*(1), 45–53. doi:10.1109/TSG.2014.2329846

Wu, D., Lesieutre, B., Ramanathan, P., & Kakunoori, B. (2016). Preserving privacy of AC optimal power flow models in multi-party electric grids. *IEEE Transactions on Smart Grid*, 1–11.

Xiao, X. (2015). Basic problems of the new complex AC-DC power grid with multiple energy resources and multiple conversions. *Trans China Elect Rotech Soc, 30*, 1–14.

Xie, J., Liang, C., & Xiao, Y. (2018). Reactive power optimization for distribution network based on distributed random gradient-free algorithm. *Energies, 11*(3), 534. doi:10.3390/en11030534

Yanjun, H., Hong, W., Amir, K., Bin, L., Jie, J., Kegang, Z., & Chuan, H. (2018). A review of power management strategies and component sizing methods for hybrid vehicles. *Renewable & Sustainable Energy Reviews, 96*, 132–144. doi:10.1016/j.rser.2018.07.020

Yarlagadda, S., Hartley, T. T., & Husain, I. (2013). A battery management system using an active charge equalization technique based on a DC/DC converter topology. *IEEE Transactions on Industry Applications, 49*(6), 2720–2729. doi:10.1109/TIA.2013.2264794

Yazie, C., Akalu, T., Yalfal, T., & Baye, B. (2017). *Characterization of potato production, marketing and utilization in North Western Amhara* Region, Ethiopia. *Journal of Horticulture and Forestry, 9*(3), 17-25.

Yenesew, M. (2017). Enhanced particle swarm optimization (PSO) algorithm for reactive power optimization in the distribution system [IJERME]. *International Journal of Engineering Research and Modern Education*, 2(2), 56–64.

Yeshalem, M. T., & Khan, B. (2017). Design of an off-grid hybrid PV/wind power system for remote mobile base station: A case study. *AIMS Energy, 5*(1), 96–112. doi:10.3934/energy.2017.1.96

Youssef, K, Krim, S, Mimouni, M. F. (2016). Control of a wind farm connected to the grid at a frequency and variable voltage. *International Journal of Renewable Energy Research, 6*(3), 747-758.

Youssef, K., Dhaker, A., Saber, K., & Mohamed, F. M. (2018). Power management and second-order sliding mode control for standalone hybrid wind energy with battery energy storage system. In *Proceedings of the Institution of Mechanical Engineers, Part I: Journal of Systems and Control Engineering, 232*(10), 1-22.

Youssef, K., Dhaker, A., Saber, K., & Mohamed, F. M. (2019a). A second-order sliding-mode control for a real time emulator of a wind power system synchronized with electrical network, *International Transactions on Electrical Energy Systems,* 1-20.

Youssef, K., Dhaker, A., Saber, K., & Mohamed, F. M. (2018a). Classical vector, first-order sliding mode and high-order sliding-mode control for a grid-connected variable speed wind energy conversion system: A comparative study. *Wind Engineering, 42*(1), 1–22.

Youssef, K., Dhaker, A., Saber, K., & Mohamed, F. M. (2019). Intelligent droop control and power management of active generator for ancillary services under grid instability using fuzzy logic technology. *Control Engineering Practice*, *81*, 215–230.

Yu, X. (2011). *Distribution system reliability enhancement,* (MSc thesis), Elec. and Comp. Eng., Georgia Institute of Technology.

Zabihollah, N. A., Masoud, G., Misagh, I. S., Erfan, K., & Ahmad, S. (2018). Evaluation of hydrogen production from harvesting wind energy at high altitudes in Iran by three extrapolating Weibull methods. *International Journal of Hydrogen Energy*, *43*(6), 3110–3132. doi:10.1016/j.ijhydene.2017.12.154

Zamani, R., Golshan, M. E. H., Alhelou, H. H., & Hatziargyriou, N. (2019). A novel hybrid islanding detection method using dynamic characteristics of synchronous generator and signal processing technique. *Electric Power Systems Research*, *175*, 105911. doi:10.1016/j.epsr.2019.105911

Zamani, R., Hamedani-Golshan, M. E., Haes Alhelou, H., Siano, P., & Pota, R., H. (2018). Islanding detection of synchronous distributed generator based on the active and reactive power control loops. *Energies*, *11*(10), 2819. doi:10.3390/en11102819

Zhang, H. (2018). Distribution grid state estimation based on smart meter data.

Zhang, J., Welch, G., Bishop, G., & Huang, Z. (2011, June). Reduced measurement-space dynamic state estimation (remedyse) for power systems. In 2011 IEEE Trondheim PowerTech (pp. 1-7). IEEE.

Zhanga, X., Bieb, Z., & Lib, G. (2011). Reliability assessment of distribution networks with distributed generations using Monte Carlo method. *Energy Procedia*, *12*, 278–286. doi:10.1016/j.egypro.2011.10.038

Zhang, L., & Sun, M. Research on the type of load of accessing to microgrid based on reliability. In *Proceedings 2nd International conference on electrical, computer engineering and electronics*, Jinan, China. pp. 1665–70. 10.2991/icecee-15.2015.314

Zhao, B., Song, Q., & Liu, W. (2014). Experimental comparison of isolated bidirectional DC – DC converters based on all-Si and all-SiC power devices for next-generation power conversion application. *IEEE Trans Ind Electron*, *61*, 13, 89–93.

Zhao, Z., Lee, W., Shin, Y., & Song, K. (2013, September). An optimal power scheduling method for demand response in home energy management system. *IEEE Transactions on Smart Grid*, *4*(3).

Zhou, H., Bhattacharya, T., Tran, D., Siew, T. S. T., & Khambadkone, A. M. (2011). Composite energy storage system involving battery and ultracapacitor with dynamic energy management in microgrid applications. *IEEE Transactions on Power Electronics*, *26*(3), 923–930. doi:10.1109/TPEL.2010.2095040

Zhou, N., Meng, D., Huang, Z., & Welch, G. (2014). Dynamic state estimation of a synchronous machine using PMU data: A comparative study. *IEEE Transactions on Smart Grid*, *6*(1), 450–460. doi:10.1109/TSG.2014.2345698

Zineb, C., Mohammed, O., & Mohamed, M. (2017). Battery and supercapacitor for photovoltaic energy storage: A fuzzy logic management. *IET Renewable Power Generation*, *11*(8), 1157–1165. doi:10.1049/iet-rpg.2016.0455

Zobaa, A. F., & Aleem, S. H. E. A. (2017). Power quality in future electrical systems, IET, 2017.

About the Contributors

Baseem Khan received his Bachelor of Engineering degree in Electrical Engineering from the Rajiv Gandhi Technological University, Bhopal, India in 2008. He received his Master of Technology and Doctor of Philosophy degrees in Power systems and Electrical Engineering from the Maulana Azad National Institute of Technology, Bhopal, India, in 2010 and 2014, respectively. Currently, he is working as an Foreign Expatriate in the Department of Electrical and Computer Engineering, Institute of Technology, Hawassa University, Hawassa, Ethiopia. His research interest includes power system restructuring, power system planning, smart grid technologies, meta-heuristic optimization techniques, reliability analysis of renewable energy system, power quality analysis, renewable energy integration and fault analysis. He received various scholarships and awards in his career.

Hassan Haes Alhelou (S'15) is a Faculty Member at Tisheen University, Lattakia, Syria. He is also a Ph.D. Researcher at the Isfahan University of Technology (IUT), Isfahan, Iran. His major research interests are power systems, power system dynamics, power system operation and control, dynamic state estimation, frequency control, smart grids, microgrids, demand response, load shedding, and power system protection.

Ghassan Hayek was born in Lattakia in 1958, received a BSc degree in Electrical Power engineering from the Aleppo University, Aleppo, Syria, in 1985 and the Ph.D. degree in Electrical Engineering - Automation and Control from TU Dresden, Dresden, Germany, in 1989, respectively. He was the recipient of the Outstanding Reviewer Award ISA Transactions Journal in 2018 and many other Awards form prestigious journals and publishers. Also, He has been an editor of several special issues and books in the field of power systems. His main research areas are Electric drive, Mico-grid, Power system operation and control, Power system dynamic, control and stability, intelligent control, fuzzy control, automation and process control, preventive maintenance, condition monitoring, Fault diagnosis.

* * *

Degarege Anteneh is working as research scholar in Department of Electrical and Computer Engineering, Hawassa University. His research interests include smart grid, reliability assessment, Power quality issue and distribution subsystem reconfiguration.

Tesfaye Belay currently working as MSc research scholar at the Department of Electrical and Computer engineering, Hawassa University, Ethiopia. He completed his bachelors from Mettu University, Ethiopia in the year of 2016. His research interests are Micro grid planning, Smart Grid technologies, and Reliability assessment and enhancement.

Samuel Degarege working as research scholar in Electrical and Computer Engineering, Hawassa University, Ethiopia. His research interests include Micro grid planning, reliability, Smart grid technology.

Mesfin Fanuel completed his MSc from Addis Ababa Institute of Science and Technology. His area of interest includes renewable energy integration, power quality, electrical machines, Micro controllers etc.

Youssef Krim was born in Kairouan, Tunisia. He received the degree in electrical engineering from the national engineering school of Monastir, University of Monastir, Tunisia, in 2015, and the Ph.D. degree in electrical engineering from the Monastir University, Tunisia, in February 2019. His current research interests include control systems for the integration of distributed renewable generation and storage into power networks.

Abirami Manoharan is currently working as an Assistant Professor in the Department of Electrical and Electronics Engineering, Government College of Engineering, Srirangam. Her research contributions include 10 SCI articles. She is a Senior Member in IEEE (SM 93333628) and also a Member in Institution of Engineers (India) (M 1558689).

Fsaha Mebrahtu completed his MSc in Power and Energy Engineering from Hawassa University Ethiopia. His research interests are Wind energy generation, micro grid integration, Smart grid technologies.

Tesfahun Molla received the B.Sc. and M.Sc. degrees both from Hawassa University, Ethiopia, in 2013 and 2019, respectively. He is currently working as a Lecturer at Hawassa University. His research interests include smart grid energy management system, optimization and integration of renewable energy systems, power system operation and intelligent control, and application of artificial intelligence in power system.

Suriya Ponnambalam is a research scholar at Annamalai University in the department of Electrical and Electronics Engineering. Her research interest is mainly focused on power system expansion planning.

C. Sharmeela (CS) holds a B.E. in Electrical and Electronics Engineering, M.E. in Power Systems Engineering from Annamalai University, Chidambaram, India and a Ph.D in Electrical Engineering from Anna University, Chennai, India in 1999, 2000 & 2009 respectively. She did P.G Diploma in Electrical Energy Management and Energy Audit from Annamalai University, Chidambaram in 2010. At present, she holds the post of Associate Professor in department of EEE, CEG campus, Anna University, Chennai, India. She has 20 years of teaching experience and taught various subjects to undergraduate and post graduate students. She did number of research projects and consultancy work in renewable energy, power quality & design of PQ compensators for various industries. She published more than 30 research publications in peer reviewed journals, more than 50 research papers in international and national conferences in India and abroad. CS talks importance of power quality more than 10 years. She guides 14 PhDs (5 completed and 9 ongoing), 58 Master's level, 10 Undergraduate's level and 4 AMIE students. She co-ordinate and organized short term course on power quality for Tamilnadu Electricity Board Engineers in various years. She received best paper award in IEEE POWERCON 2004 conference held in Singapore. She is a Life-member of the IEEE, CBIP, Institution of Engineers (India), ISTE and SSI. She is co-authored a book chapter "Solar Micro Inverter" in IGI global.

Bawoke Simachew completes his MSc from Hawassa University. His area of interest includes distribution system reconfiguration, smart grid, heuristic optimization techniques.

Ganesan Sivarajan is presently working as Assistant Professor in the Department of Electrical & Electronics Engineering, Government College of Engineering, Salem, Tamil Nadu, India. He has 16 years / 11 years of teaching / research experience. He has authored 12 SCI research articles. He is an active member in various professional bodies like the Institution of Engineer (India) and Indian Society for Technical Education.

P. Sivaraman was born in Vellalur, Madurai district, Tamilnadu, India. He completed schooling in Govt. Higher Secondary School, Vellalur, B.E. in Electrical and Electronics Engineering, M.E. in Power Systems Engineering from Anna University, Chennai, India in 2012 & 2014 respectively. He has more than five years of experience in the field of power quality studies & harmonic assessments, trouble shooting for various power quality problems, power system analysis and studies, micro grid planning and design, renewable energy integration studies, solar PV system, providing the techno economical solution to various power quality problems for industries across all over India. He has trained more than 200 personnel on renewable energy and power quality & carried out power quality assessments for over 200 sites across all over India in last 5 years and has validated over 200 power quality reports. He is expert in power system simulation software's like ETAP, PSCAD, DIGSILENT POWER FACTORY and MATLAB and also speaker who is well versed on both National and International Standards. He had published several papers in national and international conferences. He is authored a book chapter "Solar Micro Inverter" in IGI global. He is a member of The European Energy Center (EEC), Member of the International Association of Engineers (IAENG), Member of International Association of Online Engineering (IAOE), Member of South Asia Institute of Science and Engineering (SAISE), Member of Science and Engineering Institute (SCIEI) and Member of Universal Association of Computer and Electronics Engineers (UACEE).

S. Subramanian is working as Professor in the Department of Electrical Engineering, Annamalai University, Annamalainagar, India. His contributions in the research include 40 SCI articles and 19 Ph.Ds. awarded under his guidance. He is a senior member in IEEE (SM 90389250) and also a Fellow of Institution of Engineers (India) (F1130982).

Index

N

P

R

S

T

U

V

W

Premier Reference Source
Emerging GIS Applications for Emergency and Disaster Management

Premier Reference Source
Managerial Strategies and Green Solutions for Project Sustainability

Premier Reference Source
Comparative Approaches to Using R and Python for Statistical Data Analysis

Premier Reference Source
Solutions for High-Touch Communications in a High-Tech World

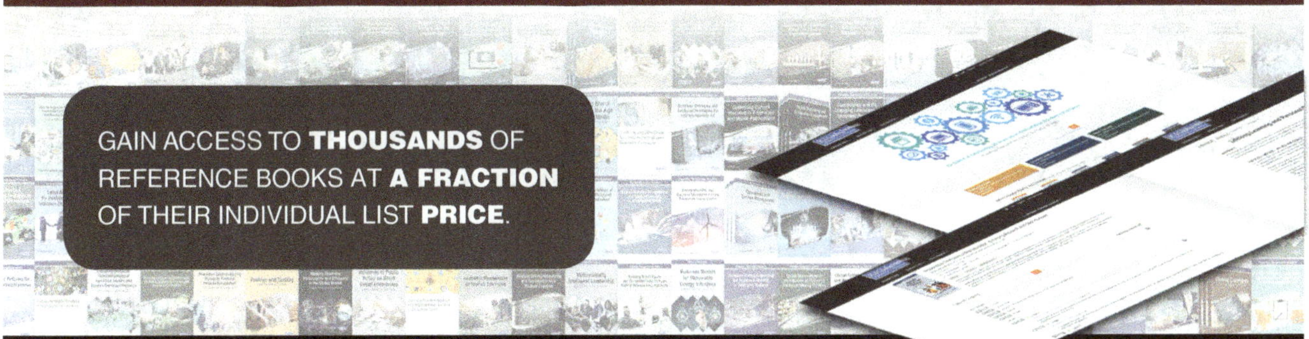
GAIN ACCESS TO **THOUSANDS** OF REFERENCE BOOKS AT **A FRACTION** OF THEIR INDIVIDUAL LIST **PRICE**.

www.ingramcontent.com/pod-product-compliance
Lightning Source LLC
LaVergne TN
LVHW060258011026
849161LV00011B/2735
* 9 7 8 1 7 9 9 8 1 2 3 0 2 *